Different Paths Towards Sustainable Biofuels?

A Comparative Study of the International, EU, and Chinese Regulation of the Sustainability of Biofuels

Taotao Yue

Different Paths Towards Sustainable Biofuels?

A Comparative Study of the International, EU, and Chinese Regulation of the Sustainability of Biofuels

intersentia

METRO

Ius Commune Europaeum

Intersentia Ltd
Sheraton House | Castle Park
Cambridge | CB3 0AX | United Kingdom
Tel.: +44 1223 370 170 | Fax: +44 1223 370 169
Email: mail@intersentia.co.uk
www.intersentia.com | www.intersentia.co.uk

Distribution for the UK and Ireland:
NBN International
Airport Business Centre, 10 Thornbury Road
Plymouth, PL6 7PP
United Kingdom
Tel.: +44 1752 202 301 | Fax: +44 1752 202 331
Email: orders@nbninternational.com

Distribution for Europe and all other countries:
Intersentia Publishing nv
Groenstraat 31
2640 Mortsel
Belgium
Tel.: +32 3 680 15 50 | Fax: 32 3 658 71 21
Email: mail@intersentia.be

Distribution for the USA and Canada:
International Specialized Book Services
920 NE 58th Ave. Suite 300
Portland, OR 97213
USA
Tel.: +1 800 944 6190 (toll free) | Fax: + 1 503 280 8832
Email: info@isbs.com

Different Paths Towards Sustainable Biofuels? A Comparative Study of the International, EU, and Chinese Regulation of the Sustainability of Biofuels

ISBN 978-1-78068-413-0
D/2016/7849/93
NUR 828

British Library Cataloguing in Publication Data. A catalogue record for this book is available from the British Library.

In loving memory of my grandfathers

ACKNOWLEDGEMENTS

People in China say that the world consists of men, women, and female PhDs. This social prejudice is perhaps even more relevant for a female PhD student pursuing her study abroad. With some doubt and hesitation started I the venture of PhD at Maastricht University in 2011, uncertain whether I could ever finish it and what I would achieve. More than four years afterwards, I am glad to have finally accomplished my PhD research while maintaining still a lively and cheery mind. At this moment, it is high time to formally express my sincere gratitude to all the people who have given me support and help in these years.

This dissertation is the outcome of PhD research at Maastricht University with the support of the China Scholarship Council (CSC). I deeply appreciate the Faculty of Law of Maastricht University and CSC for making my PhD research possible. In particular, my heartfelt gratitude goes to my supervisors, Professor Marjan Peeters and Professor Ellen Vos, who gave me the chance of starting the research and who have guided me with enormous patience, inspiration, and knowledge. I would like to express my profound gratitude to Marjan for leading me to the research field of environment and climate law, as well as helping me to find the way, start from scratch, and move gradually forwards. I deeply appreciate her insightful and provoking questions and comments that not only inspire me to accomplish this dissertation but also teach me how to be a researcher. Beyond work, Marjan has continuously encouraged and helped me to be integrated in the life of Maastricht, via inviting me for a number of times to join her family events and local activities. I also extend special thanks to Ellen, my co-supervisor in Maastricht University, for her invaluable comments and guidance with my research as well as her unconditional support towards my career. Their extremely professional approach towards work and sincere commitments dedicated to legal scholarship have set excellent examples for me along the journey of my PhD research and will keep on enlightening me in the future.

I am deeply grateful to the members of the assessment committee: Professor Fons Coomans, Professor Michael Faure, Professor Elisa Morgera, and Professor Zhiping Li. I really appreciate the time and patience they have spent in reading my thesis, in particular their kind encouragement and insightful comments and suggestions which incentivised me to improve my research from various perspectives. Special gratitude goes to Michael who is an 'old friend' of Chinese people and has provided generous support and inspiring comments for my research within the METRO and the Ius Commune research school. I extend additional thanks to Fons for his efforts in chairing the assessment committee. I also want to express my whole-hearted thanks to all the experts who agreed to sit in the supervisory committee and participate in my defence.

My sincere thanks also go to my colleagues and friends in the Faculty of Law of Maastricht University, as it would not be possible for me to finish this research without their support and help. Sincere thanks are owed to the unconditional help from the METRO team, Yleen Simonis, Chantal Kuypers, Elke Hundhausen, Marina Jodogne, and Marjo Mullers, for their professionalism of dealing with details in both daily work and big events. My profound gratitude also goes to Licette Poll, Ingrid Loontjens, Monique Laenen, Pauline Arends, and Peggy van Wersch who always responded to my questions with great patience and helped me to deal with documents or procedures in each stage of my research. I am especially grateful for the stimulating discussions with Professor Peter Van den Bossche, Professor Menno Kamminga, Professor Chris Backes, Dr. Niels Philipsen, Dr. Laura Tilindyte, Dr. Denise Prévost, Dr. Sarah Schoenmaekers, and Dr. Jing Liu. During the stay in Maastricht, a great deal of friends within the law faculty helped me whenever I faced difficulties in life or at work, whose friendship enriched my life, brought me beautiful memories, and made me feel as if doing PhD research is not a solo task. Heartfelt thanks go to Sandra Nobrega, Ancui Liu, Huizhen Chen, Yuan Gao, Wenqing Liao, Jie Wang, Tianxiang He, Xiahong Chen, Tian Lv, Huilan Zhu, Xiaowei Yu, Jiangqiu Ge, Liang Yu, Liuhu Luo, Yufeng Wu, Bo Chen, Guang Shen, Mengmeng Shi, Mengxing Lv, Yu Zhao, Xi Chen, Yingying Jing, Yaojin Peng, Yu Yan, Yang Gao, Jingjing Fu, Joanne Faggin, Jaqueline Da Silva Albino, Kevine Kindji, Emma Dermawati, Viorelia Gasca, Mehdi Piri Damagh, and Tobias Heldt. Special thanks go to Ancui Liu who has given me enormous help when I was in or out of Maastricht.

I am greatly thankful to the experts and friends outside the Maastricht University for their willingness to share their immense knowledge and brilliant ideas which broadened and deepened my perspectives in the research. Particular thanks go to Professor Tianbao Qin, Dr. Seita Romppanen, Professor Joanne Scott, Professor Kati Kulovesi, Professor Nicolas De Sadeleer, Professor Xi Wang, Professor Carmen G. Gonzalez, Professor Ben Boer, Dr. Wybe Douma, Ignacio Vazquez Larruscain, Dr. Carlo Hamelinck, Dr. Reinier de Man, Max Salomon Janssen, Dr. Yoshiko Naiki, Dr. Lorenzo Squintani, Dr. Federico Ortino, Dr. Leonie Reins, Natascha Trennepohl, Haomiao Du, Dr. Jinshan Zhu, and Yelena Gordeeva. I would like to express additional gratitude to Professor Tianbao Qin, Professor Joanne Scott, Ignacio Vazquez Larruscain, Carlo Hamelinck, and Dr. Reinier de Man for agreeing upon my request of face-to-face interviews. Many thanks go to the UNESCO-UNEVOC and UNFCCC Secretariat for giving me the precious internship opportunities.

Last but not least, my deepest thanks go to my beloved family for fully supporting me throughout the PhD research. I am wholeheartedly indebted for their unconditional support and considerate understanding over the years when I stayed a long distance away and could not spend more time with them, for which I may never be able to repay in a single lifetime. I sincerely hope that this dissertation would become their pride and also serve as the starting point for my future career!

CONTENTS

LIST OF ABBREVIATIONS

AFOLU	Agriculture, forestry and other land use
AFTF	Alternative Fuel Task Force
AIA	Advance Informed Agreement
ASEAN	Association of Southeast Asian Nations
BEFS	Bioenergy and Food Security
CAEP	Committee on Aviation Environmental Protection
CAP	Common Agricultural Policy
CBD	Convention on Biological Diversity
CCS	Carbon Capture and Storage
CDDA	Common Database on DesignatedAreas
CDM	Clean Development Mechanism
CER	Certified Emission Reduction
CESCR	Committee on Economic, Social, and Cultural Rights
CFS	Committee on World Food Security
CITES	Convention on International Trade in Endangered Species
CJEU	Court of Justice of the European Union
CMP	Conference of the Parties serving as the Meeting of the Parties
CMS	Convention on the Conservation of Migratory Species in Wild Animals
CN	Combined Nomenclature
CNPC	China National Petroleum Corporation
COP	Conference of Parties
CPC	Communist Party of China
DNA	Designated National Authority
DOE	Designated Operational Entity
EAG	Environmental Advisory Group
EB	Executive Board
ECOSOC	Economic and Social Council
EEA	European Environment Agency
EEA	European Economic Area
EFTA	European Free Trade Association
EIA	Environmental Impact Assessment

EPA	Environmental Protection Agency
ETS	Emission Trading Scheme
EU	European Union
FAME	Fatty Acic Methyl Esters
FAO	Food and Agriculture Organization of the United Nations
FQD	Fuel Quality Directive
FRA	Forest Resources Assessment
GATT 1994	General Agreement on Tariffs and Trade 1994
GBEP	Global Bioenergy Partnership
GHG	Greenhouse gases
GMTF	Global Market-based Measure Technical Task Force
GPG	Good Practice Guidance
GPG-LULUCF	Good Practice Guidance for Land Use, Land-use Change and Forestry
HS	Harmonised System
ICJ	International Court of Justice
ICAO	International Civil Aviation Organization
ICESCR	International Covenant on Economic, Social and Cultural Rights
ICRW	International Convention on the Regulation of Whaling
IEA	International Energy Agency
IFAD	International Fund for Agricultural Development
ILUC	Indirect Land Use Change
ILA	International Law Association
ILO	International Labour Organisation
IMO	International Maritime Organization
IPCC	Intergovernmental Panel on Climate Change
ISO	International Organisation for Standardization
ITPGR	International Treaty on Pland Genetic Resources for Food and Agriculture
IUCN	International Union for the Conservation of Nature
LMO	Living Modified Organisms
LULUCF	Land Use, LandUse Change and Forestry
MARPOL	International Convention for the Prevention of Pollution from Ships
MEA	Multilateral Environmental Agreement
MEF	Major Economies Forum
Mtoe	Million tonnes of oil equivalent
NBSAP	National Biodiversity Strategies and Action Plans
NEA	National Energy Administration
NDRC	National Development and Reform Commission
NGO	Non-governmental Organization
NUTS	Nomenclature Territorial Units for Statistics
NWFP	Non-Wood Forest Product
OECD	Organization for Economic Cooperation and Development
OHCHR	Office of the High Commissioner for Human Rights
PDD	Project Design Documents

PPM	Process and Production Method
RSB	Roundtable on Sustainable Biomaterials
SBSTTA	Subsidiary Body on Scientific, Technical and Technological Advice
SEA	Strategic Environmental Assessment
SIA	Sustainability Impact Assessment
Sinopec	China Petrochemical Corporation
STRP	Scientific & Technical Review Panel
TBT	Technical Barriers to Trade
TEU	Treaty on European Union
TFEU	Treaty on the Functioning of the European Union
UK	United Kingdom
UN	United Nations
UNCCD	United Nations Convention to Combat Desertification
UNCED	United Nations Conference on Environment and Development
UNEP	United Nations Environment Programme
UNFCCC	United Nations Framework Convention on Climate Change
UNIDO	United Nations Industrial Development Organization
US	United States
VAT	Value Added Tax
WCED	World Commission on Environment and Development
WCO	World Customs Organisation
WFP	World Food Programme
WHC	World Heritage Convention
WTO	World Trade Organisation

LIST OF TABLES

Chapter 1

INTRODUCTION

1. Research Background

1.1. Sustainability Concerns about Biofuels

Biofuel is distinguished from fossil fuel by the fact that its raw materials (feedstocks) are sourced from biomass such as agricultural crops and other biodegradable matters.[1] Fuels were produced from biomass, such as starch crop and vegetable oil, even before fossil fuels were popularized – Rudolf Diesel's first engine was designed to run on peanut oil.[2] Biofuels regained popularity since the mid-1970s when Brazil and the United States began to use sugarcane and corn to produce ethanol as transport fuels for energy security and rural development purposes.[3] More recently, reducing greenhouse gas (GHG) emissions in the transport sector has become the key objective of promoting biofuels as a substitute for fossil fuels.[4] Based on the three-fold objective in energy security, GHG emission

1 There are similar terminologies used such as 'bioenergy' and 'agrofuel.' This thesis uses the term 'biofuel' to refer to the fuel produced from biomass for transport use. The definition of the term 'biofuel' is to be further discussed in section 2.1 of this chapter.

2 Biofuels began to be produced in the late 19th century when ethanol was derived from vegetable oil and biofuels were seen as viable transport fuels until the 1940s. IEA 2011, p. 10.

3 Commercial interest in the production of biofuels for transport rose in the 1970s, when the 1973 oil crisis and 1979 energy crisis broke out. During the 1990s, the industrialized economies of North America and Europe actively pursued policies in support of domestic biofuel industries for the sake of energy security and rural economies. More countries followed and have since then launched biofuels programmes. IEA 2011, p. 10; Elbehri, Segerstedt & Liu 2013, p. 13.

4 It is observed that the reduction of GHG emissions and reducing use of fossil fuels rank among the main drivers for biofuel development, particularly in countries belonging to the Organization for Economic Cooperation and Development (OECD). IEA 2011, p. 10; Elbehri, Segerstedt & Liu 2013, p. 13.

reduction, and rural development, governmental support has played a major role in the increased production and consumption of biofuels worldwide.[5]

In effect, instead of being applauded for the assumed contribution to energy security, rural development, and GHG emission reduction, biofuels are increasingly subject to criticism for their potential negative impacts on environmental and socio-economic sustainability (see Table 1). From the *environmental* perspective, the criticism concentrates on the ecological and climate change impacts during the biofuel production process, particularly with regard to the effects of the changes in land use. It has been pointed out that using a large area of land to grow biofuel feedstocks, which are mainly starch- and oil-rich crops, may threaten biodiversity by deforestation, monoculture, and invasive species,[6] and may also cause air, water, and soil pollution and degradation in the processes of crop cultivation.[7] Moreover, although biofuels are promoted for reducing GHG emissions, the actual contribution to GHG emission reduction has been seriously doubted,[8] especially when emissions during the life cycle of biofuels from direct[9] and indirect[10] land use changes are considered. Apart from environmental concerns, biofuels are also criticised for causing negative *social* effects particularly for people in poverty, since the feedstocks used for producing biofuels are mostly food sources and their diverted use for biofuel production may give rise to food availability and accessibility problems,[11] as well as land acquisition for biofuels that affects the land

5 Over the ten years between 2000 and 2009, the production of biofuels has increased dramatically: fuel ethanol output experienced an increase from 16.9 to 72.0 billion liters while biodiesel grew from 0.8 to 14.7 billion liters. The increased biofuel production has been driven by governmental interventions: for example, over 50 countries have adopted financial incentives or blending mandates to promote biofuels. With this growing trend, the International Energy Agency (IEA) predicted that biofuels would provide up to 27 per cent of world transportation fuel by 2050. Elbehri, Segerstedt & Liu 2013, p. 13; Sorda, Banse & Kemfert 2010, p. 6977; IEA 2011, p. 7 and 11.

6 See for example Wiens, Fargione & Hill 2011, p. 1085-1095; DeHaan *et al.* 2010, p. 33-38; Ottinger 2009, p. 256.

7 Biofuels may increase the need for higher yields and the intensification of production, which thus place greater stress on resources and generate unintended consequences by way of pollution and soil degradation. See for example Elbehri, Segerstedt & Liu 2013, p. 1 and 13; Gerbens-Leenes *et al.* 2012; Fargione, Plevin & Hill 2010.

8 See for example Larson 2006; von Blottnitz & Curran 2007; Gnansounou *et al.* 2009; Johnson 2009.

9 It has been warned that increasing biofuels production and use may even cause, rather than reduce, GHG emissions if deforestation and land clearing take place. Searchinger *et al.* 2008; Fargione *et al.* 2008.

10 In addition to emissions from direct land use changes of forested land or grassland, there may be indirect land use change (ILUC) effects if biofuels are produced on existing agricultural land but the demand for food and feed crops remains the same and may lead to more food and feed production elsewhere. See for example Searchinger *et al.* 2008, p. 1238-1240; Sanchez *et al.* 2012, p. 1106.

11 The concern is that increased biofuel production may increase food costs and food insecurity. See for example Molony & Smith 2010; Sexton *et al.* 2008; Runge & Senauer 2007.

use rights of local and indigenous communities.[12] In *economic* terms, biofuels were assumed to bring in new economic opportunities for people in rural areas by increasing the demand for and prices of agricultural products,[13] but some cost-benefit analyses reveal that the economic viability and benefits may have been overestimated,[14] and many studies demonstrate resource efficiency problems with regard to certain types of biofuels when the input of energy and resources for producing biofuels is considered.[15] Furthermore, the environmental, social, and economic facets can be interconnected - the environmental impacts might incur socio-economic problems and vice versa.[16] Although there are still debates about sustainability concerns and whether they should all be attributed to biofuels,[17] the Intergovernmental Panel on Climate Change (IPCC) has admitted in its reports[18] that biofuels can result in both positive and negative effects and their deployment needs to balance a range of environmental, economic, and social objectives which are not always compatible.[19] Moreover, as noted by the IPCC, the potential institutional, social, environmental, economic, and technological implications of bioenergy (including biofuels) options may range from a local scale to the global scale,[20] which may call for solutions or regulation at different levels.

12 The other social sustainability concerns associated with increased biofuel production include increased poor labour conditions and poor economic remuneration for local farmers. See for example Konopacky 2012, p. 403-404; Cotula, Dyer & Vermeulen 2008.

13 Demirbas 2009, p. S108.

14 See for example De Gorter & Tsur 2010.

15 Resource efficiency concentrates on the efficiency of using natural resources while decreasing associated environmental impacts, so as to decouple economic growth from resource use and environmental degradation. Berndes & Hansson 2007, p. 5965-5979; Prabhakar & Elder 2009; Börjesson & Tufvesson 2011, p. 108-120; EEA 2013.

16 Robertson *et al.* 2008, p. 49.

17 There is discussion over whether the negative environmental or socio-economic impacts should be completely attributed to biofuels. Murphy *et al.* 2011; Rajagopal & Zilberman 2007; Vuichard, Ciais & Wolf 2009; Rathmann *et al.* 2010.

18 Chum *et al.* 2011; Smith *et al.* 2014.

19 Smith *et al.* 2014, p. 79.

20 For example, biodiversity may have global and local implications; food security may be impacted on a global level and national level. In another report, the IPCC also illustrates the impacts and performance of biomass production and use according to region- and site-specific, or region-oriented and global-oriented. Smith *et al.* 2014, p. 884; Chum *et al.* 2011, p. 308.

Table 1. The major sustainability concerns with regard to biofuels

Environmental	Social	Economic
biodiversity (e.g. deforestation, monoculture, and invasive species)	food availability and accessibility	economic viability
air, water, soil pollution, and degradation	land use rights of local and indigenous communities	efficiency of using natural resources
GHG emissions from direct land use changes (e.g. land use change from forested land or grassland)	poor labour conditions or poor economic remuneration for local farmers	
GHG emissions from indirect land use changes (if biofuels are produced on existing agricultural land but the demand for food and feed crops remains and may lead to more food and feed production elsewhere)		

1.2. *Research on Regulating the Sustainability of Biofuels*

The sustainability concerns give rise to increased discussions in legal literature on how biofuels are and can be regulated to avoid unsustainable consequences. Firstly, authors have discussed regulation of the overall or certain negative impacts of biofuels. For instance, Ottinger analysed the potential benefits and problems of biofuels and argues that legally binding and enforceable standards should be considered for the cultivation and processing of biofuels with respect to the risks.[21] In view of the multiple impacts on land use and GHG mitigation, Khanna called for the development of a mechanism to attach a score or index to each land plot.[22] Gonzalez examined the impact of the biofuel boom on food security in the global South and suggested specific measures to promote food security through international law and regulation.[23] Secondly, authors have examined the regulation of biofuels in a domestic or international legal system. For example, Sorda *et al.* provided an overview of individual countries' biofuel policies on each continent,[24] and Morgera *et al.* carried out comparative case studies on bioenergy policy and law

21 Ottinger 2009, p. 259.

22 Khanna *et al.* 2011, p. 581.

23 Gonzalez 2010, p. 478-479.

24 Sorda, Banse & Kemfert 2010.

at both the international and national levels.[25] Nevertheless, most authors focus on the regulation of biofuel in one or two legal systems, among which the EU's regulatory approach to biofuel sustainability has been much discussed. To name a few, Johnson analysed the EU sustainability criteria for biofuels in terms of their relationship with the EU energy strategy and the resulting effects on energy, climate, and development policy.[26] Romppanen has discussed whether biofuel regulation by the EU is adequate to ensure the sustainable production of biofuels in a global scenario.[27] Harrison *et al.* examined the external impacts of EU biofuel regulation on developing countries,[28] and Afionis and Stringer discussed the possibility of EU provide normative leadership in the regulation of biofuels in the world.[29] Furthermore, the regulation of biofuels at the international level has been explored, for instance from the perspectives of the relationship between environment and trade,[30] the relationship between developing and developed countries,[31] or the application of international legal principles.[32] Thirdly, many authors have discussed the governance approach involving private standards or certifications. For instance, the 'meta-standard' approach, based on certification schemes, has been discussed by Endres[33] and Lin,[34] and Schleifer explored biofuel governance from the perspective of the 'orchestration' of private certification schemes and public regulatory frameworks.[35] Romppanen raised the question of whether certification schemes provided as a central implementation mechanism can fully guarantee the sustainability of biofuels,[36] and Lee *et al.* questioned the impact of a certification approach on small-scale farmers.[37] While Echols discussed the compatibility of certification with WTO law,[38] Scarlat and Dallemand argued for a need for international harmonisation on monitoring and a control of certification schemes.[39]

This review reveals some research gaps. Firstly, literature has frequently concentrated on one sustainability concern such as biodiversity or GHG emissions, while the other concerns are not equally addressed in the meantime. Secondly, literature has usually focused on the regulation of developed legal systems such as the EU, while regulation in less developed legal systems such as China is less

25 Morgera, Kulovesi & Gobena 2009.
26 Johnson 2011.
27 Romppanen 2012a.
28 Harrison *et al.* 2010, p. 197-211.
29 Afionis & Stringer 2012, p. 114-123; Afionis & Stringer 2014, p. 47-64.
30 Switzer 2007.
31 Mathews 2007.
32 Lima 2009.
33 Endres 2010.
34 Lin 2011a; Lin 2012a.
35 Schleifer 2013.
36 Romppanen 2012b; Romppanen 2013.
37 Lee *et al.* 2011.
38 Echols 2009.
39 Scarlat & Dallemand 2011.

examined.[40] Thirdly, while the governance approach of private standard or certification is hotly discussed, the regulation involving state activities still deserves a close examination, which will be carried out in this thesis and thereby the term 'regulation' is understood in the following way. As Black writes:

> 'The main textbooks on regulation identify three definitions. In the first, regulation is the promulgation of rules by government accompanied by mechanisms for monitoring and enforcement, usually assumed to be performed through a specialist public agency. In the second, it is any form of direct state intervention in the economy, whatever form that intervention might take. In the third, regulation is all mechanisms of social control or influence affecting all aspects of behaviour from whatever source, whether they are intentional or not.'[41]

The three definitions actually reflect divided views of regulation. The first two definitions are 'clearly "centred" definitions', for which regulation is 'a state activity'[42] or 'state control or governance or direction by rule(s).'[43] The third one represents 'a broader definition'[44] and 'decentred regulation'[45] that 'breaks the connection with the state'[46] and could be 'any mechanism of control or direction.'[47] Although agreeing on the importance of the third definition,[48] this thesis will concentrate on the 'centred' definition of regulation which is promulgated by government or inter-governmental international regimes, for the reason that governmental support or mandate has played a critical role in the increased production and consumption of biofuels.

2. Research Questions

In view of discussions about sustainability concerns and the regulation of biofuels, this thesis is dedicated to answering the research question of 'how the sustainability of biofuels is regulated in the international, EU, and Chinese legal systems', which is analysed according to the following three steps:

40 Wang has examined current legislation on biofuels in China, while many other existing literature usually made analysis from policy-making perspective. Wang 2011; Gan & Yu 2008; Yang, Zhou & Liu 2009; Wang, Xiong & Liu 2009; Wang & Tian 2011; Huang *et al*. 2009.

41 Black 2002, p. 11; Baldwin, Scott & Hood 1998, p. 3; Baldwin & Cave 1999.

42 Black 2002, p. 11.

43 Lee 2011, p. 124.

44 *Ibid.*

45 Black 2002, p. 11.

46 *Ibid.*

47 Lee 2011, p. 124.

48 As commented by Chowdhury and Wessel, 'regulation' has been an area of social science research from the disciplines of not only law but also economics and political science, and literature has observed that the process of regulation has undergone dramatic change from being a limited political activity of the state to a more open-ended process characterized by interaction between public authority and private actors in both soft and hard form. Chowdhury & Wessel 2012, p. 335 and 337.

(1) What are the regulatory approaches to the sustainability of biofuels in the international, EU, and Chinese legal frameworks?
(2) Do they address sustainability concerns in relation to all the environmental, social, and economic dimensions?
(3) Does the concept of 'global environmental law' offer a prospect for an inclusive approach for the regulation of the sustainability of biofuels to avoid negative environmental and socio-economic impacts?

Hence, this thesis will examine the regulation of biofuels in the international, EU, and Chinese legal frameworks, analyse whether the regulations address sustainability concerns in relation to the environmental and socio-economic dimensions, and finally explore the prospect for an inclusive regulatory approach based on the concept of 'global environmental law', if the examined regulations fail to ensure the sustainability of biofuels on a domestic or global scale. The following sub-sections will set forth the scope of the research question and its three-step analysis in detail.

2.1. *Definition of Biofuels*

The term 'biofuels' in this thesis refers to the fuels derived from biomass[49] that are used for road or non-road transport, irrespective of its physical form as a liquid or gas. In fact, there is not yet a unified definition of 'biofuels'[50] and its scope may be interpreted broadly or narrowly. A broad definition of biofuels refers to any type of fuel, as a kind of energy carrier,[51] derived from biological sources, regardless of its physical form (liquid, gaseous, or solid), its way of use (traditional[52] or modern), or

49 There is also no unified definition of 'biomass.' In one report the IPCC defines biomass as 'the total mass of living organisms in a given area or volume; dead plant material can be included as dead biomass', which may include products, by-products, and waste of biological origin (plants or animal matter), but excluding materials embedded in geological formations and transformed to fossil fuels or peat. Verbruggen *et al.* 2011, p. 955.

50 The term 'biofuel' *per se* is controversial, as some argue that the prefix of 'bio-' may wrongly imply ecological or environmental beneficial senses and they prefer the term of 'agro-fuels.' The terms 'biomass fuel' can also be used interchangeably with biofuels. Altieri 2009, p. 236; Levidow & Paul 2010, p. 64; White & Dasgupta 2010, p. 593-607; McMichael 2010, p. 609.

51 There is a difference between 'fuel' and 'energy'; while energy refers to 'the amount of work or heat delivered', fuel is an 'energy carrier intended for energy conversion' and an energy carrier is 'a substance for delivering mechanical work or transfer of heat.' Verbruggen, Moomaw & Nyboer 2011, p. 958; FAO, Bioenergy and Food Security (BEFS), Defintions of key bioenergy and food security terms, available at: <www.fao.org/energy/befs/definitions/en/> (last accessed 24 February 2015).

52 In an IPCC report, traditional biomass refers to biomass such as fuelwood, charcoal, agricultural residues, and animal dung used with the so-called traditional technologies such as open fires for cooking, rustic kilns and ovens for small industries. Traditional biomass is still widely used in developing countries and may lead to indoor or outdoor air pollution and thus can hardly be regarded as clean or renewable energy. Allwood *et al.* 2014, p. 1253.

its end use (power, heating, or transportation).[53] A less broad definition characterises biofuels with a certain physical form (liquid) or a certain method of production, regardless in which sector the biofuel is used.[54] In a narrower sense, biofuels merely refer to the biomass-derived fuel for use in the *transport* sector,[55] which mainly consists of liquid bioethanol[56] and biodiesel[57] that respectively substitute gasoline and diesel. With each definition, biofuel can be divided into three generations according to its raw material (feedstock) and conversion technology: first-generation biofuels are derived from grains, oilseeds, animal fats, and waste vegetable oils with mature conversion technologies; second-generation biofuels use non-traditional conversion processes with feedstocks mostly derived from the cellulosic fractions of agricultural and forestry residues, municipal solid waste; third-generation biofuels can be derived from feedstocks such as algae through advanced processes under development.[58]

53 For example, the IPCC 2011 special report on renewable energy, the UNFCCC, and the FAO adopt a broad interpretation of biofuels as 'any liquid, gaseous or solid fuel produced from biomass', or 'a fuel produced from dry organic matter or combustible oils produced by plants', or 'fuel[s] produced directly or indirectly from biomass.' Verbruggen, Moomaw & Nyboer 2011, p. 955; UNFCCC, Glossary of climate change acronyms, available at: <http://unfccc.int/essential_background/glossary/items/3666.php> (last accessed 24 February 2015); FAO, Bioenergy and Food Security (BEFS), Definitions of key bioenergy and food security terms, available at: <www.fao.org/energy/befs/definitions/en/> (last accessed 24 February 2015).

54 For example, in an IEA biofuel report, the term of biofuels is defined as 'liquid and gaseous fuels produced from biomass, which thus excludes solid forms of biofuels.' In a more recent report of the IPCC Working Group III, biofuels are defined as 'fuels, generally in liquid forms, produced from organic matter or combustible oils produced by living or recently living plants.' A UNEP report on biofuels addressed only the modern use of biomass for energetic purposes, while under the United Nations Industrial Development Organization (UNIDO) biofuels are defined as energy carriers derived from the conversion of biomass to provide sustainable inputs for heat, power, and transport applications. IEA 2011, p. 8; Allwood *et al.* 2014, p. 1253; United Nations Environment Programme (UNEP), Biofuels Vital Graphics: Powering a Green Economy, 2011, available at: <www.unep.org/climatechange/-mitigation/Portals/93/documents/Bioenergy/VBG_Ebook.pdf> (last accessed 24 February 2015), p. 7; United Nations Industrial Development Organization (UNIDO), 'UNIDO's Biofuels Strategy Sustainable Industrial Conversion and Productive Uses of Biofuels', available at: <www.unido.org/fileadmin/user_media_upgrade/Media_center/2013/-Feature_stories/68441_FINAL_DRAFT_UNIDO_BIOFUEL_STRATEGY.pdf> (last accessed 2 February 2016).

55 Bessou *et al.* 2011, p. 5.

56 Bioethanol refers to ethanol produced from starch- or sugar-rich biomass such as sugar cane or corn. Allwood *et al.* 2014, p. 1253.

57 Biodiesel refers to fuel derived from oil-rich biomass such as vegetable oils or animal fats. However, it may be argued that, when produced with methanol as one component, biodiesel is only approximately 95 percent 'bio', as methanol is most commonly derived from non-renewable natural gas. Knothe 2010, p. 364.

58 The second- and third-generation biofuels produced through new processes are also referred to as 'next-generation' or 'advanced' biofuels. Another method of classification divides biofuels into three types totally based on raw materials: woodfuels, agrofuels, and biofuels from municipal wastes. Allwood *et al.* 2014, p. 1253; FAO, Bioenergy and Food Security →

This thesis adopts a narrow definition and confines biofuels to those for use in the transport sector, based on the consideration that biofuels have a relatively more important role in the transport sector than in the other sectors. According to the IPCC, transportation accounts for approximately 23 per cent of the total energy-related carbon emissions, and GHG emissions in transport sector have been increasing due to the continuing growth in passenger and freight activity,[59] which makes it an urgent priority to find alternative energy for transportation. Moreover, compared to the other alternative energy options in the transport sector, biofuels have the advantage that they can be blended with current oil products, which may spare the costs of changing engines and fuel supply facilities.[60] This character makes biofuels listed among the few commercially viable transport decarbonisation options,[61] particularly for heavy vehicles, aircraft, and perhaps ships which can hardly rely on electricity or other alternative fuel in the short- and medium-term.[62]

2.2. *Inclusion of the Three Dimensions of Sustainability*

An essential element of the research question is to analyse whether the examined regulatory approaches fully integrate concerns regarding the environmental, social, and economic dimensions to ensure the sustainability of biofuels, which is grounded in the meaning of 'sustainability' or 'sustainable development.'

To study regulations that address the 'sustainability of biofuel', there is a need to firstly look into what 'sustainability' entails. Defining 'sustainability' is complex[63] due to its multi-faceted and interdisciplinary[64] nature, so there is rarely a uniformly

(BEFS), Definitions of key bioenergy and food security terms, available at: <www.fao.org/energy/befs/definitions/en/> (last accessed 2 February 2016).

59 Sims *et al.* 2014, p. 603.

60 It is estimated that liquid biofuels provided 2.7 per cent of all global road transport fuels in 2009. Biofuels are consumed by being blended with oil products, with an average bioethanol blend rate of 5 per cent (E5) to 10 per cent (E10) and biodiesel blend rate of 2 per cent (B2) to 5 per cent (B5). However, while a low rate of biofuel blends are adaptable to current engine designs, high rate of blends needs a special design of vehicle engines and other facilities such as flexible-fuel vehicles. United Nations Environment Programme (UNEP), Biofuels Vital Graphics: Powering a Green Economy, 2011, available at: <www.unep.org/-climatechange/mitigation/Portals/93/documents/Bioenergy/VBG_Ebook.pdf> (last accessed 2 February 2016), p. 12; IEA 2011, p. 10.

61 It is considered that the competitiveness of biofuels compared to other low-carbon energy sources for transport (such as natural gas, electricity, or hydrogen) stems from the fact that no new distribution networks are needed and that conventional petrol or diesel engines can be fuelled with biofuels, at least up to a certain blend rate. Elbehri, Segerstedt & Liu 2013, p. 65.

62 IEA predicts that biofuels could play a significant role in replacing liquid fossil fuels suitable for planes, marine vessels and other heavy transport modes that cannot be electrified. IEA 2011, p. 7.

63 Bosselmann highlights the similarity between defining sustainability and defining justice, arguing that sustainability is both simple and complex. Sustainability is complex because it is difficult to categorically say what sustainability is, and one cannot define it without a further reflection on values and principles. Bosselmann 2013, p. 9.

64 The idea of sustainability can be traced back to theories of environmental economists or sociologists such as Thomas Malthus, Garrett Hardin, Herman Daly, and the publication of →

accepted definition.[65] 'Sustainability' is at times used interchangeably[66] with 'sustainable development', which has also been considered to be a vague and broad concept,[67] and whether the two terms have the same meaning is disputable.[68] Despite the interdisciplinary and vague features, sustainability or sustainable development has gradually appeared in law-making processes and outcomes on the international and domestic levels since the 1992 Rio Earth Summit,[69] and its meaning, legal status, and implementation have been much discussed by lawyers especially in the field of environmental law.[70] Discussions on the meaning of sustainability or sustainable development cannot avoid tracing it back to the Brundtland Report where sustainable development was formally defined and well accepted politically,[71] and the concept has rarely been defined in international instruments since then.[72] The Brundtland Report defined sustainable development

The Limits to Growth. Environmental economists define sustainability in terms of non-depletion of capital, but there is disagreement on the extent to which advancing technology enables human-made capital to replace natural capital and how far the idea of non-depletion of natural capital should be taken. French admits that the inevitable interdisciplinary nature of sustainable development has ensured that these issues range across the entire disciplinary spectrum relating to both the natural and social sciences. Dresner 2008, p. 3 and 7-27; French 2010, p. 51.

65 Bosselmann 2013, p. 9.

66 The term 'sustainability' is used interchangeably with 'sustainable development' on some occasions. For example, in Agenda 21 'sustainability' and 'sustainable development' are used interchangeably, because, according to Dresner, it is politically important for many people to avoid making a distinction. Dresner 2008, p. 65.

67 For instance, some authors criticised the term 'sustainable development'as a meaningless concept; because different people use the term in different ways - some emphasising development through economic growth and others emphasising sustainability through environmental protection. Montini 2008, p. 523; Dresner 2008, p. 2.

68 Some authors draw a distinction between the two terms, arguing that sustainable development gives priority to development while sustainability is primarily about the environment, or they regard sustainable development as a 'means' to the 'end' of sustainability. Dresner 2008, p. 71; Voigt 2009, p. 4.

69 See for example Segger 2008, p. 88-107, 142-155, and 167-174; Boyle 2008, p. 206-214; French 2010, p. 53 and 63-64; De Sadeleer 2014, p. 13-18; Voigt 2009, p. 21-24.

70 See for example Voigt 2009; French 2010, p. 51-68; Sands & Peel 2012, p. 206-217; Bosselmann 2013, p. 667-680.

71 In 1987, the World Commission on Environment and Development (WCED), which had been formed by the UN General Assembly in 1983, published its report *Our Common Future* which is also known as the Brundtland Report. The Brundtland Report defines sustainable development as 'development that meets the needs of the present without compromising the ability of future generations to meet their own needs.' Such a definition has been well accepted politically and it facilitated the recognition of sustainable development in the follow-up international processes such as the 1992 Rio Earth Summit and the 2002 World Summit. Segger 2008, p. 95-97 and 115.

72 Sustainable development was not defined in the 1992 Rio Declaration and Agenda 21 or the 2002 Johannesburg Declaration and Plan of Implementation or the 2012 Rio+20 outcome of *The Future We Want*. The only international treaty that contains a definition of sustainable development is the *2002 Convention for Cooperation in the Protection and Sustainable Development of the Marine and Coastal Environment of the Northeast Pacific*, and it does not differ much from the Brundtland Report's definition. Vogler 2014, p. 430-446.

in an innovative way by drawing links between the economy, society, and the environment, and presented an optimistic view of integrating the environmental protection and socio-economic development and assuming that the three dimensions - economy, society, and the environment - can be compatible and mutually supportive.[73] Although such a definition has been criticised for a lack of clarity,[74] it nevertheless has political significance[75] and lays the foundation for extracting normative elements from the meaning of sustainable development, such as integration, sustainable use, inter-generational and intra-generational equity.[76] Among the identified elements, integration of the three dimensions embeds the very fundamental and generic essence of defining sustainability and sustainable development.[77]

Based on the definition in the Brundtland Report, sustainable development is first and foremost a concept *bringing together* the environmental protection and socio-economic development, in particular with regard to their interrelationships.[78] The environmental, social, and economic dimensions, which respectively stand for

73 Baker 2006, p. 20, 22, and 24; Dresner 2008, p. 33-34.

74 Some authors have argued that the ambiguity of the definition severely diminishes its usefulness; there is a concern that the lack of clarity in the definition allows anything to be claimed as 'sustainable.' On the other hand, some argue that the lack of clarity has also been politically advantageous, because it has allowed groups with different and often conflicting interests to reach some common ground upon which concrete policies can be developed. Baker 2006, p. 25-27; Dresner 2008, p. 31 and 35; Kane 1999, p. 17-18; Bugge 2008, p. 19; French 2010, p. 54; Bosselmann 2013, p. 670-671.

75 According to Dresner, the Brundtland report sets much of the subsequent agenda for both academic debate about sustainability and international political debate about environment and development. French states that it was the publication of the Brundtland Report that convinced the international community to taking the issues seriously. Dresner 2008, p. 32 and 65, 68 and 36; French 2010, p. 52-53.

76 For example, the Brundtland report's definition identifies the crucial elements of sustainable development as meeting the basic needs, recognising environmental limits, and the principles of inter-generational and intra-generational equity, and, in that sense, sustainable development is not such a vague idea as it seems. Sands concluded four recurring elements which comprise the legal elements of the concept of 'sustainable development' as intergenerational equity, sustainable use, equitable use or intragenerational equity, and integration. French argued for the following principles as the basic framework for sustainable development: integration, sustainable use, equity and the right to sustainable development, and the duty to co-operate. Dresner 2008, p. 67; Bugge 2008, p. 20; Sands & Peel 2012, p. 207; French 2010, p. 58-62; Boyle & Freestone 1999, p. 8-16; Magraw & Hawke 2007, p. 619; Schrijver 2008, p. 229-230; Resolution 3/2002, *Sustainable Development New Delhi Declaration Of Principles Of International Law Relating To Sustainable Development*, The 70th Conference of the International Law Association, held in New Delhi, India, 2-6 April 2002.

77 French noted that sustainable development will only be realised when the principle of integration is properly and fully implemented, since this principle holds the other principles together and it has played a pivotal role in the discussions on sustainable development. Voigt argued that the principle of integration, as a central aspect of sustainable development, provides a legal tool for resolving normative conflicts. According to ILA committee on sustainable development, integration incorporates the very essence of sustainable development. French 2010, p. 59; Voigt 2009, p. 3; ILA 2012, p. 34.

78 Voigt 2009, p. 15; French 2010, p. 51.

the biosphere and natural resources,[79] the human relationships and institutions, and the allocation and distribution of scarce resources, are frequently referred to as the three[80] 'pillars' of sustainable development.[81] In addition to bringing together the three dimensions, sustainable development moreover seeks to *integrate* the economic, social, and environmental dimensions which represent different values and interests.[82] According to Voigt, the strength of sustainable development rests in the necessity of seeking reconciliation and integration between the institutionally disconnected but intrinsically interlinked dimensions.[83]

However, divergence exists as to how to understand the principle of integration. On the one hand, the principle of integration has been largely formulated to emphasise the need to integrate environmental considerations into economic and other social development plans, since the purpose of Principle 4 of the Rio Declaration is to ensure that development decisions do not disregard environmental considerations.[84] On the other hand, as the 2002 Johannesburg Plan of Implementation stresses 'the integration of the three components of sustainable development - economic development, social development and environmental protection - as interdependent and mutually reinforcing pillars',[85] integration can be interpreted as meaning that the pursuit of pure environmental values is not what

79 According to Kiss, the term 'environment' can describe a limited area, the entire planet, or even include a part of the outer space that surrounds the Earth, and the term 'biosphere' is used in particular by the United Nations Educational, Scientific, and Cultural Organization to designate the totality of the human environment or the part of universe where all life is concentrated according to present knowledge. Kiss 2005, p. 6.

80 Although a fourth 'pillar'of culture or institution has been argued, such an effort has been commented as an 'ever-expanding number of self-serving interpretations' since the original three-dimension formulation is regarded as sufficient. Gaines 2014, p. 9.

81 The 'pillars' image in an architectural style depicts that sustainability consists of at least the economic sustainability, social sustainability, and environmental sustainability, and the whole edifice cannot stand if any one pillar is weak. The three dimensions can also be termed as environmental sustainability, social sustainability, and economic sustainability. However, Winter argues for a 'two pillars' diagram that emphasises natural resources as the fundament: instead of the edifice with the three pillars of equal status, the more appropriate illustration should be that the two pillars of economy and society stand on the fundament of natural resources, with the ultimate purpose of 'future generation' as the ceiling. Baker 2006, p. 7 and 26; Drexhage & Murphy 2010, p. 6; Winter 2008, p. 27-28.

82 As commented, the objectives of environmental protection, economic development, and social justice often pull in different directions, and if dealt separately they would run counter to each other. Voigt 2009, p. 4-5 and 15.

83 Voigt 2009, p. 4-5 and 15.

84 Principle 4 of Rio Declaration provides that 'environmental protection shall constitute an integral part of the development process and cannot be considered in isolation from it.' Boyle and Freestone stated that the real implications of Principle 4 are more to be found in its impact on developing countries for not neglecting environmental consideration in development planning. Rio Declaration On Environment And Development, available at: <www.un.org/documents/ga/conf151/aconf15126-1annex1.htm> (last accessed 27 February 2015); Boyle & Freestone 1999, p. 10-11.

85 Para. 2 of the World Summit on Sustainable Development Plan of Implementation, available at: <www.johannesburgsummit.org/html/documents/summit_docs/2309_planfinal.htm> (last accessed 27 February 2015).

sustainable development intends to serve.[86] Sands and Peel interpreted the integration as not only the integration of environmental consideration but also the need to take account of development in applying environmental objectives.[87] Segger pointed out that the focus of sustainable development is on the need to reconcile and integrate environmental protection and social development with economic development by taking all three pillars into account in a balanced way.[88]

Moreover, ambiguity and difficulty exist in how to apply the principle of integration and how to reconcile and balance the three dimensions, particularly when their values and interests conflict. There is concern that the reconciliation and balance would *de facto* subordinate environmental policies to economic imperatives,[89] so it is argued that the economic and social development goals must be defined in terms of natural resource limits.[90] In practice, although the principle of integration has been affirmed in case law, the question of how to reconcile and balance the three dimensions and the result of reconciliation and balance have rarely been decided by courts. For instance, the International Court of Justice (ICJ) in the Gabčíkovo-Nagymaros Case stated that '[T]his need to reconcile economic development with protection of the environment is aptly expressed in the concept of sustainable development',[91] but meanwhile ruled that 'it is not for the Court to determine what shall be the final result' and it is for the Parties themselves to find an agreed solution.[92] The Court made no judgment as for whether the project at stake was or was not sustainable, but left the adjustment of balance between environmental protection and economic development to the parties themselves, thus avoiding debates about its substantive content while focusing on improving the processes of management.[93] In the same vein, in the Pulp Mills Case, the ICJ reaffirmed 'the balance between economic development and environmental protection that is the essence of sustainable development',[94] but the reference to reconciliation between economy and environment in the case was not included into the Court's assessment of the law and facts and only the procedure of

86 Boyle & Freestone 1999, p. 11.
87 Sands & Peel 2012, p. 208.
88 Segger 2008, p. 115.
89 Boyle & Freestone 1999, p. 11.
90 For instance, Dresner argues that the goals of economic and social development must be defined in terms of sustainability and the environment's limit, irrespective of whether it is in developing or developed countries, as the 'sustainable' part is based on the idea of limitations imposed by the state of technology and social organisation on the environment's ability to meet present and future needs. Voigt states that it is impossible to ignore the fact that ultimate limits to human conduct are set by securing the functioning ecological basis on which all life and all human activities depend. Dresner 2008, p. 67 and 76; Voigt 2009, p. 5.
91 ICJ Case Gabcikovo-Nagymaros Judgment 1997, para. 140.
92 ICJ Case Gabcikovo-Nagymaros Judgment 1997, para. 141.
93 However, the Court did not expressly refer to the Environmental Impact Assessment (EIA), while Judge Weeramantry in his separate opinion argued that a continuing obligation to carry out EIA was imported to the joint development treaty between the litigants. ILA 2012, p. 9.
94 ICJ Case Pulp Mills 2010, para. 177.

environmental impact assessment was stressed by the Court.[95] One reason could be that, as stated in Vice-President Weeramantry's opinion in the Gabčíkovo-Nagymaros Case, whether development is sustainable as for its impact on the environment is a question to be answered in the context of the particular situation involved.[96]

Accordingly, it might be difficult for the law to decide how the three dimensions can be reconciled and balanced or what the substantive result of reconciliation and balance would be, which should be best examined in specific cases. However, the integration principle has the value of providing a foundation for *questioning* whether the substance of the law or policy includes a consideration of the values and interests in all the environmental, economic, and social dimensions to the extent that it avoids or mitigates potential negative impacts in any of the dimensions. According to Gaines, weaving together environmental, economic, and social conditions constitutes one core element of sustainable development, so there is a need for an integrated evaluation of the social, economic, and environmental consequences of policy choices and interventions, because any policy that does not take into consideration all the three factors is unlikely to lead to sustainable development on a global scale.[97]

Based on the afore-mentioned analysis of the meaning of sustainability, the regulation of the sustainability of biofuels will be examined from the perspective of whether they inclusively address concerns in the environmental and socio-economic dimensions. In fact, such an analysis of biofuels policy can be traced back to as early as the Brundtland Report, in which the Brundtland Commission already noted the social, environmental, and economic impacts of biofuels.[98] Thus, this thesis concentrates on the question of whether the examined biofuel regulation includes and addresses environmental, economic, and social sustainability concerns with a view to mitigating or avoiding their negative impacts, instead of making a judgement on how to reconcile and balance the concerns. When the regulation on biofuels in a legal system dismisses any dimension and the legal system involves different levels of regulatory authority, such as the EU, this thesis examines further whether the regulation leaves a possibility for the authority of a lower level to adopt *complementary* requirements to address the dismissed sustainability dimension or to set *stricter* requirements for the sustainability concern already addressed. Analysis from the perspective of inclusiveness is meaningful because dismissing one dimension in the regulation of biofuels would lead to unsustainable consequences

95 Tladi 2012, p. 8.

96 ILA 2012, p. 34.

97 Gaines 2014, p. 10 and 14.

98 The Brundtland Commission noted that biofuels would become *uneconomical* with low oil prices and have their own special *social* and *environmental* hazards given the competition with food crops for agricultural land, organic waste effluent, water pollution caused by fertilizers, but on the other hand biofuels would provide additional benefits of rural development, employment generation, and reduced vulnerability to crises in the world oil markets. World Commission on Environment and Development, 'Our Common Future', 1987, paras. 79 and 83.

on a domestic or global scale.[99] Furthermore, as the regulation of biofuels involves multiple goals for climate change mitigation, energy security, and rural economy, it is difficult to find a way to find a balance between the different goals and values, but it is more viable and necessary to examine whether both the environmental and socio-economic impacts are inclusively considered to avoid or minimise possible adverse effects.

2.3. *Potential for a Global Inclusive Approach*

Should the examined regulatory approaches at either the domestic or international level prove to be deficient in addressing the sustainability concerns of the three dimensions, a further question is whether an inclusive approach can be expected for regulating the sustainability of biofuels. As mentioned in section 1 of this Chapter, the sustainability concerns in relation to biofuels not only have local and national implications but they also may have global relevance, which might call for a regulatory approach that goes beyond the international or domestic legal systems. This thesis will explore this question by drawing on the conclusive observations from examining and comparing the current international, EU, and Chinese regulation, and furthermore on the conceptions that call for evolution of 'global environmental law', for example in the essay of Yang and Percival.[100]

Recent years have seen a flood of literature addressing the limits of regulating the emerging and complex global environmental issues, such as climate change, through a 'top-down' approach by national or international public authorities.[101] Discussions on unconventional regulatory and governing approaches to the complex global environmental issues have led to arguments for 'multi-level',[102] 'transnational',[103] 'polycentric',[104] 'regime complex',[105] 'network',[106] 'orchestrator',[107] or 'transnational'[108] governance, which can also be linked to the situation governing biofuels. For example, Scott cited the EU's biofuel sustainability scheme as an

99 Gaines has applied a holistic analysis for examining the EU's biofuel policy, concluding that 'the matter of the biofuels mandate in Europe is a vivid example of how shortcomings in holistic analysis may have led to an unsustainable policy', because the policy fails to prevent sustainable development drawbacks such as the effect on food and feed production of shifting some agriculture to crops for fuel, land conversion to agricultural uses, and an incomplete assessment of full life-cycle of biofuel cultivation, refining, and fuel distribution. Thus, it is argued that a more thorough holistic analysis of biofuels before committing to their use should have anticipated these sustainability problems and allowed for more calibrated decisions about whether or under what conditions the promotion of biofuels would further sustainable development. Gaines 2014, p. 17-18.

100 Yang & Percival 2009.

101 See for example Hoffmann 2014, p. 433-434.

102 Kern & Bulkeley 2009; Scott 2011a.

103 Andonova, Betsill & Bulkeley 2009; Pattberg & Stripple 2008; Bulkeley *et al.* 2012.

104 Cole 2011; Ostrom 2014.

105 Keohane & Victor 2011; Abbott 2012.

106 Bäckstrand 2008; Juhola & Westerhoff 2011.

107 Abbott & Snidal 2010; Hale & Roger 2014.

108 Pattberg & Stripple 2008; Andonova, Betsill & Bulkeley 2009; Bulkeley *et al.* 2012

example of multi-level governance in a global dimension;[109] De Beer analysed the regulation of biofuels in a setting of multi-level and network governance;[110] Romppanen discussed the role of private certification schemes in the EU's biofuel regulation and the use of 'new governance' and 'co-regulation' approaches.[111] In the meantime, some authors emphasise the effect of globalisation in legal developments and argue for the convergent or common values and principles as components of 'global law' applicable worldwide.[112] For instance, arguments for the emergence of 'global administrative law' emphasise the increasing 'reach and forms of transgovernmental regulation and administration designed to address the consequences of globalized interdependence.'[113] In the environmental law area, Wiener has used the term 'global environmental law' to describe the 'vertical' borrowing between national law and international law,[114] and Hey stresses strengthening the justice between developing and developed states and the exercise of public power at a global level on the basis of proper administrative principles in 'global environmental law.'[115] The 'global environmental law' argued by Yang and Percival is defined as 'the set of legal principles developed by national, international, and transnational environmental regulatory systems to protect the environment and manage natural resources', which is supported by the evolving process of transplantation, convergence, harmonisation, and integration.[116]

This thesis draws on the conceptual framework of 'global environmental law' defined by Yang and Percival for exploring the prospect of achieving an inclusive regulatory approach to the sustainability of biofuels for the following reasons. Firstly, 'global environmental law' is characterised by blurring the boundary between national law and international law,[117] which coincides with the feature of the regulation on sustainability of biofuels that involves both international and domestic law. Secondly, Yang and Percival argued that global environmental law evolves from the processes of 'harmonization', 'integration', 'transplantation', and 'convergence',[118] which has particular implications for the question of whether an inclusive regulation for biofuel' sustainability may evolve based on the existing

109 Scott 2011b, p. 30-31.

110 De Beer 2011.

111 Romppanen 2013, p. 340.

112 The idea of 'global law' could be traced back to as early as Kant's 'eternal peace' by uniform globalisation. Ehrlich, however, found that the centre of gravity of globalisation lay in society itself and argued for 'Global Bukowina' – meaning that civil society itself will globalise its legal orders, since *lex mercatoria* might be the most successful example of global law without the state. On the basis of *lex mercatoria*, Teubner developed a broader theory of global legal pluralism with the hypothesis that law is not generated by the states but instead creates itself and that the center of lawmaking moved away from the state but into the periphery of transnational actors. See for example Teubner 1996.

113 Kingsbury *et al*. 2005, p. 16.

114 Wiener 2001.

115 Hey 2009.

116 Yang & Percival 2009, p. 623.

117 Yang & Percival 2009, p. 616, 623, 664.

118 Yang & Percival 2009, p. 616, 623, 664. See Chapter 1, section 3.5.

domestic and international regulating efforts.[119] 'Transplantation' is hereby defined as the *deliberate* copying and adaptation of significant portions of statutes or particular doctrines of law into one country from another.[120] Different from the deliberate/purposeful copying, the term 'convergence' describes *unintended* similarities between disparate legal systems as a response to common/similar external pressures rather than deliberate copying.[121] 'Integration' and 'harmonization' both refer to *joint multi-country* efforts or cooperation to achieve some level of uniformity with respect to particular laws or legal issues, which are thus different from uni-directional transplantation or convergence.[122] Meanwhile, a distinction can be drawn between 'integration' and 'harmonization': integration is defined as the process of *linking* separate legal systems,[123] whereas harmonisation refers to the *adjusting and conforming* of disparate standards and requirements of different legal systems to an international system or to each other.[124]

119 For instance, Yang and Percival observed the evolvement of global environmental law from the impact of the UNFCCC and national environmental law. Yang & Percival 2009, p. 626-635.

120 This definition follows the description by Watson. Yang and Percival cite examples of the spread of environmental impact assessments from the United States to many other countries and international treaties, the borrowing from the US Clean Air Act to Kyoto Protocol and to air quality regulation in China and Thailand. Yang & Percival 2009, p. 626; Watson 1993, p. 22-24.

121 The common or similar external pressures include common functional goals, governance considerations, and ecological and public health constraints have driven the design, implementation, and operation of regulatory systems in similar directions. The examples of convergence include convergence in the corporate law context discussed by Hansmann & Kraakman; the greater involvement of civil society in environmental concerns; more stringent environmental protection measures adopted by US states to tackle global climate change problems. Yang & Percival 2009, p. 627, and 631-635; Hansmann & Kraakman 2001, p. 468.

122 Such multi-country cooperation can most easily be seen in formal international treaties and institutions, such as the WTO. Integration and harmonisation are designed to coordinate and facilitate cooperation in order to achieve an environmental objective. Yang & Percival 2009, p. 627 and 636.

123 The author identifies climate change as the 'most visible driver of regulatory integration', particularly with regard to the emission trading scheme and CDM projects. The international emission transaction log and CDM registrar have facilitated international and transnational regulation by promoting the integration of emission trading regulatory schemes worldwide. National and regional implementation schemes of the Kyoto Protocol, especially the European emission trading system, have become closely integrated with each other and with the requirements of the Kyoto Protocol. Yang & Percival 2009, p. 636-638.

124 This definition of harmonisation is similar to the one adopted by Faure that harmonisation 'refers to a legislative process whereby various pieces of environmental legislation are either brought together in one new document, or are at least coordinated.' Yang and Percival take the example of international standard formulation by ISO: ISO standards and certification are reinforcing the idea within multinational corporations that use of uniform operating standards and practices with respect to pollution, worker safety, and other matters. Furthermore, the International Maritime Organization (IMO) has taken on a significant leadership role in coordinating national marine pollution standards and other marine environmental protection efforts around the world. Yang & Percival 2009, p. 635-636, 638-640; Faure 2000, p. 174.

3. Methodology

This thesis will reach general concluding observations through desktop research, text analysis, and a comparative study of the regulation of biofuels in the international, EU, and Chinese legal frameworks. This section aims to set forth reasons for the following questions: why the international, EU, and Chinese legal systems are studied; how the comparative analysis was conducted; what analytical model was used in the analysis; and how the literature and information were collected for this desktop research.

3.1. *Study of the International, EU, and Chinese Legal Systems*

This thesis studies the regulatory approaches in the international, EU, and Chinese legal frameworks. Relevant rules in the international framework are examined to present the international setting against which biofuels are regulated. The sustainability of biofuels is not a completely domestic issue and may have international ramifications, but there is not yet a multi-lateral treaty for biofuels. Mapping the international setting may reveal what elements exist for the regulation of the sustainability of biofuels within the current international treaty regimes, thus clarifying the 'room' for domestic regulators to deal with sustainability of biofuels.

Among domestic legal systems, this thesis chooses to conduct case study of the regulatory approaches in the EU and China for the following reasons. Firstly, both the EU and China have become the leading consumers and producers of biofuels. The biofuel penetration rate in the EU's transport energy mix reached 4.9 per cent in 2015, which amounted to a total biofuel consumption of 14 million tonnes of oil equivalent (Mtoe),[125] and the EU remains the top producer of biodiesel in the world.[126] China is among the largest bioethanol producing countries in the world after the US, Brazil, and the EU.[127] Secondly, both the EU and China have adopted regulatory measures that attempt to tackle certain sustainability concerns about biofuels. The EU has established a binding sustainability scheme for biofuels which is declared by the European Commission as being 'the most comprehensive and advanced' scheme in the world.[128] Although China has not set out a similar binding sustainability scheme as the EU, the concerns regarding food security and land use impacts have been highly emphasised in China's biofuel regulation and policy.[129]

125 EurObserv'ER, Biofuels Barometer 2015, available at: <www.eurobserv-er.org/biofuels-barometer-2015/> (last accessed 14 February 2015).

126 According to estimates, the sum of biodiesel produced by Germany and France exceeds the US' production. The world's biggest biodiesel producers in 2014, available at: <www.statista.com/statistics/271472/biodiesel-production-in-selected-countries/> (last accessed 12 February 2016); Sorda *et al.* 2010, p. 6977.

127 Renewable Fuels Association, World Fuel Ethanol Production, available at: <http://ethanolrfa.org/pages/World-Fuel-Ethanol-Production> (last accessed 2 February 2016).

128 Commission Communication (2010/C160/02).

129 Renewable Energy Medium- and Long-Term Plan 2007. See discussion in Chapter 4, section 2.

Finally, as there is not yet a unified definition of the sustainability of biofuels, the examination and comparison of the EU and Chinese biofuel regulations may provide insight into how the two different legal systems integrate different sustainability concerns related to biofuels.

3.2. *Comparative Analysis*

Based on the study of the regulation of biofuels in the international, EU, and Chinese legal systems, this thesis will draw its concluding observations based on a comparative law methodology. Comparative law traditionally follows an approach of setting specific 'objects' of at least two jurisdictions against each other, determining their similarities and differences, explaining the causes, and evaluating the solutions.[130] In addition to the discovery, explanation, and evaluation of similarities and differences, comparative law may also have the purpose of exploring the influence between different legal systems and developing a general theory of law.[131] Comparative study in this thesis generally follows the approach of discovering the similarities and differences (interchangeably called 'convergence' and 'divergence')[132] of regulation of the sustainability of biofuels (as the specific 'object') in the 'different jurisdictions' of the international, EU, and Chinese legal systems.

Furthermore, the comparative analysis in this thesis may be distinguished from the traditional approach of comparative study as it is *problem-oriented.* Comparative law traditionally takes national 'legal systems' or 'families of law' as the starting point,[133] but a generalisation about national traits may become more difficult nowadays because hybridisation in law tends to blur the boundaries of legal systems and families. Therefore it is difficult to expose generalised contrasts between legal systems where there is a high rate of hybridisation.[134] In this context,

130 Kokkini-Latridou explained comparison as 'scholarly process in which specific "objects" of at least two jurisdictions are set against each other (1) to determine their similarities and differences; (2) to explain the causes of the similarities and differences and (3) to evaluate the solutions.' Kokkini-Latridou 1986, p. 155; Boele-Woelki 2008, p. 7; De Cruz 2010, p. 36.

131 Michaels 2012, p. 298.

132 Knill identified that there is a broad consensus on the definition of convergence as 'the tendency of societies to grow more alike, to develop similarities in structures, processes, and performances' and stated that 'the most basic way of assessing policy convergence is to analyse the extent to which the policies of countries have become more similar to each other over time.' Based on this definition, the thesis uses the terms of 'convergence' and 'divergence' interchangeably with 'similarities' and 'differences.' Knill 2005, p. 765-768.

133 Wiener 2011, p. 552.

134 According to Wiener, diffusion and hybridisation are the two powerful forces to make national legal systems open to changes and thus blur the legal 'systems' or 'families.' Hybridisation is not understood by Wiener as the same as convergence but as entailing mixed exchanges of diverse ideas that overlap with diversity at the level of particular rules and policies. Based on this observation, Wiener argued that where legal systems are involved in high rates of borrowing, diffusion, and hybridisation, it will be increasingly difficult and unhelpful to expose stark generalised contrasts between them. Wiener 2011, p. 553-554; Wiener 2004, p. 75-77.

a comparative study might be conducted in a 'pragmatically applicable' way by starting with major public policy questions faced by legislators and regulators,[135] and thus even the intrinsically different legal systems can be compared.[136] In the same vein, instead of generalising stark contrasts between different legal systems, the comparative study in this thesis starts from and finally comes to conclusions about the public policy problem of how to regulate the sustainability of biofuels.

First, a problem-oriented approach provides for the starting point of a comparison of the intrinsically different international, EU, and Chinese legal systems. Comparative law primarily addresses legal systems belonging to the same level or echelon - either among different legal systems at the national level (e.g. a comparison between US and China) or among various legal regimes at the international level (e.g. a comparison between international trade and environmental law).[137] More recently, the vertical comparison of different levels or echelons has been noted,[138] despite the assumption of inherent incompatibility between the national and international legal systems.[139] However, the horizontal and vertical classification could not completely depict the comparison among the international, EU, and Chinese legal systems, since the EU and China are neither of the same horizontal level (i.e. supranational and national) nor on the same vertical line (i.e. between the EU and China). The problem-oriented comparative approach, in this sense, may fall under the scope of a *functional* comparison that identifies problems and effects as the yardsticks and views the even doctrinally different

135 Markesinis 2003, p. 53; Whitman 2007; Wiener 2011, p. 554.

136 Wiener argued against the view that fragility of generalisations means an end of comparing national legal systems or the view that comparative law is impossible because legal systems are intrinsically different from each other. Wiener argued that comparisons can and must be made, but on the basis of a much more systematic empirical study and an improved approach will involve a comparison of rules and institutions as modules or memes that can be exchanged across interconnected legal systems, rather than of categorical generalisations about national legal systems or legal origins. Wiener 2011, p. 553-554.

137 A comparative study can be generally classified into one of two types: horizontal and vertical, based on which Momirov and Fourie divided horizontal types into two sub-categories: among legal systems at the national level and among legal systems or regimes at the international level. Another type of classification divides a comparison into internal and external: the former refers to vertical levels or horizontal branches of a single legal system, while the latter refers to different legal systems and different research fields. Momirov & Fourie 2009, p. 295; Monateri 2012, p. 146-149.

138 Momirov and Fourie classified a vertical comparison into the 'top-down' type and 'bottom-up' type: the former typically refers to the internalisation of international norms and regulation by a national legal order whereby national law is required to incorporate international concepts into the national legal system; the latter refers to the transposition of legal concepts, or the ideas behind them, from national to international level. Wiener also noted that there are many examples of trans-echelon vertical legal borrowing between national levels and especially pointed out the phenomenon of borrowing from the national level to the international legal system. Momirov & Fourie 2009, p. 296; Wiener 2001, p. 1297-1301.

139 This kind of legal comparison has been viewed skeptically due to the assumption of an inherent incompatibility between the national and international legal orders. Momirov & Fourie 2009, p. 296; Dannemann 2006, p. 390.

institutions as functionally equivalent and comparable.[140] Such an approach may invite criticism for example for 'methodological eclecticism',[141] but pragmatism[142] can serve as a justification for carrying out a comparative study on the international, EU, and Chinese systems.[143]

Second, the problem of regulating the sustainability of biofuels is also where the comparative study concludes. The observations from comparative analysis will provide the basis for answering the question of whether there is prospect for a global inclusive approach to ensuring the sustainability of biofuels. Conventionally, a comparative study leads to a conclusion that informs national law-makers of 'better' law or arguments for legal unification.[144] This would also be the case for a functional comparative study.[145] The comparative analysis in this thesis, however, does not aim to determine which law is 'better' nor does it attempt to argue for the international legal unification of the regulation of biofuels, but it focuses on the problem of how the sustainability of biofuels is regulated to avoid or minimise the potential negative impacts. Bearing this in mind, this thesis will examine the implications of 'global environmental law' for the comparative analysis and the problem of the regulation of the sustainability of biofuels. Global environmental law is depicted as a process which is evolving dynamically through the trends of 'convergence', 'transplantation', 'harmonization' and 'integration',[146] which makes it different from the sole argument for legal unification. Percival and Yang already outlined the implications of global environmental law for comparative environmental law: an effective understanding of global environmental law requires 'looking beyond superficial similarities to gain a good grasp of principles, practice, and outcomes of different environmental governance systems and their

140 Functional comparison was popularised by Konrad Zweigert and Hein Kötz and starts from the premise that the function of law lies in responding to social problems and that all societies face in essence the same problems. This makes it possible to compare legal institutions, even if they display different doctrinal structures, as long as they fulfil the same function, because in this case they are functionally equivalent. Michaels 2012, p. 299; Michaels 2006, p. 342; Zweigert & Kötz 1998, p. 34.

141 The functional method may meet with criticism, for example for methodological eclecticism, but such criticism could be justified by pragmatism or methodological pluralism. Michaels 2006, p. 362.

142 Michaels 2006, p. 362.

143 For example, a number of authors have contributed to the book on EU, China, and global governance, covering a wide range of topics such as trade, climate change and energy, and financial and monetary systems. Wouters *et al.* 2012.

144 Comparative law generally has the following purposes: (1) to inform national lawmaking, (2) to assist judges in the resolution of difficult questions, (3) to provide a basis for legal unification or harmonisation, or (4) to simply increase knowledge and extend awareness. In other words, the functions of a comparative study may focus on either understanding the legal rules and institutions to adopt better laws or as a critique of the law, or formalising functions of system building and preparing for legal unification. Michaels 2012, p. 298; Michaels 2006, p. 363.

145 It is stated that understanding legal norms as responses to problems supposedly also makes it possible to designate which law is better and, on this basis, to reform domestic law or to create an international uniform law. Michaels 2012, p. 299.

146 Yang & Percival 2009, p. 626-636.

functional components.'[147] Thus, this thesis will not only outline the similarities and differences using the comparative analysis, but it also addresses what the similarities and differences imply for the development of a global inclusive regulatory approach to sustainability of biofuels.

3.3. *Analytical Model for Text Analysis*

In analysing and comparing regulatory approaches in the international, EU, and Chinese legal systems, one problem that arises is how to map the 'instrument mix' or even 'instrument mess'[148] in relation to biofuels, since the sustainability of biofuels concerns both environmental and social-economic dimensions and has an impact of various scales, which may involve regulations in different sectors and at different levels.[149] In this view, this thesis adopts a 'demand-supply-trade' analytical model in the international, EU, Chinese, and comparative chapters, so as to provide a systematic and coherent analysis of the regulatory approaches to the sustainability of biofuels.[150]

The terms 'demand' and 'supply' are used in economics to describe consumers' or producers' desire, willingness, and ability to buy or to produce specific goods or services at specified prices,[151] while public authorities can exert

147 In this sense, the authors may have agreed with the functional comparative method, because they admitted that 'provisions that appear to be similar might function rather differently and have little in common; conversely, seemingly disparate schemes might be functionally equivalent because of their substantive effect or analogous operation.' Yang & Percival 2009, p. 660-661.

148 Peeters questioned whether the legislative package which supports EU climate policy is so vast and diverse that it has become an 'instrument mix' or even an 'instrument mess.' For a good understanding of specific provisions and related potential legal conflicts, see Peeters 2014, p. 173.

149 It has been commented that the issue of biofuels is exceedingly complex in economic and political terms, which makes its regulation complex and diffused. Specifically, the forward and backward linkages with other highly volatile and price-distorted markets such as oil and agriculture make the economics of the biofuels industry highly complex. Diffused decision-making and policy implementation adds political complexity, because authority over policy making and associated regulatory rule making is dispersed among different actors and at different levels. De Beer 2011, p. 1-2.

150 Biofuels are usually analysed through the supply chain or value chain from crop production to biofuel processing to final consumption, but the demand-supply-trade analysis was once used for analysing bioenergy prospects in the Europe. UNEP, 'Towards Sustainable Production and Use of Resources: Assessing Biofuels', 2009, available at: <www.unep.fr/scp/rpanel/pdf/assessing_biofuels_full_report.pdf> (last accessed 23 November 2012), p. 51; ICTSD, 'Biofuel Production, Trade and Sustainable Development', International Centre for Trade and Sustainable Development, 2008, available at: <http://ictsd.org/i/publications/46253/> (last accessed 27 November 2012), p. 9; Karin Ericsson, 'Prospects for Bioenergy in Europe – Supply, Demand and Trade', 2006, Lund University Thesis, available at: <www.lunduniversity.lu.se/o.o.i.s?id=24732&postid=-547515> (last accessed 11 December 2012).

151 O'Sullivan & Sheffrin 2003, p. 79.

'facultative constraints'[152] on human behaviour in society and thus influence the 'aggregate' demand and supply[153] through regulatory instruments such as command-and-control or market-based mechanisms.[154] The influence of regulatory instruments is evident for the demand and supply of biofuel's, because the biofuel industry has hinged on governmental support policies and market interventions that stimulate production and consumption.[155] In view of sustainability concerns, the question is whether and how such regulatory instruments incorporate sustainability concerns and constrain the demand and supply of unsustainable biofuels. Trade measures related to biofuels will be discussed separately in the chapters because of their external influence and international law implications.

Demand-side regulation to be discussed in the thesis includes the instruments that create and influence the demand for biofuels, such as a biofuel consumption target. Biofuel consumption targets are usually presented in the form of setting a minimum amount or percentage of biofuels in the overall transport fuel consumption, either as a certain percentage of biofuel in the transport fuel sold to consumers (also called blend mandate) or as a certain volume/quantity of biofuels consumed within a fixed period.[156]

Supply-side analysis in this thesis is focused on the regulation of the sustainability of the production of biofuels. As biofuels are currently mainly produced from agricultural crops,[157] the regulation on agriculture and land management is particularly relevant for the sustainable production of biofuels. Therefore, this thesis will concentrate on examining the regulation of biofuel feedstock production to see whether environmental and socio-economic dimensions are considered.

In the trade-side section this thesis explores the regulation of the sustainability of biofuels in international trade. For example, if the domestic sustainability

152 'Facultative constraints' refer to restrictions that are designed by humans and therefore are open to both change and violation and can be thought of in terms of property rights, obligations, and/or liabilities, and a more generally combination of these forms. Weigel 2008, p. 9-10.

153 In macroeconomics, aggregate demand or supply is the total demand or supply for final goods and services in an economy at a given time. It specifies the amounts of goods and services that will be purchased or sold at all possible price levels, which is the demand or supply for the gross domestic product of a country. Sexton 2015, p. 630; O'Sullivan & Sheffrin 2003, p. 307.

154 The command-and-control method is reflected in setting performance standards and permits or licenses, while market-based mechanisms could be tax levies and emission/pollution trading. Keohane, Revesz & Stavins 1998, p. 313-314.

155 Market intervention may include policies designed to stimulate domestic production through subsidies, R&D investment, import tariffs, budgetary support, and demand by mandating biofuel uses, and, as a result, the biofuel market is heavily price-distorted. De Beer 2011, p. 1; Charles 2013, p. 3-4.

156 De Gorter & Just 2009, p. 738.

157 There is no agreed-upon common definition of 'agriculture' and 'crop', and the ambiguity lies in whether tree stands in agricultural production systems, such as fruit plantations and agro-forestry, are identified as forestry or agricultural crops. FAOSTAT, available at: <http://faostat.fao.org/site/375/default.aspx> (last accessed 5 March 2015).

requirement for biofuels is also applied to biofuels imported from third countries, such a requirement has an impact on the international trade of biofuels. A separate discussion of trade-related regulation is necessary because the demand for biofuels is at times met by a supply from third countries and whether the biofuels are sustainably produced in third countries should be considered in the regulation of biofuels in order to achieve the objective of sustainable production.

3.4. *Literature and Information Collection*

Academic literature provides the major source of information and arguments in this thesis. Such literature was sourced in the legal databases provided by Maastricht University's library, such as the HeinOnline, SSRN, Westlaw, Picarta, and Google Scholar. Chinese literature can be accessed through the databases of China Academic Journals (CAJ) and Chinese Journal Full Text Database (CJFD). The key words used for searches included but were not limit to: 'biofuel', 'bioenergy', 'biomass energy', 'sustainability', 'EU biofuel/bioenergy', 'China biofuels/-bioenergy', or 'international biofuels/bioenergy', and the key word of 'biofuel' generated the majority of results. In addition to academic literature, third-party reports and studies conducted by international organisations (e.g. the Food and Agriculture Organisation and the International Energy Agency), non-governmental institutes (e.g. the International Institute for Sustainable Development), and scientific bodies (e.g. the IPCC and the European Environment Agency Scientific Committee) form another source of information.[158]

Regulatory documents and instruments were primarily accessed on the websites of the issuing authorities. International instruments were obtained on the official websites of the issuing international treaty regimes and the United Nations Treaty Collection database.[159] EU's legal documents are published in the Official Journal of European Union and were acquired via EUR-Lex,[160] and other official documents or inter-institutional decision-making processes were accessed on the official website of the EU[161] or the European Parliament's Legislative Observatory.[162] Information on the related legislation of the EU Member States was accessed on EUR-Lex,[163] and specific legislation regarding biofuels could be found and even

158 Hirschl *et al.* 2012 p. 160.

159 The database was established in accordance with Art. 102 of the UN Charter: treaties and international agreement entered into by any UN member shall be registered with the Secretariat and published by it. United Nations Treaty Collection, available at: <https://treaties.un.org/Home.aspx?lang=en> (last accessed 3 March 2015).

160 EUR-Lex, Access to European Union Law, available at <http://eur-lex.europa.eu/oj/direct-access.html> (last accessed 3 December 2015).

161 European Union, Official website of the European Union, available at: <http://europa.eu/-index_en.htm> (last accessed 3 December 2015).

162 European Parliament, Legislative Observatory, available at: <www.europarl.europa.eu/-webnp/cms/pid/17> (last accessed 3 December 2015).

163 EUR-Lex, Access to European Union Law, National Law, available at: <http://eur-lex.europa.eu/collection/n-law.html> (last accessed 3 December 2015).

compared in the database of 'RES LEGAL.'[164] Chinese official documents are compiled in the Gazette published by the central and provincial governments which were mostly accessed on the official websites of the Central People's Government or local government.[165] Chinese legal instruments were also sourced via a database provided by the Legislative Affairs Office of the State Council[166] or via unofficial databases for lawyers.[167] As most Chinese instruments are not officially translated into English in full text,[168] this thesis relied on non-official translations provided in the databases[169] or literature,[170] and when multiple non-official versions of translations exist, this thesis chose the most relevant version or translated the text from the original text with *pinyin* attached. Furthermore, face-to-face interviews, email correspondence, and news from media provided an additional source of information for this thesis. Interviews were carried out with biofuel policy officers from the European Commission DG Climate Action,[171] Chinese expert on environmental and climate law,[172] and biofuel consultant from ECOFYS

164 RES LEGAL, Legal sources on renewable energy, available at: <www.res-legal.eu/> (last accessed 3 December 2015).

165 State Council Gazette (国务院公报) [Guo Wu Yuan Gong Bao], available at: <www.gov.cn/gongbao/2015/issue_4583.htm> (last accessed 3 December 2015).

166 Legislative Affairs Office of the State Council, the Full Text Search System of Laws and Regulation, available at: <http://search.chinalaw.gov.cn/search.html> (last accessed 3 December 2015).

167 *Fa Lv Tu Shu Guan Xin Fa Gui Su Di* (法律图书馆新法规速递) [New Chinese Laws and Regulation], available at: <www.law-lib.com/law/> (last accessed 3 March 2015); *Bei Da Fa Yi Fa Lv Jiao Xue Yan Jiu Ping Tai* (北大法意法律教学研究平台) [Beijing University Fayi Legal Education and Research Platform], available at: <www.lawyee.org/Act/Act.asp> (last accessed 3 December 2015).

168 For the major Chinese legislative documents, English summaries of titles, applicable scope, and major contents are provided by the State Council. A limited number of pieces of legislation, regulations, and policy documents with national importance (e.g. the constitutional law, national civil law, administrative laws, criminal law, procedural law, national social and economic plans of the Communist Party of China) or of relevance for external relations (e.g. China's Policies and Actions for Addressing Climate Change) have been officially translated in full text. See Legislative Affairs Office of the State Council, China Legislative Information Network System, available at: <www.chinalaw.gov.cn/article/-english/> (last accessed 3 December 2015); The Database of Laws and Regulation, available at: <www.npc.gov.cn/englishnpc/Law/Frameset-page8.html> (last accessed 3 December 2015); Central Compilation & Translation Bureau, available at: <www.cctb.net/bygz/> (last accessed 3 December 2015); National Development and Reform Commission, available at: <http://qhs.ndrc.gov.cn/zcfg/201111/t20111123_467502.html> (last accessed 3 December 2015).

169 *Bei Da Fa Bao* (北大法宝), available at: <http://en.pkulaw.cn/> (last accessed 3 December 2015); The World Law Guide, Legislation China, available at: <www.lexadin.nl/wlg/legis/-nofr/oeur/lxwechi.htm> (last accessed 3 December 2015).

170 The translation may draw on literature either on general the legal and political system of China or on specific issues related biofuels.

171 Interview with Ignacio Vazquez Larruscain (Brussels, 17 December 2013). A record of the interview is on file with the author.

172 Interview with Tianbao Qin (Maastricht, 9 February 2015). A record of the interview is on file with the author.

respectively.[173] Access to EU's information was also acquired by sending questions about EU biofuel legislation via the EUROPE DIRECT Contact Centre.[174] News and reports from the media, especially the EurActiv, provided a source for updated information.

4. Structure

This thesis consists of six chapters including the general introduction and conclusion in Chapters 1 and 6.

Chapter 2 examines and analyses the regulation of biofuels in the international legal system. In the absence of an international agreement on biofuels, the question is whether the related international regimes, such as climate change, biodiversity, and human rights, provide for the regulation of the sustainability of biofuels. This chapter examines not only general commitments relating to biofuels in treaties, but also decisions, resolutions, and guidelines on biofuels adopted by treaty bodies. It was found that, although these international regimes contain rules or have adopted decisions that can be applied to certain sustainability concerns, they are too fragmented and weak to provide for inclusive and binding international regulation for biofuel's sustainability.

Chapter 3 carries out an analysis on the regulation of biofuels in the EU. The EU has adopted a binding sustainability scheme for biofuels, so the question is how the regulatory approach is set out and ensures sustainability of biofuels. According to the 'demand-supply-trade' analysis, this chapter examines the related provisions in a series of legal instruments such as consumption targets, fuel quality, passenger car emission standards, preferential energy taxes, emission trading rules, and common agricultural policy. It was found that the EU's approach still has limits in ensuring the sustainability of biofuels, and, moreover, the exhaustive harmonisation based on the market functioning objective restricts EU Member States to adopt complementary or stricter sustainability requirements for biofuels.

Chapter 4 probes into the regulation of biofuels in China. China has not established binding sustainability criteria for biofuels in law, but policy principles and administrative rules have been adopted to avoid certain negative impacts of biofuels. This chapter examines the regulatory approaches concerning the demand, supply, and trade of biofuels, particularly noticing the uniqueness of the Chinese legal and political context, and in conclusion, it finds that the current national regulation is too general and vague to ensure the sustainability of biofuels, but the central government does not restrict local (provincial) governments from adopting complementary or stricter sustainability requirements.

Chapter 5 compares the regulatory approaches at the international level and in the EU and China. The comparison follows the analytical model of 'demand-supply-trade', identifies the similarities and differences, and analyses the possible

[173] Interview with Carlo Hamelinck (Utrecht, 13 February 2015). A record of the interview is on file with the author.

[174] Emails of questions and answers via the EUROPE DIRECT Contact Centre are on file with the author.

reasons underlying the similarities and differences. The comparative analysis reveals the differences in regulatory approaches and the existing gaps in including sustainability concerns of all the environmental and socio-economic dimensions, but meanwhile it recognises the convergences in relation to the protection of certain essential environmental and social interests. Based on the comparative observations, this chapter attempts to frame the observations in global environmental law to argue for a global inclusive approach to regulate the sustainability of biofuels.

Chapter 2

THE INTERNATIONAL REGULATION OF THE SUSTAINABILITY OF BIOFUELS

1. Introduction

This chapter aims to set forth the international context in which the sustainability of biofuels may be regulated. The international context is firstly explored for the following reasons. Above all, the sustainability concerns about biofuels are largely about environmental and socio-economic impacts on a global scale, especially regarding GHG emissions, biodiversity loss, and food prices. Moreover, the increasing international trade in biofuels[1] has resulted in biofuel production in developing countries where environmental and socio-economic sustainability impacts are most evident.[2] Next, mapping the international terrain may assist in exploring the question that to what extent the rules on the sustainability of biofuels have been internationally harmonised[3] and what room is left for domestic law to regulate.[4] This exploration is necessary particularly when there is no multilateral treaty[5] on biofuels[6] and sustainability concerns such as GHG emissions, biodiversity

1 Junginger *et al.* 2011, p. 2028.

2 Ottinger & Miller 2010, p. 23.

3 Harmonisation hereby refers to the adjusting and conforming of disparate standards and requirements of different legal systems to an international system. Yang & Percival 2009, p. 635- 636, 638-640; Faure 2000, p. 174.

4 According to Kamdem, 'harmonisation' is different from 'unification' in that it builds a level of legislative or regulatory similarity leaving space for certain national specificity: although harmonisation can eliminate certain divergences, it allows for the existence of national legislative or regulatory specificity. In contrast, the unification and uniformanisation denote a greater level of unity or homogeneity in regulatory or legislative content. Kamdem 2008, p. 718-721; Faria 2009, p. 6-8.

5 The term 'treaty' is herein used as a generic term, referring to the international instrument: (1) concluded by states or international organisations with treaty-making capacity, (2) binding upon contracting parties based on their express consent, (3) governed by international law, and (4) presented in written form, regardless of the label of convention, agreement, protocol, pacts, covenant or declaration. In this thesis it can be used interchangeably with the terms 'convention' or 'agreement.' United Nations Treaty Collection, Overview, Definition of key terms used in the UN Treaty Collection, available at: <https://treaties.un.org/- →

loss, food and land use right are respectively subject to different international regimes[7] on climate change, biodiversity, and socio-economic rights.[8]

Following the analytical model of 'demand-supply-trade', this chapter respectively examines the biofuel sustainability-related norms[9] under the international regimes of climate change, biodiversity, and socio-economic rights. On the demand side, although international law hardly mandates the use of biofuels, the GHG emission limitation norms under the international climate change regime may constitute an essential reason for states to promote the consumption of biofuel.[10] As for supply, the international biodiversity conventions and international socio-economic treaties contain states' commitments to the protection of biodiversity and socio-economic rights, so that states may need to refrain from producing biofuels which would be detrimental to these commitments. Finally, international trade law is relevant where the sustainability requirement of one state is applied to the biofuel produced in a third country.

Pages/Overview.aspx?path=overview/definition/page1_en.xml#treaties> (last accessed 7 February 2014).

6 A search with the key word 'biofuel' in the UN Treaty Collection database yielded zero results. Nevertheless, bilateral 'treaty-contracts' on cooperation in biofuel development can be found between certain countries, such as those between Brazil and the US and between Brazil and Benin. United Nations Treaty Collection, Databases, Full text search, available at: <https://treaties.un.org/Pages/Result.aspx?searchText=biofuel&dir=&file=&query=All&tab=SEARCH> (last accessed 7 February 2014); Memorandum of Understanding Between The Government of The Federative Republic of Brazil And The Government of The United States of America to Advance Cooperation on Biofuels (03 September 2007), available at: <www.brazilcouncil.org/sites/default/files/MOUtoAdvanceCooperationonBiofuels%20-Mar092007.pdf> (last accessed 9 February 2016).

7 As expounded by Krasner, 'international regimes' refer to the implicit or explicit principles, norms, rules, and decision-making procedures around which states' expectations converge in a given area. It has been observed that, with the emergence of 'self-contained regimes' and geographically or functionally limited treaty-systems, the 'general international law' has become the field of operation for specialist systems such as 'trade law', 'human rights law', and 'environmental law' regimes. Krasner 1983, p. 1-2; ILC 2006, paras. 6-10; Hasenclever 1997, p. 8-9; Keohane 1982, p. 325.

8 It has already been observed by Morgera, Kulovesi and Gobena that 'there are no intergovernmental mechanisms or international agreements specifically addressing bioenergy, although several multilateral instruments impose obligations with implications for the way in which bioenergy is regulated at national level', such as the UN Framework Convention on Climate Change, Kyoto Protocol, Convention on Biological Diversity, United Nations Convention to Combat Desertification, WTO law, as well as international human rights treaties. Morgera, Kulovesi & Gobena 2009, p. 15-17.

9 The term 'norm' is used to include the 'law' that is legally binding and has formal legal sources and the other instruments that have normativity and can affect states' behavior. International law is conventionally perceived as a cluster of norms having a legally binding effect on states, but a 'norm' is more open-textured in its content and can even be seen as soft. Krasner 1983, p. 1-2; Boyle 2010, p. 122-123; Shelton 2010, p. 142.

10 It is observed that the reduction of GHG emissions and reducing the use of fossil fuels rank among the main drivers for biofuel development, particularly in countries belonging to the Organization for Economic Cooperation and Development (OECD). IEA 2011, p. 10; Elbehri, Segerstedt & Liu 2013, p. 13.

This chapter is structured as follows. In section 2, norms under international climate regime, especially those addressing the GHG emission reduction obligation and calculation, are examined to assess their relevance for the demand for biofuels. Section 3 examines the international biodiversity regime, in particular the Convention on Biological Diversity (CBD) and the Ramsar Convention, in order to explore the norms relating to the production of biofuels. Furthermore, the provisions addressing the right to food and land under the International Covenant on Economic, Social, and Cultural Rights (ICESCR) are examined in relation to the production of biofuels. Section 4 sets out the key international trade rules relating to the regulation of the sustainability of biofuels, aiming to demonstrate under which conditions international trade law may restrict a state's regulation of the sustainability of biofuels. Finally, section 5 draws conclusions about whether the examined international regimes can provide for an inclusive regulatory approach to the sustainability of biofuels at the international level.

2. Demand for Biofuels: the International Climate Regime

At the international level, the United Nations Framework Convention on Climate Change (UNFCCC) serves as the 'central forum'[11] where international negotiation and law making on climate change take place, with 196 participants including the EU and China.[12] This section will examine the relationship between commitments relating to GHG emissions under the UNFCCC and the demand for biofuels, and it will furthermore look into whether the norms developed under the treaties provide for requirements for the sustainability of biofuels, in particular regarding the GHG emission in biofuels' life cycles.

2.1. Commitments Relating to GHG Emissions

The UNFCCC enshrines the objective of stabilising GHG concentrations at a level that would prevent dangerous anthropogenic interference with the climate system,[13] and it follows a framework convention approach by setting out the ultimate objectives, guiding principles, and institutional arrangements.[14] It does not specify concrete obligations or measures to be taken for reducing GHG emissions,[15] but

11 Admittedly, there is a trend of 'bottom-up' growth of international climate change governance, including the Major Economies Forum on Energy and Climate Change (MEF) launched in 2009, which could serve as another forum of energy and climate talks among the 17 major economies. However, the UNFCCC is still the most important multilateral process with a nearly universal participation of 196 states. Hoffmann 2014, p. 433; Bodansky & Rajamani 2012, p. 1-2.

12 United Nations Framework Convention on Climate Change (UNFCCC), Status of Ratification of the Convention, available at: <http://unfccc.int/essential_background/convention/-status_of_ratification/items/2631.php> (last accessed 29 Februaruy 2016).

13 UNFCCC, Art. 2.

14 Bodansky & Rajamani 2012, p. 15.

15 The UNFCCC follows the 'framework' convention approach which 'guide[s] rather than determine[s]' the measures to be taken. Bodansky & Rajamani 2012, p. 17.

generally it provides that all Parties shall formulate domestic measures to mitigate climate change by *addressing* anthropogenic GHG emissions[16] and Annex I Parties, including the EU, shall mitigate climate change by *limiting* anthropogenic GHG emissions and protecting GHG sinks.[17] The Kyoto Protocol concretised the general UNFCCC provisions by means of quantified emission limitation/reduction commitments for Annex I Parties.[18] To assist the implementation of reduction goals, Annex I Parties are allowed to use flexibility mechanisms, namely international emission trading, the clean development mechanism, and joint implementation, to achieve their GHG reduction commitments.[19] Towards the end of the first commitment period of the Kyoto Protocol in 2012, Parties agreed on an eight-year second commitment period (known as 'Doha Amendment') from 2013 to 2020 with the flexibility mechanisms maintained,[20] which has, however, not yet entered into force.[21] The Paris Agreement[22] as adopted at the twenty-first session of Conference of the Parties to the UNFCCC sets forth long-term goals and actions for the post-Kyoto Protocol era. None of the UNFCCC, the Kyoto Protocol, or the new Paris Agreement mentions the use of biofuels in their texts, and they contain no quantified GHG emission reduction target specifically set out for the transport sector. However, the treaty commitments relating to GHG emissions may affect the demand for biofuels in the following ways.

First, the transport sector constitutes an import emission resource and the use of biofuels has been recognised as an important means, *inter alia*, to reduce GHG emissions in the transport sector. According to the recent report of the IPCC, the transport sector accounts for approximately 23 per cent of the total energy-related carbon emissions, and GHG emissions have continued to rise in this sector due to the continuing growth in passenger and freight activities.[23] Therefore, it will be critical and challenging to reduce GHG emissions from the transport sector especially for the Annex I countries that undertake binding reduction commitments. Although the use of biofuels is not mentioned in the climate treaties, the Kyoto Protocol obliges Annex I Parties to implement climate change mitigation policies

16 It is provided that 'All Parties … shall: … (b) Formulate, implement, publish and regularly update national and, where appropriate, regional programmes containing measures to mitigate climate change by addressing anthropogenic emissions by sources and removals by sinks of all greenhouse gases not controlled by the Montreal Protocol.' UNFCCC, Art. 4(1)(b).

17 It is stated that the Annex I Parties 'shall adopt national policies and take corresponding measures on the mitigation of climate change, by limiting its anthropogenic emissions of greenhouse gases and protecting and enhancing its greenhouse gas sinks and reservoirs.' UNFCCC, Art. 4(2)(a).

18 Kyoto Protocol, Arts. 2(1) and 3.

19 Kyoto Protocol, Arts. 17, 12, and 6.

20 Decision 1/CMP.8.

21 The second commitment period began on 1 January 2013 and will end in 2020. However, the second commitment period has not entered into force up to date (1 January 2016) due to insufficient submission of ratification. Status of the Doha Amendment, available at: <http://unfccc.int/kyoto_protocol/doha_amendment/items/7362.php> (last accessed 21 December 2015).

22 Paris Agreement, paras. 2(1) and 4(1).

23 Sims *et al*. 2014, p. 603.

and measures, such as an increased use of renewable forms of energy[24] which can include biofuels. The IPCC also stresses the importance of a large-scale transformation of energy systems by 'tripling to nearly a quadrupling of the share of zero- and low-carbon energy supply from renewables.'[25] Moreover, it is expected that there will be a continued reliance on liquid and gaseous fuels, supported by an increase in the use of biofuels until 2030, due to technology constraints in reducing the carbon intensities of transport sector in the short term.[26] The IPCC in its 2007 assessment report identified biofuel as one of the key mitigation technologies and practices commercially available for the transport sector.[27] In the IPCC 2014 assessment report, although Working Group II alerted to the negative impacts of biofuels,[28] Working Group III mainly recommends biofuel as a main sectoral mitigation measure in the transport sector that offers significant potential for climate change mitigation[29] if produced and converted sustainably.[30] While sustainability of biofuels remains to be an issue for which no scientific consensus exists within the IPCC,[31] the recognition and recommendation of the use of biofuels as an emission mitigation strategy may exert certain influence on domestic biofuel policies, considering that there is no clear rule regarding biofuel under the

24 Kyoto Protocol, Art. 2(1)(a)(iv).

25 Edenhofer *et al.* 2014a, p. 12.

26 It is estimated that substantial shares of electricity and/or hydrogen will be used to fuel electric and fuel-cell light-duty vehicles in the second half of the 21st century, while biofuels have the advantage of being compatible with the existing petroleum fuel infrastructure and engines. Edenhofer *et al.* 2014b, p. 73; Sims *et al.* 2014, p. 615.

27 IPCC 2007, p. 60.

28 The Working Group II on adaptation to and impact of climate change mentions that land-use change to produce bioenergy can affect food crops, biodiversity, and ecosystem services. Working Group II offers a subtle but significant caveat to the IPCC's past view of biofuels as one of the 'key mitigation technologies' for reducing fossil fuel use and greenhouse gas emissions. It warned that an increased cultivation of crops for biofuel may boost water consumption and 'exacerbate the already serious water scarcity' in countries such as China and Spain. Additionally, land acquisition for biofuels production may have 'negative impacts on the lives of poor people' by depriving them of land and increasing food prices. Moreover, reductions in greenhouse gas emissions from using biofuels may be cancelled out by the effects of deforestation and the draining of wetlands. Field 2014, p. 50 and 91.

29 It is recommended that one way of reducing direct (tank-to-wheel) GHG emissions from passenger and freight transport is by substituting oil-based products with natural gas, bio-methane, or biofuels, electricity or hydrogen produced from low GHG sources. Edenhofer *et al.* 2014b, p. 68 and 73; Sims *et al.* 2014, p. 603.

30 The Working Group III on the mitigation of climate change concluded that bioenergy can play a critical role in mitigation, but there are issues to consider such as the sustainability of the practices and the efficiency of bioenergy and its impacts on environmental and socio-economic objectives and concerns other than climate change. Smith *et al.* 2014, p. 817; Edenhofer *et al.* 2014b, p. 77-78, and 89; Sims *et al.* 2014, p. 603.

31 The Working Group III also admitted that some biofuel technologies could help reduce the carbon intensity of fuels, but their total mitigation potential is very uncertain. It is also mentioned that commercially available liquid and gaseous biofuels already provide benefits together with mitigation options that can be increased by technology advances, particularly drop-in biofuels for aircraft. Edenhofer *et al.* 2014b, p. 68, 73, 77-78, and 89; Sims *et al.* 2014, p. 603; Beck *et al.* 2014a, p. 83.

UNFCCC. Furthermore, the IPCC's reports have gained epistemic authority in climate-related knowledge matters,[32] as they are reviewed and approved by scientists and governments.[33]

Second, the Clean Development Mechanism (CDM) makes it possible for biofuel projects in the non-Annex I countries to generate GHG emission offsetting credits. The CDM provided for in the Kyoto Protocol reflects the 'where flexibility'[34] for Annex I countries to achieve GHG emission reduction commitments - the mechanism allows emission reduction projects in developing countries to earn certified emission reduction (CER) credits which are equivalent to tonnes of carbon dioxide and can be traded and used by Annex I countries to meet their targets under the Kyoto Protocol.[35] Driven by the joint benefits for Annex I and non-Annex I countries, more than seven thousand CDM projects have been registered with the CDM Executive Board (EB), including projects that use biomass to generate energy.[36] The biomass projects are to a large extent generating power or heating instead of biofuels for transport,[37] but the CDM leaves open possibility for biofuels projects to be registered and to acquire CER credits. As yet, seven monitoring

32 Beck *et al.* 2014b; Mbengue 2011, p. 97.

33 The IPCC assessment reports have to go through a series of reviews by governments and experts, and finally the full reports are to be accepted at the Working Group Sessions and approved, particularly regarding the summary for policymakers, by IPCC member governments at the Working Group Session and then accepted at a Session of the Panel. IPCC, Principles and Procedures, available at: <www.ipcc.ch/organization/-organization_procedures.shtml> (last accessed 1 April 2015); IPCC, Activities, available at: <www.ipcc.ch/activities/activities.shtml> (last accessed 1 April 2015).

34 'Where flexibility' means that the Kyoto flexibility mechanism allows countries to decide where emission reductions should happen with lower costs. Bodansky & Rajamani forthcoming, p. 22.

35 The CERs are issued when the projects satisfy the required modalities and procedures through the processes of validation, registration, verification, and certification. In these processes, the CDM Executive Board (EB), the 'designated operational entities' (DOEs), and the 'designated national authority' (DNAs) play different and important roles. Parties to the Kyoto Protocol adopted an amendment to the Kyoto Protocol by Decision 1/CMP.8 in Doha, Qatar, 2012, including modalities and access to the CDM credits, but a discussion on the amendment is outside the scope of this thesis. UNFCCC, Clean Development Mechanism (CDM), available at: <https://cdm.unfccc.int/> (last accessed 1 April 2015).

36 Up to 28 February 2008, there are 7,598 registered CDM projects, which include biomass energy projects noting the graphic 'trend of types of projects registered and registering.' UNFCCC, About CDM, CDM Insights, Project Activities, available at: <https://cdm.unfccc.int/Statistics/Public/CDMinsights/index.html> (last accessed 1 April 2015).

37 A search on the CDM website with the key words 'biofuel', 'biodiesel', 'bioethanol', and 'ethanol' produces very few results, among which only the Shandong Province Jinjiang Biodiesel Production Project (Project 8727) in China produces biofuels as defined in the thesis. In contrast, a search with the key word 'biomass' produces 433 results, among which most are projects for generating power/electricity and heating, and it was estimated that biomass energy projects take a share of more than 13 per cent of all registered projects of CDM. UNFCCC, CDM, Project Search, available at: <https://cdm.unfccc.int/Projects/-projsearch.html> (last accessed 1 April 2015); Tuerk, Cowie & Leopold 2011, p. 5.

methodologies have been approved by the CDM EB for biofuel-related projects,[38] which may help to change the situation that there is only a minimal number of biofuel CDM projects.[39] Moreover, the CDM EB has agreed to develop new methodologies in the areas such as aviation and renewable energy to broaden the applicability of the CDM, including the methodologies for biofuels.[40]

Based on the afore-mentioned two points, the commitments relating to GHG emissions under the UNFCCC and the Kyoto Protocol may incentivise Parties, particularly the Annex I Parties, to see biofuel as a means to achieve the GHG emission reduction goals in transport. Nevertheless, there is distinction to be made between emissions from international and domestic transport. According to the IPCC Guidelines, emissions from international aviation and maritime transport (also known as international bunker fuel emissions) are excluded from the national totals and are reported separately,[41] and activities resulting in emission reductions from reduced consumption of international bunker fuels are also not eligible to collect offset credits under the CDM;[42] but domestic aviation and maritime transport

38 This result is according to a search through the approved large or small scale baselines and monitoring methodologies on the CDM website, including AMS-I.H, AMS-III.T., AMS-III.AK., AMS-III.AQ., AM0057, AM0089, ACM0017. UNFCCC, CDM, Methodologies, Approved Baseline and Monitoring Methodologies for Large Scale CDM Project Activities, available at: <https://cdm.unfccc.int/methodologies/PAmethodologies/approved> (last accessed 1 April 2015); UNFCCC, CDM, Methodologies, Approved SSC methodologies, available at: <https://cdm.unfccc.int/methodologies/SSCmethodologies/approved> (last accessed 1 April 2015).

39 An analysis attributed one reason for the minimal number of biofuel CDM projects to the lack of a monitoring methodology for biofuel projects. Monitoring methodology is a necessary element to be included in the project design document for project validation, and it is also an element that needs to be included in the monitoring report for project verification, certification, and the issuance of the Certified Emission Reduction (CERS). Bakker 2006, p. 5; Decision 3/CMP.9, paras. 11, 13-14.

40 The EB requested the secretariat to elaborate concept notes on each specific area for its consideration before the development of the methodologies. In the Annex 2 and 7, it is specified that the broadly applicable and highly useable methodology for biofuel applications is to be developed, and a concept note is to be prepared at EB 85 before 2016. As for aviation, a methodology is to be developed regarding renewable energy applicable to the aviation sector, such as the use of solar power for at-gate aircraft, which might also include biofuels for aviation and energy efficiency methodologies applicable to the aviation sector. CDM-EB82 Meeting Report, para. 40, Annexes 2 and 7.

41 In accordance with the IPCC Guidelines for the preparation of greenhouse gas (GHG) inventories and the UNFCCC reporting guidelines on annual inventories, emissions from international aviation and maritime transport should be calculated as part of the national GHG inventories of Parties, but they should be *excluded* from national totals and reported separately, which means that these emissions are not subject to the limitation and reduction commitments of Annex I Parties under the Convention and the Kyoto Protocol. UNFCCC, Emissions from fuel used for international aviation and maritime transport (international bunker fuels), available at: <http://unfccc.int/methods/emissions_from_intl_transport/-items/1057.php> (last accessed 5 October 2015).

42 EB 25 Report (extract) Paragraph 58 'Guidance On Bunker Fuels' (Extract of the report of the twenty-fifth meeting of the Executive Board, para. 58).

fuels shall be included. In this sense, the emission reduction commitments under the UNFCCC mainly exercise influence on domestic transport.

2.2. *GHG Emissions Accounting*

In view of the possible incentive for biofuels provided by the GHG emission commitments, the critical question is whether the international climate regime correspondingly provides rules that require an assessment of whether biofuels indeed save GHG emissions compared to the replaced fuels, particularly considering life-cycle GHG emissions from land use and land use changes.[43] As yet, the UNFCCC or the Kyoto Protocol does not provide specific rules for biofuels, except for the UNFCCC online glossary which defines biofuels as renewable energy only when the vegetation producing it is maintained or replanted.[44] Nevertheless, the IPCC Guidelines for National Greenhouse Gas Inventories (IPCC Guidelines), which will be discussed below, aims to completely cover GHG emissions from all sources and removals and will have implications for the accounting of life-cycle emissions of biofuels.

2.2.1. The Role of the IPCC Guidelines

The UNFCCC and Kyoto Protocol respectively set out the requirement of communicating national GHG inventories by emission sources and sink[45] removals, while the methodologies for calculating the inventories are to be agreed upon by the Conference of Parties (COP) or Conference of the Parties serving as the meeting of

43 Life-cycle assessment may involve an assessment of the type and management of raw materials, conversion technologies, end-use technologies, system boundaries and reference energy systems with which the biofuel chain is compared. It was pointed out that whether biofuels offer carbon savings depends on how they are produced. For example, converting rainforests, peatlands, savannas, or grasslands to produce crop-based biofuels creates a 'biofuel carbon debt' by releasing more CO_2 than the annual greenhouse gas (GHG) reductions that these biofuels would provide by displacing fossil fuels. Cherubini *et al.* 2009, p. 434; Fargione *et al.* 2008, p. 1235; von Blottnitz & Curran 2007, p. 607; Larson 2006, p. 109.

44 The glossary defines 'biomass fuels or biofuels' as 'a fuel produced from dry organic matter or combustible oils produced by plants' and supplements that '[t]hese fuels are considered renewable as long as the vegetation producing them is maintained or replanted, such as firewood, alcohol fermented from sugar, and combustible oils extracted from soy beans. Their use in place of fossil fuels cuts greenhouse gas emissions because the plants that are the fuel sources capture carbon dioxide from the atmosphere.' UNFCCC, Glossary of climate change acronyms, available at: <http://unfccc.int/essential_background/glossary/-items/3666.php> (last accessed 24 February 2015).

45 'Sink' refers to 'any process, activity or mechanism which removes a greenhouse gas, an aerosol, or a precursor of a greenhouse gas from the atmosphere. Forests and other vegetation are considered sinks because they remove carbon dioxide through photosynthesis.' UNFCCC, Glossary of climate change acronyms, available at: <http://unfccc.int/essential_background/glossary/items/3666.php> (last accessed 2 April 2015).

the Parties to the Kyoto Protocol (COP/CMP).[46] The calculating methodologies are substantially adopted on the basis of guidelines as issued by the IPCC. The reliance on the IPCC's scientific expertise regarding the GHG inventory estimation is explicitly authorised in the Kyoto Protocol: '[M]ethodologies for estimating anthropogenic emissions ... shall be those accepted by the Intergovernmental Panel on Climate Change ....'[47]

The IPCC has published a number of guidelines to provide methodology guidance for measuring, monitoring, and reporting on GHG emission sources and removals,[48] which were later mandated by COP or COP/CMP as methodologies for national communications, Biennial (Update) Reports, and for the calculation of legally-binding targets.[49] For example, it was decided by the first meeting of UNFCCC COP in 1995 that the IPCC Guidelines *should* be used by Annex I Parties in preparing their national communications and by non-Annex I Parties, as appropriate and to the extent possible, in the fulfilment of their commitments.[50] At the third meeting of the UNFCCC COP in 1997, it was reaffirmed that the Revised 1996 IPCC Guidelines for National Greenhouse Gas Inventories should be used as methodologies for estimating anthropogenic emissions in the calculation of legally-binding targets during the first commitment period.[51] According to the more recent decisions, Annex I Parties shall use the 2006 IPCC Guidelines in preparing their

46 The UNFCCC provides that Parties shall, in accordance with the principle of common but differentiated responsibilities, develop, periodically update, publish, and make available to the COP national inventories of anthropogenic emissions using comparable methodologies to be agreed upon by the COP. Under the Kyoto Protocol, Annex I Parties undertake GHG emission limitation and reduction commitments, and shall estimate anthropogenic emissions according to methodologies agreed upon by COP/CMP. The COP/CMP shall regularly review and, as appropriate, revise such methodologies and adjustments, taking fully into account any relevant decisions by the COP. UNFCCC, Arts. 4(1)(a), 4(2)(c), 12(1)(a), and 7(2)(d); Kyoto Protocol, Arts. 5(1), (2) and 13(4)(e).

47 Kyoto Protocol, Art. 5(2).

48 The IPCC Guidelines were first accepted in 1994 and published in 1995. The publication includes the 1996 IPCC Guidelines for National Greenhouse Gas Inventories, the 2006 IPCC Guidelines for National Greenhouse Gas Inventories, as well as the 2000 Good Practice Guidance and Uncertainty Management in National Greenhouse Gas Inventories, the 2003 Good Practice Guidance for Land Use, Land-Use Change and Forestry, the 2013 Revised Supplementary Methods and Good Practice Guidance Arising from the Kyoto Protocol, the 2013 Supplement to the 2006 IPCC Guidelines for National Greenhouse Gas Inventories: Wetlands. IPCC, Task force on National Greenhouse Gase Inventories, Publications, available at: <www.ipcc-nggip.iges.or.jp/public/gl/invs1.html> (last accessed 12 May 2015).

49 Following the updates of IPCC Guidelines, the COP and COP/CMP have adopted decisions relating to the methodological issue such as decision 3/CP.1, 4/CP.1, 9/CP.2, 2/CP.3, 11/CP.4, 3/CP.5, 18/CP.8, 14/CP.11, 1/CP.16; 15/CMP.1, 2/CMP.7, 4/CMP.7, 2/CMP.8, and 6/CMP.9. UNFCCC, Arts. 4(1)(a), 4(2)(c), 12(1)(a), and 7(2)(d); Kyoto Protocol, Arts. 5(1), (2) and 13(4)(e); Decision 4/CP.1 1995; Decision 2/CP.17 2011; Ellis & Moarif 2015, p. 11.

50 However, Parties that already have an established and comparable methodology could continue to use it, provided that they include sufficient documentation to back up the data presented. Decision 4/CP.1, para. 1(a) and (b).

51 Decision 2/CP.3, para. 1.

national inventories communications,[52] and the methodologies for estimating GHG emissions under Kyoto Protocol shall be consistent with the 2006 IPCC Guidelines for the second commitment period (2013-2020).[53] Annex I Parties may use national methodologies, but only when they are better able to reflect national situations, compatible with the 2006 IPCC Guidelines, well documented, and scientifically based.[54] The inventories information reported in national communications is to be reviewed by expert teams, as part of Kyoto Protocol compliance mechanism, to assess its conformity with the IPCC Guidelines and supplementary guidance as adopted by the COP and COP/CMP.[55] Non-Annex I Parties undertake differentiated inventories communication obligations, which will be discussed later, but they are also guided by the IPCC Guidelines and their supplementary method.[56] Moreover, to promote consistency, standardised formats are provided for Parties in their submissions, such as the Common Reporting Format (CRF) and the National Inventory Report (NIR), in accordance with the IPCC Guidelines as adopted under the UNFCCC and Kyoto Protocol.[57]

Hence, the IPCC Guidelines have been identified and endorsed by the COP to UNFCCC and COP/CMP to the Kyoto Protocol as the methodologies that Parties use, at a minimum, to develop their GHG inventories.[58] The Paris Agreement also states that 'each Party shall provide a national inventory report … prepared using good practice methodologies accepted by the IPCC and agreed upon by the COP/CMP to the Paris Agreement.'[59] In this vein, the methodologies for estimating GHG inventories by emission sources and sink removals under the UNFCCC are to a certain extent harmonised based on the IPCC Guidelines.

2.2.2. IPCC Guidelines for Complete GHG Emissions Accounting

The IPCC Guidelines aim to estimate GHG emissions and removals in a complete and accurate manner: completeness requires that the estimates are reported for all relevant categories of sources and sinks, and accuracy means that the national

52 The Annex I country national inventories communications are due by 15 April each year since 2015. Decision 24/CP.19, para. 3; Decision 22/CP.19; Decision 6/CMP.9.

53 Decision 4/CMP.7; Decision 6/CMP.9.

54 Decision 24/CP.19, Annex I, para. 10.

55 It has been observed that the review of the Parties' annual reports is conducted mostly in the form of 'centralised' reviews, in Bonn, by expert review teams. A team conducting a centralised review is made up of about ten sectoral and 'generalist' experts from both Annex I and non-Annex I countries. Zahar 2010a, p. 411; Decision 22/CMP.1, para. 50.

56 Decision 17/CP.8; Decision 2/CP.17.

57 UNFCCC, Reporting Requirements, available at: <http://unfccc.int/national_reports/-annex_i_ghg_inventories/reporting_requirements/items/2759.php> (last accessed 3 April 2015); UNFCCC, Greenhouse Gas Inventory Software for non-Annex I Parties (NAIIS), available at: <http://unfccc.int/national_reports/non-annex_i_national_communications/-non-annex_i_inventory_software/items/7627.php> (last accessed 3 April 2015).

58 Decision 6/CMP.9 2013, para. 8; Paris Agreement, Art. 13(7)(a); UNFCCC Secretary 2014, p. 16.

59 Decision 24/CP.19 2013, para. 4.

greenhouse gas inventory contains neither over- nor under-estimates so far as can be judged.[60]

For completeness, the 2006 IPCC Guidelines include five volumes to estimate human-induced[61] GHG emissions and removals from the sectors of 'energy', 'industrial processes and product use' (IPPU), 'agriculture, forestry and other land use' (AFOLU), and 'waste and other',[62] which thus could cover the life-cycle emissions of biofuels – from biomass cultivation, industrial conversion, and finally to end use. It is particularly noticeable that the previously separate sectors of agriculture, land use and land use change, and forestry are for the first time integrated into one volume as AFOLU for more completeness and coherence.[63] Inventory estimation in this volume is based on division of six land use categories – forest land, cropland, grassland, wetlands, settlements, and other. The estimation includes both the circumstances of 'land remaining in that category' (e.g. cropland remaining cropland) and 'land converted from one category to another (e.g. forest land converted to cropland).[64] Such an integration and categorisation signifies the recognition that GHG emissions and carbon stock changes[65] can occur across all types of managed land,[66] including *inter alia* carbon stock changes from biomass,[67]

60 IPCC Guidelines 2006, Volume 1, Chapter 1, p. 1.8.

61 Inventories should result from human activities. The distinction between natural and anthropogenic emissions and removals follows straightforwardly from the data used to quantify human activity. IPCC Guidelines 2006, Volume 1, Chapter 1, p. 1.4.

62 Each volume is further divided into individual categories and sub-categories. For example, the energy volume has a separate chapter for emissions from mobile combustion, and mobile combustion is further divided into the sub-categories of road and off-road transport. GHG emissions and removals are firstly calculated at the sub-category level and then are summed up to form the national total emissions and removals for each greenhouse gas in all the sectors.

63 The revised guidelines regarding AFOLU reflect the elaborations in the Good Practice Guidance and Uncertainty Management in National Greenhouse Gas Inventories (GPG 2000) and the Good Practice Guidance for Land Use, Land-Use Change and Forestry (GPG-LULUCF). IPCC Guidelines 2006, Volume 1, Chapter 1, p. 1.4.

64 IPCC Guidelines 2006, Volume 1, Chapter 1, p. 1.9.

65 Carbon stock refers to the quantity of carbon in a 'pool' – a reservoir or system which has the capacity to accumulate or release carbon, and examples of carbon pools are living biomass (including Above and below-ground biomass); dead organic matter (including dead wood and litter); soils (soils organic matter). The carbon stock in a pool can change due to the difference between additions of carbon and losses of carbon, which results in carbon stock changes. When the losses are larger than the additions, the carbon stock becomes smaller and thus the pool acts as a *source* to the atmosphere; when the losses are smaller than the additions, the pools acts as a *sink* to the atmosphere. See IPCC, Good Practice Guide for LULUCF, Annex A Glossary, 2003, available at: <www.ipcc-nggip.iges.or.jp/-public/gpglulucf/gpglulucf_files/GPG_LULUCF_FULL.pdf> (last accessed 3 April 2015); FAO, Terms and Definitions for the National Reporting tables for FRA 2005, available at: <http://www.fao.org/docrep/007/ae156e/AE156E03.htm#P284_12333> (last accessed 3 April 2015).

66 In the AFOLU sector, human-induced GHG emissions and removals are defined as all those occurring on 'managed land' where human interventions and practices have been applied to perform production, ecological, or social functions. IPCC Guidelines 2006, Volume 1, Chapter 1, p. 1.4 and 1.5.

soils, and managed wetlands.[68] Therefore, GHG emissions in a biofuel's life cycle, including those from the cultivation of the biofuel feedstock and the induced land use change from non-cropland or food cropland could be completely accounted for.

Meanwhile, the IPCC Guidelines seek to ensure accuracy that avoids both over- and under-estimates. To avoid double estimates, CO_2 emissions from combusting the biomass-derived fraction of biofuels[69] are not counted in the inventory total of energy volume, for the reason that CO_2 emissions and removals from biomass are counted in the AFOLU volume.[70] The excluded CO_2 emissions in energy volume, nonetheless, have to be reported as an information item for quality assurance and control purposes.[71] Meanwhile, non-CO_2 GHG emissions such as N_2O and CH_4 from biofuel combustion, as well as all the GHG emissions from processing and transporting the biomass shall be counted in the inventory totals.[72] Thus, to exclude CO_2 emissions of biofuel combustion from the inventory calculation in the energy volume does not mean that there is no need to count the GHG emissions from biofuels.

It is hence crucial to see whether and how CO_2 emissions from biomass combustion are counted in the AFOLU volume. The default Tier 1[73] estimation is that CO_2 emissions from the combustion of short-lived biomass or annual crops, including biomass for energy use, are *balanced* by carbon uptakes prior to harvest, so the net CO_2 emission from combusting such biomass is zero within the uncertainties of the estimates.[74] However, when methods of a higher tier (Tier 2 or Tier 3) are

67 Biomass is defined as including *a)* the total mass of living organisms in a given area or of a given species usually expressed as dry weight, and *b)* Organic matter consisting of or recently derived from living organisms excluding peat, and includes products, by-products and waste derived from such material. IPCC Guidelines 2006, Glossary, p. G.3.

68 IPCC Guidelines 2006, Volume 1, Chapter 1, p. 1.5.

69 In the IPCC Guidelines, the term biofuel is defined broadly as including 'any fuels derived from biomass, either deliberately grown or from waste products', but 'peat is not considered a biofuel in these guidelines due to the length of time required for peat to re-accumulate after harvest.' IPCC Guidelines 2006, Glossary, p. G.3.

70 IPCC Guidelines 2006, Volume 2, Chapter 3, p. 3.12.

71 IPCC Guidelines 2006, Volume 2, Chapter 3 Mobile Combustion, p. 3.12; IPCC Guidelines 2006, Volume 4, Chapter 1, p. 1.5.

72 IPCC Guidelines 2006, Volume 2, Chapter 3 Mobile Combustion, p. 3.13.

73 In the IPCC Guidelines, a Tier represents a level of methodological complexity. Usually three tiers are provided: Tier 1 is the basic and default method, Tier 2 is the intermediate, and Tier 3 is the most demanding in terms of complexity and data requirements. Tiers 2 and 3 are sometimes referred to as higher tier methods and are generally considered to be more accurate. In AFOLU Volume, Tier 1 methods are designed to be the simplest to use. IPCC Guidelines 2006, Volume 1, Chapter 1, p. 1.6; IPCC Guidelines 2006, Volume 4, Chapter 1 Introduction, p. 1.11.

74 For annual crops, the IPCC Guidelines assume that biomass carbon stock lost through harvest and mortality equal biomass carbon stock gained through regrowth in that same year and therefore there are no net CO_2 emissions or removals from biomass carbon stock changes. However, for perennial crops CO_2 emissions or removals resulting from changes in biomass stocks due to harvest and regrowth will still be counted. IPCC Guidelines 2006, Volume 1, Chapter 1, p. 1.6; IPCC Guidelines 2006, Volume 4, Chapter 5, p. 5.7; IPCC, Task Force on →

used,[75] the CO2 emissions may not be balanced by carbon uptakes from the atmosphere and thus the net emissions may not be zero.[76] Moreover, the inventory accounting in AFOLU volume should cover GHG emissions from harvesting and the regrowth of crops, changes in soil carbon and organic matter, the production and use of fertilisers, and land use changes caused by biomass production. Thus, the carbon balance estimation exists only for certain biomass when the simplest Tier 1 method is applied and, as it has been stated by the IPCC, the Guidelines do not automatically consider biomass used for energy as 'carbon neutral' even if the biomass is thought to be produced sustainably.[77]

2.2.3. The Difficulty in Ensuring Complete GHG Emissions Accounting

However, the completeness and accuracy may be difficult to ensure in applying the IPCC Guidelines to assessing the life-cycle GHG emissions of biofuels, because a complete estimation and reporting of life-cycle emissions from biofuels is based on the premise that all countries follow the IPCC Guidelines to account for and report their national inventories. As admitted by the IPCC, growing crops for biofuels can result in land use changes, both directly through the conversion of land to biofuel production and indirectly by driving land use change elsewhere.[78] While GHG emissions due to direct land use changes within one country are reported in the AFOLU sector, emissions from indirect change can occur many years after the land conversion.[79] Hence, a complete estimate is only possible *provided* all countries follow the IPCC Guidelines and report to the UNFCCC.[80] However, actually Parties undertake different obligations and methods with regard to inventory accounting and reporting - the monitoring, reporting and verification requirements under the UNFCCC currently differentiate among Parties in terms of timeline, frequency, content, and level of flexibility.[81]

Greenhouse Gases Inventories, FAQs, Q2-10, available at: <www.ipcc-nggip.iges.or.jp/faq/faq.html> (last accessed 16 October 2014).

75 In AFOLU volume, Tier 2 can use the same methodological approach as Tier 1 but it applies emission and stock change factors that are based on country- or region-specific data. At Tier 3, higher order methods are used, including models and inventory measurement systems tailored to address national circumstances, repeated over time, and driven by high-resolution activity data and disaggregated at sub-national level. IPCC Guidelines 2006, Volume 4, Chapter 1 Introduction, p. 1.11.

76 Where higher testimation shows that this emission is not balanced by a carbon removal from the atmosphere, this net emission or removal should be included in the emission and removal estimates for AFOLU Sector through carbon stock change estimates. IPCC Guidelines 2006, Volume 1, Chapter 1 Introduction, p. 1.6.

77 IPCC, Task Force on Greenhouse Gases Inventories, FAQs, Q2-10, available at: <http://www.ipcc-nggip.iges.or.jp/faq/faq.html> (last accessed 16 October 2015).

78 IPCC, Task Force on Greenhouse Gases Inventories, FAQs, Q2-10, available at: <http://www.ipcc-nggip.iges.or.jp/faq/faq.html> (last accessed 16 October 2015).

79 *Ibid.*

80 *Ibid.*

81 Ellis & Moarif 2015, p. 9-10; UNFCCC Secretary 2014, p. 11 and 17.

Firstly, while it is mandatory, which is indicated by the use of 'shall' without a qualifier, for Annex I Parties to report on the total GHG emissions, metrics and methods, and national inventory systems and arrangements, such a requirement is not mandatory for non-Annex I countries.[82] Secondly, Annex I Parties shall provide national GHG inventories covering all the six volumes following the 2006 IPCC Guidelines and its 2013 supplements, but non-Annex I Parties may communicate national inventories following 1996 IPCC Guidelines and 2000 and 2003 Good Practices about land use and land-use change and forestry (LULUCF).[83] Thirdly, reporting on the approach to GHG inventory accounting of LULUCF is not mandatory to non-Annex I Parties.[84] Reporting of LULUCF inventories may even vary among Annex I Parties, as GHG emissions from cropland management, grazing-land management, revegetation, wetland drainage and rewetting are *optional* items to be reported under the Kyoto Protocol.[85] Finally, Parties may choose between different methods (tiers) contained in the 2006 IPCC Guidelines.[86] The simplest Tier 1 method is applied by default, which assumes that CO_2 emissions from the combustion of short-lived biomass (including biomass for energy use) is *balanced* by carbon uptake prior to harvest and the net emissions are thus zero within the uncertainties of the estimates.[87] Annex I Parties are only *encouraged* to apply higher tier methods which need more specific data and are considered more accurate,[88] but, as it is admitted by IPCC, applying the IPCC Guidelines to

82 It is furthermore observed that most reporting requirements for developing countries are couched in non-mandatory terms such as 'shall, to the extent possible', 'should', and 'encouraged', which indicate a less mandatory extent. Ellis & Moarif 2015, p. 13-15.

83 Decision 17/CP.8, Annex; Decision 2/CP.17.

84 Ellis & Moarif 2015, p. 14-15; Kyoto Protocol, Art. 3(3) and (4); Decision 6/CMP.9 2013, paras. 8 and 10.

85 As stipulated by the Kyoto Protocol, carbon stock changes and GHG emissions and removals from the LULUCF sectors shall be used to meet the Annex I Parties' commitments under the Protocol. However, the Kyoto Protocol restricts the accounting of the LULUCF sector to emissions and removals from specific activities that are defined under Art. 3(3) and (4) of the Protocol. Annex I Parties undertake different obligations with regard to Art. 3(3) and Art. 3(4): while reporting on activities under Art. 3(3) is mandatory, namely about afforestation, reforestation and deforestation occurred since 1990, reporting is elective on activities under Art. 3(4), namely forest management (mandatory in the second commitment period), revegetation, cropland management, and grazing land management. UNFCCC, Reporting and accounting of LULUCF activities under the Kyoto Protocol, available at: <http://unfccc.int/land_use_and_climate_change/lulucf/items/4129.php> (last accessed 4 April 2015).

86 Decision 24/CP.19, Annex I, para. 9.

87 Plant biomass, including above-ground and below-ground parts, is recognised as 'the main conduit for CO_2 removal from the atmosphere', as it is assumed that increases in total carbon stocks over time are equated with a net removal of CO_2 from the atmosphere. IPCC Guidelines 2006, Volume 1, Chapter 1 Introduction, p. 1.6; IPCC Guidelines 2006, Volume 4, Chapter 5 Cropland, p. 5.7; IPCC, Task Force on Greenhouse Gases Inventories, FAQs, Q2-10, available at: <www.ipcc-nggip.iges.or.jp/faq/faq.html> (last accessed 16 October 2014).

88 Parties are *encouraged* to refine estimates of anthropogenic emissions and removals in the land use, land-use change and forestry (LULUCF) sector through the application of Tier 3 methods. Decision 15/CP.17, Annex I, paras. 10 and 13; IPCC Guidelines 2006, Volume 1, Chapter 1 Introduction, p. 1.6-1.7; IPCC Guidelines 2006, Volume 4, Chapter 1, p. 1.10-1.11.

accurately estimate CO_2 emissions from biofuels[89] may require specific data to ensure that the relevant emissions are comprehensively considered at the appropriate level.[90]

It has been observed in the literature that the majority of national systems compile GHG inventories that favour the economic interest of the reporting state, and a large and growing proportion of global anthropogenic emissions are not reported or independently reviewed.[91] Moreover, many scientists have expressed the concern that the IPCC Guidelines might even be applied erroneously if emissions from the combustion of biofuels are excluded from the accounting, without sufficiently considering land use changes.[92] It has been pointed out that, although the IPCC Guidelines applied under the UNFCCC have a complete and accurate accounting system, because emissions are reported from both land and energy sectors worldwide, these conditions do not apply to any treaties or regulations that seek to limit emissions from energy use but do not limit emissions from land-use, or do so only weakly and do not apply worldwide.[93] Therefore, now that it is not possible at the moment to ensure all countries in the world follow the IPCC Guidelines and report to the UNFCCC, particularly with regard to the GHG emissions caused by land use and land use changes, the application of the IPCC Guidelines may not ensure completeness and accuracy in accounting for a biofuel's life-cycle emissions.

2.3. *Other Sustainability Concerns*

In addition to the concern about GHG emissions, a further question is whether the other sustainability concerns, such as biodiversity and socio-economic rights, are considered in the commitments and guidelines relating to biofuels.

Within the texts of the UNFCCC and Kyoto Protocol, sustainable development is enshrined as a guiding principle. In the UNFCCC, it is affirmed that '[T]he Parties have a right to, and should, promote sustainable development',[94] which contains both 'sustainable social and economic development'[95] and the 'sustainable

89 It is admitted that the IPCC approach does not include the emissions from individual facility or industry in the energy volume, while individual methodologies and emission factors provided in the IPCC Guidelines may be relevant for estimating CO_2 emissions from the use of bioenergy in an individual facility or industry. IPCC, Task Force on Greenhouse Gases Inventories, FAQs, Q2-10, available at: <www.ipcc-nggip.iges.or.jp/faq/faq.html> (last accessed 16 October 2014).

90 IPCC, Task Force on Greenhouse Gases Inventories, FAQs, Q2-10, available at: <www.ipcc-nggip.iges.or.jp/faq/faq.html> (last accessed 16 October 2014).

91 Zahar 2010a, p. 424; Zahar 2010b, p. 322.

92 Searchinger *et al.* 2009, p. 527; Haberl 2012, p. 18.

93 EEA Scientific Committee 2011, p. 5-6.

94 UNFCCC, Art. 3(4).

95 The preamble of the UNFCCC affirms that 'responses to climate change should be coordinated with social and economic development in an integrated manner with a view to avoiding adverse impacts on the latter, taking into full account the legitimate priority needs of developing countries for the achievement of sustained economic growth and the eradication of poverty.' UNFCCC, preamble, Art. 3(5), and Art. 4(2)(a).

management' of GHG sinks and reservoirs such as biomass and forestry.[96] In the Kyoto Protocol, Annex I Parties shall implement or elaborate policies and measures to achieve their emission reduction commitments as well as to promote sustainable development, such as sustainable forest management practices, afforestation and reforestation and sustainable forms of agriculture.[97] The CDM provided in the Kyoto Protocol has the purpose of assisting non-Annex I Parties in achieving sustainable development,[98] which requires that, for the project's validation and registration, confirmation from the host country shall be obtained to attest that the project activities assist sustainable development in that country.[99] Furthermore, documentation on environmental impacts is frequently required to be included in the project design document prepared for the validation and registration of a project.[100] Actually, many authors have observed the link between climate change and sustainable development: climate change policies could have significant ancillary benefits for the local environment, while local and national policies to address many sustainable development problems (i.e. air quality or energy access and diversity) may also limit GHG emissions.[101]

However, the international climate regime has rarely specified what 'sustainable development' entails for the obligations of the Parties and it has been usually considered as ignoring other sustainability concerns. In the UNFCCC text, the need to achieve sustainable social and economic development and to consider specific circumstances of economic growth, particularly for developing countries, is set out to balance the climate objectives,[102] but the text mentions little about how to define the 'sustainable' development. In the specific circumstances of CDM under the Kyoto Protocol, little guidance has been provided to specify what 'sustainable development' means for hosting countries. The COP/CMP of the Kyoto Protocol has adopted decisions on the modalities and procedures on CDM operation, but the decisions do not further provide guidelines on how to define and assess a project's contribution to sustainable development and substantial discretion is left to the hosting countries.[103] For example, an environmental impact assessment of the CDM projects is needed only *if the host country so requires*, and a determination of which impacts are 'significant' is completely up to the host country and project participants; where an environmental impact assessment is conducted, it needs to be carried out 'in accordance with the procedures required by the host Party.'[104] The lack of guidance on how Parties should establish standards for environmental impact assessment or sustainable development assessment could be problematic, since environmental interests and sustainability could be outweighed by economic

96 UNFCCC, Art. 4(1)(d).
97 Kyoto Protocol, Art. 2(1)(a)(ii) and (iii).
98 Kyoto Protocol, Art. 12(2).
99 Decision 3/CMP.1, para. 40(a).
100 Decision 3/CMP.1, Appendix B(e).
101 Swart, Robinson & Cohen 2003; Beg *et al*. 2002; Banuri 2009.
102 UNFCCC, preamble, Art. 3(4) and (5), and Art. 4(2)(a).
103 See the discussion about sustainable development and CDM in e.g. Olsen 2007; Sutter & Parreño 2007; Olsen & Fenhann 2008; Boyd *et al*. 2009.
104 Decision 3/CMP.1, Appendix B(e); Van Asselt 2011a, p. 1235.

incentives for host countries and project participants.[105] Moreover, the international climate regime has been criticised for being 'single-minded'[106] on the goal of carbon reduction while ignoring the unintended negative consequences for biodiversity or human rights.[107] For instance, CDM projects on afforestation and reforestation have raised concerns about the possible negative impacts on biodiversity,[108] and the international climate regime as a whole does not make an adequate effort to accommodate biodiversity considerations in climate change rules.[109] Particularly, the afforestation and reforestation projects may be developed for the purpose of providing raw materials such as wood and wood residues to generate energy. For example, there is already a reforestation project hosted in Brazil that uses the harvested biomass to replace fossil fuels in an iron and steel factory.[110] The literature has cited the case of biofuels as an example of the conflict between the pursuit of the climate change mitigation objective and biodiversity or human rights.[111]

Nevertheless, some efforts have been made to strengthen the 'sustainability' part of CDM projects: the CDM EB has been evaluating and reporting on the use of 'voluntary sustainable development tools' (such as the CDM Gold Standard[112]),[113]

105 This is problematic because participants have an economic incentive to abstain from conducting an environmental impact assessment in order to keep the costs of a project low, and CDM host countries have incentives (financial or other) to attract CDM investments. For example, Lin points out the risks of rent-seeking activities in CDM certification by designated operational bodies as well as by host country agencies. Van Asselt 2011a, p. 1235; Rousseaux 2005, p. 10; Lin 2012a, p. 43.

106 Roht-Arriaza 2010a, p. 213.

107 The unintended negative consequences include violations of rights of farmers or forest peoples, especially indigenous peoples, massive displacement or evictions, effects in food, water, and energy security, increasing the mal-distribution of land and resources and further impoverishing the already poor. Roht-Arriaza 2010a, p. 213; Morgera 2013, p. 359-390; Pedersen 2011, p. 403; Cameron 2009, p. 699.

108 Rousseaux 2005; Sagemuller 2006; Van Asselt 2011a; Savaresi 2011.

109 Van Asselt 2011a, p. 1211-1212.

110 Project Design Document Form For Afforestation And Reforestation Project Activities (Cdm-Ar-Pdd) - Version 04, CDM - Executive Board, UNFCCC, available at: <https://cdm.unfccc.int/filestorage/R/N/4/RN4YPQ1628K03HCISXFDEZJLVWATBO/PDD.pdf?t=RVN8bnk2MjhwfDCq3qz7QnG2SppQIfk9Vk4B> (last accessed 21 November 2015).

111 For instance, Roht-Arriaza cited biofuels as one of the most notorious contexts in which climate change policies cause human rights violations. It has also been cautioned that even biofuel projects using 'waste' materials - second generation biomass - may have unanticipated negative impacts. Roht-Arriaza 2010a, p. 220-221.

112 It is commented that voluntary initiatives such as the CDM Gold Standard have become important instruments for promoting sustainable development and for the integration of environmental and social concerns into the CDM. Van Asselt 2011a, p. 1272; Levin, Cashore & Koppell 2009.

113 At COP/CMP 7, the Executive Board was requested to 'develop appropriate voluntary measures to highlight the co-benefits brought about by clean development mechanism (CDM) project activities and programmes of activities (PoAs), while maintaining the prerogative of the Parties to define their sustainable development criteria.' The COP/CMP requested at the COP/CMP 8 that the Executive Board evaluate the use of the voluntary sustainable development tool and to report on its findings to the COP/CMP. Decision 8/CMP.7, para. 5; Decision 5/CMP.8, paras. 21 and 22.

and is requested by the COP/CMP to develop guiding tools that assist host Parties, upon their request and on a voluntary basis, in monitoring the CDM project's sustainable development benefits.[114] The Paris Agreement emphasises the long-term climate goals are achieved 'in the context of sustainable development',[115] and determines to establish 'a mechanism to contribute to the mitigation of greenhouse gas emissions and support sustainable development.'[116] Thus, how the objective of reducing GHG emissions could be met while fostering sustainable development is to be seen in the international processes and mechanisms in progress under the UNFCCC.

2.4. *Interim Conclusion*

It is observed in this section that the commitments to address or limit GHG emissions under the international climate regime may drive the demand for biofuels, as long as biofuels are recognised as contributing to reducing GHG emissions in the transport sector. Hence, the essential question is whether the climate regime correspondingly provides rules or guidance for assessing the contribution of biofuels to reductions in GHG emissions. Biofuels have not been especially addressed among a range of topics under the UNFCCC or Kyoto Protocol, but the IPCC Guidelines for national inventory communication, as adopted by COP or COP/CMP, have provided implications for accounting for the life-cycle GHG emissions of biofuels. The IPCC Guidelines aim to estimate GHG inventories including emissions and removals from all sources and sectors in a complete and accurate manner, which could account for all the human-induced emissions caused by biofuels in their life cycle if it would be followed by all the states. However, since states in practice undertake different obligations and methods in communicating and reporting their national inventory totals, the application of the IPCC Guidelines may be limited in ensuring a complete and accurate accounting of the life cycle GHG emissions for biofuels, in particular when land use changes occur. Furthermore, although the international climate regime commits to the principle of sustainable development, it has not provided reliable measures to sufficiently assess and integrate other environmental and socio-economic sustainability considerations, such as biodiversity and human rights, in the pursuit of climate change mitigation goal and it may even lead to conflicting consequences.

114 The COP/CMP requested at the 9th session that the Executive Board develop guiding tools to assist designated national authorities, upon the request of the host Party and on a voluntary basis, in monitoring the sustainable development benefits in its territory of clean development mechanism project activities and programmes of activities, recognising that the use of such guiding tools is the prerogative of Parties and subject to the availability of funds from Parties included in Annex I. At the 10th session, the COP/CMP expressed satisfaction that 16 sustainable development co-benefits description reports had been published using the voluntary sustainable development tool. Decision 3/CMP.9, para. 8; Decision 4/CMP.10, para. 3(h).

115 Paris Agreement, Arts. 2(1) and 4(1).

116 Paris Agreement, Art. 6(4).

3. Supply of Biofuels: International Biodiversity and Socio-Economic Rights Regimes

The regulation of the supply of biofuels, and in particular the production of biofuel feedstocks, mainly falls within national jurisdiction due to the close relationship with agriculture and land use policies. However, the potential impacts on biodiversity or food availability and prices during the production processes of biofuels give rise to important sustainability concerns, which will have an effect on international biodiversity and socio-economic rights.[117] This section explores the international regimes on biodiversity and socio-economic rights to map the rules relevant for the concerns related to the sustainability of biofuels in the production stage.

3.1. *The International Biodiversity Regime and the Supply of Biofuels*

Approximately one third of multilateral environmental agreements address biodiversity either in full or in part, and they form the biodiversity regulatory regime at the international level.[118] Among the major biodiversity treaties,[119] the Convention on Wetlands of International Importance especially as Waterfowl Habitat (Ramsar Convention) and the Convention on Biological Diversity (CBD) are of the greatest relevance with regard to biofuels, because these two Conventions aim to protect wetlands or biodiverse areas/ecosystems[120] and biofuel feedstock production may pose a potential threat to these areas/ecosystems. While the Ramsar Convention concentrates on a specific ecosystem and the CBD takes a 'comprehensive and cutting-edge' approach,[121] they both boast wide membership[122]

117 See e.g. Ottinger 2009, p. 256; Koh & Ghazoul 2008, p. 2450-2451; Fletcher Jr. *et al.* 2010; Wiens, Fargione & Hill 2011; Witt 2010; Altieri 2009; Naylor *et al.* 2007; Runge & Senauer 2007; Sexton *et al.* 2008; Molony & Smith 2010.

118 McGraw 2002, p. 20.

119 The major biodiversity treaties include but are not limited to the Convention on International Trade in Endangered Species (CITES), the Convention on the Conservation of Migratory Species of Wild Animals (CMS), the World Heritage Convention (WHC), International Treaty on Plant Genetic Resources for Food and Agriculture (ITPGR), and the International Convention on the Regulation of Whaling (ICRW).

120 The mission of the Ramsar Convention is 'the conservation and wise use of all wetlands through local and national actions and international cooperation' in order to achieve sustainable development throughout the world.' The CBD sets forth its principal objectives as threefold: conservation of biodiversity, sustainable use of its components, and fair and equitable sharing of benefits from the utilisation of genetic resources. Ramsar Convention Secretariat 2013, p. 7; CBD, Art. 1.

121 The CBD can be distinguished from other biodiversity treaties such as the Ramsar Convention due to its comprehensive and global approach, which seeks to fill the 'biodiversity gap' and consolidate existing efforts, for it addresses nearly all aspects of biodiversity and cuts across existing species and habitat instruments. McGraw 2002, p. 21-22; Bowman, Davies & Redgwell 2010, p. 593-594, and 625.

122 The Ramsar Convention entered into force in 1975 and has 168 Contracting Parties. China is a Party to the Convention; while EU is not a Party, all the 28 European Union Member States →

and have a similar institutional structure.[123] Moreover, they both contain general commitments for the wise or sustainable use of the components of biodiversity as well as specific decisions focusing on the relationship between biodiversity and biofuels.

3.1.1. General Commitment to Conservation and Sustainable/Wise Use of Biodiversity

The CBD enshrines conservation and the sustainable use of the components of biodiversity or biological diversity[124] as its objective and core commitment, and Parties have an obligation to develop and integrate national strategies, plans, or programmes for conservation and the sustainable use of biodiversity.[125] The core commitment for contracting Parties of the Ramsar Convention is to conserve and make wise use of all the wetlands within their territories as defined in the Convention,[126] and there is a general obligation for the Parties to formulate and implement national land-use planning to include wetland conservation and to promote the wise use of wetlands in their territories.[127] Hence, the essential question

have ratified the Convention. EU as a supranational body is not eligible to join the Convention in accordance with Art. 9.2 of the Convention, but the EU may develop bilateral working agreements with the Convention Secretariat. The Convention on Biological Diversity (CBD) was opened for signature during the 1992 United Nations Conference on Environment and Development (UNCED) and entered into force 18 months thereafter. Until now, the CBD accomplished virtually universal participation, with 193 parties including China and the EU, but the US is not a party to the Convention although it signed in 1993. Ramsar Convention on Wetlands, available at: <www.ramsar.org/sites/default/files/documents/library/-contracting_parties_list_20150312_e.pdf> (last accessed 7 April 2014); Convention on Biological Diversity, 'List of Parties', available at: <www.cbd.int/convention/parties/list/> (last accessed 7 April 2014).

123 The Ramsar Convention has the Conference of the Contracting Parties (COP) which meets every three years, a Secretariat responsible for day-to-day administration and a subsidiary body of the Scientific and Technical Review Panel to provide scientific and technical guidance. The CBD has a similar structure of treaty bodies such as the Conference of Parties (COP), the Secretariat, and the Subsidiary Body on Scientific, Technical, and Technological Advice (SBSTTA). Ramsar, The Bodies of the Convention, available at: <www.ramsar.org/about/the-bodies-of-the-convention> (last accessed 7 April 2015); CBD, Conference of Parties (COP), available at: <www.cbd.int/cop/> (last accessed 7 April 2015).

124 The CBD adopts a broad definition of 'biological diversity': biological diversity means the variability among living organisms from all sources including, *inter alia*, terrestrial, marine and other aquatic ecosystems and the ecological complexes of which they are part; this includes diversity within species, between species and of ecosystems. CBD, Art. 2.

125 CBD, Art. 6.

126 Art. 1(1) of the Ramsar Convention defines wetlands as 'areas of marsh, fen, peatland or water, whether natural or artificial, permanent or temporary, with water that is static or flowing, fresh, brackish or salt, including areas of marine water the depth of which at low tide does not exceed six metres.' This is an 'exceptionally broad definition', since habitats such as peat bogs, floodplains, coastal waters and beaches, mangrove swamps, coral reefs are also included. Bowman, Davies & Redgwell 2010, p. 405.

127 Ramsar Convention, Art. 3.1.

is what the general commitments about conservation and sustainable/wise use entail for the production of biofuels.

The CBD distinguishes between *in-situ* and *ex-situ* conservation, which respectively refers to the conservation of ecosystems, habitat, and populations in their natural surroundings or the conservation of biological diversity components outside their natural habitats.[128] The CBD provides for the establishment of protected areas as an important measure for conserving biological diversity,[129] but it does not require the formulation of an international list of the protected areas as the Ramsar Convention does.[130] Sustainable use is explicitly defined as 'the use of components of biological diversity in a way and at a rate that does not lead to the long-term decline of biological diversity, thereby maintaining its potential to meet the needs and aspirations of present and future generations.'[131] According to these definitions, the CBD does not exclude human use or intervention from conservation and sustainable use of biodiversity,[132] so it is important to measure what constitutes the 'sustainable' use. Guidance on the implementation of the sustainable use of biodiversity has been set out by the CBD Secretariat in the Addis Ababa Principles and Guidelines for the Sustainable use of Biodiversity (Addis Ababa Principles and Guidelines), which consists of fourteen practical principles, operational guidelines, and a few instruments that provide a framework of general relevance that assists governments and communities.[133] However, it has been commented that the Convention text is too broad and vague, and the language used for setting out obligations is imbued with qualifying words such as 'in accordance with its particular conditions and capabilities' or 'as far as possible and as appropriate.'[134]

Under the Ramsar Convention, Parties undertake a commitment to conserve wetlands designated as internationally important and to make wise use of all the wetlands.[135] As for the wetlands of international importance which are designated

128 CBD, Art. 2.

129 Other measures may include the formulation and implementation of national protective legislation and regulations, risk control of living modified organisms and alien species, as well as the protection of indigenous peoples' knowledge and equitable sharing. CBD, Art. 8.

130 The measure of identification of protected areas is to identify and monitor components of biological diversity such as ecosystems and habitats containing high diversity, threatened species and communities, or genomes and genes of social, scientific or economic importance, as well as to identify processes and categories of activities having significant adverse impacts on conservation and the sustainable use of biodiversity. CBD, Art. 7 and Annex I.

131 CBD, Art. 2.

132 It has been commented that the reference to 'sustainable use' is a 'clear indicator' that human use and benefit to humans are the reasons for conserving biodiversity and thus are different from the preservationist concept. Bowman, Davies & Redgwell 2010, p. 598.

133 Addis Ababa Principles and Guidelines 2004, para. 1.

134 In addition, the broad commitments intend to be implemented through reporting in the form of National Biodiversity Strategies and Action Plans (NBSAPs) and national reports with no non-compliance sanction prescribed in the Convention, and the form and means of implementation are largely up to contracting parties. Bowman, Davies & Redgwell 2010, p. 626-627; Adam 2010, p. 123.

135 Ramsar Convention, Arts. 2 and 3.

by Parties and included in the Ramsar List,[136] Parties commit to maintaining their ecological characters[137] and are obliged to inform the Ramsar Secretary if the ecological character changes.[138] As for other wetlands, Parties undertake a commitment to promote 'as far as possible the wise use of wetlands in their territory.'[139] Nevertheless, the commitment to wise use is applied to both the wetlands on Ramsar List and the other wetlands,[140] and the meaning of 'wise use' has been interpreted as *synonymous* to 'sustainable use': it entails the maintenance of the ecosystems and the continued delivery of ecosystem services for human well-being,[141] which implies that human uses on a sustainable basis is entirely compatible with the Ramsar Convention and wetland conservation in general.[142] In this vein, human activities on wetlands are permissible but may vary depending on the merits of each case, and the activities may range from small to large-scale exploitation.[143]

136 To distinguish the wetlands of international importance, Parties are required to designate at least one wetland at the time of accession for inclusion in the Ramsar List of Wetlands of International Importance, and to continue to designate suitable wetlands within its territory for the List. Parties have great freedom in deciding which sites should be included in the List: the Convention allows for a unilateral designation of sites without prejudice to the exclusive sovereignty of the Contracting Parties; and Parties may extend the boundaries of the sites as well as add, delete and restrict the sites. Ramsar Convention, Arts. 2(1) and 2(4); Bowman, Davies & Redgwell 2010, p. 406.

137 Ecological character is the combination of the ecosystem components, processes and benefits or services that characterise the wetland at a given point in time. Ramsar Convention Secretariat 2010, p. 14.

138 To a certain extent, the wetlands on the Ramsar List may enjoy a higher priority in relation to preserving the ecological characters and may attract more public attention. The underlying rationale is that wetlands included in the List acquire a new national and international status: they are recognised as being of significant value not only for the country or the countries in which they are located, but for humanity as a whole. Ramsar Convention, Art. 3(2); Ramsar, Wetlands of International Importance, available at: <www.ramsar.org/about/wetlands-of-international-importance> (last accessed 7 April 2015); Bowman, Davies & Redgwell 2010, p. 410 and 420.

139 Ramsar Convention, Art. 3(1).

140 Bowman, Davies & Redgwell 2010, p. 414-419; Ramsar Convention Secretariat 2013, p. 15 and 46.

141 How to define and interpret 'wise use' has been a core issue under the Ramsar Convention. A definition of 'wise use' was first adopted by the Contracting Parties at COP3 in 1987, Canada, as 'the wise use of wetlands is their sustainable utilization for the benefit of humankind in a way compatible with the maintenance of the natural properties of the ecosystem.' This definition was later revised in Annex A of Resolution IX.1 in 2005 as '[W]ise use of wetlands is the maintenance of their ecological character, achieved through the implementation of ecosystem approaches, within the context of sustainable development.' 'Sustainable utilization' was defined in 1987 as the '[H]uman use of a wetland so that it may yield the greatest continuous benefit to present generations while maintaining its potential to meet the needs and aspirations of future generations.' Ramsar Convention Secretariat 2010, p. 6.

142 Ramsar Convention Secretariat 2013, p. 46.

143 In practice, it is observed that the key ecological characteristics of many wetlands, such as the blanket peat bogs of upland Britain, have been shaped by human activities over the centuries, and some wetlands such as rice paddies and reservoirs have been specifically created for human purposes. Bowman, Davies & Redgwell 2010, p. 415; Ramsar Convention Secretariat 2013, p. 14 and 49.

Like the CBD, guidance on the implementation of 'wise use' has been issued in the form of guidelines by the Ramsar Secretariat,[144] but it is the Parties that decide how to make wise use of wetlands.[145]

Therefore, the general commitment to conservation and sustainable/wise use of the components of biodiversity does not necessarily exclude human activities or natural resource exploitation. In this vein, biofuel feedstock production activities may be compatible with the CBD or Ramsar Convention if they are in line with the definition of and guidance on sustainable/wise use.

3.1.2. Specific Guidance on Biofuels and Biodiversity

Although the texts of the CBD and the Ramsar Convention do not provide for specific rules for biofuels, the relationship between biofuels and biodiversity has been noted in instruments adopted by Conference of the Parties (COP) of both the Conventions. The CBD COP undertakes the function of keeping the implementation of the Convention under review, as well as considering and adopting protocols, amendments, annexes, and additional actions for achieving the Convention's objectives.[146] Under the Ramsar Convention, the COP has the competence to adopt resolutions, recommendations, and guidelines, to set out guidance on a range of issues relating to wetlands, and further to spell out the interpretation of the Parties' commitments and responsibilities.[147] Due to the potential impact of the production

144 Bowman, Davies & Redgwell 2010, p. 419; Ramsar Convention Secretariat 2013, p. 47.

145 The Convention text only provides for the obligation of Parties to notify the Secretariat of adverse ecological changes of wetland sites on the Ramsar List and to respond to the Secretariat's inquiries about such reports received from third parties. Moreover, Parties are requested to consult with each other with respect to the implementation of their obligations. Until the COP in 1984, Parties were urged to submit detailed National Reports to the Secretariat prior to each ordinary meeting of the Conference, and this tradition has continued unbroken to this day. National reports are submitted to the Secretariat prior to each COP meeting, and the Secretariat submits regional overviews of implementation to the Conference based on study and summary of the national reports. Ramsar Convention, Arts. 3.2 and 5; Ramsar Convention Secretariat 2013, p. 14 and 34; Bowman, Davies & Redgwell 2010, p. 416.

146 The CBD COP convenes every two years and functions as a governing body of the Convention. To date the COP has held 12 ordinary meetings and 1 extraordinary meeting, during which more than 2 protocols and more than 300 decisions were adopted. It is noted that the Cartagena Protocol on Biosafety is an international treaty that intends to protect biodiversity from transboundary movements of living modified organisms (LMOs) out of modern biotechnology and has established an advance informed agreement (AIA) procedure to ensure that the necessary information is provided for importing countries to make informed decisions before agreeing to import such organisms. CBD, Art. 23(4); CBD, Convention bodies, Conference of the Parties (COP), available at: <www.cbd.int/cop/> (last accessed 1 March 2016).

147 Every three years (sometimes at an interval of four years), representatives of the governments of each of the Contracting Parties meet as the Conference of the Contracting Parties to agree on a work programme and budgetary arrangements for the next triennium and to consider guidance on a range of ongoing and emerging environmental issues. Art. 6 of Ramsar Convention; Ramsar Convention on Wetlands, The Conference of the Contracting Parties, →

of biofuels on wetlands and biodiversity, decisions or resolutions have been respectively adopted under CBD and the Ramsar Convention to focus on the issue of biofuels. The following passages will analyse the decisions or resolutions according to their substantive guidance regarding biofuels and will summarise their synergies in substance and legal status.

3.1.2.1. Decisions of the CBD COP

Decisions adopted under CBD COP are generally divided into seven thematic areas according to different ecosystems and biomes,[148] among which the topic of biofuels and biodiversity has been continuously addressed since the 9th meeting in 2008. Generally, the decisions declare that 'biofuel production and use should be sustainable in relation to biological diversity'[149] by promoting the positive and minimising the negative impacts of biofuels on biodiversity and the livelihood of indigenous and local communities,[150] and biofuel's sustainability 'must necessarily take into account the environmental, economic and social pillars of sustainable development.'[151] Furthermore, the following specific guidance has been set out.

a). Impact Assessment

CBD COP Decision X/37 encourages Parties and other governments to undertake impact assessments for minimising or avoiding the negative impacts of biofuels on biodiversity, as well as impacts on biodiversity that affect related socio-economic conditions.[152] Thus, the impact assessment on biofuels under CBD entails not only an environmental impact assessment but also an impact assessment of the socio-economic aspects of biodiversity, such as the impact on indigenous and local communities.

Actually, prior to the 9th meeting in 2008, the Body on Scientific, Technical and Technological Advice (SBSTTA) of CBD had conducted an assessment on the potential impacts of biofuels on biodiversity.[153] The adverse effects include not only environmental impacts, such as habitats, biodiversity components, ecosystem services, GMOs and invasive alien species, local environment, but also socio-economic effects, such as competition for land, indigenous and local communities, small-holder farmers, and food security.[154] CBD COP Decision IX/2 recognises the

available at: <www.ramsar.org/about/the-conference-of-the-contracting-parties> (last accessed 8 April 2015); Ramsar Convention Secretariat 2013, p. 15.

148 They include marine and coastal biodiversity, agricultural biodiversity, forest biodiversity, inland waters biodiversity, and dry and sub-humid lands biodiversity, mountain biodiversity and island biodiversity. Bowman, Davies & Redgwell 2010, p. 602; CBD, Convention, Convention bodies, Conference of Parties (COP), available at: <www.cbd.int/cop/> (last accessed 10 February 2014).

149 Decision IX/2, para. 1.

150 Decision IX/2, para. 2.

151 Decision IX/2, preamble.

152 Decision X/37, para. 10.

153 CBD SBSTTA Recommendation XII/7, paras. 3(b) and 3(c).

154 CBD SBSTTA Recommendation XII/7, para. 3(c)(i)-(vi).

impacts as biodiversity-related factors in the production and use of biofuels[155] and calls for further investigation and monitoring of the impacts of biofuels[156] in view of the knowledge gaps and uncertainties.[157] Decision X/37 continues to recognise the impacts of biofuels on the ecological character of biodiversity, food and energy security, and land tenure and resource rights (e.g. water), in particular with regard to indigenous and local communities.[158] Thus, Parties are invited to assess both the direct and indirect effects and impacts of biofuels on biodiversity in their full life cycle as compared to other types of fuels.[159]

b) Sound Policy Framework

CBD COP Decision IX/2 (2008) urges Parties, and invites other governments, to develop and apply 'sound policy frameworks' for the sustainable production of biofuels.[160] The Decision does not set out what 'sound' means, but the following characteristics can be identified for developing such policy frameworks.

First, while acknowledging national diversity, the policy framework should take into account the impact of biofuels during their full life cycle as compared to other fuel types.[161] Parties and other governments are encouraged to develop and implement land use and water policies, in particular to address direct and indirect land and water use changes affecting the areas with high value of biodiversity or of cultural, religious and heritage interest.[162]

Second, the policy framework should minimise the negative impacts on biodiversity that would affect socio-economic conditions or the food and energy

155 The Decision recognises the current preliminary views of Parties that the biodiversity-related factors of biofuel production and use that should be considered are outlined in paras. 3(b) and 3(c) of Recommendation XII/7 of the Subsidiary Body on Scientific, Technical and Technological Advice. Decision IX/2, para. 9.

156 The CBD COP decision calls upon Parties, other Governments, and the research community, as well as invites other relevant organisations to 'continue to investigate and monitor the positive and negative impacts of the production and use of biofuels on biodiversity and related socio-economic aspects, including those related to indigenous and local communities.' The Executive Secretary is requested to 'further compile this evidence and to make it available through the clearing-house mechanism of the Convention and other appropriate means.' Decision IX/2, para. 5.

157 SBSTTA 12 Recommendation XII/7 also recognised that there are gaps in knowledge and information associated with the impacts of biofuel production and use on biodiversity and the production and use of biofuel from second-generation feedstocks, including the application of the precautionary approach. CBD SBSTTA Recommendation XII/7, para. 3(d) and (e).

158 Parties and other Governments are urged to ensure that the sustainable agricultural practices and food and energy security of indigenous and local communities are addressed and respected, subject to national legislation, taking into account the customary laws of indigenous and local communities, when assessing the impacts of biofuel production and use. Decision X/37, paras. 2 and 4.

159 Decision X/37, para. 6.

160 Decision IX/2, paras. 3(c) and 4.

161 Decision IX/2, para. 3(c).

162 Decision X/37, para. 9.

security.[163] In accordance with an impact assessment on the ecological character as well as socio-economic conditions of biodiversity, Parties are invited to develop and implement policies that minimise or avoid the negative impacts of biofuel production on biodiversity and related socio-economic conditions.[164]

Third, the policy framework should make use of relevant tools and guidance already developed under CBD, such as the precautionary approach[165] and the guiding principles on alien invasive species.[166] Particularly, Decision X/37 urges Parties and other governments to apply the precautionary approach when genetically modified organisms are introduced into the environment,[167] since it is set out in the preamble of the CBD that 'where there is a threat of significant reduction or loss of biological diversity, lack of full scientific certainty should not be used as a reason for postponing measures to avoid or minimize such a threat.'

Fourth, supportive measures for biofuels should promote the positive and minimise the negative impacts of the production of biofuels.[168] Decision XI/27 underscores the importance of national incentive measures[169] as drivers for biofuels expansion, and invites governments to evaluate their incentive measures by using the Aichi Biodiversity Targets[170] given the national socio-economic conditions.[171]

c) National Inventory and National Report

Parties and other governments are invited to develop national inventories and identify 'go' and 'no go' areas or ecosystems for biofuel production.[172] Decision X/37 invites Parties to develop national inventories so as to identify areas of high biodiversity value and critical ecosystems as well as areas important to indigenous and local communities.[173] It also invites Parties to assess and identify areas and

163 Decision IX/2, para. 3(b).

164 Decision X/37, para. 6.

165 Decision IX/2, para. 3(c)(i).

166 Decision VI/23.

167 Decision X/37, para. 16.

168 Decision IX/2, para. 4.

169 According to Art. 11 of CBD, 'Each Contracting Party shall, as far as possible and as appropriate, adopt economically and socially sound measures that act as incentives for the conservation and sustainable use of components of biological diversity.' Thus, incentive measures hereby refer to the measures that introduce economic and social incentives for the conservation and sustainable use of the components of biodiversity. CBD, Art. 11.

170 The Aichi Biodiversity Targets include 5 strategic goals and 20 targets. The Target 3 sets the goal that '[B]y 2020, at the latest, incentives, including subsidies, harmful to biodiversity are eliminated, phased out or reformed in order to minimize or avoid negative impacts … consistent and in harmony with the Convention and other relevant international obligations, taking into account national socio economic conditions.' See CBD, Strategic Plan 2011-2020, Aichi Biodiversity Targets, available at: <www.cbd.int/sp/targets/> (last accessed 11 February 2015).

171 Decision XI/27, para. 5.

172 It is recorded that the development of biodiversity inventories for biofuels was an area of contention in the discussion on biofuels. The COP eventually agreed to *invite* governments and relevant organisations to develop national inventories. Jungcurt *et al.* 2010, p. 22.

173 Decision X/37, para. 7(a).

ecosystems that could be used for biofuels' production (as 'go' areas) or protected from biofuels' production (as 'no-go' areas).[174] The purpose is to assist policy makers in applying conservation measures and identifying areas that are inappropriate for the production of biofuel feedstock.[175] However, the meaning of 'high biodiversity value' and 'critical ecosystems' is not explicitly set out in the decision, and it is unclear whether the exercise of identifying biodiverse areas for biofuels will support the protection of biodiversity or go beyond a business-as-usual planning and zoning.[176]

Moreover, Parties are invited to report on the ways and means of minimising or avoiding the negative impacts of biofuels through the CBD implementation mechanisms. Decision X/37 recognises the need to include such ways and means in national biodiversity strategies and action plans (NBSAPs) and national development plans,[177] and invites Parties to report as part of their fifth national reports under CBD as appropriate. Decision XI/27 invites Parties to consider biofuel issues continuously in updating and implementing their national biodiversity strategies and action plans (NBSAPs) and the fifth national reports.[178] The content to be contained in the fifth national report includes progress and information about the designation of national inventories and 'go' or 'no-go' areas,[179] as well as voluntary tools that are used in strategic environment and socio-economic assessment regarding the impacts of biofuels on biodiversity and integrated land-use planning.[180]

3.1.2.2. Resolutions of the Ramsar COP

Resolution X.25 on the relationship between wetlands and biofuels was adopted by the Ramsar COP meeting in 2008 and set out the potential impacts of biofuels on wetlands: concerns are expressed with regard to the pressure from the development of biofuels on wetlands conversion, potential GHGs released from wetlands carbon stock, possible ignorance of the full range of ecosystem services, and the threat of irrigated agriculture expansion.[181] Hence, the Resolution states that biofuel

174 Decision X/37, para. 7(b).

175 Decision X/37, para. 7.

176 Although Parties are invited to develop national inventories and areas, it is not clear whether this exercise will support the protection of biodiversity or 'go beyond a business-as-usual planning and zoning exercise', and there is doubt that this exercise may undermine area-based efforts to protect biological diversity. Morgera 2011, p. 106.

177 Decision X/37, para. 5.

178 Decision XI/27, para. 2(a).

179 The decision invites Parties and other Governments to report on biofuel matters, voluntary tools, and national inventories and designated areas in their fifth national reports, where feasible. Decision XI/27, para. 2(b) and (c) and para. 4.

180 Decision XI/27, paras. 2(b) and 4.

181 Meanwhile, there is recognition that biofuels crops differentiate in impact on wetlands and some crops grown on degraded lands can in certain cases benefit the rehabilitation of wetlands. Resolution X.25, paras. 5, 7, 8, 11, 12, and 13.

production should be sustainable in relation to all wetlands[182] and calls for measures to be taken to address biofuels and wetlands.

a). Impact Assessment

The Resolution calls upon Parties to assess the potential impacts, benefits, and risks of proposed biofuel crop production schemes affecting wetlands - included in the Ramsar List or not, particularly with regard to implications for surface and groundwater resources.[183] To assess the impacts, Environmental Impact Assessment (EIA) and Strategic Environmental Assessment (SEA) are to be applied.[184] However, the Resolution further states that, when avoidance of the assessed negative impacts is not feasible, appropriate mitigation/compensation/offset actions (e.g. through wetlands restoration) will be used as far as possible.[185] Such a statement implies that, biofuel crops could still be grown, even when an environmental impact assessment reveals that there might be adverse impacts on the wetlands, if mitigation and/or compensation/offset measures are taken 'as far as possible.'

In fact, at the COP XI, Resolution XI.10 notes that the production and harvesting of biofuels may have direct and indirect adverse impacts[186] on the ecological character of wetlands, which are factors to be considered in carrying out impact assessments and developing integrated sectoral policy for the energy sector.[187] However, considering that 'at present there is no scientific consensus regarding quantification and measurement of indirect impact',[188] Parties are encouraged to conduct deeper studies and analyses about the potential impacts, benefits, and risks of biofuel production for wetlands.[189]

b) Appropriate Land Use Policies

The Resolution urges Parties to consider formulating appropriate land use policies for the sustainable production of biofuels feedstocks and to accelerate the implementation of policies that promote the positive and minimise the negative impacts on wetlands.[190] For example, Parties are encouraged to consider the cultivation of biomass on rewetted peatlands, by using the land management

182 Resolution X.25, para. 14.

183 Resolution X.25, para. 15.

184 The COP has adopted Resolutions on impact assessments in 1999 and 2008: Resolution VII.16, 'The Ramsar Convention and impact assessment: strategic, environmental and social', and Resolution X.17, 'Environmental Impact Assessment and Strategic Environmental Assessment: updated scientific and technical guidance.' Resolution X.25, para. 15; Resolution VII. 16; Resolution X.17.

185 Resolution X.25, para. 15.

186 For example, there are indirect impacts of atmospheric emissions, including water quality impacts from emissions like particulate materials, sulphur or nitrogen compounds, and those due to climate change effects resulting from greenhouse gas emissions. Resolution XI. 10, Annex, para. 5(vii).

187 Resolution XI.10, paras. 5(vi) and (vii), 13(iii), 17(i) and 19(i).

188 Resolution XI.10, fn. 1.

189 Resolution X.25, para. 21.

190 Resolution X.25, para. 16.

technique of paludiculture that helps stop peat oxidation and simultaneously provide sustainable harvests from peatlands.[191] Parties are also encouraged to promote sustainable forest and agricultural practices that mitigate the adverse impacts of biofuel production.[192] However, the Resolution does not further elaborate on what 'appropriate land use' and 'sustainable forest and agricultural practices' entail.

c) Ecosystem Approach

The ecosystem approach is a 'strategy for the integrated management of land, water, and living resources that promotes conservation and sustainable use in an equitable way',[193] which has also been emphasised under the CBD as the framework for analysis and implementing CBD's objectives.[194] The ecosystem approach has been recognised as being particularly important for wetlands, considering the past loss and degradation of wetlands ecosystems and their ecosystem services.[195] In this context, Resolution X.25 strongly urges Parties to ensure that biofuel crop production policies should consider the full range and value of ecosystem services provided by wetlands, such as carbon storage, flood protection, the production of food and fibres, the biodiversity they support, and groundwater recharge,[196] as well as the livelihoods provided by wetlands.[197] In this vein, Parties should consider the trade-offs among the ecosystem services alongside the cost-benefit analysis, and should make use of the precautionary approach as defined in 1992 Rio Declaration on Environment and Development.[198] The ecosystem approach may entail the core meaning of 'wise use' that identifies the critical linkages between people and the sustainable development of natural resources, as well as community engagement in

191 Paludicultures are land management techniques that cultivate biomass from wet and rewetted peatlands under conditions that maintain the peat body, facilitate peat accumulation and sustain the ecosystem services associated with natural peatlands. Resolution X.25, para. 17; Wetlands International, 'Paludiculture: sustainable livelihood options', available at: <www.wetlands.org/Whatwedo/Savingpeatlands/Paludiculture/-tabid/3468/Default.aspx> (last accessed 9 May 2015).

192 Resolution X.25, para. 17.

193 Bowman, Davies & Redgwell 2010, p. 603.

194 Under the CBD, at COP II in 1995, the Parties adopted the ecosystem approach as the framework for analysis and the implementation of the objectives of the Convention. At COP V, Decision V/6 was adopted to focus on the ecosystem approach and to contain an annex to set out operational guidance and 12 common principles, with the requirement to integrate these principles within national plans and priorities, and COP VII adopted implementation guidelines for each of the 12 common principles. Bowman, Davies & Redgwell 2010, p. 603; Decision IX/2, para. 3(c)(iii).

195 Finlayson *et al.* 2011, p. 178.

196 Resolution X.25, para. 13.

197 Resolution X.25, para. 19.

198 Principle 15 of Rio Declaration states that '[I]n order to protect the environment, the precautionary approach shall be widely applied by States according to their capabilities. Where there are threats of serious or irreversible damage, lack of full scientific certainty shall not be used as a reason for postponing cost-effective measures to prevent environmental degradation.' Rio Declaration on Environment and Development 1992, principle 15.

negotiating trade-offs and determining equitable outcomes for conservation.[199] However, the Resolution on biofuels and wetlands does not specify further how to conduct the trade-offs among ecosystem services in the case of biofuel crop production, particularly between the potential impacts on wetlands and possible gains for livelihood.

3.1.3. Interim Conclusion

According to the afore-mentioned general commitments and specific guidances, it is observed that the CBD and Ramsar Convention have some synergies in addressing the issue of biofuels and biodiversity. Both the CBD and Ramsar COPs have adopted instruments, in the form of decisions or resolutions, to raise awareness of the potential impacts of the production of biofuels on biodiversity. Moreover, both the CBD and Ramsar COPs recognise the need to formulate 'appropriate' or 'sound' national policies to minimise and avoid negative impacts, as well as the importance of conducting impact assessments, designating biodiversity inventories and 'go' or 'no-go' areas, or adopting a precautionary and an ecosystem approach. Moreover, the two Conventions intend to maintain close institutional cooperation between their treaty bodies. On the part of Ramsar Convention, Resolution X. 25 'acknowledges' the CBD COP Decision IX/2 and X/37 and invites its subsidiary body – the STRP – to take into account the findings and conclusions of the CBD subsidiary body SBSTTA.[200] As for the CBD, its Executive Secretary is instructed to contribute to and assist with the on-going work regarding the sustainability of biofuels in other international organisations and processes, which should include those within the Ramsar Convention.[201] Moreover, the secretariats of the two Conventions have channels to cooperate, such as through the Joint Work Plan[202] between the two Conventions, or the Joint Liaison Group[203] among treaties agreed at the 1992 Rio Summit.

However, the approach of the two Conventions in setting out general commitments and specific guidelines reveals a remaining deficiency for addressing

199 Finlayson *et al.* 2011, p. 177.

200 Resolution X.25, para. 23.

201 Decision X/37, paras. 11(b) and 13.

202 The secretariats of the Ramsar Convention and the CBD signed Memorandum of Cooperation since January 1996. In November of that year, the CBD's COP3 invited Ramsar 'to cooperate as a lead partner' in implementing CBD activities related to wetlands. Shortly thereafter an innovative Joint Work Plan was put in place for 1998-1999. The two conventions are still working closely together, presently under a 5th Joint Work Plan. The secretariats of other organisations and conventions, such as FAO and CITES, are present at the meetings of SBTTA and COP of CBD. Ramsar Convention Secretariat 2013, p. 18; Bowman, Davies & Redgwell 2010, p. 625.

203 The Joint Liaison Group comprises the secretariats of the CBD and the UNFCCC and the UNCCD (UN Convention to Combat Desertification), in order to enhance coordination and exchange of information among the multilateral environmental agreements reached as fruits of 1992 Rio Summit. CBD takes the lead in the Biodiversity Liaison Group, which also includes the World Heritage Convention and the Ramsar, CITES and Bonn Convention. Bowman, Davies & Redgwell 2010, p. 625; Van Asselt 2011a, p. 1265.

concerns relating to biofuels and biodiversity. In the first place, they do not provide sufficient guidance for evaluating the 'wise' or 'sustainable' use of biodiversity in the case of biofuel crop production. The general commitment to sustainable/wise use does not necessarily preclude human intervention and the utilisation of natural resources, including the activities of biofuel feedstock production, so it is important to set out how to make sustainable or wise use. Although the COP decisions or resolutions set forth the need for and importance of establishing appropriate or a sound national policy framework, they do not define the standards or criteria for 'appropriate' or 'sound' policy. To the contrary, it is even implied from Resolution X.25 that biofuel crop production may continue when the avoidance of the assessed negative impact on wetlands is not feasible, if the mitigation/compensation/offset actions are applied as far as possible.

In fact, certain gaps have been identified in developing universally accepted definitions of the terms essential for assessing the sustainability of biofuels relating to biodiversity. As requested by Decision XI/27,[204] the CBD Executive Secretary has compiled information on relevant definitions of the key terms associated with the CBD COP decisions on biofuels.[205] In this document, the Executive Secretary identifies the key terms used in CBD COP decisions like 'full life cycle', 'direct and indirect land use', 'sustainable and unsustainable production and use of biofuels', etc. It notes the lack of universally accepted definitions of the key terms, as well as the difficulty in developing criteria or setting thresholds of numerical limits for applying the key terms.[206] In the opinion of the Executive Secretary, the definition and interpretation of these terms are largely matters of national systems, and it is up to states to determine whether and how to apply the definitions at a national level.[207] Meanwhile, the existing non-binding initiatives engaged by non-governmental organisations have been noticed. For example, the CBD Executive Secretary attaches attention to the role of International Union for Conservation of Nature (IUCN)[208] in consolidating scientific criteria and methodology to identify

204 The Decision requests the Executive Secretary, as part of his ongoing work with regard to Decision X/37, in collaboration with Parties, other Governments and relevant organisations, and considering ongoing work, to compile information on relevant definitions of relevant key terms to enable Parties to implement Decisions IX/2 and X/37 (both referring to biofuels and biodiversity) and to report on progress to a meeting of the Subsidiary Body on Scientific, Technical and Technological Advice prior to the 12th meeting of the Conference of the Parties. Decision XI/27, para. 10.

205 CBD/SBSTTA/18/15 2014.

206 For example, the use of the term 'sustainable' is widespread and there are various official definitions of the term, but difficulties can be encountered when trying to operationalise the term in practice. CBD/SBSTTA/18/15 2014, paras. 2-4.

207 CBD/SBSTTA/18/15 2014, para. 29.

208 Since 2009, an IUCN World Commission on Protected Areas and IUCN Species Survival Commission joint task force was established to convene a consultation process to consolidate scientific criteria and methodology to identiy sites of 'Key Biodiversity Areas' of global significance. These criteria are developed to support national and regional processes in identifying important sites within their jurisdiction and intend to help government agencies as well as the private sector and donor agencies. CBD/SBSTTA/18/15 2014, paras. 25, 28 and 29.

'Key Biodiversity Areas' of global significance, and the Roundtable on Sustainable Biomaterials (RSB)[209] in providing sources of definitions of relevant terms and glossaries.[210] Moreover, the CBD COP decisions have also encouraged the development of 'voluntary initiatives'[211] and utilising 'voluntary tools' for assessing the impact on biodiversity.[212] The resort to voluntary initiatives may to a certain extent reflect the constraint of developing normative work on biofuels under the CBD.[213] The constraint is particularly evident in addressing the issue of life-cycle GHG emissions related to biofuels, as the CBD or Ramsar COP instruments have not mentioned GHG emissions from biofuel and certain states explicitly rejected addressing climate change and biofuels under the CBD.[214]

In the second place, the general commitments to biodiversity conservation and sustainable/wise use are regarded as weakly binding on Parties, and the specific guidance set out in COP decisions or resolutions has ambiguous binding effect. The environmental agreements to address biodiversity are generally regarded as 'weak' or 'softly binding'[215] - the original attempt to impose binding substantive obligations to reduce biodiversity loss has been frequently reduced to ambiguous and weak commitments.[216] The reason for this weak bindingness can be attributed to the lack of effective implementation and compliance mechanisms of the biodiversity conventions,[217] and may in addition lie in the difficulty to undertake

209 The Roundtable on Sustainable Biomaterials (RSB) is an international initiative to bring together multiple stakeholders from business, civil society, government and research who are concerned with the sustainability of biomaterials production, conversion, processing and use. See Roundatable on Sustainable Biomaterials, available at: <http://rsb.org/> (last accessed 1 May 2015).

210 CBD/SBSTTA/18/15 2014, para. 3.

211 Decision IX/2, para. 10.

212 Decision X/37, para. 11(a); Decision XI/27, para. 2(b).

213 It has been commented that the limited agreement reached under CBD may represent a midway solution that arguably allows the CBD to continue with normative work on biofuels, while the work with CBD might simply be useful input for Parties involved in normative work on biofuels in other fora. Morgera 2011, p. 103.

214 It was recorded that during a meeting of SBSTTA, while the EU called for references to full life-cycle assessments and climate-change related impacts of biofuels, biofuels producing countries, represented by Argentina, repeatedly requested that references to action on biofuels going beyond the collection and dissemination of information be deleted, with Argentina noting that climate change and biofuel issues are already addressed in other fora. IISD 2008, p. 8.

215 Generally, the multilateral environmental agreements addressing environmental degradation and destruction have been considered weak and inadequate to achieve their declared goals. For example, the CBD is a multilateral framework agreement consisting primarily of general obligations that could evolve into binding commitments by means of protocols. Adam 2010, p. 124 and 142.

216 Adam 2010, p. 137.

217 Apart from suffering from weak implementation, the CBD stands out for the lack of a compliance mechanism. Adam 2010, p. 141-144.

strong biodiversity agreements that deal with the root causes underlying drivers of biodiversity loss.[218]

The specific guidelines of COPs as adopted in the form of resolutions or decisions have ambiguous and uncertain legal status. Although the institutional arrangement with the COPs as the governing body has become a common pattern for environmental treaties, there is no definitive conclusion about whether the COP can be viewed as an international organisation and whether the outcomes of COP decisions or resolutions are sources of international law and thus have binding effect on Parties.[219] The ambiguity and uncertainty particularly arises in the occasions where the COP activities are not specifically authorised in the constitutive treaty[220] or do not require the states' formal *consent* to come into effect,[221] so they cannot fit easily within the traditional categories of international law sources.[222] To

218 It has been commented that, as the root causes of biodiversity loss may entail changing the status quo for 'special interest' groups, instead of undertaking critically necessary measures to address th eunderlying factors, governments have preferred to address biodiversity loss by adopting weak agreements of dubious effectiveness. Adam 2010, p. 134, 163, and 165.

219 See the discussion about COP law making in: Churchill & Ulfstein 2000; Brunnée 2002; Brunnée 2005; Handl 2005; Wiersema 2009; Ulfstein 2007; Lavranos 2002; Gehring 2007; Drumbl 2010, p. 9-10 and 15-16; Ulfstein 2010, p. 29-34; Sand 2011; Röben 2013.

220 Some decisions are taken with explicit empowerment in the treaty, such as decisions regarding treaty amendment, protocol adoption, annex amendment or adoption. These activities may fit easily within traditional categories of the sources of international law, as they give rise to new obligations for states and are identifiable as hard law. Certain treaties explicitly provide that the COP is authorised to adopt binding decisions with delegated power to COP, but there is a rare example of such a provision. For example, in the Montreal Protocol on Substances that Deplete the Ozon Layer, Art. 2.9 allows the adoption of certain new obligations with binding effect for all parties by a double majority of the developing and developed states. However, this article has never been used. Another example is the internal-level rules which are addressed to other treaty bodies or administrative agencies, so they are regarded as internally binding on the addressed treaty bodies. Ulfstein 2007, p. 881-882.

221 International law is largely consent-based, which requires the state parties' consent before it will be binding on them. However, some activities by COPs are secondary rules requiring only consensus or even some form of majority voting instead of formal consent by Parties, although they may have substantive implications for Parties, such as COP resolutions or decisions that interpret provisions of the treaty. Decisions adopted based on consensus are different from the consent-based activities, because a state does not have to consent formally to be bound by a decision reached by consensus. Wiersema termed such activities as 'consensus-based' COP activities. Wiersema 2009, p. 231-233, 245.

222 Primary formal sources of international law are identified as 'international conventions', 'international custom', and 'the general principles of law.' However, instruments which are not legally binding but may have certain normativity or substantive commitments, frequently generalised as 'soft law', are gaining momentum and their legal status and functions in law making and implementation have been discussed extensively in international law studies. It is generally acknowledged that declarations, guideline, principles, and resolutions adopted by COPs, although non-binding, to some extent contribute to the interpretation and development of the binding law of the treaties and applicable customary law, so they fall within the scope of soft law. However, classifying them as soft law is perhaps generalised and inadequate to capture the complex relationship between the activities and underlying treaty obligations. Some may classify them as 'informal law-making' or '*de facto* law-making.' →

address the ambiguity, Wiersema proposed to determine the legal status by analysing the scale of consent from four axes: *a)* treaty authorisation, *b)* consent of Parties in voting, *c)* intent contained in language, and *d)* effect in implementation.[223] The legal status of COP resolutions and decisions on biofuels is hereby explored based on these axes (see Table 2). It is found that, the COP decisions and resolutions on biofuels under the CBD and Ramsar can hardly be regarded as binding rules of international law, given the general authorisation,[224] consensus-based procedural rule,[225] and the non-binding intent reflected in language.[226] For instance, the CBD COP decisions on biofuels were adopted based on the general authorisation provided in Article 23(4)(i) of the Convention, in contrast with the explicit delegation of law-making power in the cases of, *inter alia,* adoption of or amendment to protocols.[227] Similarly, the Ramsar COP resolution on biofuels was adopted based on the general authorisation of 'adopting other recommendations or resolutions to promote the functioning of the Convention' provided in Article 6(2) of the Convention, in contrast with the explicitly delegated power in adopting, for example, 'changes with regard to the Ramsar List of wetlands of international importance.'[228] In terms of the rule of procedure, both the CBD and Ramsar COPs

Wiersema 2009, p. 232 and 245; Churchill & Ulfstein 2000, p. 631-643, 655, 658; Brunnée 2005, p. 118.

223 Wiersema argues that the legal status of the COP activities could focus on the degree of consent achieved from voting, the level of authorisation for the COP activity that represents consent for delegated rule making, the intent expressed by the parties in decision or resolution texts, or the effect of the COP activity on the parties. Wiersema 2009, p. 249-257.

224 The level of authorisation represents a level of consent at the time of treaty adoption. General authorisation provides open-ended powers rather than the explicit delegation of law-making power, so it was argued that generally authorised activities could be regarded as less directly authorised and thus of a lesser legal status than specifically authorised activities. Wiersema 2009, p. 252.

225 Voting within the COP on a specific activity represents the degree of consent required and achieved for the activity in question. There are generally three types of voting in decision making within international organisations: unanimity, majority, and consensus. Consensus-based voting is different from unanimous voting because consensus means that *there is a general agreement with the absence of objections.* In most instances, resolutions and decisions are passed by consensus; in the absence of consensus, a treaty may require a majority or super-majority vote, usually among Parties present and voting. Thus, unless objection or majority vote is marked, the decisions are adopted by consensus. Schermers & Blokker 2011, p. 535-546; Wiersema 2009, p. 251.

226 According to Wiersema, the level of obligation and intent can be measured by two criteria: the form or category of the COP activity – 'resolution', 'recommendation', 'notification' or 'decision'; and the language used in the decisions which may range from 'shall' to 'encourage', 'recommends', 'urge' or 'invite.' Wiersema argues that the form of 'resolution' and the language of 'shall' are more likely to imply harder legal status than the others. Wiersema 2009, p. 252-253; Schermers & Blokker 2011, p. 767-768.

227 Art. 23(4)(i) authorises CBD COP to 'undertake any additional action that may be required for the achievement of the purposes of this Convention in the light of experience gained in its operation', which serves as a general authorisation to provide open-ended powers. CBD, Art. 23(4).

228 Art. 6(2) provides that the COP shall be competent regarding Convention implementation, changes with regard to the Ramsar List of wetlands of international importance, making →

adopt decisions or resolutions on substantive matters by consensus rather than unanimity of Parties;[229] if a consensus cannot be achieved, decisions or resolutions can be passed by a majority vote, which has rarely taken place.[230] Moreover, the language used in the CBD and Ramsar COP decisions and resolution is featured with words of 'urge', 'invite', 'call upon', or 'encourage' instead of 'shall' which implies a stronger normative force.[231] As for the effect of implementation, although the CBD decisions request Parties to report on the regulation of biofuels in the fifth national reports, the submitted national reports, which serve as the main means of implementation, thus far remain a limited source of information on biofuels due to inadequate reporting.[232] However, when a similar analysis is applied to COP decisions about GHG emission assessment adopted under the UNFCCC and Kyoto Protocol, it is found that these COP decisions intend to have a more harmonising

recommendations about wetlands to contracting parties, requesting other international bodies to prepare reports and statistics, and adopting other recommendations or resolutions to promote the functioning of the Convention. Ramsar Convention, Art. 6(2).

229 According to rule of procedure established by the CBD COP, decisions on substantive matters are taken by consensus of Parties. The Rule of Procedure adopted by Ramsar COP in 2012 provides that the Parties shall make every effort to reach agreement on all substantive matters by consensus. Schermers & Blokker 2011, p. 540; Gardner 2003, p. 2-9; The Ramsar Convention Secretariat 2013, p. 33.

230 It is provided by the rule of procedure that if all efforts to reach consensus have been exhausted and no agreement reached, the decision, except that about certain financial mechanisms, shall be taken by a two-thirds majority vote of the Parties present and voting unless otherwise provided. While there is no record about voting procedure, an exceptional case (Decision VI/23 in 2002) is marked to indicate that the decision could be adopted in spite of a formal objection of Parties and subsequent disagreement about the adoption. This exceptional case and the raised disagreement about the meaning of consensus also had an impact on the other COPs such as the Ramsar COP. Moreover, to solve the issue, the Executive Secretary of CBD requested a legal opinion from the UN Office of Legal Affairs, which set forth a definition implying that there can be no consensus by definition with formal objection. Under the Ramsar Convention, if all efforts to reach consensus have been exhausted and no agreement reached, the decision shall, as a last resort, be taken by a simple majority vote of the Parties present and voting, unless otherwise provided by the Convention. However, it is admitted that in fact there has not yet been a vote on any substantive matter, and all decisions of the COP, in the form of resolution and recommendation, have in the end been made by consensus. Schermers & Blokker 2011, p. 540; Gardner 2003, p. 2-9; The Ramsar Convention Secretariat 2013, p. 33.

231 According to Wiersema's argument, the terms such as 'encourage' and 'strongly urge', although containing normative force to different degrees, would be regarded as reflecting political rather than legal commitments. Wiersema 2009, p. 252-253.

232 At the 12th meeting of the Conference of Parties, the progress in implementation of Decision IX/2 and Decision X/37 was reviewed, particularly regarding the activities within the CBD (by Executive Secretariat and the subsidiary body) and by the CBD Parties. It is estimated that, among the 92 fifth national reports available, almost half (44 reports) make no mention of biofuels or bioenergy; and among the remainder, most provide limited information to evaluate progress in the implementation of Decisions IX/2 and X/37. Nevertheless, it is admitted that more empirical study is needed to conclude how the Parties implement and accord significance to the COP decisions and resolutions. CBD/COP/12/23, p. 9; Wiersema 2009, p. 255.

effect upon Annex I Parties,[233] considering the specific authorisation[234] and the language of 'shall',[235] which reflects that the COP decisions with vague legal status may have differential normative significance.[236]

Table 2. Analysis of the legal status of COP decisions from three axes

	Authorisation	Voting	Language
Ramsar Convention COP Resolutions	General authorisation	Consensus	'urge', 'invite', and 'encourage', etc.
CBD COP Decisions	General authorisation	Consensus	'strongly urge', 'urge', 'call upon', and 'encourage', etc.

233 Ulfstein argues that COP/CMP of the Kyoto Protocol is explicitly empowered to undertake legally binding activities, since the COP/CMP is authorised in the Kyoto Protocol to develop guidelines, modalities, procedures, and particularly 'rules' for the key provisions of the Protocol such as CDM and emissions trading scheme. Ulfstein argues that the term 'rule' is different from a non-binding 'principle' or 'guideline', and such an interpretation is supported by substantive considerations. Kyoto Protocol, Arts. 6, 12, 17; Ulfstein 2010, p. 32; Wiersema 2009 p. 252; Brunnée 2005, p. 110-111.

234 According to Art. 7(2)(d) of the UNFCCC, the COP is competent to develop and periodically refine comparable methodologies for preparing inventories of GHG emissions and for evaluating the effectiveness of national measures. Furthermore, Arts. 5(2) and 13(4)(e) of the Kyoto Protocol provides that the methodologies agreed by the COP serving as CMP of the Kyoto Protocol is used for assessing and evaluating the implementation of this Protocol. UNFCCC, Art. 7(2)(d); Kyoto Protocol, Arts. 5(2) and 13(4)(e).

235 Decision 24/CP.19 states that Annex I Parties 'shall' use the revised guidelines in preparing their inventories which are due by 15 April each year since 2015, and it is further stated in the guidelines that Annex I Parties 'shall' use the methodologies provided in the 2006 IPCC Guidelines. Decision 6/CMP.9 states that Annex I Parties shall apply the 2006 IPCC Guidelines and IPCC 2013 Revised Supplementary Methods for providing information in the second commitment period of Kyoto Protocol under Art. 3(3) and (4) of Kyoto Protocol. Decision 24/CP.19, para. 3; Decision 6/CMP.9, paras. 8 and 11.

236 For a further discussion regarding normativity and informal law-making, see Weil 1983; Boyle 2010; Shelton 2010; Goldmann 2010; Wessel 2011.

	Authorisation	Voting	Language
UNFCCC/Kyoto Protocol COP/CMP Decisions	Specific authorisation	Consensus[237]	'shall'

3.2. *International Socio-economic Rights Regime and Biofuels*

The sustainability of biofuels is not only reflected in concerns about environmental impacts, but also in concerns about the socio-economic impacts on food availability and land grab, which are most relevant with the International Covenant on Economic, Social and Cultural Rights (ICESCR).[238] The ICESCR has been recognised as an essential part of the 'international bill of rights' which refers to a cluster of major international human rights treaties.[239] International human rights law presents a regime different from the afore-mentioned international climate change or biodiversity law, since it consists of norms that grant rights directly to the individuals.[240] Moreover, human rights treaties are deemed to have a substance inherently of high normative worth, which may go beyond states' consent to be

237 It is noted that the UNFCCC COP has not reached an agreement on voting rules for substantive matters, but there are two alternative propositions (Alternative A and Alternative B) provided. Nonetheless, by either alternative voting method, consensus is the voting rule in principle for adopting decisions on matters of substance, and the difference resides in the procedure when consensus could not be reached. Alternative A provides that 'Parties shall make every effort to reach agreement on all matters of substance by consensus. If all efforts to reach consensus have been exhausted and no agreement has been reached, the decision shall, as a last resort, be taken by a two-thirds majority vote of the Parties present and voting'; while Alternative B suggests 'Decisions on matters of substance shall be taken by consensus, except that decisions on financial matters shall be taken by a two-thirds majority vote.' See Rule 42, Adoption of The Rules of Procedure, FCCC/CP/1996/2 22 May 1996, available at: <http://unfccc.int/resource/docs/cop2/02.pdf> (last accessed 6 May 2014).

238 The International Covenant on Economic, Social and Cultural Rights (ICESCR) was adopted by the United Nations General Assembly on 16 December 1966 and entered into force from 3 January 1976. The ICESCR has 164 State Parties (up to April 2015) including China and all the 28 EU Member States (not including the EU). United Nations Human Rights, Status of Ratification Interactive Dashboard, available at: <http://indicators.ohchr.org/> (last accessed 10 April 2015).

239 The 'international bill of human rights' consists of thefollowing treaties: Universal Declaration of Human Rights, the International Covenant on Economic, Social and Cultural Rights, the International Covenant on Civil and Political Rights, and their instruments on rights of certain groups such as the Convention on the Rights of the Child, the International Convention on the Protection of the Rights of All Migrant Workers and Members of Their Families, the Convention on the Elimination of All Forms of Discrimination against Women. Von Glahn & Taulbee 2007, p. 446 and 451.

240 Human rights treaties differ in one respect from other treaties: they do not create a system of rights, duties, and obligations between states but seek to protect the rights of individuals. Von Glahn & Taulbee 2007, p. 446 and 451.

bound after the initial consent of treaty commitments has been given by states.[241] Another characteristic is the interdependence and indivisibility of all human rights,[242] even between the civil and political rights and economic, social and cultural rights.[243] This section attempts to explore the implications from the ICESCR and other international instruments on socio-economic rights for addressing the sustainability of biofuels in relation to the socio-economic concerns about food and land.

3.2.1. Food Availability

One major sustainability concern regarding biofuels, particularly crop-based biofuels, is that they may threaten food price stability and eventually food security[244] if crops are used to produce energy instead of food,[245] although there is uncertainty regarding the causal link between biofuel production and food security.[246] By all means, impact on food availability and security, if there is an impact, will affect access to adequate food, particularly for people in poverty.[247]

Article 11 of the ICESCR provides the right to an adequate standard of living, including the right to adequate food and the right to be free from hunger.[248] The meaning of the right to food has been further interpreted in 'General Comment 12'

241 Human rights lawyers think that human rights treaties have a substance that is inherently of a high normative value, above and beyond the states' consent to be bound by them - human rights obligations have a life of their own as soon as states have manifested their initial commitment to be bound. Connors & Schmidt 2010, p. 128-129.

242 The interdependence and indivisibility are to some extent reflected in the institutional arrangement. At the international level, the UN provides for the forum in which human rights instruments are negotiated, drafted, adopted, and monitored. Treaty bodies are established under the UN Human Rights Council to monitor the implementation of the core treaties through state reporting and in some cases individual communication and the treaty bodies have independent experts rather than states representatives. Connors & Schmidt 2010, p. 394-395 and 418; Carlarne 2008, p. 461.

243 The interdependence and indivisibility of all human rights, whether or not classified in various categories, has become a leading axiom in international human rights discourse. Connors & Schmidt 2010, p. 178-179.

244 FAO defines food security as the 'situation that exists when all people, at all times, have physical, social and economic access to sufficient, safe and nutritious food that meets their dietary needs and food preferences for an active and healthy life' by the FAO. The food insecurity can be attributed to the rising food prices. Eide 2008, p. 7; Gonzalez 2010, p. 472;

245 FAO found that new demand for biofuels was the major driver of dramatic increases in food prices. The report by Doornbosch and Steenblik, prepared for the OECD, argued that the pro-biofuels policies would cause a 'food-versus-fuel' battle majorly due to the impact on food prices. FAO, 'State of Agriculture Commodity Markets', 2009, available at: <ftp://ftp.fao.org/docrep/fao/012/i0854e/i0854e.pdf> (last accessed 1 April 2014), p. 19-21;

246 Biofuels may not be the only reason for the increase in food prices, and the other underlying causes include weather conditions, speculative investment, as well as some structural factors such as rising energy costs, stagnation in crop productivity, policy failures that constrain agricultural development, climate change, etc. Some biofuel proponents argue that biofuels have received a disproportionate amount of the blame. See Koh & Ghazoul 2008, p. 2455.

247 Msangi & Ewing 2008; Gonzalez 2010, p. 471-473; Runge & Senauer 2007.

248 ICESCR, Art. 11.

by the Committee on Economic, Social and Cultural Rights (CESCR) which consist of 18 independent experts who monitor the implementation of the ICESCR.[249] It has become common practice for treaty bodies of the UN human rights treaties to issue 'General Comments' which are adopted by consensus of the body of independent experts, with the purpose of providing authoritative guidance on the general treaty obligations of State Parties or setting out how the experts interpret the scope of substantive provisions.[250] Although the general comments are not legally binding on State Parties, they have the value of offering interpretative guidance or dealing with wider thematic/cross-cutting issues, particularly in view of the fact that human rights treaties are usually drafted in vague or ambiguous wording.[251] According to the General Comment 12, the right to food entails the meaning that every person 'has physical and economic access at all times' to adequate food in quality and quantity.'[252] Such an interpretation may imply that, if biofuel production results in food price increases and food supply shortages, it may thus impair the right to food by reducing food accessibility and quantitative adequacy.

As provided in the ICESCR, State Parties undertake obligations with regard to the right to adequate food. According to the General Comment 12, Parties undertake a three-level obligation with regard to the right to adequate food: the obligation to respect, protect, and fulfil.[253] States hence commit not only to the 'negative' obligation of refraining from interrupting or depriving access to adequate food, but also to the 'positive' obligation to facilitate and provide the access to adequate food. It is important to notice that the 'negative' obligation requires states to refrain from certain behaviours with immediate effect, instead of a gradually

249 The Committee was established under the United Nations Economic and Social Council (ECOSOC) Resolution 1985/17 of 28 May 1985 to carry out the monitoring functions assigned to the United Nations Economic and Social Council (ECOSOC) in Part IV of the Covenant. United Nations Human Rights, Office of the High Commissioner for Human Rights, Committee On Economic, Social and Cultural Rights, available at: <www.ohchr.org/en/hrbodies/cescr/pages/cescrindex.aspx> (last accessed 10 May 2015).

250 The term 'general recommendation' is used to refer to the comments of the Committee on the Elimination of Racial Discrimination and the Committee on the Elimination of Discrimination against Women. Connors & Schmidt 2010, p. 408-409.

251 It has been noted that the general comments or recommendations are frequently invoked by states and complainants in the context of reporting and complaints procedures and sometimes by national courts in their judgments. They are thus called 'a new species of soft law.' Connors & Schmidt 2010, p. 408-409.

252 It is stated that the core content of the right to adequate food implies food availability and food accessibility. Economic and physical accessibility respectively refers to the financial costs associated with the acquisition of food and access to food for physically vulnerable individuals. CESCR General Comment No. 12, paras. 6 and 8.

253 The three-level categorisation was developed by Asbjørn Eide in his study on the right to adequate food and is used widely as a framework for analysing state duties with regard to human rights generally. The obligation to respect implies that states should not take any measures that result in *preventing* such access. The obligation to protect requires a state to ensure that enterprises or individuals do not *deprive* individuals of their access to adequate food. The obligation to fulfil means states must proactively engage in activities intended to *strengthen* people's access to food, as well as to provide such access in case of natural or other disasters. General Comment 12, para. 15; Eide 1987, paras. 112-114.

'progressive' implementation.[254] Moreover, the CESCR distinguishes 'core' obligations regarding the socio-economic rights set out in the Covenant, stating that the core obligations – satisfaction of the *minimum essential level* of the socio-economic rights – have an immediate and non-derogable effect upon a State Party.[255] As for the right to food, the core obligation is the access to the minimum essential food to ensure freedom from hunger.[256] The circumstances in which a State Party violates the right to food include, *inter alia, a)* failure to ensure the minimum essential level required to be free from hunger (either by direct action or omission),[257] *b)* adoption of legislation or policies which are manifestly incompatible with pre-existing legal obligations relating to the right to food, *c)* failure to regulate activities of individuals or groups so as to prevent them from violating the right to food of others, and *d)* failure to take into account right to food when entering into international agreements.[258] Accordingly, in the case of biofuels, State Parties of the ICESCR may violate the right to food when their biofuel legislation/policy causes hunger or is 'manifestly incompatible' with the right to food or fails to prevent individuals or groups undertaking biofuel-related activities from violating the right of others, or entering into international agreement without consideration of the right to food.

A further question is whether a State Party violates the right to food if the consequences take place in a third country. General Comment 12 does not provide a

254 Economic and social rights are frequently regarded as 'second generation' rights or 'relative' standards, since they require 'positive' action by states and can only be implemented gradually due to the potential great cost for states. For example, the ICESCR provides that Parties can achieve the full realisation of the rights 'progressively' or make a detailed plan of action for the 'progressive implementation', and it is also specifically stated in the General Comment 12 that 'the right to adequate food will have to be realized progressively.' However, not all social and economic rights impose a positive obligation and may be realised progressively, and some obligations contained in ICESCR require action of immediate effect, such as the duty to exercise social and economic rights without discrimination. The obligation on Parties to refrain from preventing or depriving access to adequate food also requires states to take action with immediate effect, as 'some measures at these different levels of obligations are of a more immediate nature.' ICESCR, Arts. 2(1) and 14; General Comment 12, paras. 6 and 16; General Comment 3, para. 1; Gordon 1998, p. 711-712.

255 In General Comment 15 on the right to water, the CESCR further set forth a distinction between a violation of a 'core obligation' and of other obligations: the 'core obligation' refers to the satisfaction of the minimum essential level of the right to water, and is 'incumbent upon every State party' and is non-derogable. It is stated in General Comment 3 that 'a minimum core obligation to ensure the satisfaction of, at the very least, minimum essential levels of each of the [ICESCR] rights is incumbent upon every State party', which means that States Parties cannot justify any non-compliance with such core obligations. General Comment 15, paras. 37 and 40; General Comment 3, para. 10.

256 General Comment 12 states that State Parties are obliged to 'ensure for everyone under its jurisdiction access to the minimum essential food which is sufficient, nutritionally adequate and safe, to ensure their freedom from hunger', and it emphasises that a violation of the obligation occurs when states fail to ensure the 'minimum essential level required to be free from hunger.' General Comment 12, paras. 14 and 17.

257 The State has to demonstrate that every effort has been made to use all the resources at its disposal in an effort to satisfy, as a matter of priority, those minimum obligations. General Comment 12, para. 17.

258 General Comment 12, para. 19.

direct answer to the question, but it mentions that State Parties should take steps to respect the right to food in other countries and to provide the necessary aid when required;[259] and State Parties should refrain from food embargoes or similar measures which endanger conditions for food production and access to food in other countries.[260] This implies that Parties also undertake certain commitments regarding the right to food of people in third countries, particularly when the minimum essential level of food is endangered. Another argument is that the Covenant *per se* does not contain jurisdictional constraint, and the term 'jurisdiction' should not be interpreted in a narrow way that would limit it to territorial jurisdiction, but it could be extended to be interpreted to mean 'economic effective control.'[261] There is a similar dispute over whether non-ratifying states of the ICESCR have an obligation to respect, protect, and fulfil the right to food, which largely boils down to the question of whether the right to food becomes customary international law or not.[262] As it has been observed, one spectrum of the argument is that the right to food has achieved customary international law status or even that of a peremptory norm of general international law, because the right to food has been referred to in a number of international and regional human rights instruments, humanitarian law, UN resolutions, and declarations.[263] The other spectrum of the argument is to regard the right to food (and other social and economic rights) as 'soft' and non-binding, since their primary function is to provide guidelines that states may choose to follow or not.[264] The argument in between is that the minimum core component of right to food - the right to be free from hunger - may have already achieved the status of international customary law.[265] It is noticed that the Maastricht Principles on the Extraterritorial Obligations of States in the Area of Economic, Social, and Cultural Rights, the outcome of an expert meeting convened in September 2011, declare that '[A]ll States have obligations to respect, protect and fulfil human rights, including civil, cultural,

259 The wording of 'take steps' implies that the full realisation of the relevant rights may be achieved progressively, steps towards that goal must be taken within a reasonably short time after the Covenant's entry into force for the states concerned. General Comment 12, para. 36; General Comment 3, para. 2.

260 General Comment 12, para. 37.

261 For example, General Comment 12 states that 'Every State is obliged to ensure for everyone under its jurisdiction access to the minimum essential food which is sufficient, nutritionally adequate and safe, to ensure their freedom from hunger.' The 'jurisdiction' could hereby be interpreted as 'within economic control' of that sate. Narula 2006, p. 735-737.

262 Narula 2006, p. 771-772; Steiner, Alston & Goodman 2008, p. 264-270.

263 Narula 2006, p. 775.

264 This point of view is in line with the 'human rights hierarchy' according to which economic, social and cultural rights, as 'relative' standards, are subordinated to civil and political rights as 'absolute' standards. Narula 2006, p. 775.

265 It was argued that the right to food provided in Art. 11 of ICESCR has a twofold meaning: the right to adequate food and the right to be free from hunger. The right to adequate food is a 'relative' standard since it is defined as sustainable access to food in quantitative and qualitative sufficiency, while the right to be free from hunger is 'absolute' and a fundamental right and may make a strong case to achieve international customary law status. Narula 2006, p. 780.

economic, political and social rights, both within their territories and extraterritorially.'[266] In sum, there is no definitive conclusion on whether the right to food can be enforced for people affected by biofuel activities in the non-ratifying countries, but at least ICESCR Parties should take steps to minimise the impact of biofuels on the right to food in third countries and take immediate measures if there is an impact on the minimum essential level of food availability and accessibility.

Although the CESCR, in its General Comment, does not address the link between biofuels and the right to food, other human rights bodies within the UN have paid attention to the impact of biofuels on food availability. For instance, the UN Special Rapporteur on the Right to Food[267] has been engaged in investigating and calling for attention for the negative impacts of biofuels on right to food as well as on land acquisition and environment.[268] The UN Human Rights Council held a special session on the food crisis from a human rights perspective in May 2008,[269]

266 The Principles indicate that states have extraterritorial obligations that exist alongside their territorial obligations corresponding to their effective control. Extraterritorial obligations differ from territorial obligations, however, in that such obligations can be shared with other states. Meanwhile, the extraterritorial obligations are subject to the conditions and in the circumstances set out in the principles. Maastricht Principles, p. 2; De Schutter *et al.* 2012, p. 1097.

267 The Special Rapporteur is an independent expert appointed by the Human Rights Council to examine and report on a country's situation or a specific human rights theme. This is an honorary position and the expert is not a staff member of the United Nations nor paid for his/her work. Since 1979, special mechanisms have been created by the United Nations to examine specific country situations or themes from a human rights perspective. The United Nations Commission on Human Rights, replaced by the Human Rights Council in June 2006, has mandated experts to study particular human rights issues, which constitutes a system of special procedures. The mandate of the Special Rapporteur on the right to food was originally established by the Commission on Human Rights in April 2000 by Resolution 2000/10. Subsequent to the replacement of the Commission by the Human Rights Council in June 2006, the mandate was endorsed and extended by the Human Rights Council by its Resolution 6/2 of 27 September 2007. United Nations Human Rights, Special Rapporteur on the right to food, available at: <www.ohchr.org/EN/Issues/Food/Pages/FoodIndex.aspx> (last accessed 11 April 2015).

268 The former UN Special Rapporteur on the Right to Food, Jean Ziegler, examined the impact of agro-fuels on hunger and food security. In his interim report presented in August 2007, he expressed serious concern about biofuels and warned that 'biofuels would bring hunger in their wake' and described 'the sudden, ill-conceived, rush' to convert food into fuels as 'a recipe for disaster.' His successor, Olivier de Schutter, continued to raise concerns about the impact of biofuels on food prices and the right to food and underlined the impacts of agrofuels on food prices, concentration of land, and the environment, and criticised that the current sustainability approaches such as the EU's sustainability criteria are not sufficient to solve these concerns. The current Special Rapporteur, Hilal Elver, also addressed the negative impact of biofuels on food, land, carbon emissions, and environment, and called for EU legislators to take action to bridge the gap. Report on the right to food 2007, p. 8-15; De Schutter 2008, p. 9; 'MEPs take note: Biofuels are synonymous with high, volatile food prices', 23 February 2015, available at: <www.euractiv.com/sections/energy/meps-take-note-biofuels-are-synonymous-high-volatile-food-prices-312347> (last accessed 11 April 2015).

269 The 7th Special session of the Human Rights Council on 'The negative impact on the realization of the right to food of the worsening of the world food crisis, caused inter alia by the soaring food prices', 22 May 2008, Geneva, available at: <www.ohchr.org/EN/- →

and adopted resolutions that call for states and international organisations to review the impact of biofuel policies and projects on food before they are implemented.[270] The political attention to the relationship between food and biofuels culminated in the High-level Conference on World Food Security with the theme of 'the Challenges of Climate Change and Bioenergy' in June 2008.[271] At this Conference, the participants explored a range of topics about climate change, bioenergy, and food security,[272] and sought to identify *international consensus* on sustainable biofuels through reviewing the existing international instruments and identifying bioenergy policy options.[273] Finally, a declaration was produced at the Conference, which states that 'it is essential to address the challenges and opportunities posed by biofuels, in view of the world's food security, energy and

HRBodies/HRC/SpecialSessions/Session7/Pages/7thSpecialSession.aspx> (last accessed 4 April 2014).

270 The Human Rights Council emphasised in its Resolution that states and relevant multilateral institutions should review any policy or measure which could have a negative impact on the realisation of the right to food before adopting such a policy or measure. The policy or measure that should be reviewed includes biofuels policy and projects. The UN Human Council's Resolution calls upon state members of the UN and invites UN High Commissioner for Human Rights and the Special Rapporteur on the right to food to participate in the High-level Conference on World Food Security with the theme of 'the Challenges of Climate Change and Bioenergy' in 2008. Resolution adopted by the Council at its Seventh Special Session S-7/1, Human Rights Council Seventh Special Session 22 May 2008, 'Report of the Human Rights Council on its Seventh Special Session', A/HRC/S-7/2 17 July 2008, available at: <http://daccess-dds-ny.un.org/doc/UNDOC/GEN/G08/-147/03/PDF/G0814703.pdf?OpenElement> (last accessed 4 April 2014), para. 3; Eide 2008, p. 12.

271 On 3 June 2008, the High-Level Conference opened in Rome and was attended by 42 Heads of State and Government, 100 high-level ministers, and 60 non-governmental and civil society organisations. The participants included the governmental delegates of China, the European Community and the European Parliament. High-Level Conference on World Food Security: The Challenges of Climate Change And Bioenergy Revised List of Participants, Rome, 3-5 June 2008, available at: <www.fao.org/fileadmin/user_upload/foodclimate/HLCdocs/-HLC2008-lop.pdf> (last accessed 11 April 2015).

272 These include challenges and opportunities of climate change adaptation and mitigation for food security, financial mechanisms for climate change adaptation to and mitigation in the food and agriculture sectors, and the perspective from the civil society and private sector. Bioenergy, food security and sustainability - towards an international framework. Rome, 3 -5 June 2008, available at: <ftp://ftp.fao.org/docrep/fao/meeting/013/k2498e.pdf> (last accessed 5 April 2014); Climate Change, Bioenergy and Food Security: Options for Decision Makers Identified by Expert Meetings, available at: <ftp://ftp.fao.org/docrep/fao/-meeting/013/k2396e.pdf> (last accessed 5 April 2014).

273 Five essential areas of action were identified and concrete suggestions about the five areas were given for further consideration. Particularly, to build a framework on 'sustainability principles' is one of the five areas, and it is suggested that a common reference framework of sustainability principles in environmental, social and institutional dimensions could be discussed and agreed at the international level. Bioenergy, food security and sustainability - towards an international framework. Rome, 3 -5 June 2008, available at: <ftp://ftp.fao.org/-docrep/fao/meeting/013/k2498e.pdf> (last accessed 5 April 2014).

sustainable development needs.'[274] In this sense, although the UN human rights framework has not yet taken legal action to address biofuels and the right to food, it has sought to raise political awareness and coordinate international efforts.

3.2.2. Land Grab

Biofuels production has raised serious concern about 'land grab' in developing countries.[275] There is not yet an agreed definition of 'land grab', but it generally refers to the situation in which land acquisition of a certain scale, through purchase or lease, takes place without properly consulting or compensating local people.[276] Land grab would infringe on human rights in the form of forced eviction and displacement or deprival of land tenure[277] and land access.[278] Although it is not

274 The declaration identifies two major aspects of addressing the biofuels challenges and opportunities: firstly, in-depth studies are necessary 'in accordance with the three pillars of sustainable development'; secondly, it is deemed desirable to exchange experiences not only on biofuels technologies but also on norms and regulation. To foster the exchange, the declaration calls upon national governments and other stakeholders to enhance coherent, effective and result-oriented international dialogues on biofuels. Declaration of the High-Level Conference on World Food Security: The challenges of climate change and bioenergy, available at: <www.fao.org/fileadmin/user_upload/foodclimate/HLCdocs/declaration-E.pdf> (last accessed 5 April 2014).

275 Some civil society organisations such as ActionAid, Oxfam, Grain, IIED, and International Land Coalition have expressed grave concerns about the linkage between the demand for biofuels in developed countries, such as the EU, and land grab and people eviction in developing countries. For example, Cotula *et al.* 2009; Anseeuw *et al.* 2012; Broadhurst 2011; Deininger & Byerlee 2011; 'Fuel for thought: Addressing the social impacts of EU biofuels policies', ActionAid, April 2012, available at: <www.actionaid.org/sites/files/actionaid/-fuel_for_thought.pdf> (last accessed 11 April 2015); 'Land and Power: The growing scandal surrounding the new wave of investments in land', 151 Oxfam Briefing Paper, 22 September 2011, available at: <www.oxfam.org/sites/www.oxfam.org/files/bp151-land-power-rights-acquisitions-220911-en.pdf> (last accessed 11 April 2015).

276 Whether or not a transaction can be classified as 'land grab' depends on its context. Land deals and transactions are not necessarily 'land grabs' and have negative impact on human rights. If the lease is properly structured and access to land is obtained in a responsible way, local communities would benefit from the lease. Negative impacts will be result if the community is not consulted properly and people lose resources to support their life. PANGEA, 'Land Grab Refocus Roots and possible demise of land grabbing', December 2011, available at: <www.ascension-publishing.com/BIZ/LandgrabStudy.pdf> (last accessed 11 April 205).

277 According to the FAO, land tenure is defined as 'the relationship, whether legally or customarily defined, among people, as individuals or groups, with respect to land', and rules of tenure define how property rights to land are to be allocated within societies – the rights to use, control, and transfer lands as well as responsibilities and retraints. Put simply, according to the FAO, land tenure systems determine who can use what resources for how long and under what conditions. FAO land tenure studies (2002) 'Land tenure and rural development' available at: <ftp://ftp.fao.org/docrep/fao/005/y4307e/y4307e00.pdf> (last accessed 6 April 2014).

278 The underlying human rights violation prompts UN Special Rapporteurs on the Right to Food to denounce land grabbing driven by biofuels. De Schutter, O., 'Biofuels: animal feed a →

completely clear what the extent is of the contribution of biofuels to the land grab issue,[279] the link between biofuels and land grab cannot be denied.

In international human right treaties, instead of being codified as a general 'right to land', the right regarding land access and land tenure is provided for certain groups such as indigenous peoples and women. The Convention 169 on Indigenous and Tribal Peoples, which was adopted by the International Labour Organization in 1989,[280] serves as the 'only binding international instrument' on the right of indigenous peoples.[281] The Convention on the Elimination of all forms of Discrimination against Women (CEDAW) focuses on equal treatment in rights to land.[282] Besides, the voluntary guidelines on land tenure[283] have been endorsed by the Committee on World Food Security,[284] in which guiding principles of responsible tenure governance are set out for both states and non-state actors including business enterprises. Land can also be seen as a condition for the enjoyment of the right to an adequate standard of living and even some civil-political rights such as the right to respect of home.[285] In the case of biofuels, the focus is on the right of land access and tenure for indigenous peoples, and land access may act as a condition for enjoying the right to an adequate standard of living.

minor factor' (Financial Times, 19 June 2013), available at: <www.srfood.org/images/-stories/pdf/press_cuts/20130620_biofuels_ft.pdf> (last accessed 11 April 2014).

279 There is a report revealing that the link between biofuels and land grad could have been exaggerated based on the 'Land Matrix' data. The 'Land Matrix' is a global land monitoring initiative consisting of a database for large-scale land deals, including transaction of land use rights, control and ownership. Carlo Hamelinck, 'Land grabs for biofuels driven by EU biofuels policies', ECOFYS, BIENL13469, July 2013, available at: <www.ecofys.com/files/files/ecofys-2013-report-on-land-grabbing-for-biofuels.pdf> (last accessed 11 April 2014).

280 The Convention entered into force in 1991, but until now there are only 22 ratifying countries. China and the major EU countries such as Germany and France are still not State Parties. ILO, Ratifications of C169 - Indigenous and Tribal Peoples Convention, 1989 (No.169), available at: <www.ilo.org/dyn/normlex/en/f?p=1000:11300:0::NO:11300:P11300_INSTRUMENT_ID:31 2314b> (last accessed 5 April 2014).

281 Wickeri & Kalhan 2010, p. 18.

282 Wickeri & Kalhan 2010, p. 16-19.

283 Voluntary Guidelines on Tenure of Land, p. 3-4.

284 The Committee on World Food Security (CFS) is an international and intergovernmental platform, consisting of members of all the Member States of the Food and Agricultural Organization (FAO), the International Fund for Agricultural Development (IFAD), the World Food Programme (WFP), and non-Member States of FAO that are Member States of the United Nations. The Committee reports to the UN General Assembly through the Economic and Social Council (ECOSOC) and to FAO Conference. Committee on World Food Security, CFS Structure, available at: <www.fao.org/cfs/cfs-home/about/structure/en/> (last accessed on 12 February 2016).

285 The right to respect of home privacy, family, home and correspondence, and protection of honour and reputation is established in Art. 17 of Internaitonal Covernant on Civil and Political Rights. This right is required to be guaranteed against all arbitrary or unlawful interferences and attacks whether they emanate from state authorities or from natural or legal persons. 'Home' is to be understood to indicate the place where a person resides or carries out his usual occupation. Wickeri & Kalhan 2010, p. 16.

3.2.2.1. *Land Rights for Indigenous Peoples*

The Convention 169 on Indigenous and Tribal Peoples provides that State Parties shall identify lands traditionally occupied by indigenous peoples[286] and take measures to safeguard their rights to land to which they have traditional access for subsistence and conventional activities.[287] Accordingly, indigenous peoples have rights not only to the lands traditionally occupied by them, but also the lands to which they traditionally had access, not necessarily owned or occupied, for subsistence and other traditional activities. To protect such rights, State Parties have an obligation to identify such lands, as well as to legal procedures to resolve potential land claims,[288] protect against forced removal,[289] compensate for lost land,[290] and establish rights to natural resources.[291] Moreover, indigenous peoples shall be consulted when resources on the identified lands are explored, in order to ascertain whether and to what degree their interests are prejudiced and can be compensated.[292] Relocation for exploring resources on the lands is allowed only if it is necessary and free and informed consent of indigenous peoples at stake is obtained.[293]

Hence, according to the Convention, if the lands to which the indigenous peoples have traditional access are retained or exploited by states or third parties for biofuel feedstock production, indigenous peoples shall be *consulted* so as to identify their degree of interest, and relocation cannot take place without obtaining their *free* and *informed consent*. Such procedural safeguards were confirmed in the 2007 Declaration on the Rights of Indigenous Peoples[294] which declares that free, prior, and informed consent shall be obtained if land confiscation, taking,

286 The Convention includes a section on land and requires State Parties to identify lands traditionally occupied by indigenous peoples and guarantee their ownership and protect their rights. Convention 169, Art. 14.

287 The lands do not need to be exclusively occupied by indigenous peoples. Convention 169, Art. 15.

288 Convention 169, Art. 14.

289 Convention 169, Art. 19.

290 The compensation could be either in form of money or land of at least equal quality and quantity. Convention 169, Art. 16.

291 Convention 169, Art. 15.

292 Convention 169, Art. 15(2).

293 Where their consent cannot be obtained, such relocation shall take place only following appropriate procedures established by national laws and regulation, including public inquiries where appropriate, which provide the opportunity for effective representation of the peoples concerned. Convention 169, Art. 16(2).

294 The Declaration was approved after 143 states voted in favour, 11 abstained (e.g. Russia) and 4 (i.e. Australia, Canada, New Zealand and the United States) voted against. The declaration sets out the individual and collective rights of indigenous peoples and their rights to culture, identity, language, employment, health, education, ect. It is stated that 'indigenous peoples have the right to the lands, territories and resources which they have traditionally owned, occupied or otherwise used or acquired', and 'States shall give legal recognition and protection to these lands, territories and resources.' Declaration on the Rights of Indigenous Peoples, Art. 26.

occupation, use, or damage takes place;[295] and indigenous peoples shall have the right to participate in the process of recognition and adjudication that establishes and implements their right to lands.[296] However, Convention 169 may have limited influence since it has only 22 ratifying Parties excluding China and most EU Member States;[297] although the Declaration on the Rights of Indigenous Peoples has achieved wide approval among states, it is after all a declaration rather than a binding convention.

3.2.2.2. Land Rights as Condition for the Enjoyment of the Right to an Adequate Living Standard

Access to land is often necessary for the enjoyment of the socio-economic rights, such as right to food and housing provided for in Article 11 of ICESCR, as land is an important resource for producing food and building accommodations.

Firstly, the protection of land access and tenure is essential for ensuring the right to food. It is set out in General Comment 12 that being able to feed oneself directly from productive land is one aspect of food availability.[298] Furthermore, respect for and the protection of food accessibility requires economic and physical access to land by landless persons and indigenous peoples.[299] For instance, the former Special Rapporteur on the Right to Food made land access a priority in his mandate, and recommended that '[G]overnments investing in farmland abroad should ensure that they do so in accordance with their human rights obligations', and '[I]nternational human rights bodies should consolidate the right to land and take land issues fully into account when ensuring respect for the right to adequate food.'[300]

Secondly, land access and tenure are also significant for the enjoyment of the right to adequate housing and the prevention of forced eviction.[301] The relationship between land rights and the right to housing and the prevention of forced eviction

295 Declaration on the Rights of Indigenous Peoples, Art. 27.

296 Declaration on the Rights of Indigenous Peoples, Art. 28.

297 The Convention has been ratified by several EU Member States such as Denmark, the Netherlands, and Spain. ILO, Ratifications of C169 – Indigenous and Tribal Peoples Convention, 1989 (No. 169), available at: <www.ilo.org/dyn/normlex/en/-f?p=1000:11300:0::NO:11300:P11300_INSTRUMENT_ID:312314b> (last accessed 5 April 2014).

298 General Comment No. 12, para. 12.

299 General Comment No. 12, para. 13.

300 Olivier De Schutter, 'Access to Land and the Right to Food', Report of the Special Rapporteur on the right to food presented at the 65th General Assembly of the United Nations [A/65/281], 21 October 2010, available at: <www.srfood.org/images/stories/pdf/-officialreports/20101021_access-to-land-report_en.pdf> (last accessed 5 April 2014), p. 22.

301 According to Art. 11(1) of ICESCR, the right to adequate housing, together with the right to adequate food and clothing, is one element of achieving an adequate standard of living and ratifying states must take appropriate steps to ensure the realisation of this right. This right is also provided for some special groups, such as Art. 5(e)(iii) of the International Convention on the Elimination of All Forms of Racial Discrimination, Art. 14(2)(h) of the Convention on the Elimination of All Forms of Discrimination Against Women, and Art. 27(3) of the Convention on the Rights of the Child. ICESCR, Art. 11(1).

has been noted, as 'land is often a necessary and sufficient condition on which the right to adequate housing is absolutely contingent for many individuals and even entire communities.'[302] The security of tenure and protection against forced eviction are regarded as a 'touchstone' for identifying whether the right to adequate housing is protected and are also necessary with regard to access to land.[303] Particularly, the prohibition on forced eviction is an 'absolute' obligation with immediate effect,[304] so State Parties have an obligation not only to refrain from such action, but also to ensure their law is enforced 'against its agents or third parties who carry out forced eviction.'[305] In defining circumstances of forced eviction,[306] the General Comment 7 mentions that forced eviction may occur in the name of development, such as for the construction of dams, energy projects, or land clearing for agricultural cultivation, when these instances are connected to conflict over land rights and land acquisition measures.[307]

In this vein, although the ICESCR does not include the right to land, the impact on land tenure and access in the process of biofuel feedstock production may also violate the right to adequate food or housing, if the production impairs economic and physical access to land for food or housing. Forced eviction by states or other entities due to biofuel feedstock production cannot be justified by any economic or climate change purpose.

302 Commission on Human Rights, Report of the Special Rapporteur on Adequate housing as a component of the right to an adequate standard of living, Millon Kothari, UN. Doc. E/CN.4/2005/48, 3 March 2005.

303 Wickeri & Kalhan 2010, p. 24; FAO Land Tenure Studies, 'Land Tenure and Rural Development', 2002, available at: <www.fao.org/3/a-y4307e/> (last accessed 11 April 2015), p. 7; General Comment No. 4, para. 8(a).

304 General Comment 7 states that '[I]t is to be noted that the State's obligation to ensure respect for that right is not qualified by considerations relating to its available resources.' In General Comment 4, it is also mentioned that 'States parties should consequently take *immediate* measures aimed at conferring legal security of tenure upon those persons and households currently lacking such protection…' General Comment No. 7, para. 8; General Comment No. 4, para. 8(a).

305 According to General Comment 7, the state itself must refrain from forced evictions and ensure that the law is enforced against its agents or third parties who carry out forced evictions (as defined in para. 3 above). General Comment 7, para. 8.

306 In General Comment 7, the CESCR defines 'forced eviction' as 'the permanent or temporary removal against their will of individuals, families and/or communities from the homes and/or land which they occupy, without the provision of, and access to, appropriate forms of legal or other protection.' Nonetheless, lawful eviction with appropriate justification and procedures is permissible. A justifiable eviction includes cases of persistent non-payment of rent, or damage to rented property without any reasonable cause, but shall be conducted in a manner compatible with international human rights law and legal safeguards in resources, procedure and remedies. General Comment 7, paras. 3 and 11.

307 The CESCR identifies the cases in which forced evictions take place, such as forced population transfer and internal displacement, and in the many instances of forced eviction that involve military or ethnic violence. General Comment 7, paras. 5-7.

3.2.3. Interim Conclusion

According to the analysis in section 3.2, State Parties of ICESCR may violate the right to food if they adopt biofuel legislation and policies that manifestly affect the quantity (availability) and price (economic accessibility) of food supply, or if they fail to regulate biofuel-related activities of individuals or groups that infringe on the right to food. State Parties undertakes the obligation with immediate effect to prevent the production of biofuels from infringing on the minimum essential level of food, and State Parties may even have an obligation to do so when the consequence takes place in third countries. Unlike the right to food, the ICESCR does not set out the 'right to land', but the protection of land tenure and access has been particularly provided for indigenous peoples and can be contended to have basis on the right to adequate living - precisely the right to food and housing provided in Article 11(1) ICESCR, and forced eviction by states or any other entities for biofuel feedstock production is prohibited. Although the General Comments have no legal force on State Parties, they may have a practical influence on State Parties and provide helpful interpretative guidance to states and other stakeholders.[308]

Moreover, it is observed that the features of interdependence and indivisibility of international human right rules might provide for a chance to address the socio-economic impacts of biofuels in an integrative way. As afore-mentioned, the Human Rights Council and the Special Rapporteur on Human rights have made an effort in promoting international political awareness and consensus with regard to the relationship between biofuels and the right to food as well as land access. Moreover, the international human rights regime is characterised by a relatively integrated institutional arrangement under the UN framework,[309] and in this sense the socio-economic right concerns related to biofuels, such as the right to food, access to land, and right of indigenous peoples, might be integrated and coordinated under Human Rights Council and the UN.[310] Nevertheless, while the overlap between climate change and human rights has been noted,[311] an

308 Connors & Schmidt 2010, p. 179-180.

309 It has been commented that the UN provides for the forum in which human rights instruments are negotiated, drafted, adopted, and monitored. Moreover, treaty bodies are established under the UN Human Rights Council to monitor the implementation of core treaties through state reporting and in some cases individual communication. The Human Rights Council and the Office of the UN High Commissioner for Human Rights can play a role respectively in coordinating and assisting the development and implementation of human rights law. Connors & Schmidt 2010, p. 394-395 and 418.

310 Although Convention 169 on Indigenous Peoples was negotiated and adopted under the International Labour Organisation (ILO), the International Labour Office in a General Survey argued that the ILO's fundamental rights and conventions form 'an integral part of the Unites Nations' overall human rights framework' and are broadly recognised in the other international and regional texts. International Labour Office 2012, para. 19.

311 For example, the Office of the High Commissioner for Human Rights (OHCHR) and the Committee on Economic, Social and Cultural Rights (CESCR) have recognised the connection between demand for biofuels in developed countries and extensive human rights problems in biomass exporting countries. The OHCHR wrote in a report that 'Whereas agro-fuel →

international instrument that integrates environmental protection and climate change with the socio-economic rights regime still lacks.

4. International Trade of Biofuels: The International Trade Regime

Global trade in biofuels (e.g. liquid bioethanol and biodiesel) and biofuel feedstocks (e.g. vegetable oil and agricultural residues) has grown rapidly.[312] Meanwhile, due to the consideration of the sustainability of biofuels, a number of domestic regulatory frameworks have set out various sustainability or environmental requirements for biofuels.[313] The regulation of the sustainability of biofuels in importing countries hence has an influence on exporters and exporting countries, which gives rise to potential trade disputes and involves dispute settlement by international trade law. In contrast with international environmental law, an umbrella organisation like the World Trade Organization (WTO) has been formed to manage international trade issues with established institutional, law-making, and judicial mechanisms.[314]

The WTO is committed to the overriding objective of helping trade flow smoothly, freely, fairly, and predictably. It undertakes the functions of administering trade agreements, acting as a forum for trade negotiations, settling trade disputes, reviewing national trade policies, assisting developing countries in

production could bring positive benefits for climate change and for farmers in developing countries', agrofuels have also contributed to food price increases 'because of the competition between food, feed and fuel for scarce arable land.' The CESCR in 2008 urged states to implement strategies to combat global climate change that 'do not negatively affect the right to adequate food and freedom from hunger.' The Special Rapporteur on Right to Food has also addressed the theme of 'climate change and food security.' Roht-Arriaza 2010a, p. 221-222.

312 Compared to 2000 when practically no liquid biofuels were traded, 2009 has seen a net trade of liquid biofuels amounting to 120 to 130 PJ (1 petajoule = 23884.58 tonne of oil equivalent). It was expected that larger quantities of these products to be traded internationally in the future, with Latin America and sub-Saharan Africa as potential net exporters and North America, Europe and Asia expected to become net importers. In 2008 around 9 per cent of global biofuel production was traded internationally and a net amount of 40 to 51 PJ of fuel ethanol was traded in 2009, but data related to fuel bio-ethanol trade are imprecise because of the various potential end uses of ethanol (i.e., fuel, industrial and beverage use) and the lack of proper codes for biofuels in global trade statistics. World biodiesel production started below 20 PJ in 2000 and reached about 565 PJ in 2009. Chum *et al.* 2011, p. 251-252; Junginger *et al.* 2011; Lamers *et al.* 2011; Heinimö & Junginger 2009.

313 Such regulatory measures include, *inter alia*, the Low Carbon Fuel Standard of California, the Renewable Fuel Standard of the US, the Biofuels Life Cycle Assessment Ordinance of Switzerland, the Selo Combustíve Social - Social Fuel Seal of Brazil, the Biomass Sustainability Ordinance of Germany, the Renewable Transport Fuel Obligation of the UK, and Directive 2009/28/EC on the promotion of the use of energy from renewable sources of the EU. A review of these regulatory measures has been made under the FAO BEFS approach. Food and Agriculture Organization (FAO), Bioenergy and Food Security (BEFS), A Compilation of Bioenergy Sustainability Initiatives, available at: <www.fao.org/-energy/befs/compilation/en/> (last accessed 10 February 2015).

314 Carlarne 2008, p. 461.

trade policy issues through technical assistance and training programmes, and cooperating with other international organisations.[315] While a multilateral trade agreement on biofuels has not been negotiated or reached under the WTO, four trade disputes about biofuels have thus far been brought before the WTO dispute settlement body.[316] These disputes respectively centre on the compatibility of domestic measures for biofuels with WTO law, such as the sustainability criteria,[317] mandated targets,[318] supported mechanisms,[319] or anti-dumping tariffs.[320] As for the

315 WTO, WTO in Brief, available at: <www.wto.org/english/res_e/doload_e/inbr_e.pdf> (last accessed 13 April 2015).

316 The cases are sorted through the index of dispute issues provided on WTO website, according to the subjects of WTO disputes. The issue of 'biodiesels' is among the subjects listed therein. WTO, Dispute Settlement: the Disputes, Index of disputes issues, available at: <www.wto.org/english/tratop_e/dispu_e/dispu_subjects_index_e.htm#selected_subject> (last accessed 13 April 2015).

317 In dispute DS459, Argentina requested consultations with the European Union and its Member States regarding certain measures that allegedly affect the importation and marketing of biodiesel, including the measures to introduce a mechanism to control and reduce greenhouse gas emissions. WTO, Dispute Settlement: Dispute DS459 European Union and Certain Member States – Certain Measures on the Importation and Marketing of Biodiesel and Measures Supporting the Biodiesel Industry, available at: <www.wto.org/english/tratop_e/dispu_e/cases_e/ds459_e.htm> (last accessed 12 April 2015).

318 In dispute DS443, Argentina requested consultations with the European Union and Spain concerning certain measures affecting the importation of biodiesels for accounting purposes with regard to the compliance with the mandatory targets for biofuels. The key measure challenged by Argentina is the Spanish Ministerial Order regulating the allocation of quantities of biodiesel needed to achieve the mandatory target of renewable energy. This measure is the national implementation of the European Union regulatory framework for energy from renewable sources. On 6 December 2012, Argentina requested the establishment of a panel, but at its meeting on 17 December 2012, the DSB deferred the establishment of a panel. WTO, Dispute Settlement: Dispute DS443 European Union and a Member State – Certain Measures Concerning the Importation of Biodiesels, available at: <www.wto.org/english/tratop_e/dispu_e/cases_e/ds443_e.htm> (last accessed 12 April 2015).

319 In Dispute DS459, Argentina also requested a consultation on measures to establish support schemes for the biodiesel sector in the EU and its Member States. WTO, Dispute Settlement: Dispute DS459 European Union and Certain Member States – Certain Measures on the Importation and Marketing of Biodiesel and Measures Supporting the Biodiesel Industry, available at: <www.wto.org/english/tratop_e/dispu_e/cases_e/ds459_e.htm> (last accessed 12 April 2015).

320 In the Dispute DS473, Argentina requested consultations with the European Union regarding (a) provisional and definitive anti-dumping measures imposed on biodiesel originating in, *inter alia*, Argentina, as well as the investigation underlying the measures; and, (b) a provision in Council Regulation (EC) No 1225/2009 of November 2009, which refers to the adjustment or establishment of costs associated with the production and sale of products under investigation in the determination of dumping margins. On 23 June 2014, the Director-General established the Panel. The Panel expects to issue its final report to the parties by the end of 2015. In Dispute DS480, Indonesia requested consultations with the European Union on: (a) provisions of Council Regulation (EC) No 1225/2009 on protection against dumped imports from countries not members of the European Community; and (b) anti-dumping measures imposed in 2013 by the European Union on imports of biodiesel originating in, *inter* →

complaint about sustainability criteria for biofuels, the most relevant WTO rules are *inter alia* Article I, III, and XX of the General Agreement on Tariffs and Trade 1994 (GATT 1994), with the focal point on whether a unilateral sustainability requirement constitutes discrimination against the imported biofuels and becomes 'unnecessary obstacles to international trade.'[321] While the case regarding sustainability criteria for biofuels is still in the stage of consultation,[322] scholars have extensively discussed the consistency of sustainability criteria for biofuels with WTO rules.[323] In this view, the following paragraphs of this section aim to demonstrate what and how the WTO rules relate to the regulation of the sustainability of biofuels, concentrating on Article I, III, and XX of GATT 1994 which are the most relevant provisions for the regulation of the sustainability of biofuels.

4.1. *Articles I and III of GATT 1994*

Article I and III of GATTT 1994 respectively establish the 'most-favoured-nation' and the 'national treatment' principles, and the message underlying both the principles is prohibition of discrimination between 'like' products. Article I[324] prohibits WTO Members from providing border or internal measures[325] that

alia, Indonesia. WTO, Dispute Settlement: Dispute DS473 European Union – Anti-Dumping Measures on Biodiesel from Argentina, available at: <www.wto.org/english/-tratop_e/dispu_e/cases_e/ds473_e.htm> (last accessed 12 April 2015; WTO, Dispute Settlement: Dispute DS480 European Union – Anti-Dumping Measures on Biodiesel from Indonesia, available at: <www.wto.org/english/tratop_e/dispu_e/cases_e/ds480_e.htm> (last accessed 12 April 2015).

321 Admittedly, there are more provisions and questions to be discussed, such as Art. XI of GATT 1994 and the Agreement on Technical Barriers to Trade. For instance, in dispute DS459, Argentina contended that the EU's biofuel sustainability requirement with regard to the threshold of 35 per cent GHG emissions saving is incompatible with the Arts. I:1, III:1, III:2, III:4 and III:5 of the GATT 1994, Arts. 2.1, 2.2, 5.1, 5.2 of the TBT Agreement (Agreement on Technical Barriers to Trade), and Art. XVI:4 of the Marrakesh Agreement Establishing the World Trade Organization WTO, Dispute Settlement: Dispute DS459 European Union and Certain Member States – Certain Measures on the Importation and Marketing of Biodiesel and Measures Supporting the Biodiesel Industry, available at: <www.wto.org/english/-tratop_e/dispu_e/cases_e/ds459_e.htm> (last accessed 12 April 2015).

322 The case is in the consultation phase, which means that a complainant requested consultations with the respondents and a panel has not yet been established, nor has withdrawal or an agreed solution been notified. WTO, Dispute Settlement: Dispute DS459 European Union and Certain Member States – Certain Measures on the Importation and Marketing of Biodiesel and Measures Supporting the Biodiesel Industry, available at: <www.wto.org/english/tratop_e/dispu_e/cases_e/ds459_e.htm> (last accessed 2 March 2016).

323 Deshpande 2006; Dufey 2006; Van den Bossche, Schrijver & Faber 2007; de Vera 2007; Rancourt 2009; Mitchell & Tran 2010; Lydgate 2012a; Lydgate 2012b; Lydgate 2013; Kent 2013; Douma 2010; Puddle 2007; Payne 2008; Switzer 2007; Switzer & McMahon 2011, p. 713-736; Motaal 2008; Swinbank & Daugbjerg 2013, p. 813-834; Machado-Filho 2008; Taylor 2013.

324 GATT 1994, Art. I(1): General Most-Favoured-Nation Treatment.

325 Art. 1(1) applies to measures including *a)* customs duties and charges of any kind imposed on or in connection with importation or exportation, or imposed on the international transfer of →

discriminate among like products originating in or destined for different countries, by granting advantage, favour, privilege, or immunity to any one party or product of any one country. Article III[326] prohibits WTO Members from providing internal measures[327] to afford less favourable treatment to imported products, so that imported products can enjoy equal competitive conditions with domestic 'like' products.[328] Thus, if a sustainability requirement is invoked and applied to biofuels of foreign origin, the focal questions regarding compatibility with WTO law might be: *a)* whether the sustainability measure at issue falls under the scope of Article I and III; *b)* whether the 'sustainable' and 'unsustainable' biofuels, as distinguished by the measure at issue, is a 'like' product; *c)* whether the sustainability measure leads to 'advantage' or 'less favourable treatment.'[329]

The first question amounts to whether the sustainability measure affects internal sale, offering for sale, purchase, transportation, distribution, or use of biofuels in the domestic market within the scope of Article III(4).[330] The word of 'affecting' implies a meaning broader than 'regulating' or 'governing',[331] so the sustainability requirement that may have the effect of constraining or incentivising sales or use of biofuels can fall within its scope.[332]

Identifying the 'likeness' of sustainable and unsustainable biofuels becomes a key issue in determining a biofuels sustainability measure is discriminatory,[333] since whether a biofuel is sustainable or not largely depends on its production process

payments for imports or exports, *b)* the method of levying such duties and charges, and with respect to all rules and formalities in connection with importation and exportation, *c)* to all matters referred to in paras. 2 and 4 of Art. III, such as laws, regulations and requirements affecting their internal sale, offering for sale, purchase, transportation, distribution or use.

326 GATT 1994, Art. III(1): National Treatment on Internal Taxation and Regulation.

327 Art. III applies to internal measures, instead of border measures relating to exportation or importation, that apply after imported products enter into domestic market, such as customs duties and charges in connection with importation or exportation, and internal measures may include internal taxes and other charges as well as laws, regulations, and requirements affecting internal sale, purchase, distribution or use.

328 Rancourt 2009, p. 92.

329 Mitchell & Tran 2010, p. 35 and 37.

330 As WTO case law suggests, if the measure falls outside the scope of Art. III(4), it will also fall outside the scope of Art. I(1). Mitchell & Tran 2010, p. 35 and 37.

331 The word 'Affect' can be interpreted to cover not only laws and regulations which directly govern the conditions of sale or purchase, but also laws or regulations which 'might adversely modify the conditions of competition between domestic and imported products.' Mitchell & Tran 2010, p. 38; Douma 2010, p. 398.

332 However, consideration would be also given to whether the biofuel sustainability requirement falls within the scope of Art. III of GATT at all, since WTO case law has held that measures based on the production and processing methods of products fall outside the scope of Art. III. Mitchell & Tran 2010, p. 38-39; Douma 2010, p. 396-397.

333 'Likeness' is essential for applying both Arts. I and III and it appears in many different provisions in WTO law, but the approach to 'likeness' is broadly similar under the GATT provisions. Mitchell & Tran 2010, p. 36.

and method.[334] This matter proves to be the most contentious in the literature on trade in biofuels, as the WTO has not adopted a definitive conclusion on this matter. Therefore, an interpretation can only be proposed based on the WTO case law.[335] On the one side, the GATT panel reports once held that products are not 'unlike' under Article III(4) of GATT 1994 by virtue of different production methods, as the sole difference of production methods has no impact on the 'physical characteristics' of the final product.[336] On the other side, 'physical characteristics' of products should be just one aspect of the criteria to determine the likeness. Other aspects could include, according to the Appellate Body, the extent to which consumers treat the products as alternative means of satisfying the same demand.[337] In this vein, supporters for sustainability measures related to biofuels may gain insights from the EC-Asbestos case in which the products having different environmental and health impacts were identified as 'unlike' products, and the Appellate Body emphasised that consumer perceptions were central to 'likeness' under the GATT 1994.[338] For example, Douma concludes that sustainable and non-sustainable biofuels, as distinguished by EU sustainability criteria, should be regarded as unlike products.[339] Rancourt argues that whether there is 'likeness' should depend on whether in a specific market the sustainability of the biofuels production process has a significant impact on consumer preferences and tastes and thus on the nature and the extent of the competitive relationship between differently produced biofuels.[340] However, Mitchell and Tran argue that the EC-Asbestos case does not necessarily lead to a justification of sustainability criteria, because carcinogenicity in that case was itself a physical characteristic of the product, whereas land-related

334 The production method can be generally termed as the 'process and production method' (PPM) under WTO law, which can be further classified into product-related PPM and non-product-related PPM. Rancourt 2009, p. 93.

335 The WTO has no firm answer to the question of what constitutes a 'like' product, and each case is judged on its merits. Swinbank 2009, p. 498-499.

336 The panel reports were produced in cases of *a)* United States – Restrictions on Imports of Tuna, Report of the GATT Panel, GATT Doc. DS21/R, 3 September 1991, para. 5.14, and *b)* United States – Restrictions on Imports of Tuna, Report of the GATT Panel, GATT Doc. DS/29/R, 16 June 1994, para. 5.8. These two panel reports were not adopted and have no legal status, but it has been argued they can provide 'useful guidance' and the principle stated in these panel reports has not been explicitly overturned. Meanwhile, these cases have been subject to considerable academic criticism. Mitchell & Tran 2010, p. 36.

337 The other criteria include: the extent to which they share end uses and the tariff classification of the products. The criteria are not exhaustive, but the Appellate Body has yet to extend the factors further. The criteria have been drawn from the following cases: Border Tax Adjustments, Report of the Working Party, GATT Doc. L/3464, 2 December 1970; Japan – Taxes on Alcoholic Beverages, Report of the Appellate Body, WTO Doc. WT/DS8/AB/R, WT/DS1 0/AB/R, WT/DS1 1/AB/R, 1 November 1996, p. 21; European Communities – Measures Affecting Asbestos and Asbestos-Containing Products, Report of the Appellate Body, WTO Doc. WT/DS1 35/AB/R, 5 April 2001. Mitchell & Tran 2010, p. 36.

338 European Communities – Measures Affecting Asbestos and Asbestos-Containing Products, Report of the Appellate Body, WTO Doc. WT/DS1 35/AB/R, 5 April 2001; Mitchell & Tran 2010, p. 36; Rancourt 2009, p. 93-95.

339 Douma 2010, p. 405.

340 Rancourt 2009, p. 95.

sustainability criteria do not affect the final physical characteristics of biofuel - biofuel products that are only different in terms of the land use are probably 'like' products.[341] Swinbank mentions that it would be extremely difficult to claim that a biofuel that shows a GHG emission saving of 34 per cent is not a like product with a biofuel of 35 per cent GHG emission saving.[342]

The next question is whether identifying certain biofuels as 'sustainable' amounts to providing an advantage or favourable treatment and, if so, whether the advantage or favourable treatment is immediately and unconditionally given to the like biofuel products, and whether the advantage or favourable treatment results in less favourable treatment of the like products of foreign origin. The term 'advantage' has been interpreted as meaning 'more favourable competitive opportunities' or to 'affect the commercial relationship between products of different origins', including an exemption from fees or charges or freedom from certain requirements;[343] the term of 'less favourable treatment' is concentrated on modification of the 'condition of competition in the relevant market.'[344] While it is generally recognised that less favourable treatment occurs only when it is 'explained by the foreign origin' of the products at stake,[345] a dispute exists as to whether a biofuel sustainability measure *de facto* treats biofuels of foreign origin unfavourably. According to Mitchell and Tran, the EU sustainability criteria will *de facto* treat certain biofuels differently and unfavourably particularly by invoking the land-related sustainability criteria, because tropical countries are less likely to meet the land-related sustainability criteria if compared to EU Member States.[346] Van den Bossche *et al.* consider that certain sustainability criteria in the production of biofuels will impose a heavier burden on imported biomass from foreign countries, particularly the developing countries.[347] In contrast, Douma concludes that in the EU's case the less favourable treatment of unsustainable biofuels from third countries is not motivated by its foreign origin.[348]

4.2. *Article XX of GATT 1994*

Even if a biofuel sustainability requirement falls afoul Articles I and III of GATT 1994, it does not necessarily violate WTO law when it can be justified by Article XX of GATT 1994 which sets out the general exception for measures otherwise

341 Mitchell & Tran 2010, p. 36-37.
342 Swinbank 2009, p. 499.
343 Mitchell & Tran 2010, p. 35.
344 The term is concerned with treatment that 'modifies the conditions of competition in the relevant market to the detriment of imported products. Meanwhile, this principle does not prohibit differential treatment so long as it is not 'less favourable treatment' of imported products compared with the like domestic products. Mitchell & Tran 2010, p. 38.
345 Dominican Republic - Measures Affecting the Importation and Internal Sale of Cigarettes, Report of the Appellate Body, WTO Doc. WT/DS302/AB/R, 25 April 2005, para. 96. Mitchell & Tran 2010, p. 38.
346 Mitchell & Tran 2010, p. 37-38.
347 Van den Bossche, Schrijver & Faber 2007, p. 71-72.
348 Douma 2010, p. 400.

inconsistent with GATT.[349] A justification based on Article XX has to go through a 'two-tier test': firstly, whether the measures at stake are found to qualify under one of the circumstances listed in paragraphs (a) to (j) of Article XX; secondly, even if qualified, whether the measures 'constitute a means of arbitrary or unjustifiable discrimination between countries where the same conditions prevail' or 'a disguised restriction on international trade.'[350] To satisfy the two-tier test is not an easy task, as there have been very few cases in which the conditions in both the paragraphs and the chapeau are met.[351] To apply the two-tier test in the case of a biofuel sustainability requirement, the following focal questions can be identified: *a)* whether the sustainability requirement falls under the paragraphs of Article XX, and *b)* whether the sustainability requirement constitutes an arbitrary or unjustifiable discrimination or disguised restriction on international trade.

Before conducting the 'two-tier test', however, the preliminary question is whether Article XX contains a jurisdictional limitation and cannot be invoked to justify a domestic measure applied outside of the regulating country.[352] This question is important in the case of biofuel sustainability requirement, because concerns about the unsustainable production of biofuels in third countries may serve as an important reason for applying the sustainability requirement.[353] Article XX does not explicitly set forth a jurisdictional limitation, but it can be inferred from case law that Article XX could be invoked if there is a 'sufficient nexus'[354] between the value or interest to be protected and the regulating country, even when the protected subject is not located within the territorial jurisdiction of the regulating country. As for biofuels, a complexity lies in the fact that a biofuel sustainability requirement needs to consider various values or interests, such as GHG emission reductions, biodiversity, as well as socio-economic rights, some of which may hardly be proven to have a 'sufficient nexus' with the regulating country. There is literature arguing that the concern about the regulation of GHG emissions should have a sufficient nexus because of the global effect of climate change, while it is necessary to examine the implications of regulation for the other requirements on a case-by-case basis – for instance, the objective of protecting local biodiversity may not always have a global implication depending on the species at stake; the concern

349 Mitchell & Tran 2010, p. 40.

350 GATT 1994, Art. XX: General Exceptions; Rancourt 2009, p. 97.

351 According to Mitchell and Tran, only one environmental measure has been upheld under GATT Art. XX in the case of United States – Import Prohibition of Certain Shrimp and Shrimp Products – Recourse to Art. 21.5 of the DSU by Malaysia, Report of the Appellate Body, WTO Doc. WT/DS5S8/AB/RW, 22 October 2001. Mitchell & Tran 2010, p. 40.

352 Van den Bossche, Schrijver & Faber 2007, p. XXXII; Rancourt 2009, p. 97.

353 As Ottinger and Miller observe, unsustainable impacts are most likely to occur in developing countries. Ottinger & Miller 2010, p. 23.

354 The Appellate Body once noted that 'We do not pass upon the question of whether there is an implied jurisdictional limitation in Article XX(g), and if so, the nature or extent of that limitation. We note only that in the specific circumstances of the case before use, there is a sufficient nexus between the migratory and endangered marine populations involved and the United States for the purpose of Article XX(g).' United States – Import Prohibition of Certain Shrimp and Shrimp Products (1998), WTO Doc. WT/DS58/AB/R, para. 134; Rancourt 2009, p. 97.

for food availability may require a comprehensive assessment of the territorial scope of a possible food crisis.[355]

As the first step of the 'two-tier test', a biofuel sustainability requirement has to fall within a list of specific and limited circumstances provided in Article XX.[356] Among the circumstances listed, a biofuel sustainability requirement is most likely to satisfy the following two paragraphs: Article XX(b) regarding measures necessary to protect human, animal or plant life or health, and Article XX(g) about measures relating to the conservation of exhaustible resources.[357] As for Article XX(b), the key element is to prove that the biofuel sustainability measure is necessary to achieve its intended goal. The necessity test under Article XX(b) has two aspects: a balancing of the contribution to the sustainability goal against trade restrictiveness, in the light of the importance of the goal itself; and a comparison with any less trade restrictive measure that would attain the same goal.[358] Under Article XX(g), the term 'relating to' implies a less stringent condition than the word of 'necessary', so the measure can be 'primarily aimed at' national resource conservation and have a 'substantial relationship' to the goal.[359] Moreover, WTO case law has suggested that the exhaustible natural resources do not have to be unrenewable mineral resources, so even biological species can be considered to be 'exhaustible' resources under Article XX(g).[360]

In the second step, the conditions set out in the chapeau of Article XX aim to prevent abuse or misuse of the general exception.[361] The step of examining the compatibility with the chapeau may prove to be more complex and largely dependent upon the specific case, because the Appellate Body has given little guidance on the meaning of a 'disguised restriction on international trade',[362] and the chapeau requires more information about the actual application of the measure

355 Rancourt 2009, p. 97-98; Van den Bossche, Schrijver & Faber 2007, p. 96-97; Douma 2010, p. 408-409.

356 The list is provided in paras. (a) to (j) of Art. XX, which is considered to be an exhaustive list of limitations. Rancourt 2009, p. 96.

357 There is a discussion about whether a sustainability requirement for socio-economic concerns could invoke Art. XX(a) which provides an exception for measures necessary to protect public morals, and it is concluded that such argument may hardly stand if the necessity of such requirement is examined. Rancourt 2009, p. 96.

358 For this purpose, the regulating country needs to provide evidence for the qualitative reasoning and material/actual contribution or quantitative projections. Mitchell & Tran 2010, p. 40-41.

359 Mitchell & Tran 2010, p. 41.

360 Rancourt 2009, p. 102.

361 The Appellate Body reminded that the chapeau serves to ensure that members' rights to avail themselves of exceptions are exercised in good faith in order to protect legitimate interests, not as a means to circumvent one member's obligations towards other WTO members. WTO, WTO rules and environmental policies: GATT exceptions, available at: <www.wto.org/-english/tratop_e/envir_e/envt_rules_exceptions_e.htm> (last accessed 14 April 2015).

362 Nevertheless, there are panel reports indicating that the relevant question is whether the design, architecture and revealing structure' of the measure reveals an intention to conceal the pursuit of trade restrictive objectives. Mitchell & Tran 2010, p. 41.

at stake.[363] Actually, the second tier test of the chapeau has proved to stand in the way of many national measures, mostly because of *inter alia* the rigid and unbending manner in adopting and implementing the trade measure at stake or the lack of serious negotiations with all relevant WTO members.[364] Notwithstanding, the WTO jurisprudence has highlighted some of the circumstances which may assist in demonstrating the compatibility of a measure with the chapeau, including coordination and cooperation at the international level, flexibility in taking different situations into account, and the connection between the rationale for the discrimination and the stated objective.[365] Thus, the domestic regulation of the sustainability of biofuels may be justified by the chapeau of Article XX if it is undertaken in a cooperative and flexible manner.

4.3. *Interim Conclusion*

Section 4 presented the focal questions regarding the regulation of the sustainability of biofuels in the context of WTO law. According to the analysis on WTO rules that are most related to the sustainability of biofuels, it could be observed that, in the absence of a multilateral agreement or case law specifically on biofuels, the current WTO law may constitute a great challenge for taking unilateral measures on the sustainability of biofuels, especially for those sustainability requirements on the production and process methods for biofuels in third countries. Such a challenge may not be unique to the issue of biofuels, but it reflects the long-discussed topic of trade and environmental or social standards. Admittedly, the pursuit of free trade and the objective of promoting sustainability are not necessarily mutually exclusive. The founding agreement of the WTO refers to the objective of sustainable development and the need for environmental protection,[366] and the Principle 12 of Rio Declaration denounces the use of discriminatory or disguised trade policy for environmental purposes or the use of unilateral trade action to deal with

363 As case law suggests, the conditions set out in the chapeau concern how the measure is applied, and they do not concern the policy but are interpreted in the context of the policy. Mitchell & Tran 2010, p. 42; Echols 2009, p. 27.

364 For instance, in two cases (US-Gasoline and US-Shrimp), the national rules were accepted under Art. XX(b) and (g), but they were deemed to result in arbitrary or unjustifiable discrimination. WTO, United States-Standards for Reformulated and Conventional Gasoline, Report of the Panel of 29 January 1996, WT/DS2/R; WTO, United States – Import Prohibition of Certain Shrimp and Shrimp Products, Report of the Panel of 15 May 1998, WT/DS58/R; Douma 2010, p. 418.

365 The rationale for the discrimination needs to have some connection to the stated objective of the measure at issue. WTO, WTO rules and environmental policies: GATT exceptions, available at: <www.wto.org/english/tratop_e/envir_e/envt_rules_exceptions_e.htm> (last accessed 14 April 2015).

366 The preamble to the Marrakesh Agreement Establishing the World Trade Organization (the 'WTO Agreement') includes direct references to the objective of sustainable development and to the need to protect and preserve the environment. See more on WTO and sustainable development in Sampson 2005; Rancourt 2009; Lydgate 2012a.

extraterritorial environmental problems.[367] However, there is difficulty in interpreting international trade law in the light of the environmental law treaty obligations partly due to the alleged primacy of the free trade objective.[368] Furthermore, it is hard to find a balance between the two objectives under the WTO regime in specific circumstances, such as in the case of the sustainability of biofuels, which relates to the sensitive question about production and process methods and jurisdictional limitation. One resolution would be the development of international cooperation and internationally recognised criteria for defining 'sustainable biofuels',[369] since the WTO law supports multilateral cooperation and coordination.[370] Efforts are needed within the WTO to address and clarify the thorny questions essential for sustainability regulation, such as production and process methods,[371] jurisdictional limitation, and a general exception for environmental or socio-economic sustainability purposes,[372] or to reinforce institutional coordination with other international regimes.[373] There might even be an expectation that a dispute about the sustainability of biofuels under WTO law

367 Principle 12 reads as follows: 'States should cooperate to promote a supportive and open international economic system that would lead to economic growth and sustainable development in all countries, to better address the problems of environmental degradation. Trade policy measures for environmental purposes should not constitute a means of arbitrary or unjustifiable discrimination or a disguised restriction on international trade. Unilateral actions to deal with environmental challenges outside the jurisdiction of the importing country should be avoided. Environmental measures addressing transboundary or global environmental problems should, as far as possible, be based on an international consensus.'

368 Rancourt 2009, p. 79-81.

369 Rancourt 2009, p. 82-85.

370 Except for the Art. XX of GATT 1994 which requires demonstration of international cooperation and coordination, the Agreement on Technical Barriers to Trade (TBT Agreement) provides that technical regulations in accordance with relevant international stardandards shall be rebuttably presumed not to create an unnecessary obstacle to international trade. Agreement on Technical Barriers to Trade, Art. 2.5.

371 For an analysis of the more recent WTO case law on production and process method, see, for example, Crowley & Howse 2014, p. 325-327.

372 Zleptnig has discussed how non-economic objectives (such as the protection of public morals, life and health, environment, or human rights) can serve as a justification for trade-restrictive measures normally prohibited under WTO law and argues that only if the WTO can demonstrate that it is not just concerned about free trade, but respects non-economic objectives as well, is it likely to remain a sustainable and legitimate form of governance. Other authors have specifically discussed the 'mutual supportiveness' or 'balancing' relationship between WTO law and climate law or sustainable development. Zleptnig 2010; Condon 2009; Messerlin 2012; Lydgate 2012a.

373 One topic focused on in the Doha Round negotiation is the trade and environment, and negotiations in this area targeted three key areas: the liberalisation of environmental goods and services; negotiations on the relationship between multilateral environmental agreements (MEAs) and the WTO; and fishery subsidies. WTO, Harnessing trade for sustainable development and a green economy, available at: <www.wto.org/english/res_e/-publications_e/brochure_rio_20_e.pdf> (last accessed 14 April 2015).

would help to clarify the precise definition and legal weight of the term 'sustainable development' in the WTO context.[374]

5. Conclusion

In the absence of a binding regulatory framework for developing and applying sustainability requirements in relation to biofuels, this chapter mapped the international terrain for regulating the sustainability of biofuels by looking into the international instruments relating to the demand, supply, and trade of biofuels in different international legal regimes. This chapter has also observed the limits in addressing the environmental and socio-economic sustainability concerns of biofuels in the examined treaty regimes, as summarised as follows in section 5.2. A final reflection in view of the research question of this thesis leads to the concluding remarks that the international instruments which note and attempt to address the sustainability of biofuels are emerging but are largely fragmented and have weak legal force, thus constituting challenges for regulating the sustainability of biofuels in an inclusive manner at the international level.

5.1. *Summary of the Main Findings*

Section 2 examined the relationship between the international climate regime and the demand for biofuels, finding that the commitment to addressing or reducing GHG emissions under the UNFCCC or the Kyoto Protocol may provide an incentive for the use of biofuels, especially for Annex I Parties undertaking the GHG emission reduction obligation, if biofuels are regarded as a means to reduce GHG emissions in the transport sector. However, the international climate regime does not specify how the GHG emissions of biofuels should be calculated, and the existing GHG inventory accounting methodology provided in the IPCC Guidelines can hardly ensure a complete calculation of the GHG emissions of biofuels due to disparate communication and reporting commitments of the states. Furthermore, it is possible that the pursuit of climate change mitigation by using biofuels may ignore the other sustainability impacts such as biodiversity and human rights.

Section 3 analysed the international instruments in the field of biodiversity and socio-economic rights that relate to the supply of biofuels. It is found that the general commitment to the conservation and sustainable or wise use of the components of biodiversity does not exclude human activities and natural resource exploitation for the production of biofuels, while the CBD and Ramsar COPs have adopted decisions or resolutions that contain specific guidance for addressing the relationship between biofuel and biodiversity, particularly by means of conducting environmental impact assessments, establishing appropriate national policies, and developing a biodiversity inventory and 'go' or 'no-go' areas. The possible impact on food availability and land grab may violate State Parties obligations related to socio-economic rights (e.g. right to adequate food) provided under the international

[374] Lydgate 2012a, p. 637-638; Lydgate 2012b; Motaal 2008, p. 61.

human rights regime, in the cases where a state adopts laws or policies infringing the rights or failing to regulate individuals and groups who undertake activities infringing such rights. However, the commitments and guidance under the international biodiversity and socio-economic rights regimes remain deficient in addressing the concerns related to the sustainability of biofuels in an inclusive and integrated manner: there are identified gaps in defining the key terms for sustainable production of biofuels, and both the biodiversity and socio-economic rights regimes have limits in addressing the issues beyond their functional mandates.

Section 4 has probed into the relationship between the regulation of the sustainability of biofuels and WTO law. In the absence of an international agreement on biofuels, the question is to what extent WTO law may allow or constrain unilateral biofuel sustainability regulation that affects the international trade of biofuels. It is found that the current provisions and case law under WTO may preclude adopting unilateral sustainability requirements, particularly socio-economic requirements, on imported biofuels, due to many debatable questions about production and process methods and jurisdictional limitation.

5.2. *Reflection*

The previous sections have addressed the international regimes to map the international regulatory terrain for the sustainability of biofuels, observing that international instruments addressing the sustainability of biofuels are emerging. The sustainability of biofuels has been increasingly noted and addressed under the existing international regimes, particularly in instruments discussed by treaty bodies such as the COPs. These instruments, mainly in the form of decisions, resolutions, or comments, could form the basis for raising awareness, efforts, and synergies leading to the regulation of the sustainability of biofuels at the international level. For instance, the decision or resolution on biofuels adopted by the CBD and Ramsar COPs can provide specific guidance for states and stakeholders to be aware and to prevent negative biodiversity impacts of biofuels, and the information compiled by CBD's Executive Secretariat could contribute to identifying the gaps and progress in the area of the regulation of the sustainability of biofuels.[375] However, this section will reflect on the observations in line with the research question: whether the environmental, social, and economic sustainability concerns are inclusively addressed?

An integrated and binding international regulatory framework to develop and apply rules on the sustainability of biofuels still lacks, and the existing instruments relating to biofuels are fragmented to the extent that different sustainability concerns are addressed respectively under various international regimes. In fact,

375 The Executive Secretary of CBD is requested by COP to take into account and collaborate with other relevant organisations and multiple stakeholder initiatives in the process of compiling, analysing, and summarising information. Decision X/37, para. 13.

fragmentation in international law has been discussed extensively.[376] According to the International Law Commission, the fragmentation is inherently caused by the emergence of 'self-contained regimes' and functionally limited treaty-systems.[377] Fragmentation is particularly evident in the field of international environmental law, mainly due to treaty proliferation (congestion), functional compartmentalisation, and institutional disjunction.[378] For example, the international climate regime focuses on the issue of GHG emission reduction, and the biodiversity and socio-economic right treaties have hardly addressed GHG emission issues. Fragmentation reflects the expansion of international legal activities into new fields and the demand for legal pluralism, but meanwhile it may create conflict between regimes that undermine their effective implementation,[379] such as in the case where a single-minded focus on GHG emission would ignore the other environmental and socio-economic sustainability impacts of biofuels.[380] Nevertheless, before an integrated international regulatory framework is formed, one way to overcome fragmentation is to improve streamlining and cooperation between the different regimes[381] and to reduce 'conflictive' fragmentation.[382] In the case of biofuels, it is significant to strengthen streamlining and cooperation between the different international regimes of climate change, biodiversity, human rights, and trade. Cooperation and coordination between treaties within the same regime can be observed, such as those between the CBD and Ramsar Convention or between the human rights treaties. However, coordination between different regimes is harder: for example, although the UNFCCC and CBD are supposed to

376 See more discussions about fragmentation in international law in Biermann, Pattberg & Vam Asselt 2009; Biermann, Davies & Van der Grijp 2009; Pauwelyn 2003; Carlarne 2008, p. 456-457; Prost & Clark 2006; Rao 2004; Teubner & Fischer-Lescano 2004, p. 1000-1002; Koskenniemi & Leino 2002; Benvenisti & Downs 2007; Van Asselt 2011a; Van Asselt 2008; Van Asselt & Zelli 2014.

377 The International Law Commission has established a study group working on the subject and produced an influential report on the fragmentation of international law. International Law Commission 2006, para. 10.

378 It has been commented that, unlike international trade law, international environmental law lacks an umbrella organisation, coherent institutional, legislative, or judicial mechanism as the WTO to manage environmental issues. It is pointed out that MEA proliferation and congestion, the weak role of UNEP, and inadequacy in existing coordination make fragmentation in international environmental law is more evident than in other international law regimes. Carlarne 2008, p. 461.

379 The impact is what once appeared to be governed by 'general international law' has become the field of operation for specialist systems such as 'trade law', 'human rights law', and 'environmental law.' International Law Commission 2006, paras. 6 and 9.

380 Roht-Arriaza 2010a, p. 220-221.

381 Various ways to address fragmentation in international law have been proposed, such as making use of legal techniques like the maxim *lex specialis derogat legi generali* and *lex posterior derogat legi priori* and improving formal mechanisms for communication and cooperation or increasing dialogue about gaps and synergies between international law regimes. International Law Commission 2006; Carlarne 2008, p. 474.

382 Biermann differentiates between three types of fragmentation: synergistic fragmentation, cooperative fragmentation, and conflictive fragmentation, in order to determine and compare degrees of fragmentation of different issue areas. Biermann, Pattberg & Van Asselt 2009.

communicate through a joint liaison, there is study finding that this mechanism has not resulted in systematic or substantial collaboration,[383] and the effort is not adequate to accommodate biodiversity considerations in the international climate law.[384] Suggestions provided in literature include, *inter alia*, the incorporation of human rights concerns into the climate change regime,[385] to fill environmental and human rights gaps in the climate change regime by the CBD,[386] the improvement of institutional coordination through the catalyst of climate change,[387] or the enhancement of synergies through autonomous management by state and non-state actors.[388] The sustainability of biofuels may to some extent serve as a great opportunity to promote integration and coordination between the different international regimes due to its cross-cutting feature. For instance, the CBD COP has requested the CBD Executive Secretary to contribute and assist in the ongoing work on the sustainability of biofuels taking place in the other international organisations and processes,[389] and encouraged governments and stakeholders to share experiences on the development and application of tools concerning the sustainable production of biofuels and their full life-cycle GHG emissions.[390]

383 It has been observed that a joint work plan between Ramsar and the CBD can serve as exemplary better practice of an integrative mechanism, but a corresponding mechanism is absent in the UNFCCC and voluntary collaboration has been inadequate. Pittock 2010, p. 372-373.

384 Van Asselt 2011a, p. 1211-1212.

385 It is suggested that integration can be achieved through incorporating 'linking' language on human rights into a climate change agreement, using private standards or certifications on human rights compliance, or embedding human rights expert processes within the UNFCCC or within the existing human rights implementation architecture. Roht-Arriaza 2010a, p. 226.

386 It is argued that normative activity under the CBD provides environmentally holistic and human rights based standards that could fill gaps relating to the protection of biodiversity and human rights in climate change law both at the level of international law-making and national implementation. UNFCCC COP decisions could thus refer to relevant CBD guidance and, in this way, human rights can be incorporated into the international climate regime at different levels without the need to create new standards. The CBD can count on a virtually universal membership. Morgera 2013, p. 361; Young 2011, p. 147.

387 Carlarne 2008, p. 471-472.

388 It is argued that both legal techniques and institutional cooperation have limitations in addressing conflicts between treaty norms and it is possible to mitigate conflict and enhance synergies through autonomous management by state and non-state actors. Van Asselt 2011a, p. 1275-1276.

389 Decision X/37, para. 13.

390 The Executive Secretary of CBD is responsible for disseminating the submitted experiences through a clearing-house mechanism. The COP decisions moreover encouraged developed countries to provide technical and/or financial support to developing countries with regard to the sustainable production of biofuels. Governments are encouraged to participate in relevant efforts carried out by various bodies other than CBD and the private sector is encouraged to improve the social and environmental performance of biofuel production through voluntary initiatives such as certification and codes of conduct. Decision IX/2, paras. 8(a) and (b), 6(a), 10, and 11; Decision X/37, para. 15; Convention on Biological Diversity, 'Responses to Notifications for Biofuels and Biodiversity', available at: <www.cbd.int/agro/biofuels/responses.shtml> (last accessed 21 March 2015).

Apart from fragmentation, the examined international instruments as for biofuels have weak or uncertain legal effects. It is observed that the international commitments and guidelines examined in the previous sections vary in their legal status and binding effect, and demonstrate different scales of normativity and bindingness.[391] For instance, the general commitments under biodiversity treaties leave large room for states to decide and implement, while the Kyoto Protocol provides binding quantified targets for Annex I Parties to reduce their GHG emissions. Meanwhile, based on an analysis of authorisation, voting, and language, COP decisions or resolutions on biofuels under the CBD and Ramsar Convention do not intend to bind State Parties, while COP decisions relating to GHG emissions accounting under the UNFCCC and Kyoto Protocol have a strong sense of requiring Annex I Parties to follow the provided methodologies. On the other hand, there are arguments that non-binding international instruments might become vehicles for arriving at a consensus on rules and principles and mobilising a consistent, general response on the part of states.[392] Moreover, non-binding instruments could serve as evidence of existing law, constitute a subsequent agreement between parties regarding the interpretation of a treaty and its provisions, or gradually acquire binding legal character via being elements of a treaty-based regulatory regime.[393]

In conclusion, in the absence of a binding integrated regulatory framework, international norms that address the sustainability of biofuels are emerging but fragmented, with still weak or uncertain legal status. As Morgera *et al.* have observed, none of the international instruments as examined in their contribution 'can of themselves provide systematic guidance or create specific obligations for countries to incorporate the full interplay of connected aspects in the environmental, food, agricultural, trade and energy sectors.'[394] The fragmentation and weak legal status constitute challenges for regulating the sustainability of biofuels in an inclusive and integrated manner at the international level and call for further institutional cooperation and coordination and the formulation of international consensus. At the same time, this means that there is much room left for states to decide on domestic measures for the sustainability of biofuels in view of their international obligations.[395] In this sense, it is necessary to examine the

391 Boyle and Shelton both argue for differential scales of normativity and bindingness instead of a binary perception of law or no-law. Shelton argues for the 'relative normativity' which is a question of hierarchy of norms, which concerns the nature, structure and content of the international legal system. The relative normativity arises in determining whether a legal rule exists to govern a problem and in deciding whether priority must be given to a specific rule among several. Boyle 2010, p. 122; Shelton 2010, p. 141.

392 It has been argued that some non-binding soft law instruments are significant mainly because they are the first step in a process eventually leading to the conclusion of a multilateral treaty or because they provide the detailed rules and technical standards required for the implementation of a treaty. Boyle 2010, p. 122; Shelton 2010, p. 141.

393 Boyle 2010, p. 122; Shelton 2010, p. 141.

394 Morgera, Kulovesi & Gobena 2009, p. 16-17.

395 In view of a lack of systematic guidance for countries at the international level, Morgera *et al.* have said that the integrated implementation of different international instruments is necessary at the national level to ensure sustainability in the bioenergy sector. Morgera, Kulovesi & Gobena 2009, p. 17.

domestic regulatory approach to sustainability of biofuels at the regional or national level before proceeding to the question of whether a global inclusive approach is needed and possible.

Chapter 3

EU REGULATION OF THE SUSTAINABILITY OF BIOFUELS

1. Introduction

This chapter aims to address the EU's approach to regulating the sustainability of biofuels. Biofuels[1] were promoted in the EU firstly to reduce dependence on oil imports and to increase energy security[2] and later the EU emphasised the CO_2 emission saving benefit as the reason for substituting oil with biofuels.[3] The pro-biofuel policy was finally entrenched in legal instruments since 2003, and it is claimed that the switch to biofuels has been responsible for achieving a three-fold objective of GHG emission reduction, energy security, and rural employment and development.[4] More importantly, with mounting concerns about the negative impacts of biofuels, the EU began to stress the sustainable production of biofuels[5] and established a set of sustainability criteria particularly for biofuels (and bioliquid[6]) in the legislation,[7] which was declared by the European Commission as

1 Biofuel is defined as 'liquid or gaseous fuel for transport produced from biomass' in EU legislation. Directive 2009/28/EC, Art. 2(i).

2 Directive 85/536/EEC; Commission Communication COM(97) 599 final, p. 4.

3 Biofuels were also identified as beneficial for reduction of sulphur content in certain liquid fuels. Commission Communication COM(97) 599 final, p. 4 and 16; Commission Communication COM(97) 196 final; Commission Proposal COM(97) 88 final.

4 Directive 2003/30/EC, paras. (2), (4), (5), (7) and (15) of the preamble; Directive 2009/28/EC, paras. (1), (2), and (3) of the preamble.

5 European Commission in the 2006 'An EU Strategy for Biofuels' emphasised that it is essential to guarantee that feedstock for biofuels is produced in a *sustainable* manner, both in the EU and in third countries, particularly with regard to the protection of biodiversity, water pollution, soil degradation and the protection of habitats and species. Commission Communication COM(2006) 34 final.

6 Bioliquid refers to 'liquid fuel for energy purposes other than for transport, including electricity and heating and cooling, produced from biomass.' Hence, they are different in their utilisation purposes, but they may be the same in the form of feedstock and end products. In order to avoid the legal loophole of using bioliquids for transport, the RES Directive also applies the sustainability scheme to bioliquids. Directive 2009/28/EC, para. (67) of the preamble, Art. 2(h).

7 Directive 2009/28/EC, Arts. 17-19.

the 'most comprehensive and advanced' binding sustainability scheme of its kinds in the world.[8] Thus, it is of particular interest for this thesis to analyse the sustainability scheme, examining what sustainability criteria are set forth and how comprehensive and advanced they are to ensure the sustainability of biofuels.

To analyse the EU's approach, the regulation regarding demand, supply, and trade of biofuels is examined at the EU level, and it is questioned whether this regulation inclusively takes into account sustainability concerns in environmental and socio-economic dimensions. The EU regulator can establish biofuel measures which may directly drive or constrain the demand, supply, or trade of biofuels, within the competences stipulated in EU law. Thus, this chapter concentrates on the regulatory measures related to the demand, supply, and trade of biofuels as provided for in EU law, and examines to what extent they aim to avoid or minimise the potential adverse effects on sustainability.

Moreover, compared to national law, the EU has a 'multi-level' regulatory framework.[9] Hence, if the measure at the EU level is not inclusive for addressing both the environmental and socio-economic concerns, it will further be examined whether the EU's measure leaves room for lower authorities – the Member States – to adopt complementary or stricter sustainability criteria for biofuels. This question calls for an analysis of the extent of harmonisation (or approximation) of the sustainability criteria for biofuels. In EU's context, harmonisation is 'the creation of rules' by EU's legislative acts that change or supplement Member States' national legislation,[10] so as to reduce differentiation and increase uniformity in achieving a common purpose. In this vein, it will be examined whether the EU's sustainability criteria allow Member States to deviate by adopting complementary requirements (apart from EU's criteria) or stricter requirements (than EU's criteria).

This chapter is structured as follows. In analysing the regulation relating to the demand for biofuels, section 2 firstly looks into the sustainability criteria and their implementing scheme established in the EU legislation, and further examines the other measures relating to demand for biofuels such as the consumption target. Section 3 delves into the regulation addressing the major sustainability concerns in the production of biofuels, such as biodiversity, carbon stock, agro-environment, and socio-economic concerns. Section 4 expounds on the implications of applying the EU sustainability criteria to imported biofuels from the third countries. Section 5 concludes by reflecting on observations from the previous sections in view of the research question about whether the EU's domestic regulation of the sustainability of biofuels provides an approach to inclusively address the environmental and socio-economic sustainability concerns.

8 Communication (2010/C 160/02), section 1.

9 Hooghe & Marks 2003, p. 233-235; Chowdhury & Wessel 2012; Kern & Bulkeley 2009.

10 In this sense, 'harmonisation' in EU context is similar to that at the international level, both referring to the approximation of laws between different legal systems. The term 'approximation' has been formally used in EU law instead of 'harmonisation', such as in Arts. 67, 81, 82, 114-115, and 151 TFEU, but they are regarded as equivalent and interchangeable, and harmonisation has been more frequently used in the literature. Vos 2001, p. 147-148.

2. EU Regulation of the Demand for Biofuels

Since the 1980s, biofuels have gained momentum due to the EU's supportive policy[11] and various favourable measures.[12] After years of development, biofuels become the major form of renewable energy in road transport[13] and the share of biofuels in energy consumption in the transport sector experienced continuous growth.[14] Thus, demand for biofuels in the EU is closely linked with the support measures. This section will discuss the demand-driving measures, financial or non-financial, such as a renewable energy consumption target and preferential energy taxes. The sustainability criteria will be first and foremost analysed, since they form the basis for the other demand-side requirements.

11 Since the Council Directive 85/536/EEC on crude-oil savings through the use of substitute fuel components in petrol, biofuels have been included in a number of legislative and policy documents on supportive measures for renewable energy, such as Council Decisions on the Altener programme for supporting renewable energy, the 'Biomass Action Plan', 'An EU Strategy for Biofuels' until the adoption of a directive on biofuels promotion and later the current directive on renewable energy. Council Decision 93/500/EEC; Council Decision 98/352/EC; Council Decision 1999/21/EC; Directive 2003/30/EC; Commission Communication COM(2005) 628 final; Communication COM(2006) 34 final; Directive 2009/28/EC.

12 The favourable measures for biofuels may include market price support and tax exemptions for biofuels. The former refers to a subsidy conferred on biofuel producers due to Member States' consumption mandates that provide a guaranteed market for their product and push prices upwards. The latter refers to the full or partial non-application of excise taxes for transport fuels that are fully applied to competing products such as gasoline and diesel, and research and development grants that promote the development of biofuel projects or technologies. Another report classified four broad groups of biofuel policy measures: *a)* budgetary support, such as direct support to biomass supply and fuel tax exemptions for biofuel producers; *b)* blending or use targets ('mandates'), which impose a minimum market share for biofuels in total transport fuel; *c)* trade measures, in particular import tariffs; *d)* measures to stimulate productivity and efficiency improvements at various points in the supply and marketing chain. Charles *et al.* 2013, p. 1; Fonseca *et al.* 2010, p. 16.

13 According to an estimate, renewable energy in transport, including other kinds of renewable energy sources such as electric cars, accounted for 4.7 per cent in 2012, while biofuels accounted for 4.5 per cent, among which biodiesel wasthe most common form of biofuel consumed in the EU, with a share of 75 per cent. Another estimate shows that the incorporation rate of biofuels in EU transport was 4.7 per cent in 2013, amounting to a total biofuel consumption of 13.6 Mtoe. Commission Staff Working Document SWD (2013) 102 final, p. 10; EurObserv'ER, Biofuels Barometer, July 2014, available at: <www.energies-renouvelables.org/observ-er/stat_baro/observ/baro222_en.pdf > (last accessed 16 April 2015).

14 According to data published by EurObserv'ER, the consumption of biofuels for transport has been expanding from 2002 to 2012, but it went through the first drop from 2012 to 2013. EurObserv'ER, Biofuels Barometer, July 2014, available at: <www.energies-renouvelables.org/observ-er/stat_baro/observ/baro222_en.pdf> (last accessed 16 April 2015).

2.1. *The Sustainability Criteria*

The sustainability criteria and their implementing scheme for biofuels (and bioliquids) were established by Article 17 to 19 of Directive 2009/28/EC.[15] Article 17 sets out the sustainability criteria and forms the core of the regulatory scheme, while Article 18 and Article 19 respectively provide the implementing scheme that verifies the compliance with the criteria. They serve as a basis for the other demand-related requirements for biofuels because Article 17(1) explicitly states that the sustainability criteria shall be fulfilled for the purposes of counting towards the renewable energy consumption target,[16] financial supports,[17] and the other incentive measures.[18] The sustainability criteria can be generally divided into carbon-related and non-carbon-related.[19] Carbon-related criteria include the requirements about GHG emission savings in the biofuel's life-cycle and the conservation of areas with high carbon stocks, and non-carbon-related criteria refer to biodiversity and agro-environment protection.

2.1.1. GHG Emissions Saving

The first sustainability criterion requires that biofuels, including those produced from non-agricultural waste and residues,[20] shall achieve a minimum rate of GHG emissions savings.[21] Such a requirement consists of two critical factors: the

15 Directive 2009/28/EC on renewable energy repealed and integrated the former directives, such as Directive 2003/30/EC, respectively on different sources of renewable energy in transport (biofuel), electricity, and heating and cooling sectors. The Arts. 17-19 are almost 'copied and pasted' in Directive 2009/30/EC on fuel quality for coherence. See the discussion in section 2.3 of this chapter.

16 It refers to the targets provided in Art. 3 of Directive 2009/28/EC regarding the share of renewable energy consumed in the total (transport) energy mix in the EU. Furthermore, the sustainability criteria are also to be fulfilled for 'measuring compliance with renewable energy obligations.' The 'renewable energy obligation' is defined in the Directive as the measure that requires energy producers, suppliers, or consumers to include a given proportion of renewable energy in their production, supply or consumption. Directive 2009/28/EC, Arts. 3 and 2(l).

17 Financial support is defined as including any support scheme that promotes the use of renewable energy by reducing its cost or increasing its price, such as *inter alia* tax refunds, direct price support, or tax exemption. Directive 2009/28/EC, Art. 17(1)(b) and Art. 2(k).

18 Apart from the three application purposes provided in Art. 17(1) of Directive 2009/28/EC, the European Commission supplemented that the sustainability criteria shall be met for biofuels to be counted towards the target of the Fuel Quality Directive for reducing GHG (biofuels only); receive investment and/or operating aid in accordance with the Community guidelines on State aid for environmental protection (biofuels only); the provisions for alternative fuel vehicles of the Regulation on CO_2 from passenger cars ('E85' bioethanol only). Communication (2010/C 160/02), section 1.1.1.

19 Lin 2011a, p. 36; Romppanen 2012a, p. 126.

20 It is provided that biofuels and bioliquids produced from waste and residues, other than agricultural, aquaculture, fisheries and forestry residues, need only fulfil the GHG emission saving criterion. Directive 2009/28/EC, Art. 17(1).

21 Directive 2009/28/EC, Art. 17(2).

minimum rate of GHG emission savings or the 'threshold' and the method of calculating the GHG emission savings of a specific biofuel.

Firstly, biofuels shall reach the minimum GHG emission saving rate of 35 per cent and, according to Directive 2009/28/EC, the threshold will be raised to 50 per cent as of 1 January 2017 and to 60 per cent as of 1 January 2018 (see Table 3). It is noted that such a requirement contains a 'limited prospective grandfathering clause' in order to protect the trust of investors:[22] the 35 per cent threshold was applied to biofuels produced from installations that were in operation on 23 January 2008[23] and the 60 per cent threshold is to be applied to biofuel installations that will start operation[24] on or after 1 January 2017. In 2015, Directive (EU) 2015/1513 amended this requirement by applying the 60 per cent emission saving threshold to biofuel installations starting operation after 5 October 2015 (rather than 1 January 2017) from 5 October 2015 (rather than 1 January 2018), which in effect enhance the emission saving requirement for new installations.[25] However, for the 'old' installations in operation on or before 5 October 2015, the 35 per cent threshold shall be applied until 31 December 2017 (instead of 1 January 2017), which actually delay the rise of minimum requirement for the existing installations (see Table 4).

Table 3. GHG emission saving thresholds (Directive 2009/28/EC)

Minimum rate	35%	35%	50%	60%
Application time	June 2009[26]	1 April 2013	1 January 2017	1 January 2018
Installations	Installations in operation after 23 January 2008	All installations	All installations	Installations in operation after 1 January 2017

22 'Grandfathering' means the exclusion of biofuels facilities existing at a certain moment in time from the sustainability requirements of later biofuels law. In the comparative study between EU biofuel policy and US biofuel policy, Jansson and Kalimo observe that the EU's 'grandfathering' is minor and limited in comparison with that of the United States, because even old installations need to meet some threshold. Jansson & Kalimo 2014, p. 119-120 and 142-143.

23 2008 is the year in which the European Commission made the proposal for Directive 2009/28/EC. This means that the producers already engaged in biofuel production before 23 January 2008 did not have to comply with the GHG emission saving criterion from April 2009 (when the Directive was adopted) to April 2013. Directive 2009/28/EC, Art. 17(2).

24 An installation shall be considered to be in operation if the physical production of biofuels or bioliquids has taken place. Directive 2009/28/EC, Art. 17(2).

25 Directive (EU) 2015/1513, Art. 1(3) and Art. 2(5).

26 Directive 2009/28/EC was adopted on 23 April 2009 and entered into force on the 20th day following its publication in the Official Journal of the European Union on 5 June 2009. Directive 2009/28/EC, Art. 28.

Table 4. GHG emission saving thresholds amended in 2015

Minimum rate	35%	50%	60%
Application time	Until 31 Dec. 2017	From 1 January 2018	From 5 October 2015
Installations	Installations in operation on/before 5 October 2015	Installations in operation on/before 5 October 2015	Installations in operation after 5 October 2015

Secondly, the method for calculating the rate of GHG emissions savings is specifically provided in Article 19 and Annex V of Directive 2009/28/EC. Simply put, the emissions savings rate results from comparing a specific value of life cycle GHG emissions[27] of biofuels with the average value of GHG emissions from fossil fuels, which can thus measure the emissions to be reduced/saved by using biofuels.[28] Hence, the calculation of the GHG emissions from biofuels is crucial for measuring the achievement the emissions savings thresholds.[29] As provided by Directive 2009/28/EC, compliance with the criterion can be proved either by calculating the actual values following the methods provided in Annex V(C) or by applying pre-determined default values.

a) Calculation of Actual Values

Actual values of GHG emission savings are calculated by following the method as provided in Annex V(C),[30] which serves as the principal way to evaluate the GHG emission savings of biofuels.[31] The method aims to cover the total emissions from a

27 The greenhouse gases taken into account in the calculation are CO_2, N_2O and CH_4. For calculation convenience, the latter two can be translated into CO_2 equivalents following the fixed values provided in the Directive. The GHG emissions from fuels are expressed in terms of gCO2eq/MJ (grams of CO_2 equivalent per MJ). Directive 2009/28/EC, Annex V(C)(2) and (5).

28 The equation is expressed as: $SAVING = (EF - EB)/EF$. EB stands for the total emissions from biofuels, and EF stands for the total emissions from fossil fuel comparator. Directive 2009/28/EC, Annex V(C)(4).

29 The calculation of GHG emissions from fossil fuels is also important. It is provided that the fossil fuel emissions shall be the latest available *actual* average emissions from the fossil part of petrol and diesel consumed in EU as reported under Fuel Quality Directive. However, implementing rules on the life-cycle GHG emission of fossil fuels were not available until October 2014 when the European Commission issued a Proposal for the methodology for the GHG intensity calculation and reporting. As for biofuels, until such actual data is available, a fixed value of 83.8 gCO2eq/MJ is applied as the fossil fuel emission comparator. For bioliquids, the comparator values are respectively 91 in electricity use, 77 in heating use, and 85 in cogeneration use. Directive 2009/28/EC, Annex V(C)(19) of RES Directive; Commission Proposal COM(2014) 617 final, Annex I.

30 Actual values refer to the means the greenhouse gas emission saving for some or all of the steps of a specific biofuel production process calculated in accordance with the methodology laid down. Directive 2009/28/EC, Art. 2(m).

31 It is also stated in the preamble of Directive 2009/28/EC that in calculating the greenhouse gas impact of land conversion, economic operators should be able to use actual values for the →

biofuel's life cycle and meanwhile to avoid double-counting. A calculation of the total emissions thus takes into account emissions as well as removals of greenhouse gases from the production and final consumption of biofuels.[32] GHG emissions from the following sources are to be added up: the extraction or cultivation of biofuels raw materials; the annualised emissions from carbon stock changes caused by land-use changes;[33] the processing of biofuels feedstocks; the transportation and distribution of biofuel products; and the end use of biofuels.[34] GHG removals are to be subtracted from the added emissions, which may happen when improved agricultural management accumulates soil carbon, carbon capture and storage (CCS) or carbon capture and replacement is applied, or excess electricity is produced from cogeneration.[35]

Apart from the GHG removals, GHG emissions from certain sources are to be exempted or deducted from the total emissions accounting. Firstly, emissions from biofuels in use (combustion) are taken as zero.[36] Secondly, in calculating emissions from the extraction or cultivation of raw materials, capture of CO_2 in the cultivation of raw materials is excluded from the total emissions. Thirdly, in accounting the annualised emissions from carbon stock changes caused by land use change, biofuels that obtain raw materials from 'degraded land' can subtract 29 gCO2eq/MJ from the annualised emissions, as a 'bonus' for utilising severely degraded land or heavily contaminated land.[37] It is important to note that the European Commission has adopted a decision to provide guidelines for the calculation of land carbon

carbon stocks associated with the reference land use and the land use after conversion. Directive 2009/28/EC, para. (7) of the preamble.

32 The general equation is: the total emissions from the use of the fuel are equal to emissions from biofuel production and final use with *subtraction* of the emission saving/removal from soil carbon or carbon capture and storage. Directive 2009/28/EC, Annex V (C)(1).

33 Annualised emissions from carbon stock changes caused by land-use change shall be calculated by dividing total emissions equally over 20 years. Directive 2009/28/EC, Annex V(C)(7).

34 Emissions from the manufacture of machinery and equipment shall not be taken into account. Directive 2009/28/EC, Annex V(C)(1).

35 Directive 2009/28/EC, Annex V(C)(1), (14), (15), and (16).

36 Directive 2009/28/EC, Annex V(C)(13).

37 The degraded land hereby refers to the land that was not in use for agriculture or any other activity in January 2008, and falls into the category of 'severely degraded land' or 'heavily contaminated land.' 'Severely degraded land' refers to land that, for a significant period of time, has either been significantly salinated or presented significantly low organic matter content and has been severely eroded. 'Heavily contaminated land' means land that is unfit for the cultivation of food and feed due to soil contamination. The Commission may decide whether the land under national or regional recovery programme (that aims to improve severely degraded or heavily contaminated land) fulfills the criteria of restored degraded land. The bonus of 29 gCO2eq/MJ shall apply for a period of up to 10 years from the date of conversion of the land to agricultural use, provided that steady increase in carbon stocks as well as a sizable reduction in erosion phenomena for land falling is ensured and soil contamination for land falling is reduced. Directive 2009/28/EC, Annex V(C)(8) and (9), Art. 18(4).

stocks,[38] which largely draws upon the AFOLU volume of 2006 IPCC Guidelines for National GHG Inventories.[39]

However, this accounting method has been criticised for excluding CO_2 emissions from biofuels in use without sufficiently taking land use changes into consideration, which has been attributed by many scientists as the 'misapplication' of the IPCC Guidelines under the UNFCCC.[40] Moreover, GHG emissions caused by indirect land use changes, when existing food cropland is used to grow biofuel feedstocks, are not included in calculating the GHG emission savings.[41] To address the issue, the European Commission has issued reports, as mandated by Directive 2009/28/EC,[42] regarding indirect land use changes. In the 2010 report on indirect land use change,[43] the Commission acknowledged that indirect land use change can have an impact on GHG emissions savings of biofuels and considered that this issue should be addressed under a precautionary approach if action is required.[44] Based on the policy options proposed in 2010,[45] the Commission made a proposal to amend Directive 2009/28/EC in 2012 to mitigate the risk of indirect land use change mainly by limiting the share of conventional biofuels in fulfilling the renewable energy consumption targets.[46] The amendments prescribe indirect land use change

38 The Commission's decision provides guidelines on carbon stock and carbon stock values from soil carbon (mineral and organic soil) and above and below ground vegetation. Commission Decision 2010/335/EU.

39 It is stated that the Commission should draw its guidelines on carbon stock on the 2006 IPCC Guidelines for National Greenhouse Gas Inventories, while it is also appropriate to draw upon other scientific sources of data where the necessary information on biofuel production lack or is not accessible. Commission Decision 2010/335/EU, paras. (2)-(4) of the preamble.

40 According to many scientists, CO_2 emitted by vehicle tailpipes from the use of biofuels can be ignored only if and to the extent the effects on land use and land use change are beneficial. Searchinger *et al.* 2009, p. 527; Haberl 2012, p. 18; EEA Scientific Committee 2011, p. 3.

41 The European Commission has also admitted that the methodology defined in the EU Directives currently does not include emissions associated with indirect changes in land use. EEA Scientific Committee 2011; Commission Staff Working Document SWD (2012) 343 final, p. 8; Commission Report COM(2010) 811 final.

42 Art. 19(6) provides that, by 31 December 2010, the Commission shall submit a report to the European Parliament and to the Council reviewing the impact of indirect land-use change on greenhouse gas emissions and addressing ways to minimise that impact. Art. 17(9) provides that the Commission shall report on requirements for a sustainability scheme for energy uses of biomass, other than biofuels and bioliquids, by 31 December 2009. Directive 2009/28/EC, Art. 19(6) and Art. 17(9).

43 Commission Report COM(2010) 811.

44 Commission Report COM(2010) 811, p. 14.

45 The Commission came up with four policy options: *a)* take no action for the time being, while continuing to monitor; *b)* increase the minimum GHG saving threshold for biofuels; *c)* introduce additional sustainability requirements on certain categories of biofuels; *d)* attribute a quantity of GHG emissions to biofuels reflecting the estimated indirect land use impact. Commission Report COM(2010) 811, p. 14.

46 In the Proposal for an amendment, the Commission proposed to mitigate indirect land-use change effects mainly through the following ways. Firstly, a 5 per cent limit is set for the share of food crops-based biofuels in calculating the numerator of the 10 per cent transport renewable energy target. Secondly, in calculating the GHG emission savings related to biofuels, the minimum rates are to be raised, and the condition that biofuels of foreign origin →

emission factors to certain biofuels, and fuel suppliers are obliged to *report* on emissions of biofuels by considering indirect land use changes based on the predetermined factors.[47] These proposed amendments have gone through lengthy negotiation among EU legislators from October 2012 to April 2015[48] before Directive (EU) 2015/1513 was finally adopted, largely because the European Parliament and the Council had divergence regarding the exact limit on food crop-based biofuels.[49] However, amongst the amendments, indirect land-use change on GHG emissions are still not taken into account in estimating compliance with the life-cycle GHG emission saving criterion for biofuels. Meanwhile, the effectiveness of using approaches of 'capping' the proportion of first-generation biofuels and 'multiple counting' of second-generation biofuels to address indirect land-use changes has been doubted, since biofuels with non-food or second-generation feedstock is not necessarily more environmentally or socio-economically beneficial.[50]

b) Use of Default Values

In order to reduce the 'disproportionate administrative burden' caused by calculating actual values case by case, Directive 2009/28/EC lays down a list of pre-

can use default values is to be abolished, and the biofuel feedstocks grown on degraded lands will no longer enjoy a favourable 'bonus.' Thirdly, designated advanced biofuels are to be further promoted by double or four-time counting. Commission Proposal COM(2012) 595 final; Commission Staff Working Document SWD (2012) 344 final, p. 4-5.

47 The Commission should review the methodology for estimating land-use change emission factors included in Annexes in the light of adaptation to technical and scientific progress. The Commission can introduce factors of disaggregation at further levels and include additional values for new biofuel feedstocks. Commission Proposal COM(2012) 595 final, p. 8.

48 The European Parliament and the Council were finally able to reach agreement in the second reading in April 2015. Ecaterina Casinge, 'Parliament rubber stamps EU biofuels reform amid final controversy' (EurActiv, 29 April 2015), available at: <www.euractiv.com/sections/-transport/parliament-rubber-stamps-eu-biofuels-reform-amid-final-controversy-314196> (last accessed 1 May 2015).

49 While the Commission proposed a 5 per cent limit, the European Parliament backed a slightly higher cap of 6 per cent, and in the Council the EU energy ministers debated a new compromise of 7 per cent. The MEPs finally approved a compromise agreement on 14 April and gave in to the Council of Ministers' request to set a cap at 7 per cent, following extensive negotiations with the Council of Ministers. The legislators also agreed to have fuel suppliers and the European Commission report on emissions deriving from ILUC, but they failed to have these emissions included in the carbon accounting. MEPs accepted the Council's proposal to set an optional, instead of mandatory, target of 0.5 per cent for advanced biofuels. See Ecaterina Casinge, 'Lawmakers agree to limit food-based biofuels' (EurActiv.com, 14 April 2015), available at: <www.euractiv.com/sections/sustainable-dev/meps-agree-limit-food-based-biofuels-313760> (last accessed 21 April 2015).

50 It was found in one study that second generation biofuels perform better with indirect land use change factors and the other investigated sustainability aspects, on the condition that they are not sourced from dedicated plantations directly competing for agricultural land. Otherwise, efficient first generation systems are preferable. Since no clear technology champion for all situations exists, the study recommends targeting policy instruments directly at the positive and negative effects of biofuel production rather than at the production itself. Havlík *et al.* 2011, p. 5690.

determined 'default values' to be used in place of actual values in circumstances specified in the Directive.[51] Firstly, only biofuels with the production pathways prescribed in the Directive can use default values.[52] Secondly, default values can be used when there are no net carbon emissions from land-use changes, calculated in line with the method provided in the Directive.[53] Thirdly, the default values may only be used for the biofuels when their raw materials are cultivated outside the EU,[54] or in designated areas within the EU,[55] or are waste or residues other than agricultural, aquaculture and fisheries residues.

The prescribed biofuel production pathways cover the major types of conventional and advanced biofuels on the market, such as sugar cane ethanol, soybean biodiesel, or waste vegetable oil biodiesel. Default values may differentiate according to the production or processing method even with the same feedstock materials. For example, palm oil biodiesel with methane capture at the oil mill has a saving rate of 56 per cent, while the value is 19 per cent for palm oil biodiesel if the production process is not specified. Generally, the advanced biofuels with raw materials from waste, cellulosic or ligno-cellulosic materials (e.g. straw or wood) can easily achieve a GHG emissions saving above the 60 per cent threshold. However, when the default value as prescribed in the Directive cannot reach the minimum emission saving rate, such as the soybean biodiesel with a default value of 31 per cent, the actual value has to be used to show compliance.[56]

51 Default value means a value derived from a typical value, an estimate of the representative GHG emission saving for a particular biofuel production pathway, by the application of pre-determined factors for common biofuel production pathways, and the default value is to be used in place of an actual value in circumstances specified in this Directive. The list of default values should be updated and expanded when further reliable data is available. Directive 2009/28/EC, Art. 2(n) and (o), para. (82) of the preamble.

52 Directive 2009/28/EC, Annex V(A), (B), (D) and (E).

53 That means the annualised emissions from carbon stock changes caused by land-use changes, as calculated in accordance with Annex V(C), is equal to or less than zero. Directive 2009/28/EC, Art. 19(1)(a) and Annex V(A).

54 Directive 2009/28/EC, Art. 19(3)(a).

55 The areas refer to a list of areas submitted by Member States that are classified as nomenclature territorial units for statistics (NUTS) according to Regulation No 1059/2003 on the establishment of NUTS. These areas are identified because the typical GHG emissions from the cultivation of agricultural raw materials in these areas can be expected to be lower than or equal to the emissions prescribed in Part D of Annex V of RES Directive. Directive 2009/28/EC, Art. 19(2).

56 Where the default value for greenhouse gas emission savings from a production pathway lies below the required minimum level of greenhouse gas emission savings, producers wishing to demonstrate their compliance with this minimum level should be required to show that actual emissions from their production process are lower than those that were assumed in the calculation of the default values. Nevertheless, default values can be used in combination with actual values by using disaggregated default values as provided in the Directive. Directive 2009/28/EC, para. (82) of the preamble, Art. 19(1)(c).

2.1.2. Land with High Biodiversity Value

Directive 2009/28/EC admits that the incentives for using biofuels should have the effect of encouraging the destruction of biodiverse lands, so it is necessary to ensure that the incentivised biofuels do not obtain raw materials from land with high biodiversity value.[57] The Directive does not give a general definition for 'land with high biodiversity value',[58] but it identifies the three types of land which have a high biodiversity value: *a)* primary forest and other wooded land, *b)* areas designated for nature or eco-system protection by national or international law, and *c)* natural or non-natural highly biodiverse grasslands. Biofuels with raw materials obtained from these lands will not be recognised as 'sustainable', unless otherwise provided. Moreover, as there is no territorial limit, this criterion includes highly biodiverse lands having that status in all countries.[59]

a) Primary Forest and Other Wooded Land

Primary forest and other wooded land means 'forest and other wooded land of native species, where there is no clearly visible indication of human activity and the ecological processes are not significantly disturbed.'[60] This provision does not set out in more detail how to identify land with such a status especially as for the primary forest or wooded land in the third countries,[61] but the Directive mentions in its preamble that a forest is biodiverse when it is a primary forest in accordance with the definition in the Global Forest Resource Assessment.[62] The Global Forest Resources Assessment is produced by the United Nations Food and Agriculture Organisation (FAO) every five years since 1948, in an attempt to describe the statuses and changes of the world's forests based on data from country reports and

57 Directive 2009/28/EC, para. (69) of the preamble, Art. 17(3).

58 The Directive does not define the terms 'biodiversity', 'value', and 'high', probably due to the complex nature of 'biodiversity' and the consequent variations in the widely accepted definitions of biodiversity, not to mention the difficulty in defining 'value' and 'high.' ECOFYS 2011, p. 25.

59 However, there is a temporal limit for such types of land to fall within the scope 'land with a high biodiversity value': land shall be qualified for one of these statuses in or after January 2008, whether or not the land continues to have the status. Thus, even if the land loses the status after January 2008 for other reasons than biofuel cultivation, the land still falls within the scope of 'land with high biodiversity value' of this Directive. Directive 2009/28/EC, Art. 17(3).

60 Directive 2009/28/EC, Art. 17(3)(a).

61 Most European forests can be qualified as 'semi-natural', while there are few undisturbed forest areas remaining in the EU. According to an assessment of FAO, primary forests are mainly distributed in Brazil, Russia, and Canada. The undisturbed forest areas in the EU, if there are, are protected under Natura 2000 and the EU Forestry Strategy. European Commission, Sustainable Forestry and The European Union Initiatives of the European Commission, available at: <http://ec.europa.eu/agriculture/publi/brochures/forestry/-full_en.pdf> (last accessed 18 April 2015).

62 Directive 2009/28/EC, para. (69) of the preamble.

remote sensing.[63] The Directive relies on the Global Forest Resources Assessment largely because countries use it worldwide to report on the extent of primary forest protected by national law.[64] Nevertheless, different from the definition of the Global Forest Resources Assessment, the EU Directive explicitly excludes 'modified natural forests' and 'semi-natural forest and plantations' from the scope of primary forests.[65] Moreover, once the forest gains the status of primary forest in or after January 2008, it will retain that status up to date even if the land no longer meets the definition of 'primary forest.'[66]

b) Areas Designated for Nature or Ecosystem Protection

The second type of land refers to the areas designated for nature protection purposes, or for the protection of rare, threatened or endangered ecosystems/species. These areas include those designated by law or by the relevant competent authority, as well as those recognised by international agreements or included in lists drawn up by intergovernmental organisations or the IUCN. These areas may include the protected areas or sites designated by law within or outside of the EU. Within the EU, protected areas mainly include the sites under Natura 2000 network, which consist of the Special Protected Areas classified under the Birds Directive 2009/147/EC and Special Areas of Conservation designated under the Habitats Directive 92/43/EEC.[67] Protected areas can be designated by relevant competent authorities of the EU Member States and non-EU countries depending on the relevant national laws and regulation.[68] Moreover, areas protected by international agreements, inter-governmental organisations or the International

63 FAO, Global Forest Resources Assessments (FRA), available at: <www.fao.org/forestry/-fra/en/> (last accessed 18 April 2015).

64 Directive 2009/28/EC, para. (69) of the preamble.

65 FAO's definition of primary forest is: 'forest/other wooded land of native species, where there are no clearly visible indications of human activities and ecological processes are not significantly disturbed.' FAO notes that the definition of primary forest 'includes: areas where collection of NWFPs (Non-Wood Forest Products) occurs, provided that the human impact is small. Dehue *et al.* 2011, p. 29; Directive 2009/28/EC, para. (69) of the preamble.

66 This is to prevent a reclassification of 'primary forest' by governments to 'other naturally regenerated forests' because of human interventions such as selective logging. Dehue *et al.* 2011, p. 30.

67 The EU boasts of more than 26,000 protected sites designated in 52 countries, and the creation of protected areas has long been the only way to preserve remarkable natural assets from adverse land use. The size of Europe's designated areas varies greatly. In addition to the protection of birds and wild habitats, the EU has also taken action specifically to protect water and marine environments and their ecosystems. Apart from the Natura 2000 at the EU level, the Nationally Designated Areas or the Common Database on Designated Areas (CDDA) is a European inventory of protected areas designated at the national level. European Environmental Agency, Protected Areas, available at: <www.eea.europa.eu/themes/biodiversity/protected-areas/introduction> (last accessed 18 April 2015).

68 Dehue *et al.* 2011, p. 45.

Union for Conservation of Nature (IUCN)[69] may also be included, provided that they are formally recognised by the European Commission,[70] although as yet no such decision of recognition has been made. However, different from the absolute exclusion of raw materials from primary forests, an exception can be made if evidence is provided to show that the production of raw materials for biofuels does not interfere with nature protection and biodiversity values, such as the protected area managed mainly for the sustainable use of natural ecosystems purposes.[71]

c) Highly Biodiverse Grassland

The highly biodiverse grassland can be both natural and non-natural.[72] Natural grassland refers to the grassland that, in the absence of human intervention, maintains the natural species composition and ecological characteristics and processes.[73] Non-natural grassland is species-rich and not degraded but would cease to be grassland in the absence of human intervention.[74] The non-natural grassland is not excluded from the production of raw materials for biofuels, if evidence is provided that the harvesting of raw materials is necessary to preserve its grassland status.[75]

Different from the other two types of land, Directive 2009/28/EC provides that the criteria and geographic ranges of the highly biodiverse grassland shall be established by the European Commission, in order to determine the grassland to be covered under this criterion in accordance with best available scientific evidence and international standards.[76] In December 2014, the European Commission finally issued the criteria and geographic ranges of highly biodiverse grassland[77] based on

69 The IUCN has been working on the IUCN protected area management categories which are used to classify protected areas to define, record, and classify the wide variety of specific aims and concerns. Under IUCN, protected areas are 'clearly defined geographical space, recognised, dedicated and managed, through legal or other effective means, to achieve the long-term conservation of nature with associated ecosystem services and cultural values', and there are six management categories for protected areas which vary in levels of protection and on-site management. Dehue *et al.* 2011, p. 44-45.

70 The Commission may recognise areas for the protection of rare, threatened or endangered ecosystems or species recognised by international agreements or included in lists drawn up by intergovernmental organisations or the International Union for the Conservation of Nature for the purposes of Art. 17(3)(b)(ii). Directive 2009/28/EC, Art. 18(4).

71 For example, some biomass collection activities can bring benefits to nature or eco-system sustainability. Directive 2009/28/EC, Art. 17(2); Dehue *et al.* 2011, p. 44-45; Jansson & Kalimo 2014, p. 117.

72 It is declared biofuels made from raw materials originating in grasslands, both temperate and tropical, including highly biodiverse savannahs, steppes, scrublands and prairies, should not qualify for the incentives provided for by this Directive. Directive 2009/28/EC, Art. 17(3)(c), and para. (69) of the preamble.

73 Directive 2009/28/EC, Art. 17(3)(c)(i).

74 Directive 2009/28/EC, Art. 17(3)(c)(ii).

75 Directive 2009/28/EC, Art. 17(3)(c)(ii); Commission Regulation (EU) No 1307/2014, Art. 3.

76 Directive 2009/28/EC, Art. 17(3), para. (69) of the preamble.

77 Commission Regulation (EU) No 1307/2014.

available information.[78] The European Commission defines grassland as 'terrestrial ecosystems dominated by herbaceous or shrub vegetation for at least 5 years continuously' including 'meadows or pasture that is cropped for hay but excludes land cultivated for other crop production and cropland lying temporarily fallow.'[79] Within the EU, grasslands falling within the geographic ranges of habitats as provided in the Habitat Directive 92/43/EEC and Birds Directive 2009/147/EC are always regarded as highly biodiverse grasslands,[80] but this does not exclude other lands if they can fulfil the definition and criteria set out by the European Commission.[81]

2.1.3. Land with High Carbon Stock

The third criterion requires that sustainable biofuels shall not obtain raw materials from lands with high carbon stock including the peatland.[82] There can be overlaps between the highly biodiverse lands and the high carbon stock lands, such as the continuously forested areas, wetlands, and peatlands.[83] Directive 2009/28/EC does not give a definition of the 'land with high carbon stock', but it identifies the types of lands which are regarded as having high carbon stocks.

a) Wetlands

Wetlands are defined in Directive 2009/28/EC as the 'land that is covered with or saturated by water permanently or for a significant part of the year.'[84] Moreover, to remain aligned with the Ramsar Convention, the Directive mentions in its preamble

78 As comprehensive information on geographic ranges of highly biodiverse grasslands is not available at the international level, the Commission provides geographic ranges only for those highly biodiverse grasslands for which information is already available. Commission Regulation (EU) No 1307/2014, para. (6) of the preamble.

79 It is important to note that grasslands may include agroforestry systems where trees are managed together with crops or animal production systems in agricultural settings. Commission Regulation (EU) No 1307/2014, Art. 1(1).

80 These are geographical ranges enlisted as habitats in Annex I and habitats of significant importance for animal and plant species of Union interest listed in Annex II and IV to Council Directive 92/43/EEC, and habitats of significant importance for wild bird species listed in Annex I to Directive 2009/147/EC. Commission Regulation (EU) No 1307/2014, Art. 2.

81 It is particularly stated that highly biodiverse grassland in the European Union is not limited to the geographic ranges referred to under (1), (2) and (3) of this Art. Other grassland might fulfil the criteria for highly biodiverse grassland set out in Art. 1. Commission Regulation (EU) No 1307/2014, Art. 2.

82 The land shall have such status in January 2008 and no longer has that status, which means that biofuel raw materials can be obtained from lands having such status in January 2008 and still have this status when the raw materials are obtained. Directive 2009/28/EC, Art. 17(4) and (5); Dehue *et al.* 2011, p. 59.

83 Thus, in consulting data sources for these land categories, an economic operator may be able to satisfy both carbon stock and biodiversity criteria with the same data sources. Dehue *et al.* 2011, p. 28.

84 Directive 2009/28/EC, Art. 17(4)(a).

that the broader definition stated in the Ramsar Convention should be taken into account,[85] since the latter also includes temporary wetlands and riparian and coastal zones adjacent to wetlands, and islands or bodies of marine water deeper than six metres at low tide lying within the wetlands.[86] In order to show compliance, the Ramsar List and the Ramsar Sites Database managed by the Ramsar Sites Information Service can be used.[87]

b) Forested Land

In addition to the primary forest protected for its biodiversity, other forested lands are protected from biofuel production for their value in conserving carbon stock. Under Directive 2009/28/EC, two categories of forested lands are regarded as having high carbon stocks according to criteria in the extension, height, canopy cover, and the ability to reach these thresholds *in situ*.[88] The first category is the 'continuously forested area' with land spanning more than *one hectare*, trees higher than *five metres*, canopy cover of more than *30 per cent*, or trees able to reach those thresholds *in situ*.[89] The second category is the so-called 'sparsely forested land' which has the same extension, height, and ability conditions as the continuously forested land but has a canopy cover of between *10 per cent and 30 per cent*.[90] Moreover, it has been clarified by the European Commission that forested land does not include 'tree stands in agricultural production systems, such as fruit tree plantations, oil palm plantations and agroforestry systems when crops are grown under tree cover.'[91] An exception is made for biofuels which obtain raw materials from the sparsely forested land, if evidence is provided that the GHG savings calculated in line with the Directive still meet the GHG emission threshold after considering the land carbon stock changes.[92]

85 Directive 2009/28/EC, para. (73) of the preamble.

86 Ramsar Convention, Arts. 1.1 and 2.1.

87 Dehue *et al.* 2011, p. 66.

88 Extension refers to the extension of the forest area, instead of the extension of the total plantation if there is an area of forest within the plantation. In forestry, canopy cover is defined as 'the proportion of the forest floor covered by the vertical projection of the tree crowns, and should be distinguished from canopy closure, which is defined as the proportion of sky hemisphere obscured by vegetation when viewed from a single point.' Dehue *et al.* 2011, p. 85-86.

89 Directive 2009/28/EC, Art. 17(4)(b).

90 The difference between continuously forested and sparsely forested land is the canopy cover threshold. Dehue *et al.* 2011, p. 86.

91 It is set out that continuously forested land does not include land that is predominantly under agricultural or urban land use. Communication (2010/C 160/02), section 4.2.1.

92 The land carbon stock change calculation must be carried out in line with the GHG methodology in the RED Annex V(C) and in the Commission Decision of 10 June 2010 on guidelines for the calculation of land carbon stocks for the purpose of Annex V to Directive 2009/28/EC. Dehue *et al.* 2011, p. 86.

c) Peatland

Sustainable biofuels shall not be produced from raw materials obtained from land that was peatland in January 2008.[93] Directive 2009/28/EC does not define peatland.[94] Generally, peatland is an area with or without vegetation with a naturally accumulated peat layer at the surface,[95] and peat itself is not considered biomass.[96] An exception can be made for raw material from peatland when the evidence demonstrates that the cultivation and harvesting do not involve the drainage of previously undrained soil, even when the soil was partially drained.[97]

2.1.4. Agro-environment Protection

Apart from the above-mentioned criteria, the impact of the production of raw materials for biofuels on the local environment, namely the local air, water and soil, is of great concern. As the majority of biofuels consumed in the EU are first-generation biofuels based on agricultural crops,[98] their local environmental impact is an overarching issue with agricultural environment management and regulation. In this sense, biofuels made from raw materials shall comply with the EU environmental requirements on agricultural activities, such as those concerning the protection of groundwater and surface water quality.[99] Nevertheless, unlike the other sustainability requirements which are applied regardless of the origin of the raw materials,[100] this criterion only applies to the biofuels with raw materials produced within the EU.

The substantive requirements about agro-environment refer to the *a)* 'statutory management requirements' under the heading 'Environment' in part A and in point

93 Directive 2009/28/EC, Art. 17(5).

94 The European Commission is requested as appropriate to develop methodologies with a view to assessing the impact of the drainage of peatlands on GHG emissions. Directive 2009/28/EC, para. (72) of the preamble.

95 According to the International Peat Society, peat is a heterogeneous mixture of more or less decomposed plant (humus) material that has accumulated in a water-saturated environment and in the absence of oxygen. Dehue *et al.* 2011, p. 96; Joosten & Clarke 2002.

96 According to IPCC, 'peat is not considered a biofuel in these guidelines due to the length of time required for peat to re-accumulate after harvest.' Communication (2010/C 160/02), section 4.2.3; IPCC Guidelines 2006, p. G.3.

97 As for the exception, the Commission later explained that even when peatland was partially drained in January 2008, a subsequent deeper drainage would constitute a breach of the criterion if it affects soil that was not already fully drained. Communication (2010/C 160/02), section 4.2.3.

98 Feedstocks of biofuels consumed in the EU mainly come from first-generation biofuel raw materials: rapeseed, soybean, palm oil for biodiesel; and wheat, maize, sugar beet and sugar cane for bioethanol. Other feedstocks include barley, rye, triticale and wine for bioethanol; and sunflower, tallow and used cooking oil for biodiesel, but they constitute a less significant source in overall biofuels feedstock profile. Van de Staaij *et al.* 2012, p. 4.

99 Directive 2009/28/EC, para. (74) of the preamble, Art. 17(6).

100 The sustainability criteria are applied generally irrespective of whether the raw materials were cultivated inside or outside the territory of the EU. Directive 2009/28/EC, Art. 17(1).

9 of Annex II to Regulation (EC) No 73/2009;[101] and *b)* the minimum requirements for 'good agricultural and environmental conditions' defined pursuant to Article 6(1) of the Regulation. The 'statutory management requirements' under the heading of the environment include EU legislation respectively on conservation of wild bird,[102] natural habitats and wild flora and fauna,[103] groundwater,[104] surface water,[105] soil quality,[106] and plant protection products.[107] In contrast with the 'statutory management requirements' which are provided at the EU level, the 'good agricultural and environmental conditions' are defined by Member States at the national or regional level so as to take account of specific characteristics of the local environmental and farming conditions.[108] Nevertheless, Member States shall define the minimum good agricultural and environmental conditions 'on the basis of the framework established in Annex III' of Regulation (EC) No 73/2009 and 'shall not define minimum requirements which are not foreseen in that framework.'[109] The framework foresees minimum requirements in the following areas: soil issues (about soil erosion, organic matter, and structure), conservation issues (about maintenance and to avoid the deterioration of habitat), and water protection and management issues (about pollution, run-off, and managing the use of water).[110]

In this vein, the criterion on the agricultural environment requires *cross-compliance* with environmental standards prescribed in the EU's Common Agricultural Policy which primarily focuses on nature conservation, soil, and water management. However, this criterion does not apply to raw materials from non-EU countries for biofuels, since otherwise it would apply EU law to agricultural activities on territories outside of EU. In view of the jurisdiction limitation, concern about the impact of biofuels on the environment in certain third countries is expressed in the preamble of Directive 2009/28/EC,[111] which is expected to be addressed by developing multilateral and bilateral agreements, voluntary schemes,

101 In this provision, The EU environmental requirement for agriculture refers to common rules for direct support scheme for farmers under the common agricultural policy (CAP) laid down in Arts. 5(1)(b) and 6(1) of Regulation 73/2009. The 'statutory management requirements' refer to the designated EU law with regard to agriculture, covering areas of 'public, animal and plant health', 'environment', and 'animal welfare.' Art. 5(1)(b) is about the 'environment', while 5(1)(a) and (c) are respectively about health and animal welfare. Art. 17(6) RES Directive only requires compliance with the requirements and standards under the 'environment' title. Regulation (EC) No 73/2009.

102 Directive 79/409/EEC, Art. 3(1), Art. 3(2)(b), Art. 4(1), (2) and (4) and Art. 5(a), (b) and (d).

103 Directive 92/43/EEC, Arts. 6 and 13(1)(a).

104 Directive 80/68/EEC, Arts. 4 and 5.

105 Directive 91/676/EEC, Arts. 4 and 5.

106 Directive 86/278/EEC, Art. 3.

107 Directive 91/414/EEC, Art. 3.

108 Regulation (EC) No 73/2009, Art. 6(1).

109 *Ibid.*

110 Regulation (EC) No 73/2009, Annex III.

111 The concern is expressed that the production of biofuels and bioliquids in certain third countries might not respect minimum environmental or social requirements. Directive 2009/28/EC, para. (74) of the preamble.

or reporting by biofuel's economic operators.[112] Furthermore, the European Commission is required to report every two years on national measures for soil, water and air protection in EU and non-EU countries which are major suppliers of biofuels or biofuel feedstocks.[113]

2.1.5. The Other Sustainability Concerns

It is observed that the EU's current sustainability criteria concentrate on environmental impacts regarding GHG emissions, biodiversity and carbon stock land, and agricultural environment, while the socio-economic concerns, such as food availability and land use rights are not included. In fact, social criteria relating to, *inter alia*, 'land rights of local communities or the fair remuneration of all workers' had been recommended by the European Parliament's Industry Committee to be included in Directive 2009/28/EC, but such a recommendation was finally dropped due to remaining doubts about whether the social criteria would be implementable worldwide in line with WTO law.[114] Moreover, the EU's sustainability criteria as currently provided do not apply to the solid and gaseous biomass used for electricity, heating, and cooling, which may incur the possibility of producing biofuels from the solid and gaseous biomass which is not yet subject to any sustainability requirement at the EU level.[115] Moreover, Article 17(8) of Directive 2009/28/EC may limit the chance for Member States to adopt sustainability requirements to address the other sustainability concerns,[116] which will be discussed in detail in section 5.2 of this chapter. Nevertheless, Directive 2009/28/EC envisages how the other sustainability concerns which are not

112 It is stated that it is therefore appropriate to encourage the development of multilateral and bilateral agreements and voluntary international or national schemes that cover key environmental and social considerations, in order to promote the production of biofuels and bioliquids worldwide in a sustainable manner. In the absence of such agreements or schemes, Member States should require economic operators to report on those issues. Directive 2009/28/EC, para. (74) of the preamble.

113 Directive 2009/28/EC, Art. 17(7).

114 The European Parliament's Committee on Industry, Research and Energy took the lead in progressing the Commission's 2008 Proposal on Renewable Energy Directive and adopted a strong stance on sustainability criteria by adding a 'food policy constraint' – use of the land for production of biofuels should not be allowed to compete with the use of land for food production – and conditions relating to labour rights. However, after much debate about whether these conditions could be implemented in practice and whether they comply with WTO rules, they were eventually removed from the final text. Interview with European Commission Policy Officer confirms that there was political discussion in 2008 about the inclusion of social criteria when the Renewable Energy Directive was developed, but difficulty was found in developing social criteria in legal terms and implementing them in areas outside the EU. Afionis & Stringer 2012, p. 117; Swinbank 2009, p. 489; Lin 2012a, p. 52-53; Interview with Ignacio Vazquez Larruscain, 19 December 2013.

115 Romppanen 2012a, p. 134.

116 Whether Member States can adopt more stringent measures for the sustainability of biofuels will be discussed in section 5 of this Chapter. Directive 2009/28/EC, Art. 17(8).

included in the binding sustainability criteria can be addressed, primarily through the approaches of monitoring/reporting and voluntary schemes.

2.1.5.1. *Monitoring and Reporting*

To fill the gap in addressing the other sustainability concerns, the EU took the approach of obliging the European Commission and EU Member States to monitor and report. The European Commission has the obligation of reporting every two years to the European Parliament and the Council on national measures regarding biofuels in the EU and non-EU countries which provide the 'significant source' of biofuels.[117] The European Commission shall report on the impact of EU's biofuels policy on social sustainability in these countries, such as the impact on the availability of foodstuffs at affordable prices in particular for people living in developing countries, the respect of land use rights, and wider development issues.[118] The report shall include a statement of whether the concerned countries have ratified and implemented the Conventions on labour rights agreed under the International Labour Organisation, as well as the Cartagena Protocol on Biosafety and the Convention on International Trade in Endangered Species of Wild Fauna and Flora.[119] It is further required that the Commission shall monitor the origin of biofuels consumed in the EU and the impact of their production, including the impact of displacement on land use in the EU and third countries.[120] EU Member States also undertake the obligation to submit a report, every two years, to the European Commission on the promotion and use of renewable energy,[121] on the basis of which the European Commission shall analyse the effectiveness of national measures and report to the Parliament and Council.[122] Thus, for the sustainability concerns which are not included in the EU's sustainability criteria, the European Commission has the major responsibility to monitor and report on the state-of-affairs and impacts not only in the EU but also in third countries.

117 Directive 2009/28/EC, Art. 17(7).

118 Moreover, the Commission shall, if appropriate, propose corrective action, in particular if evidence shows that biofuel production has a significant impact on food prices. Directive 2009/28/EC, Art. 17(7).

119 The listed labour right conventions include: Convention concerning Forced or Compulsory Labour (No 29), Convention concerning Freedom of Association and Protection of the Right to Organise (No 87), Convention concerning the Application of the Principles of the Right to Organise and to Bargain Collectively (No 98), Convention concerning Equal Remuneration of Men and Women Workers for Work of Equal Value (No 100), Convention concerning the Abolition of Forced Labour (No 105), Convention concerning Discrimination in Respect of Employment and Occupation (No 111), Convention concerning Minimum Age for Admission to Employment (No 138), Convention concerning the Prohibition and Immediate Action for the Elimination of the Worst Forms of Child Labour (No 182). Directive 2009/28/EC, Art. 17(7).

120 Directive 2009/28/EC, Art. 23(1).

121 Member States may use the typical values provided in Part A and Part B of Annex V of Directive 2009/28/EC for estimating net greenhouse gas emission saving from the use of biofuels. Directive 2009/28/EC, Art. 22(1)(j), (k), and (2).

122 Directive 2009/28/EC, Art. 23(3) and (6).

Furthermore, as mandated by Directive 2009/28/EC,[123] the Commission in 2011 made recommendations to Member States on the development of sustainability schemes on solid and gaseous biomass used for electricity, heating, and cooling.[124] It is particularly recommended to those Member States who have already developed sustainability criteria for biomass to duly integrate these recommendations, and they must at least ensure that national sustainability schemes do not constitute a means of arbitrary discrimination or a disguised restriction on trade.[125] In 2013 the European Commission made an effort to propose a Directive on sustainability criteria for solid and gaseous biomass,[126] but this effort was withdrawn in the end.[127] In 2014, the European Parliament requested the Commission to propose sustainability criteria for solid and gaseous biomass in the Resolution on the 2030 climate and energy framework.[128] From 10 February to 10 May 2016, the European Commission consults stakeholders and citizens on an updated EU policy on sustainable bioenergy for the period 2020-2030 as part of the

123 Art. 17(9) provides that the Commission shall report on requirements for a sustainability scheme for energy uses of biomass, other than biofuels and bioliquids, by 31 December 2009. Directive 2009/28/EC, Art. 17(9).

124 It was explained by the Commission that the wide variety of biomass feedstocks make it difficult to put forward a harmonised scheme at that stage. Different feedstocks present different challenges to sustainable production, GHG performance or efficient energy conversion. It is also considered that the sustainability risks relating to domestic biomass production originating from wastes and agricultural and forestry residues, where no land use change occurs, are currently low. For these reasons, the Commission did not propose binding criteria at the EU level. Moreover, as the biomass sector is fragmented and there are numerous small-scale users of biomass, it is recommended that sustainability schemes apply only to larger energy producers of 1MW thermal or 1MW electrical capacity or above. Commission Report COM(2010) 11 final, p. 8 and 9.

125 Member States are invited to take into account the recommendations for developing their sustainability criteria and for reporting and monitoring. These recommendations aim to promote the sustainable production and use of biomass, a well-functioning internal market in biomass trade and to lift barriers to bio-energy development. Commission Report COM(2010) 11 final, p. 10.

126 A leaked Proposal from the European Commission for a Directive on sustainability criteria of biomass and bioethanol injected into natural gas network was once seen, but it was not made publicly available and the Commission has since then withdrawn this effort. SWITCHBOARD, Natural Resources Defense Council Staff Blog, Debbie Hammel's Blog, European Union 'Sustainability Criteria' of Biomass Energy Fail to Protect the Climate or Our Forests, 21 August 2013, available at: <http://switchboard.nrdc.org/blogs/dhammel/-european_unions_sustainability.html> (last accessed 21 April 2015).

127 Nevertheless, in January 2014, in the Commission's Communication 'A policy framework for climate and energy in the period from 2020 to 2030', it is stated that an improved biomass policy will be necessary to deliver robust and verifiable GHG savings which should encompass the sustainable use of land, the sustainable management of forests, and address indirect land use effects. Commission Staff Working Document SWD (2014) 259 final, p. 3.

128 European Parliament resolution of 5 February 2014 on a 2030 framework for climate and energy policies 2013/2135 (INI), available at: <www.europarl.europa.eu/sides/-getDoc.do?type=TA&reference=P7-TA-2014-0094&language=EN> (last accessed 21 April 2015); Commission Staff Working Document SWD (2014) 259 final, p. 3.

EU renewable energy package.[129] Until a legislative act is finally adopted, the Commission will continue to monitor and report on the sustainability of biomass for electricity, heating, and cooling in line with Directive 2009/28/EC.

However, monitoring and reporting after all constitute a much more flexible and soft approach and its effectiveness in addressing the sustainability concerns are questionable. According to the reporting obligation required in the Directive, the European Commission issued the renewable energy progress report in 2011,[130] 2013,[131] and 2015,[132] which included the assessment of socio-economic impacts and local water, soil and air impacts in EU and non-EU countries. In the 2013 report, the Commission largely refuted a direct link between EU biofuel demand and food price increases,[133] although it admitted that the EU's biofuels production had a modest price effect on global market.[134] Concerning labour rights, the Commission noticed that a violation of labour rights occurs in lower income countries, and the enforcement of international labour Conventions is questionable in these countries even if they ratify the Conventions, but it did not further address how to resolve the labour right violation risks with regard to biofuels.[135] Finally, the Commission merely came to the conclusion that 'despite risks, no major impacts are readily

129 European Commission, Energy, Preparation of a sustainable bioenergy policy for the period after 2020, available at: <https://ec.europa.eu/energy/en/consultations/preparation-sustainable-bioenergy-policy-period-after-2020> (last accessed 2 March 2016).

130 The European Commission issued a report in 2011 on 'Renewable Energy: Progressing towards the 2020 Target.' Commission Communication COM(2011) 31 final.

131 Commission Report COM(2013) 175 final; Commission Staff Working Document SWD (2013) 102.

132 Commission Report COM(2015) 293 final; Commission Staff Working Document SWD (2015) 117 final.

133 The Commission based its argument on the following reasons: *a)* price effects caused by EU biofuels demand on global agricultural market are low or modest, which have caused price effects of 1-2 per cent and 4 per cent on relevant global market; *b)* rising food prices may be caused by many other factors such as weather and rising global demand for meat and food; *c)* uncertainty exists in the link between global food prices and domestic prices; *d)* not all local populations are impacted in similar manner. Therefore, the Commission concluded that there was no direct link between biofuel production and local food price increases, or at least is not the main driver of food price increases. Commission Report COM(2013) 175 final, p. 12; Commission Staff Working Document SWD (2013) 102, p. 23-24.

134 The Commission argued that grain use for bioethanol production formed 3 per cent of total cereal use in 2010/2011 and was estimated to have a minor price effect (1-2 per cent) on the global market price; the biodiesel consumption price effect was greater (4 per cent), which is based on analysis commissioned by the European Commission conducted by ECOFYS 'Renewable Energy Progress and biofuel sustainability.' Commission Report COM(2013) 175 final, p. 12; Commission Staff Working Document SWD (2013) 102, p. 22-23.

135 The Commission noticed that weak implementation mostly occurs in lower income countries and found that most non-EU countries which provide feedstock for EU biofuels have ratified the conventions concerned, but enforcement of the conventions is still questionable and the EU should encourage the full application of those conventions. Meanwhile, the Commission mentioned the benefits of economic development that might be brought about by the biofuel economy: mechanised harvesting gradually reduced the incidence of child labour in Argentina and Brazil. Commission Report COM(2013) 175 final, p. 12; Commission Staff Working Document SWD (2013) 102, p. 26.

apparent and current measures are sufficient', so the plan is merely to continuously assess and monitor the impacts with a view to implementing corrective measures when needed.[136] In the 2015 progress report, it is evaluated that Directive 2009/28/EC is effective and sustainability certification has proven to be an effective means of achieving the sustainability objectives, but its implementation could be improved at Member States' level.[137]

2.1.5.2. *Voluntary Schemes*

Directive 2009/28/EC provides the implementing framework to set out rules on how to demonstrate compliance with the sustainability criteria,[138] which is further specified in two Communications from the European Commission.[139] Above all, it is the obligation of 'economic operators' to show compliance with the sustainability criteria.[140] Economic operators can choose to show compliance with the EU sustainability criteria by the 'national system', 'bilateral or multilateral agreements', or a 'voluntary scheme.' The way of showing compliance with 'national system' is dependent on the Member States' implementation of the EU sustainability criteria in national law,[141] so the 'national system' may vary among one another in the information to be submitted[142] and in auditing arrangement.[143] The way of

136 Commission Staff Working Document SWD (2013) 102, p. 33.

137 Commission Report COM(2015) 293 final, p. 14 and 16.

138 Directive 2009/28/EC, Art. 18.

139 Communication (2010/C 160/01); Communication (2010/C 160/02).

140 It is up to Member States to define the scope of the economic operators and Member States shall require economic operators to show that the sustainability criteria have been fulfilled. However, farmers who supply biofuels feedstock as primary agricultural products should be excluded from the scope of 'economic operators', because compliance with agricultural environment criterion is implemented not under Directive 2009/28/EC but under Art. 22 of Regulation (EC) No 73/2009. Directive 2009/28/EC, Art. 18(1); Commission Communication (2010/C 160/02), sections 2.1 and 2.2.

141 Economic operators can show compliance with the sustainability criteria by submitting data as required by the Member States' national measures to the relevant national authority of the relevant EU Member States. In the national systems, the Member States can request information on the sustainability of biofuels in particular with information on compliance with sustainability criteria, as well as 'appropriate and relevant' information on 'measures taken for soil, water and air protection', 'the restoration of degraded land', 'the avoidance of excessive water consumption in areas where water is scarce', and information concerning measures which take into account the issues referred to in the second subparagraph of Art. 17(7) of Directive 2009/28/EC. Directive 2009/28/EC, Art. 18(3); Communication (2010/C 160/02), section 2.1.

142 An investigation reveals that not all EU Member States have finalised the implementation of the biofuel sustainability criteria, and Member States differ in their establishment and reliance on national systems. According to the investigation, the established national systems might be based on an ex-post verification of actual data provided by economic operators or it may require cross-compliance with Common Agricultural Policy requirements which is possibly combined with compliance with national nature protection legislation or introduced land zoning based on the biofuel sustainability criteria. Peters *et al.* 2012, p. 9-10.

143 Member States shall take measures to ensure that economic operators submit reliable information for showing compliance with requirements in national measures. For this, →

complying with 'bilateral or multilateral agreements' is especially provided for biofuels from third countries, so that the imported biofuels can demonstrate compliance with the EU sustainability criteria by referring to the agreement concluded by the EU and recognised by the European Commission as applicable to biofuels.[144] However, as yet no such agreements have been concluded by the EU.

Compared to the afore-mentioned ways, a 'voluntary scheme' provides for a more flexible and innovative method to show compliance with the EU sustainability criteria, since it relies majorly on national or international certification schemes developed and monitored by non-public authorities.[145] A voluntary scheme does not have to cover all the sustainability criteria,[146] and more importantly it can include the other sustainability concerns which are not addressed by the sustainability criteria.[147] By using voluntary schemes, compliance with the EU sustainability criteria is certified by existing or customised schemes which are benchmarked and recognised by the European Commission.[148] Once recognised by the European Commission, the voluntary schemes can be used to show compliance

Member States shall require economic operators to arrange for an adequate standard of independent auditing of the information submitted. To arrange standard of auditing is important in implementing national systems, because for economic operators using voluntary schemes or bilateral/multilateral agreements independent auditing is already arranged when the Commission recognises the schemes, while economic operators need not be audited before they participate in the national systems. Directive 2009/28/EC, Art. 18(3); Communication (2010/C 160/02), section 2.1.

144 It is provided that if the EU has concluded agreements containing provisions relating to matters covered by the sustainability criteria (in Arts. 17(2) to (5)), the Commission may decide that those agreements demonstrate that biofuels produced from raw materials cultivated in the countries comply with the sustainability criteria in question. Moreover, due consideration shall be given to measures taken for the conservation of areas in critical situations, basic ecosystem services (such as watershed protection and erosion control), for soil, water and air protection, indirect land use changes, the restoration of degraded land, the avoidance of excessive water consumption in areas where water is scarce and to the issues referred to in the second subparagraph of Art. 17(7). Directive 2009/28/EC, para. (77) of the preamble and Art. 18(4).

145 The standards frequently result from collaboration among NGOs, businesses, government and intergovernmental organisations. Romppanen 2012b, p. 179; Lin 2012a, p. 50.

146 Voluntary schemes are not expected to cover the criterion related to agricultural and environmental requirements and standards for EU farmers. Communication (2010/C 160/01), section 2.

147 This could include the concerns such as the conservation of areas that provide, in critical situations, basic ecosystem services (such as watershed protection and erosion control), for soil, water and air protection, the restoration of degraded land, the avoidance of excessive water consumption in areas where water is scarce and on the issues referred to in the second subparagraph of Art. 17(7). Member States may not, however, use the inclusion of such other sustainability issues in a voluntary scheme as grounds for a refusal to take into account biofuels/bioliquids that are not covered by the scheme, if those biofuels/bioliquids meet the sustainability criteria laid down in the Directive. Communication (2010/C 160/01), section 2.

148 The Commission may decide that voluntary national or international schemes setting standards for the production of biofuels contain accurate data for the purposes of Art. 17(2) or demonstrate that consignments of biofuel comply with the sustainability criteria set out in Art. 17(3) to (5). Directive 2009/28/EC, Art. 18(4).

in all the EU Member States.[149] In this way, the process of recognising and benchmarking the certification schemes against the generic sustainability criteria of the EU is critically important.[150]

The European Commission has specified the assessment and recognition procedure for voluntary schemes.[151] The voluntary scheme for recognition can be developed by government or private organisations, regardless of whether another recognised scheme already covers the same type of feedstocks or original areas. The voluntary scheme is to be benchmarked against the sustainability criteria,[152] primarily according to whether 'accurate data' and 'adequate standards of reliability, transparency and independent auditing' are provided.[153] If the assessment indicates that the voluntary scheme at issue meets the EU's requirements, the Commission can initiate the process leading to the adoption of a Commission decision for recognition. As yet, the European Commission has recognised 19 voluntary schemes,[154] and a 'voluntary scheme' serves as the major way for some Member States to implement the EU sustainability criteria.[155]

149 The national system can also include the possibility of using voluntary schemes as a way of showing compliance. However, the schemes recognised by the European Commission can be used in all national systems, while the nationally recognised schemes could not since there is no mutual recognition about the national recognised certifications. Peters *et al.* 2012, p. 9.

150 The voluntary scheme contains two layers of 'principal-agent' relations: one layer is between the European Commission and the certification schemes; the other layer is between the certification schemes and biofuel economic operators. Accordingly, there are two assessment processes: the assessment performed by the Commission to recognise the voluntary schemes; and the assessment made by the private companies to certify biofuels. Lin 2012a; Romppanen 2012b, p. 179.

151 Communication (2010/C 160/01), section 2.1.

152 As the certification schemes do not have to cover all the sustainability criteria, they should indicate which sustainability criteria they are to cover. As for the information that should be submitted by the schemes, the Commission shall establish the list of appropriate and relevant information referred to in the same article, including 'appropriate and relevant' information on 'measures taken for soil, water and air protection', 'the restoration of degraded land', 'the avoidance of excessive water consumption in areas where water is scarce', and information concerning measures on issues referred to in second subparagraph of Art. 17(7). Communication (2010/C 160/01), Section 2.1; Directive 2009/28/EC, Art. 18(4).

153 This requires that economic operators prepare information and evidence related to the claims for auditing, and economic operators have been audited before participating in the scheme. The auditing is independent, transparent, reliable and fraud-resisting, and regular audits are made after being certified. There could be 'group auditing' in particular for small-holder farmers, producer organisations and cooperatives. The auditor must be external, independent, generically skillful and specifically skillful. The ISO (International Organization for Standardization) 19011 could be relied on as an example to show auditor's compliance with the requirements. The request to the Commission for recognition should demonstrate that audits will be properly planned, conducted and reported on. Communication (2010/C 160/01), section 2.1.

154 The Commission has recognised 19 voluntary schemes by issuing assessment reports and Commission Decisions. European Commission, Energy, Voluntary Schemes, available at: <http://ec.europa.eu/energy/en/topics/renewable-energy/biofuels/voluntary-schemes> (last accessed 20 April 2015).

155 According to an investigation, some Member States only allow the use of voluntary schemes (EC-recognised or Member State recognised) to demonstrate compliance with the RED →

By benchmarking and recognising voluntary schemes, the EU adopts an approach of 'meta-standard' to implement the sustainability criteria, which means that the European Commission (as the 'principal') does not directly verify the compliance of biofuels at stake with the criteria, but it relies on other public or private actors (as the 'agent') to do the verification and certification.[156] The 'meta-standard' approach has the characteristic that its compliance is achieved through the application of existing standards and certification schemes. Therefore, 'instead of requiring producers to be certified to the meta-standard directly, compliance with the meta-standard can be achieved through certification to existing standards which have proven to provide a sufficient guarantee that (most of) the criteria of the meta-standard can be complied with.'[157] This approach has the benefits of saving time and costs,[158] increasing consistency with WTO rules,[159] and consolidating existing standards.[160] In the EU's case, the meta-standard approach leaves flexibility to economic operators in choosing between certification schemes and deciding to use the schemes separately or in combination.[161] More importantly, as voluntary schemes may cover the other sustainability concerns that are not currently addressed by the EU's sustainability criteria, they may thus provide experiences[162] or even a 'back-door' entrance for such concerns to be integrated into the binding criteria.[163] Actually, an investigation has found that most of the voluntary schemes recognised by the European Commission incorporate social sustainability standards in one way or another.[164] Furthermore, the involvement of private actors in

requirements, either achieved as biofuels pass the duty point or prior to passing of the duty point. Peters *et al.* 2012, p. 9-10.

156 Lin 2011a, p. 34; Lin 2012a, p. 43; Endres 2010, p. 77; Romppanen 2012b, p. 179.

157 Dehue, Meyer & Hamelinck 2007, p. 4.

158 This is because this approach can avoid reinventing new standards or certification schemes. Lin 2012a, p. 58; Dehue, Meyer & Hamelinck 2007, p. 5.

159 The certifications are often developed by private parties and are voluntary instead of binding, so the use of voluntary schemes could reduce the risk of falling foul of WTO rules which in principle prohibit technical regulations used as trade barriers and discrimination against imported products. However, certification does not necessarily or completely avoid the incompatibility with WTO law, and the compatibility should depend on many factors in specific cases. Lin 2012a, p. 59; Echols 2009.

160 Proliferation of voluntary standards and certification schemes would lead to confusion, overlapping, increasing costs, and forum-shopping, and the meta-standard approach can help increase consistency and avoid duplicate reinventing. Lin 2012a, p. 59; Van Dam, Junginger & Faay 2010; Dehue, Meyer & Hamelinck 2007, p. 5-6; Scarlat & Dallemand 2011.

161 Different methods can be used to show compliance with different criteria. Communication (2010/C 160/02), section 2.1.

162 To a certain extent, the different approaches with regard to social standards used in voluntary schemes can provide experiences for future legislation. Interview with Ignacio Vazquez Larruscain, 19 December 2013.

163 Kaphengst, Wunder & Krista 2012, p. 20.

164 It is found that the recognised voluntary schemes vary from having no social sustainability standard at all to having strong social sustainability indicators. Most schemes require suppliers to comply with national labour law or international conventions on labour rights. Many certification schemes impose requirements on land rights, such as showing proof of legal ownership or the prior and informed consent of land holders. Some schemes such as the →

voluntary schemes has been noted for bringing about 'dynamics' and 'modern regulatory innovation.'[165]

On the other hand, there are concerns about the stringency and transparency of the European Commission's recognition process.[166] As for stringency, although 'adequate standards of reliability, transparency and independent auditing' is required, it is dubious whether the Commission is capable of ensuring credibility and legitimacy, given that the business actors are after all driven by market incentives[167] and neither Directive 2009/28/EC nor the European Commission provides adequate guidance on the benchmarking standards[168] and procedural requirements.[169] As certification schemes are largely market-oriented, the stringency of certifying biofuels will greatly depend on the market and there might be 'race to the bottom.'[170] Moreover, since the EU-recognised certification schemes should be accepted in all Member States, this will limit the possibility for Member States to ask for higher standards.[171] The issue of transparency in the process of recognition has been highlighted by the Case T-278/11 *ClientEarth and other v. European Commission*, where non-governmental organisations challenged the Commission's implied refusal of access to certain documents of the voluntary schemes seeking recognition from the Commission.[172] The findings of the Court largely centred on the questions

Roundtable of Sustainable Bioenergy (RSB) and International Sustainability and Carbon Certification (ISCC) have standards regarding food security impacts. Kaphengst, Wunder & Krista 2012, p. 21-24.

165 It has been argued that involving the private sector in the implementation of EU Directive adds certain dynamics and constitutes modern regulatory innovation, such as 'new governance' and 'co-regulation.' However, as explained in Chapter 1, the role of private actors is not the focus of this thesis. Romppanen 2013, p. 340, 341-342.

166 For example, Lin pointed out the fundamental structural problems of reliance on the principal-agent meta-standard delegation and the possibility of rent-seeking. As the process of benchmarking plays an important role in 'gate-keeping', it has been argued that more rigorous rules and guidance from the EU are needed for the emerging governance measures. Lin 2011a, p. 34; Lin 2012a, p. 44; Romppanen 2012b, p. 184-186.

167 The sustainability criteria implementation framework places great faith in independent auditors, but problems may emerge when the businesses which audit in national systems also audit in voluntary certification schemes, and it is for the schemes themselves to arrange auditing and check the auditor's performance. Although the verifiers are delegated agents for the public authority and bound by legal principles, they are business actors driven by market incentives and competing for the same consumers, and there is a risk of a 'race-to-the-bottom.' Scott 2011a, p. 832; Romppanen 2012b, p. 181; Lin 2011a, p. 40; Pelsy 2008, p. 119; Lewandowski & Faaij 2006, p. 98-102; Kaphengst, Wunder & Krista 2012, p. 24.

168 Romppanen 2012b, p. 181.

169 Romppanen 2014, p. 239.

170 It is worried that, as certifications usually become less attractive to economic operators seeking a low level of certification obligations, and the certifications without a social requirement can also be approved by the EU and thus be automatically accepted in all EU Member States, the attractiveness and competitiveness of certifications with strong social requirements will be low. Kaphengst, Wunder & Krista 2012, p. 24.

171 Kaphengst, Wunder & Krista 2012, p. 24-25.

172 In Case T-278/11, the applicants requested the Court to declare that the Commission had infringed EU law on access to information and to annul the Commission's implied refusal decision by continuous delay. Case T-278/11, paras. 35-36.

of when the Commission made the implied decision and how to calculate the time-limit for bringing an action. The Court did not directly declare the Commission's infringement on access to information, but it asked the Commission to bear its own costs and three quarters of the costs incurred by the applicants because of the serious delay in providing the reply.[173] This case may have reflected problems in transparency and access to information in the procedure of recognising voluntary schemes, which would jeopardise the implementation and effectiveness of the sustainability criteria due to deficiency in clarity and reliability.[174] Moreover, the reliance on voluntary schemes may incur some inherent problems such as fraud, proliferation, and 'forum' (certification scheme) shopping,[175] and has limit in assessing and verifying macro-level sustainability such as competition with food crop cultivation and indirect land use change effect.[176] Concerns have also been expressed with regard to the increased complexity of the whole meta-standard and certification system and its possible consequence of leaving small-scale farmers behind.[177]

Thus, it is observed that, as for the sustainability concerns which are not included in the binding criteria, the EU attempts to address the concerns by the monitoring and reporting mechanism and voluntary schemes. These approaches may have their respective strengths and shortcomings, and therefore they may not be able to completely fill the gap of the lacking corresponding sustainability criteria to address the concerns.

2.2. *The Other Regulatory Measures Addressing the Demand for Biofuels*

The sustainability criteria as provided by law may affect the demand for biofuels when they are set as a precondition for the purposes of counting towards the renewable energy target, financial support, and the other measures that incentivise

173 There were two other cases brought before the Court about biofuels: Case T-120/10 *ClientEarth and Others v. Commission* and Case T-449/10 *ClientEarth and Others v. Commission*. The three cases are similar in the essential content in relation to application and judgment: while Case T-278/11 is about access to documents relating to the voluntary certification schemes in the Commission's recognition process, the other two are respectively about access to documents on the modeling of the indirect impact of the production of biofuels and access to documents on global trade and environmental impact study reports. Case T-278/11, paras. 35-36, 47, 48-51; Peeters & Nóbrega 2014, p. 361-362.

174 Romppanen 2012b, p. 181; Romppanen 2014, p. 239.

175 Van Dam, Junginger & Faaij 2010; Scarlat & Dallemand 2011.

176 Large-scale biomass production may cause effects that can only be assessed well at the macro level and then they are primarily a responsibility of authorities. These are often effects that cannot be directly attributed to one company but are only visible on a national or regional scale. Dehue, Meyer & Hamelinck 2007, p. 7-8; Cramer Report 2007, p. IV and V.

177 It has been commented that the meta-standard approach would increase complexity for economic operators, consumers, as well as monitoring authorities, as the implementation schemes may vary from one system to another and contain standards of different content and quality. This kind of burden may exacerbate the 'playing field' between biofuel suppliers of small scale or from developing countries and those with better economic status. Romppanen 2013, p. 341; Lima 2009, p. 480-482; Lee *et al.* 2011.

demand. In this vein, it is also important to look into the regulatory measures which can exert major influence on the use of biofuels at the EU level, focusing on the question of how they may affect the demand for biofuels and integrate the sustainability requirement.

2.2.1. Consumption Targets for Renewable Energy

As part of the EU's 2020 climate and energy package,[178] the objective of raising the share of EU renewable energy consumption was largely translated into the legal act of adopting Directive 2009/28/EC on the promotion of the use of energy from renewable sources. The Directive establishes the 'mandatory national overall targets' for the share of renewable energy in Member States' primary energy consumption by 2020. Such an approach of target setting has the following three implications for the use of biofuels.

First of all, a separate target is particularly set for the *transport sector* where biofuel constitutes an important source of renewable energy. Before Directive 2009/28/EC was adopted, the EU's legislation on renewable energy was fragmented according to sectors, with legislation and targets respectively adopted on the themes of renewable electricity, biofuels and other renewable transport fuels, and heating and cooling.[179] Directive 2009/28/EC sets up an integrated legal framework to include all the forms of renewable energy in different sectors,[180] and establishes the overall target of 20 per cent for renewable energy in the total energy mix.[181] However, a sectoral target for renewable energy consumed in transport is maintained: renewable energy in transport shall amount to at least 10 per cent of the final consumption of energy in transport in 2020.[182] The target for the transport sector is especially set out largely because of the importance of the transport sector for GHG emission reduction and the comparatively low renewable energy penetration in this sector.[183] Although this target is not a biofuel consumption target

178 At the 2007 Spring European Council, EU Heads of State and Government decided to adopt an integrated approach to climate and energy policy in order to transform the EU into a low-emission and highly energy efficient economy and thus committed to the objectives of reducing GHG emissions by 20 per cent from 1990 levels, increasing the share of renewable energy to 20 per cent, and improving energy efficiency by 20 per cent, to be achieved by the year 2020. Kulovesi, Morgera & Muñoz 2011, p. 832; Presidency Conclusions (8-9 March 2007), p. 10-12 and 19-18.

179 They respectively refer to Directive 2001/77/EC, Directive 2003/30/EC, Directive 2002/91/EC, and Directive 2004/8/EC. Kulovesi, Morgera & Muñoz 2011, p. 874.

180 Kulovesi, Morgera & Muñoz 2011, p. 875.

181 The integrated targets rather than separate sectoral targets enable Member States to coordinate their approaches to a range of planning, certification and educational issues associated with the renewable energy sector, and they could also alleviate the Member States' burden related to reporting. Kulovesi, Morgera & Muñoz 2011, p. 875; Commission Proposal COM(2008) 19 final, p. 10.

182 Directive 2009/28/EC, Art. 3(4).

183 In the European Council Conclusions of 10 and 11 December 2009, it was mentioned that significant additional efforts are needed to curb and adapt to decrease high energy consumption in the transport sector. Compared to the other sectors, renewable energy use in →

as provided in the former Directive 2003/30/EC,[184] biofuels in effect constitute an important source of renewable energy in the transport sector.[185] In this sense, the 10 per cent sectoral target for transport can provide a great incentive for the demand of biofuels and meanwhile leaves Member States the liberty to decide which kind of renewable energy is to be promoted in the transport sector.

Secondly, the Directive provides *mandatory* renewable energy targets rather than indicative targets for Member States. The 20 per cent renewable energy target is binding at the EU level, and it has devolved into mandatory national targets, ranging from 10 per cent for Malta to 40 per cent for Latvia, with considerations of 'different starting points and potentials' and 'existing level' of renewable energy in each Member State.[186] For transportation sector, moreover, each Member State shall meet the 10 per cent renewable energy target in 2020,[187] which constitutes a major change from the former indicative target as a 'reference value.'[188] Actually, the change to a mandatory target was largely motivated by the deemed 'failure' of Member States to fulfil the former indicative targets by Member States.[189] To implement the binding targets, Member States shall submit National Renewable Energy Action Plans to the European Commission, in which Member States shall

the transport sector lags behind. The Commission set out the reasons for making a specific requirement in the transport sector as follows: *a)* the transport sector presents the most rapid increase in GHG emissions of all economic sectors, *b)* biofuels could tackle the oil dependence in transport which causes the insecurity of energy supply in EU, *c)* biofuels are currently more expensive to be produced than the other renewable energies, so they would hardly be developed without a specific requirement. Council Conclusion (11 December 2009); Commission Proposal COM(2008) 19 final, p. 8.

184 According to Directive 2003/30/EC, Member States should ensure that a minimum proportion of biofuels and other renewable fuels is placed on their markets and shall set national indicative targets of 2 per cent by the end of 5.75 per cent by the end of 2010. Directive 2003/30/EC, Art. 3(1).

185 In the transport sector, hydrogen, natural gas, and biofuels are seen as alternative transport fuels, among which biofuels are considered to be most promising in the short-term as their market maturity is above those of the other options. It is also stated by the Commission that 'national targets for the use of renewable energy in land transport are addressed entirely through the use of biofuels. It is assumed that this will remain the case in future.' Wiesenthal *et al.* 2009, p. 789; Commission Report COM(2006) 845 final, p. 4.

186 Directive 2009/28/EC, para. (15) of the preamble and Annex I.

187 Nevertheless, consideration and a certain exemption are given to the Member States having a large share of aviation in their gross final consumption of energy, such as Cyprus and Malta. Directive 2009/28/EC, para. (33) of the preamble, and Art. 3(4).

188 Before Directive 2009/28/EC, the targets set in Directive 2003/30/EC were indicative as a 'reference value.' Directive 2003/30/EC, Art. 3(1).

189 The indicative targets set in Directive 2003/30/EC failed to be achieved according to evaluation made by the Commission – the share of biofuels in the transport energy in Member States was estimated only 1 per cent on average. The Commission owed the reason to the under-development of the EU regulatory framework for biofuels in the transport sector and a lack of incentives for Member States to implement the targets. Based on the conclusion that 'the biofuels directive's target for 2010 is not likely to be achieved', the Commission proposed to impose a mandatory minimum target of 10 per cent renewable energy in transport petrol and diesel by 2020. Commission Communication COM(2006) 848 final, p. 5-7; Commission Proposal COM(2008) 19 final, p. 8.

determine the share of energy from renewables in different sectors by 2020 and the measures to be taken to achieve the national targets. The Commission is to evaluate the adequacy, completeness, and credibility of the national action plans and can require an amendment if the envisaged share of renewable energy cannot reach what is expected in the indicative trajectory.[190] Nevertheless, Member States may resort to the means of 'statistical transfer' to achieve their national target,[191] akin to emission trading schemes whereby one country transfers its overachieved part of renewable energy consumption to another country.[192] As yet, it is estimated by the European Commission that the EU and the majority of Member States are advancing well towards 2020 targets,[193] but the transport sector has shown slow progress compared to the other sectors.[194]

Thirdly, as afore-mentioned, only the biofuels in compliance with sustainability criteria can be counted as fulfilling the binding renewable energy targets. Applying the sustainability requirement to the achievement of binding renewable energy targets offers a primary drive for consuming 'sustainable' biofuels within the EU.[195] Furthermore, the amendment to Directive 2009/28/EC

190 In addition to submitting national action plans, Member States shall report to the Commission every two years on the progress in renewable energy use, with the major contents prescribed in Art. 22 of the Directive. Directive 2009/28/EC, Arts. 3(2) and 4(4); Kulovesi, Morgera & Muñoz 2011, p. 875-876.

191 It is noticed that a statistical transfer mechanism applies to transfer with requirement of Arts. 3(1) and (2) of Directive 2009/28/EC, which refer to the provision on the 20 per cent target and national overall targets. Directive 2009/28/EC, Art. 6(1).

192 Statistical transfer means that renewable energy which has been produced in one Member State can be *ex-post* (after production) and virtually counted towards the national target of another Member State, according to the agreement or arrangement made between the Member States. This can serve as an incentive for overachievement in relation to the mandatory targets provided for the Member States. Directive 2009/28/EC, Art. 6; Kulovesi, Morgera & Muñoz 2011, p. 876.

193 The Commission states that it continues to assess Member States' progress in implementation and legal measures are being taken in those cases where the transposition is incomplete. Cases are open and Reasoned Opinions have been sent to Austria, Bulgaria, Cyprus, Czech Republic, Finland, Hungary, Ireland, Latvia, Luxembourg, the Netherlands, Poland and Slovenia because of failure in transposition. Commission Report COM(2013) 175 final, p. 11 and 13; Commission Staff Accompanying Document SWD (2013) 102 final, p. 17; Commission Report COM(2015) 293 final, p. 3.

194 It is estimated that only nine Member States having reached or exceeded their 2011 national targets in transport renewable energy, while 18 have underperformed. Only ten Member States had growth rates from 2010 to 2011 that would be sufficient to reach 2020 targets if they were to be maintained. The identified barriers include RES strategy and legislation shortcomings, characterised for instance by a lack of long-term vision on behalf of decision makers as well as unclear national legislative frameworks. The European Commission attributes the main reason to the uncertainty caused by delay in finalisation of the indirect land-use change policy and insufficient progress in deployment of second generation biofuels. European Renewable Energy Council 2013, p. 6; Commission Report COM(2015) 293 final, p. 3.

195 It has been commented that sustainability criteria for biofuels is a salient feature of the Package in terms of internal environmental integration. It is estimated by EurObserv'ER that certified sustainable biofuel consumption within the EU is slightly higher and increased from →

imposes a 7 per cent limit on the share of biofuels (and bioliquids) produced from food crops, precisely cereal and other starch rich crops as well as sugars and oil crops, in measuring compliance with the 20 per cent and 10 per cent renewable energy targets.[196] Meanwhile, biofuels produced from certain non-food crop feedstocks could be considered twice or four times in meeting the targets,[197] and a non-binding sub-target of 0.5 per cent is established for the designated advanced biofuels.[198]

However, the approach of setting binding renewable energy targets for each Member State will not be continued for the period during 2020-2030. The European Commission proposed an overall renewable energy target of 27 per cent by 2030 that is binding on the EU,[199] which means that there will be no binding overall or sectoral renewable energy target on individual Member States.[200] The European Parliament expressed 'regrets' especially with regard to the 'lack of national targets for renewable energy.'[201] The European Council agreed on the binding 27 per cent

11.7 Mtoe in 2012 to 11.8 Mtoe in 2013, and its share in total biofuel consumption in the EU is surging and accounts for 86.5 per cent in 2013 up from 79.8 per cent in 2012. Kulovesi, Morgera & Muñoz 2011, p. 874 and 834; EurObserv'ER, Biofuels Barometer, July 2014, available at: <www.energies-renouvelables.org/observ-er/stat_baro/observ/baro222_en.-pdf> (last accessed 16 April 2015), p. 7.

196 It is provided that for the calculation of biofuels in the numerator, the share of energy from biofuels produced from cereal and other starch rich crops, sugars and oil crops shall be no more than the set limit of the final consumption of energy in transport in 2020. In an interview with the Policy Officer of the European Commission DG Climate in 2013, it was confirmed that the limit on food-based biofuels is meaningful for both the 10 per cent target and the 20 per cent overall renewable energy target for 2020, but would not affect Member States' freedom to arrange their own trajectory and Member States can use State aid to support conventional biofuels even beyond the limit. However, if the conventional biofuels beyond the limit cannot be used for meeting EU mandatory targets, Member States would lose motivation for supporting these biofuels. Commission Proposal COM(2012) 595 final, para. (10) and Art. 2(2); Directive (EU) 2015/1513, Arts. 2(2)(a) and 2(2)(b)(iv)(d); Interview with Ignacio Vazquez Larruscain, European Commission, 17 December 2013.

197 Commission Proposal COM(2012) 595 final, Art. 2(2) and Annex IX; Council Position (10 December 2014), p. 8-9; Parliament Recommendation for Second Reading (26 February 2015), p. 71-72; Directive (EU) 2015/1513, Art. 2(2)(b)(iv)(e).

198 European Parliament Press Release, Environment committee backs compromise on cleaner biofuels, 14 April 2015, available at: <www.europarl.europa.eu/news/en/news-room/-content/20150413IPR41649/html/Environment-committee-backs-compromise-on-cleaner-biofuels> (last accessed 22 April 2015).

199 The target of 27 per cent would be fulfilled through clear commitments decided by the Member States themselves, which should be guided by the need to deliver collectively the EU-level target, and build upon what each Member State should deliver in relation to their current targets for 2020. Commission Communication COM(2014) 15 final, section 2.2.

200 According to the Commission, rather than sectoral renewable energy target, the focus of policy development should be on improving the efficiency of the transport system, further development and deployment of electric vehicles, second and third generation biofuels and other alternative, sustainable fuels as part of a more holistic and integrated approach. Commission Communication COM(2014) 15 final, section 2.2.

201 European Parliament Resolution on a 2030 Framework for Climate and Energy Policies (5 February 2014) 2013/2135(INI); European Parliament Press Release, MEPs want binding 2030 goals for CO_2 emissions, renewables and energy efficiency, 5 February 2014, available at: →

renewable energy target at the EU level, but meanwhile it invited the European Commission to further examine instruments and measures for a comprehensive and technology-neutral approach for renewable energy in transport after 2020.[202] The climate and energy policy framework for 2030, as agreed by the European Council, is to be further translated into legislation, and it is anticipated that national decision makers will thus have less pressure to promote renewable energy[203] and Directive 2009/28/EC would be substantially revised.[204] Noticing that the European Council endorsed a binding EU target of at least a 40 per cent domestic reduction in greenhouse gas emissions by 2030 compared to 1990, the question will be to what extent the 40 per cent target on GHG emissions could influence the demand for biofuels and their sustainability after 2020.

In sum, it is observed that the EU has adopted the approach of setting binding national renewable energy targets for Member States which can trigger the demand for biofuels for 2020. Particularly, by connecting the sustainability criteria with the counting of biofuels towards the targets, the targets can drive the consumption of 'sustainable' biofuels as defined in EU law. However, as the target-setting approach changes, the pressure from meeting the renewable energy targets by biofuels would diminish, and it is uncertain how the new targets as for renewable energy and GHG emissions reduction can influence the demand of 'sustainable' biofuels.

<www.europarl.europa.eu/news/en/news-room/content/20140203IPR34510/html/MEPs-want-binding-2030-goals-for-CO2-emissions-renewables-and-energy-efficiency> (last accessed 21 April 2015).

202 The European Council calls for a rapid adoption of the Directive laying down calculation methods and reporting requirements pursuant to Directive 98/70/EC of the European Parliament and of the Council relating to the quality of petrol and diesel fuels. The Council also recalls that under existing legislation a Member State can opt to include the transport sector within the framework of the ETS. The Council also notes the multiple objectives of the agriculture and land use sector, as well as the need to ensure coherence between the EU's food security and climate change objectives. The European Council invites the Commission to examine the best means of encouraging the sustainable intensification of food production, and states that policy on how to include 'Land Use, Land Use Change and Forestry' into the 2030 greenhouse gas mitigation framework will be established as soon as technical conditions allow and in any case before 2020. European Council Conclusion on 2030 Climate and Energy Policy Framework, paras. 3, 2.13 and 2.14.

203 Peeters & Schomerus 2014, p. 283.

204 It is stated by the Commission that the Directive on renewable energy sources will need to be substantially revised for the period after 2020 to give the EU the means of ensuring that the 2030 EU level target is met. An improved biomass policy will also be necessary to maximise the resource efficient use of biomass in order to deliver robust and verifiable greenhouse gas savings and to allow for fair competition between the various uses of biomass resources in the construction sector, paper and pulp industries and biochemical and energy production. This should also encompass the sustainable use of land, the sustainable management of forests in line with the EU's forest strategy and address indirect land use effects as with biofuels. Commission Communication COM(2014) 15 final, section 2.2.

2.2.2. GHG Emissions Targets for Fuels and Vehicles

Since biofuels are used to substitute petrol or diesel in transport, the GHG emission targets for fuels or vehicles and how they may influence the demand for 'sustainable' biofuels can thus have an effect on the sustainability of biofuels. Directive 2009/30/EC on fuel quality and Regulation 443/2009 on carbon emission of passenger cars were adopted together with Directive 2009/28/EC as part of the 2020 climate and energy package, and in this sense they share the common objective of reducing GHG emissions from energy use.

2.2.2.1. GHG Emission Targets for Fuels

The EU has harmonised pollutant emission limits of vehicle fuels, such as petrol and diesel, by the Fuel Quality Directive 98/70/EC for the sake of health and air quality.[205] Directive 2009/30/EC added the objective of reducing GHG emissions from vehicle fuels[206] into the Fuel Quality Directive and established target and rules on life-cycle[207] GHG emissions to fulfil this objective.[208] Directive 2009/30/EC established the GHG emission reduction target: Member States shall require fuel suppliers[209] to reduce life-cycle GHG emission intensity[210] from the supplied fuel (including fossil fuel and biofuel) by 10 per cent in 2020.[211] A 6 per cent reduction of

205 Directive 98/70/EC sets out the minimum technical specifications as to sulfur content, oxygen content, and vapour pressure for the purpose of internal market functioning, in case of a disparity in Member States' measures in the area, and international competitiveness of the European vehicle and refining industries. The Directive has been subsequently amended by Commission Directive 2000/71/EC, Directive 2003/17/EC, and Regulation (EC) No 1882/2003. Directive 98/70/EC, para. (1) of the preamble.

206 The scope of the Directive includes fuels for road vehicles, and non-road mobile machinery (including inland waterway vessels when not at sea), agricultural and forestry tractors, and recreational craft when not at sea. Directive 2009/30/EC, Art. 1(1).

207 'Life cycle greenhouse gas emissions' in the Directive means all net emissions of CO_2, CH_4 and N_2O that can be assigned to the fuel (including any blended components) or energy supplied. Directive 2009/30/EC, Art. 1(2)(ii)(6).

208 It is stated that transport fuel use makes a significant contribution to the overall EU greenhouse gas emissions, and monitoring and reducing fuel life cycle greenhouse gas emissions can contribute to helping the Community meet its greenhouse gas reduction goals through the decarbonisation of transport fuel. Directive 2009/30/EC, para. (4) of the preamble, Art. 1(1).

209 'Fuel supplier' refers to the entity responsible for passing fuel or energy through an excise duty point or, if no excise is due, any other relevant entity designated by a Member State. Directive 2009/30/EC, Art. 1(2)(ii)(6).

210 GHG emission intensity means the 'greenhouse gas emissions per unit of energy': the total mass of CO_2 equivalent GHG emissions associated with the fuel or energy supplied, divided by the total energy content of the fuel or energy supplied (for fuel, expressed as its low heating value). Directive 2009/30/EC, Art. 1(2)(ii)(7).

211 The reduction is compared to the EU-average level of life-cycle greenhouse gas emissions per unit of energy from fossil fuels in 2010, which is to be set forth by the European Commission. Directive 2009/30/EC, para. (9) of the preamble.

GHG emissions is mandatory to meet the 10 per cent target,[212] while an additional 2 per cent reduction can be achieved by deploying GHG mitigation technology such as carbon capture and storage and electric cars,[213] and another 2 per cent reduction can be achieved by using offset credits under CDM.[214] To fulfil the targets, Member States shall designate the suppliers responsible for monitoring and reporting the life-cycle GHG emission intensity of the supplied fuels including biofuels.[215]

The GHG emission target for fuels used in vehicles has a bearing on the demand for biofuels, because biofuels can be used as a means to reduce GHG emissions. It is stated in Directive 2009/30/EC that the 6 per cent reduction target is to be achieved through *inter alia* the use of biofuels.[216] Furthermore, to assist the development of the biofuel market, bioethanol can derogate from the technical specification on vapour pressure limit.[217] However, whether biofuels can reduce GHG emissions largely depends on a calculation of its GHG emission saving following the same accounting method as provided in Article 19 of Directive 2009/28/EC.[218] Moreover, for biofuels used to meet the GHG emission reduction targets, they have to comply with the same sustainability criteria as provided in Directive 2009/28/EC[219] so as to ensure policy coherence.[220] Furthermore, as provided in Directive (EU) 2015/1513 which amends Directive 2009/30/EC, fuel suppliers shall report annually to the authoritydesignated by the Member States on the biofuel production pathways volumes of biofuels derived from cereals and other starch-rich crops, sugars, and oil crops, and the life cycle GHG emissions per

212 Member States may require suppliers, for this reduction, to comply with the following intermediate targets: 2 per cent by 31 December 2014 and 4 % by 31 December 2017. Directive 2009/30/EC, Art. 1(5)(2)(a).

213 Directive 2009/30/EC, Art. 1(5)(2)(b).

214 It could be achieved through the use of credits purchased through the Clean Development Mechanism of the Kyoto Protocol. Directive 2009/30/EC, Art. 1(5)(2)(c).

215 The suppliers shall at least provide the information on: *a)* the total volume of each type of fuel or energy supplied, indicating where purchased and its origin; and *b)* life-cycle greenhouse gas emissions per unit of energy. Member States shall ensure that reports are subject to verification. The European Commission shall set out the methodology for calculating life-cycle GHG emissions from fossil fuels, but until October 2014 the Commission issued a proposal on implementing measures for the GHG emission reduction and reporting requirements. Directive 2009/30/EC, Art. 1(5)(1) and (5); Commission Proposal COM(2014) 617 final; Carbon Market Watch 2014, p. 2.

216 Directive 2009/30/EC, para. (9) of the preamble.

217 Biofuels in general must meet the same technical specifications on environmental and health ground as set out for conventional fuels, but vapour pressure limit for bioethanol blend can be increased since bioethanol blending results in a non-linear change of the vapour pressure, if designated substantive and procedural conditions can be met. Directive 2009/30/EC, paras. (28) and (29) of the preamble, Art. 1(3)(a)(4)and (5), Annex.

218 Directive 2009/30/EC, Art. 1(5)(1) and (5).

219 It is provided that energy from biofuels shall be taken into account for the purposes of Art. 7a only if they fulfil the sustainability criteria set out in paras. 2 to 6 of this Art. Directive 2009/30/EC, Art. 1(6)(1).

220 Directive 2009/30/EC, para. (10) of the preamble.

unit of energy based on he prescribed 'indirect land use change factors', and Member States shall report those data to the European Commission.[221]

However, similar to the renewable energy target in the transport sector, the GHG emission reduction target for fuels will not be continued after 2020. The European Commission proposed to discontinue the approach because it 'does not think it appropriate to establish new targets for renewable energy or the greenhouse gas intensity of fuels used in the transport sector or any other sub-sector after 2020.'[222] The European Council agreed upon the proposal, but invited the European Commission to further examine instruments and measures for the promotion of emissions reduction in transport after 2020.[223]

2.2.2.2. *Carbon Emission Target for Vehicles*

The EU has adopted binding carbon emission targets for new passenger cars[224] and vans (light commercial vehicles)[225] in an attempt to improve fuel economy and reduce emissions. It is required that the new passenger cars registered in the EU do not emit more than an average of 130 grams of CO_2 per kilometre (g CO_2/km)[226] by 2015[227] and 95 g CO_2/km[228] by 2021.[229] For new vans, the target is to limit CO_2 emissions to an average of 175 g CO_2/km by 2017 and 147 g CO_2/km by 2020.[230]

221 Directive (EU) 2015/1513, Art. 1(2)(d).

222 Communication COM(2014) 15 final, section 2.2, p. 6.

223 European Council (23 and 24 October 2014) Conclusions on 2030 Climate and Energy Policy Framework, Section 2.13.

224 New passenger car refers to the motor vehicles of category M1 as defined in Annex II to Directive 2007/46/EC ('passenger cars') which are registered in the EU for the first time and have not been previously registered outside the EU. Regulation (EC) No 443/2009, Art. 2(1).

225 Van or light commercial vehicle refers to motor vehicles of category N1 as defined in Annex II to Directive 2007/46/EC with a reference mass not exceeding 2610 kg and to vehicles of category N1 to which type approval is extended in accordance with Art. 2(2) of Regulation (EC) No 715/2007 ('light commercial vehicles') which are registered in the Union for the first time and which have not previously been registered outside the Union ('new light commercial vehicles'). Regulation (EC) No 510/2011, Art. 2(1).

226 In terms of fuel consumption, this means a fuel consumption of around 4.1 l/100 km of petrol or 3.6 l/100 km of diesel.

227 The target of 130g/km was phased in between 2012 and 2015. From 2015 onwards, all newly registered cars must comply with the limit value curve. European Commission, Reducing CO_2 Emission from Passenger Cars, available at: <http://ec.europa.eu/clima/-policies/transport/vehicles/cars/index_en.htm> (last accessed 23 April 2015); Regulation (EC) No 443/2009, Art. 1.

228 This means a fuel consumption of around 4.1 l/100 km of petrol or 3.6 l/100 km of diesel.

229 This target is to be completely complied with by 2021, but it will be phased in, depending on the mass of vehicles, from 2020, which means a shorter phase-in period. European Commission, Reducing CO_2 Emission from Passenger Cars, available at: <http://ec.europa.eu/clima/policies/transport/vehicles/cars/index_en.htm> (last accessed 23 April 2015); Regulation (EU) No 333/2014, Art. 1.

230 The 2017 target is approximately equivalent to 6.6 l/100 km of diesel. The 2020 target equates approximately to 5.5 l/100 km of diesel. The target of 175 g CO2/km will be phased in between 2014 and 2017, and the 2020 target will be effective in 2020 with no delay. Regulation (EU) No 253/2014, Art. 1; Regulation (EC) No 510/2011, Art. 2(1); European Commission, →

The influence on biofuels primarily lies in the provision on 'specific emission target'[231] provided in the regulation. To determine the manufacturer's compliance with the specific emission target,[232] the specific carbon emission of each vehicle designed to be running on a mixture of petrol with *85 per cent bioethanol* shall be reduced by 5 per cent by 31 December 2015, in recognition of the greater technological and emission reduction capability when running on biofuels.[233] This provision offers favourable treatment for vehicles which are able to run on a high blend of bioethanol and thus may incentivise the demand for biofuels. However, this favourable treatment is to be applied with the conditions that: at least 30 per cent of the filling stations in the Member State where the vehicle is registered provide this type of bioethanol, and the bioethanol shall comply with the sustainability criteria for biofuels.[234]

2.2.3. Preferential Tax Treatment

According to Directive 2009/28/EC, Member States may use 'support scheme', which could include tax exemptions or reductions, to promote the consumption of renewable energy.[235] The most relevant taxation on the consumption of biofuels (as fuels for transports) should be the energy tax[236] – an excise duty levied on the quantity of energy products at the time of release for consumption.[237] For the functioning of internal market, the EU has taken legal measure to harmonise[238]

Climate Action, Reducing CO2 Emission from Vans, available at: <http://ec.europa.eu/-clima/policies/transport/vehicles/vans/index_en.htm> (last accessed 23 April 2015).

231 'Specific emissions target' means, in relation to a manufacturer, the average of the specific emissions of CO_2 permitted in accordance with Annex I in respect of each new passenger car, taking into account the derogation permitted in the Regulation. Regulation (EC) No 443/2009, Art. 3(1)(g); Regulation (EC) No 510/2011, Art. 3(1)(i).

232 It is provided that for the calendar year commencing 1 January 2012 (2014 for vans) and each subsequent calendar year, each manufacturer of vehicles shall ensure that its average specific emissions of CO_2 do not exceed its specific emissions target determined in accordance with Annex I or, where a manufacturer is granted a derogation, in accordance with that derogation. Regulation (EC) No 443/2009, Art. 4; Regulation (EC) No 510/2011, Art. 4.

233 Regulation (EC) No 443/2009, Art. 6; Regulation (EC) No 510/2011, Art. 6.

234 *Ibid.*

235 Directive 2009/28/EC, Art. 2(k).

236 The tax is charged and calculated on the quantity of energy products and electricity at the time of release for consumption. Directive 2003/96/EC, Art. 4(2).

237 It is provided in the EU that alcohol which is not intended for human consumption is exempt from excise duties on alcohol, including, for example, alcohol used for industrial purposes, to produce foodstuffs and medicines, or to manufacture cosmetics, biofuels, paints, anti-freeze and cleaning products. In particular, alcohol not intended for human consumption must be 'denatured', which means being rendered undrinkable with the addition of foul tasting or smelling agents and (preferably) a chemical marker. Directive 92/83/EEC, Art. 27; European Commission, Taxation and Customs Union, Alcohol Not For Human Consumption, available at: <http://ec.europa.eu/taxation_customs/taxation/excise_duties/alcoholic_beverages/-not_human/index_en.htm> (last accessed 19 May 2015).

238 This legislation is adopted on the legal basis of Art. 113 TFEU (ex Art. 93 TEC), which provides for the harmonisation of indirect taxes such as turnover taxes and excise duties if it →

energy tax by setting the *minimum*[239] level of taxation.[240] Meanwhile, the directive provides circumstances in which Member States can derogate from the minimum level of taxation by exempting or reducing the tax level, applying differential tax rates,[241] or refunding all or part of the levied tax.[242]

As provided in Directive 2003/96/EC, biofuels shall be taxed at a rate equivalent with motor fuels.[243] Meanwhile, the Directive allows Member States to apply an exemption or a reduced rate of taxation on renewable energy including biofuel or biofuel blend,[244] because '[I]t is desirable ... to allow Member States to exempt or reduce excise duties so as to promote biofuels.'[245] It is worth noting that the provision does not mandate Member States to give preferential tax treatment to

is necessary for the functioning of internal market through a special legislative procedure. Directive 2003/96/EC, preamble.

239 It is provided that 'the levels of taxation which Member States shall apply to the energy products and electricity listed in Art. 2 may not be less than the minimum levels of taxation prescribed by this Directive.' This means that Member States can apply higher taxation levels, but shall not apply lower taxation levels thanstipulated by EU. Directive 2003/96/EC, Art. 4(2).

240 In the EU energy tax law, 'level of taxation' is the total charge levied in respect of all indirect taxes (except VAT) calculated directly or indirectly on the quantity of energy products and electricity at the time of release for consumption. Directive 2003/96/EC, Art. 4(2); Charles *et al.* 2013, p. 44.

241 The level of taxation is different from the rate of taxation, since the former is the total charge levied in respect of all indirect taxes (except VAT) calculated directly or indirectly on the quantity of energy products. Member States may apply differentiated taxation rates for the same products provided they comply with the minimum levels of taxation and under fiscal control. Art. 5 sets out the conditions for Member States to apply differentiated rates of taxation under fiscal control and in line with the other EU law (such as Art. 30, Arts. 110 and 111 TFEU): directly linked to product quality; quantitative consumption levels; local public goods; and differentiation between business and non-business. Directive 2003/96/EC, Art. 5.

242 In addition to the provided reduction and exemptions, Member States wishing to introduce further exemptions or reductions shall inform the Commission accordingly and provide the Commission with the relevant and necessary information. The Council, acting unanimously on a proposal from the Commission, may authorise the Member States to introduce such measures. Directive 2003/96/EC, Arts. 6 and 19.

243 Art. 2(1) lists the taxable products of the Directive according to the Combined Nomenclature codes, which may not include certain biofuel products such as pure bioethanol. However, it is provided in Art. 2(2) that 'In addition to the taxable products listed in paragraph 1, any product intended for use, offered for sale or used as motor fuel, or as an additive or extender in motor fuels, shall be taxed at the rate for the equivalent motor fuel.' Directive 2003/96/EC, Art. 2 and Annex I.

244 Art. 16 sets out the circumstances in which Member States can apply exempted or reduced rates of taxation which include biofuels, because they include the circumstance where motor fuel is made up of or contains ethanol (CN code 2207 20 00) and includes biodiesel (FAME) (CN codes 3824 90 55 and 3824 90 80 to 3824 90 99). However, Art. 16 provides that the preferential taxation rates shall cease to apply once Member States are required by the EU to comply with legally binding obligations to place a minimum proportion of the products on their markets. Directive 2003/96/EC, Art. 16(1) and (6); Commission Implementing Regulation (EU) No 1101/2014.

245 Directive 2003/96/EC, para. (26) of the preamble.

biofuels,[246] and thus it is completely at the Member State's discretion in deciding whether to adopt or cease such preferential tax treatment.[247] However, as the energy tax directive was adopted prior to the publication of Directive 2009/28/EC, there is no mention of a sustainability requirement for the biofuels that may enjoy preferential tax treatment.[248] Nevertheless, a requirement on sustainability of the biofuel could be implied in the following ways.

Firstly, the sustainability criteria as provided in Directive 2009/28/EC should be applied to biofuels eligible for the energy tax exemption or reduction. It is provided in Article 17 of Directive 2009/28/EC that only biofuels meeting sustainability criteria shall be taken into account for financial support for the consumption of biofuels. An energy tax exemption or reduction amounts to 'government revenue foregone' and would help reduce the price of biofuels, in comparison with equivalent products, so as to overcome consumers' resistance to their use,[249] which thus constitutes a kind of 'financial support for the consumption of biofuels.' In fact, in 2011 the European Commission proposed[250] to reform the currently volume-based[251] energy tax into energy content- and CO_2 emissions-based tax.[252] According to the proposal, the tax based on CO_2 emissions from biofuels can

246 In contrast, Member States shall exempt taxation levels on fuels used in air navigation and sea navigation within EU waters. This article provides a mandatory exemption from the harmonised excise duty for energy products supplied for use as a fuel for the purpose of air navigation other than in private pleasure-flying. However, it allows Member States to tax aviation fuel for domestic flights and, by means of bilateral agreements, fuel used for intra-Community flights. Directive 2003/96/EC, Art. 14(1) and (2), and para. (23) of the preamble.

247 This issue has been clarified by a case in which a biodiesel producer challenged Germany's withdrawal of a tax preference for biodiesel. The Court held that the Member States could exclude certain types of biofuels from the tax exemption scheme. The general principles of legal certainty and legitimate expectations do not prevent the Member States from withdrawing a tax exemption scheme before its expiry date, and it is for national court to decide whether it is appropriate. Case C-201/08, para. 69.

248 The tax preference measure based on Art. 16 of the Energy Tax Directive may not have to inform the Commission, as it is provided that 'Member States shall inform the Commission of measures taken pursuant to Arts. 5, 14(2), 15 and 17.' However, Art. 16 provides that the preferential taxation rates shall cease to apply once Member States are required by the EU to comply with legally binding obligations to place a minimum proportion of the products on their markets. Member States shall communicate to the Commission the schedule of tax reductions or exemptions applied by 31 December 2004 and every 12 months thereafter. Directive 2003/96/EC, Arts. 26(1), 16(6) and (7).

249 Charles *et al*. 2013, p. 44.

250 Commission Proposal COM(2011) 169/3.

251 Directive 2003/96/EC bases taxation levels on energy volume rather than energy content of the fuels consumed. The volume-based taxation level has shortcomings of creating tax benefits for certain conventional fuels such as coal, while renewable energies cannot obtain a comparatively preferential tax level. European Commission, Press Release, 'Revision of the Energy Taxation Directive – Questions and Answers', MEMO/11/238, 13 April 2011, available at: <http://europa.eu/rapid/press-release_MEMO-11-238_en.htm> (last accessed 24 April 2015).

252 Energy tax would be composed of two parts: one part is based on CO_2 emissions from energy use and the other part is based on energy content. The taxation rates set out in the Proposal →

be exempted until 2023, but the exemption is conditional on compliance with the sustainability criteria.[253] However, the Proposal is not yet adopted largely due to the requirement of unanimity within the European Council.[254]

Secondly, the energy tax exemption or reduction may constitute State aid in EU law and in that case has to comply with the rules on State aid. It is explicitly stated in the energy tax directive that measures such as tax exemptions and reductions might constitute State aid and hence have to be notified to the European Commission pursuant to EU law, the obligation of which is not exempted by a notification on basis of the energy tax directive.[255] Articles 107 and 108 of the Treaty on the Functioning of the European Union (TFEU) provide the legal framework for defining and controlling State aid by Member States, according to which national preferential energy taxes may need to be examined by the European Commission to the extent of its consistency with the internal market.[256] Nevertheless, the preferential energy taxes for biofuels on environmental grounds could be exempted from going through the State aid control procedure, if it is able to meet the 'block exemption' laid down by the European Commission. However, according to the Commission's implementing measure on the 'block exemption',[257] only the State aid for biofuels that meet the sustainability criteria and are not produced from food crops can qualify for the 'block exemption' on environmental grounds.[258] Moreover,

would be combination of the two parts and are still the EU minimum requirements. Commission Proposal COM(2011) 169/3.

253 It is proposed that biofuels would be taxed for their energy content part but not for CO_2 emission part, and since biofuels generally have lower energy content, the exemption would provide more preferential treatment to biofuels. It is meanwhile clearly stated that this exemption should be conditional on compliance with sustainability criteria defined in Directive 2009/28/EC and would be phased out by 2023. The provision on incompatibility between obligations and tax reductions was also withdrawn in the Proposal. Commission Proposal, COM(2011) 169/3, p. 5, 6, 7, and 10.

254 It is reported that the Commission has long been seeking to change energy taxation, but some Member States have repeatedly thwarted any progress and are likely to continue to do so. EurActiv, 'EU to revive debate on minimum energy tax levels' (EurActiv.com, Reuters, 15 January 2013), available at: <www.euractiv.com/energy/eu-revive-debate-minimum-energy-news-517048> (last accessed 29 October 2014).

255 Directive 2003/96/EC, Art. 26(2).

256 According to Art. 107(1) TFEU, State aid refers to the aid directly or indirectly granted by Member States or through State resources, which distort or threaten to distort competition by favouring certain undertakings and affects trade between Member States. Art. 107(2) TFEU lists measures falling within the definition but shall or may be compatible with the internal market. State aid, except for aid under the *de minimis* threshold and within exemption conditions, is normally to be notified to the Commission and subject to the Commission or the Council decisions according to procedures provided in Art. 108 TFEU.

257 The Commission has adopted Commission Regulation (EU) No 651/2014 that declares certain categories of aid as compatible with the internal market in application of Arts. 107 and 108 TFEU.

258 It is stated in the preamble para. (63) of Commission Regulation (EU) No 651/2014 that 'aid to biofuels should only be covered by this Regulation in so far as it is granted for sustainable biofuels in accordance with the Directive 2009/28/EC', and moreover 'aid for food-based biofuels should be excluded from aid under this Regulation' so as to incentivise the shift →

if the national mandate on the blend of biofuels in oil is already provided,[259] further State aid for the biofuel, such as in the form of tax reduction according to Directive 2003/96/EC,[260] should be excluded from the scope of the 'block exemption', because the blend mandate already provides a sufficient incentive for biofuel investments.[261]

2.2.4. Emissions Trading Scheme

Inspired by the carbon trading rules incorporated into the Kyoto Protocol,[262] the EU has established an EU-wide emission trading scheme, known as EU ETS, which is recognised as the 'key tool' for cutting industrial GHG emissions in a cost-effective manner.[263] Directive 2009/29/EC brings the EU ETS into a third trading period (2013-2020)[264] which covers more sectors and gases.[265] Moreover, being part of the

towards more advanced forms of biofuels. Commission Regulation (EU) No 651/2014, para. (63) of the preamble.

259 The supply or blending obligation may refer to the 'renewable energy obligation' as defined in Art. 2(l) of Directive 2009/28/EC, including national support schemes requiring energy producers to include a given proportion of energy from renewable sources in their production, supply, or consumption, as well as the schemes under which such requirements may be fulfilled by using green certificates.

260 It is further specified that aid schemes in the form of energy tax reductions fulfilling the conditions of Council Directive 2003/96/EC shall be compatible with the internal market, but aid shall not be granted for biofuels which are subject to a supply or blending obligation. Furthermore, the beneficiary shall pay at least the minimum level of taxation set by the energy taxation directive and shall be selected on the basis of transparent and objective criteria and shall pay at least the respective minimum level of taxation set by Directive 2003/96/EC. In another Commission Guidelines on State Aid for Environmental Protection and Energy (2014-2020), it is provided that the Commission can apply a simplified approach to assess the necessity and proportionality of the aid, such as in the case of Directive 2003/96/EC. If the beneficiaries pay less than the Union minimum tax level set by the relevant applicable Directive, the aid will be assessed on the basis of the conditions for non-harmonised environmental taxes. Commission Regulation (EU) No 651/2014, Arts. 44(1), (2), and (4); (2014/C 200/01), paras. 172, 173, and 175.

261 Commission Regulation (EU) No 651/2014, para. (63) of the preamble; Charles *et al.* 2013, p. 44.

262 It is noted that the EU was critical of market mechanisms during the negotiation on Kyoto Protocol between 1995 and 1997, but, following the adoption of the Kyoto Protocol, the EU weighed the pros and cons of emissions trading and in 2000 implied that a Community-wide emissions trading scheme would be established. Kulovesi, Morgera & Muñoz 2011, p. 848-849.

263 European Commission, Climate Action, The 2020 climate and energy package, available at: <http://ec.europa.eu/clima/policies/package/> (last accessed 24 April 2015).

264 The first trading period of the ETS ran from 2005 to 2007 as a 'learning-by-doing' phase, aiming to focus on setting up the necessary institutions and procedures. The second trading period ran from 2008 to 2012, in parallel with the first commitment period under the Kyoto Protocol, concentrating on ensuring that the EU and its Member States comply with their Kyoto Protocol Commitments. Kulovesi, Morgera & Muñoz 2011, p. 850.

265 In the third period, a single EU-wide cap on emissions applies in place of the previous system of national caps, and auctioning, instead of free allocation, becomes the default method for allocating allowances. The trading period after 2020 will focus on addressing the growing surplus of allowances, by the proposed ways of postponing (or 'back-load') auctioning of →

2020 Climate and Energy package, the EU ETS has an interlinkage with the renewable energy legislation and may offer economic incentives for the demand for renewable energy, in the case where the consumption of renewable energy reduces GHG emissions and thus exempts or offsets the emission allowance to be surrendered.[266] In this vein, the EU ETS rules may affect the demand for biofuels as far as the transport sector is included in the emission trading system.[267]

2.2.4.1. *Aviation Emission Trading and Biofuels*

Aviation is to date the only transport covered in EU ETS,[268] as the emission trading system has been majorly dedicated to capping GHG emissions from power and heat generation and other energy-intensive industries. The inclusion may stimulate the consumption of biofuels in aviation, because it is provided in Directive 2008/101/EC that 'the emission factor for biomass shall be zero' in the monitoring and reporting of emissions from aviation activities.[269] By attributing an emission

some allowances, establishing a market stability reserve in 2021, and implementing faster reduction rate in the emission cap (from the current 1.74 per cent to 2.2 per cent per year from 2021). European Commission, Climate Action, The EU Emissions Trading System (EU ETS), available at: <http://ec.europa.eu/clima/policies/ets/index_en.htm> (last accessed 24 April 2015); see more on the third period of EU ETS in e.g. Kulovesi, Morgera & Muñoz 2011, p. 850-863; Peeters & Weishaar 2009; Faure & Peeters 2008; De Cendra de Larragán 2008; Peeters 2011.

266 As for the relationship between EU ETS and renewable energy policy, it has been commented that the ETS can partly enable the EU's confidence in establishing a relatively ambitious renewable energy target, or measures taken in non-EU ETS areas can 'shape' the stringency of renewable energy targets. Kulovesi, Morgera & Muñoz 2011, p. 844; Abrell & Weigt 2008, p. 6; Carlarne 2010, p. 171; De Cendra de Larragán 2012, p. 55.

267 However, whether ETS can in fact incentivise the use of biomass or biofuels largely depends on a comparison between the cost of renewable energy, efficiency technology, and the cost of buying allowances from carbon market. A study shows that the option of using liquid biofuels in transport may incur much higher costs than biomass for power and heating, liquid biofuels would not benefit from integrating aviation or other transport sector into EU ETS unless the carbon prices are high. Schwaiger *et al.* 2012, p. 104-105.

268 Aviation is included in the EU ETS since the beginning of 2012 through the adoption of Directive 2008/101/EC. All flights within EU, as well as between EEA and EFTA states, and international flights to and from the EU airports are generally included in EU ETS. But in November 2012, the European Commission proposed an initiative branded as 'Stopping the Clock', which exempts the international flights to and from EU airports from the EU ETS for a fixed period; meanwhile, the intra-EU (EEA and EFTA) flights are not exempted from their obligation under EU ETS. This exemption is supposed to be temporary for the period of 2013-2016 and would resume on the condition of the international negotiation under ICAO. The EU made this change following an agreement by the International Civil Aviation Organization (ICAO) Assembly in October 2013 to develop a global market-based mechanism addressing international aviation emissions by 2016 and apply it by 2020. European Commission, Climate Action, Reducing emissions from aviation, available at: <http://ec.europa.eu/clima/policies/transport/aviation/index_en.htm> (last accessed 25 April 2015); Directive 2008/101/EC; Decision 377/2013/EU; Regulation (EU) No 421/2014.

269 It is provided that default IPCC emission factors, taken from the 2006 IPCC Inventory Guidelines or subsequent updates of these Guidelines, shall be used unless activity-specific emission factors identified by independent accredited laboratories using accepted analytical →

factor of zero, such an arrangement in fact exempts the GHG emissions from the biomass-derived part of the aviation fuel from the obligation of surrendering allowances, because the emissions related to the surrendered allowances are calculated by multiplying the 'fuel consumption' by the 'emission factor.' The zero emission factor is also provided in Directive 2003/87/EC for the monitoring and reporting of emissions from stationary installations.[270] Such a zero emission assumption is in line with the method of calculating GHG emissions saving in Directive 2009/28/EC – emissions from the use of biofuels is taken to be zero.[271] The question is whether the biofuels exempted from surrendering allowances should meet any sustainability requirement.

The legislation on EU ETS does not provide a requirement about the sustainability of biofuels, but compliance with the EU biofuel sustainability criteria can be implied from the implementing measures on EU ETS monitoring and reporting. The European Commission has adopted implementing rules on the monitoring and reporting of GHG emissions for the EU ETS,[272] in which it refers to the provisions of Directive 2009/28/EC about biofuels. Firstly, the definitions of biofuel-related terms are consistent with those defined in Directive 2009/28/EC.[273] Secondly, the Commission states that the 'preferential treatment with regard to allowance surrender obligation' pursuant to the EU ETS allowance trading rules, referring hereby to the zero emission factor, constitutes a 'support scheme' and consequently 'financial support' under Directive 2009/28/EC.[274] It is recalled that, according to Article 17(1) of Directive 2009/28/EC, biofuels eligible for financial support shall comply with the sustainability criteria. Thus, in case the exemption from surrendering allowances constitutes financial support, the biofuels qualified for such preferential treatment must conform to the sustainability criteria. Thirdly, it is provided that, for determining the biomass fraction in the fuel mix,[275] the use of biofuels for aviation shall be assessed in accordance with Article 18 of Directive 2009/28/EC which sets out the verification rules for compliance with sustainability criteria. Finally, compliance with the sustainability criteria has been emphasised in other guidance documents about aviation emission trading. For example, in the guidelines regarding EU ETS monitoring and reporting for aircraft operators, the

methods are more accurate. The emission factor for biomass shall be zero. Directive 2008/101/EC, Annex (2)(b).

270 It is provided that accepted emission factors shall be used, and the emission factor for biomass shall be zero. Moreover, the installations 'exclusively using biomass' are excluded from the scope of EU ETS. For measuring 'small installations' which are excluded from EU ETS, emissions from biomass is not included in emission counting, according to Annex I(1) and (3) of Directive 2009/29 and Art. 27 of Directive 2003/87. Directive 2003/87/EC, Annex IV; Directive 2009/29/EC; Directive 2003/87/EC, Art. 1(28), and Annex I.

271 Directive 2009/28/EC, Annex V(C)(13).

272 Commission Regulation (EU) No 601/2012.

273 It is stated that the definition of biomass should be consistent with the definition of the terms 'biomass', 'bioliquids' and 'biofuels' set out in Art. 2 of Directive 2009/28/EC. Commission Regulation (EU) No 601/2012, para. (2) of the preamble, Art. 3(20)-(22).

274 Commission Regulation (EU) No 601/2012, para. (2) of the preamble.

275 Art. 39 of the Commission Regulation has provided ways of proving the biomass fraction in the fuel mix. Commission Regulation (EU) No 601/2012, Art. 53.

European Commission states that biofuels must meet the sustainability criteria in order to be calculated based on a zero emission factor.[276] In the initiative 'European Advanced Biofuels Flight Path' which launched a collaboration between the European Commission and the major European airlines and biofuel producers, it is emphasised that only sustainable biofuels will be produced and used for the initiative.[277] Nevertheless, whether the Commission's implementing measures and interpretations regarding the zero emission factor and the sustainability of biofuels have sufficient legitimacy needs to be further discussed,[278] and the question about the sustainability of biofuels produced in third countries would arise if the EU aviation emission trading finally includes international flights.

2.2.4.2. *CDM Regulation and Biofuels*

The EU ETS allows participants to use offset credits from the Kyoto Protocol's CDM to fulfil part of their emission reduction obligations: the credits of Certified Emission Reductions (CERs) generated from CDM projects in third countries are recognised as equivalent to the allowances under the EU ETS and thus can be used to show compliance with the GHG emission reduction obligation under the EU ETS.[279] CDM projects, as discussed already in Chapter 2, are deployed in non-Annex

276 Guidance Document, The Monitoring and Reporting Regulation General guidance for Aircraft Operators MRR Guidance document No. 2 Version of 16 July 2012 European Commission DG Climate Action (2012), sections 5.4.8 and 5.4.9, available at: <http://ec.europa.eu/clima/policies/ets/monitoring/docs/gd2_guidance_aircraft_en.pdf> (last accessed 25 April 2015).

277 In 2011, the European Commission services, in coordination with major European airlines and key European biofuel producers, launched an initiative to speed up the commercialisation of aviation biofuels in Europe. The initiative is a shared and voluntary commitment to achieve annual production of two million tonnes of *sustainably* produced biofuel for aviation by 2020. It is explicitly stated on the cover of the technical paper for the initiative that '[I]t is understood that in the context of this text the term "biofuel(s) use in aviation" categorically implies "sustainably produced biofuel(s)" according to the EU legislation.' '2 million tons per year: A performing biofuels supply chain for EU aviation', August 2013. Update, available at: <http://ec.europa.eu/energy/sites/ener/files/-20130911_a_performing_biofuels_supply_chain.pdf> (last accessed 25 April 2015); 'Launch of the European Advanced Biofuels Flightpath', available at: <http://ec.europa.eu/energy/-sites/ener/files/20110622_biofuels_flight_path_launch.pdf> (last accessed 25 April 2015).

278 A study argued that the interpretation of European Commission in the para. (2) of the Monitoring and Reporting Regulation suffers from several weaknesses and is likely to be altogether erroneous. Moreover, in case that the Commission directly introduces a provision whereby conformity with the sustainability criteria is required for the 'preferential treatment' of a zero emission factor, the Commission would risk exceeding its implementing powers, since such an interpretation would *de facto* alter the scope of application of the EU ETS. Ville Inkinen 2014, p. 65.

279 In 2004 the EU adopted a 'linking directive', Directive 2004/101/EC, to connect the CDM and JI under the Kyoto Protocol with the EU ETS. Credits from the CDM and JI are recognised as equivalent to EU emission allowances and could be traded within EU ETS. Directive 2004/101/EC, Art. 1; European Commission, Climate Action, International Carbon Market, availablet at: <http://ec.europa.eu/clima/policies/ets/linking/index_en.htm> (last accessed 25 April 2015).

I Parties of the UNFCCC and may include biofuel projects.[280] By recognising CDM CERs, EU ETS may drive the generation of offset credits from CDM projects including biofuel projects in developing countries.[281] The question is whether the EU, as a market of demand for CDM credits, applies a requirement on the sustainability of the biofuels that generate the offset credits traded under the EU ETS.

The CDM credits to be traded under the EU ETS must conform to the 'quantitative limit' and 'quality control.'[282] The 'quantitative limit' refers to the constraint on how many CERs that can be used for complying with the EU ETS obligations and whether unused CERs can be carried to the next phase.[283] 'Quality control' refers to the requirement about CERs from what types or location of CDM projects can be used for compliance with the EU ETS, in addition to the 'additionality' and 'sustainability' requirements under the Kyoto Protocol,[284] since it has been concerned that many CDM projects may lack environmental integrity or efficiency.[285] As for the location, only CERs from new CDM projects starting after January 2013 in the Least Developed Countries, in contrast with advanced developing countries such as China, can be used within the EU ETS from 2013 onward.[286] Regarding the type of project, credits are not accepted from nuclear energy projects, afforestation or reforestation activities, and projects involving the destruction of industrial gases.[287] Moreover, acceptance of credits from hydroelectric projects exceeding 20 MW of installed capacity is subject to the condition of compliance with international criteria and guidelines.[288] However, similar quality

280 As discussed in Chapter 2, Section 2.1, biomass energy projects take a share of more than 13 per cent of all registered CDM projects, but the number of biofuel projects is minimal. UNFCCC, CDM, Project Search, available at: <https://cdm.unfccc.int/Projects/-projsearch.html> (last accessed 1 April 2015).

281 The actual effect of EU ETS on biofuel CDM projects may need further data and studies.

282 Scott 2011a, p. 814-815.

283 For example, since 2013, CDM credits used under the third phase of the EU ETS are limited to a quantity of no more than 50 per cent of the EU emissions reduction made between 2008 and 2020, and unused entitlements from phase 2 (2008-2012) are transferred to phase 3 (2013-2020) while the exact amount per operator for phase 3 is determined in line with the methodology set out in the EU ETS Directive. Directive 2009/29/EC, Art. 1(13); Commission Regulation (EU) No 1123/2013; European Commission, Climate Action, International Carbon Market, available at: <http://ec.europa.eu/clima/policies/ets/linking/index_en.htm> (last accessed 25 April 2015).

284 Scott 2011a, p. 815.

285 It is stated that the EU wants to see JI and CDM further reformed in order to improve their environmental integrity and efficiency, such as through more use of standardised baselines and alternative ways of assessing additionality. European Commission, Climate Action, International Carbon Market, available at: <http://ec.europa.eu/clima/policies/ets/-linking/index_en.htm> (last accessed 25 April 2015); Ruthner *et al.* 2011; Amatayakul & Berndes 2012, p. 197.

286 Directive 2009/29/EC, Art. 1(13).

287 Directive 2004/101/EC, Art. 1(2), paras. (8) and (9) of the preamble; Commission Regulation (EU) No 550/2011, Art. 1.

288 It is provided that in the case of hydroelectric power production project activities with a generating capacity exceeding 20 MW, Member States shall, when approving such project →

requirements regarding the sustainability of biofuel projects have not been specifically set forth by the EU.

2.3. *Interim Conclusion*

In section 2, it was found that the EU has employed various regulatory measures that may have an effect on the demand for biofuels and the sustainability of biofuels. The sustainability criteria are set out in Directive 2009/28/EC (and Directive 2009/30/EC), including requirements for life-cycle GHG emission saving, biodiversity, carbon stock, and agro-environment protection. However, other sustainability concerns, particularly the socio-economic concerns, are not included in the binding criteria. Although the other sustainability concerns are addressed in the ways of monitoring/reporting and voluntary schemes, these approaches have respective strengths and shortcomings and cannot have the same binding effects as the mandatory criteria. Apart from the sustainability criteria, the EU has adopted other regulatory measures that may affect the demand for biofuels. The renewable energy targets, especially the 10 per cent target for transport renewable energy, serve as the major driver for the consumption of biofuels. The other measures, such as the GHG emission reduction target for fuel and vehicle, preferential energy taxes, aviation emission trading, and the acceptance of CDM credits, can incentivise the use and demand for biofuels to different extents. The question is whether the incentivising measures integrate sustainability requirements on biofuels. The analysis in this section reveals that compliance with sustainability criteria has been explicitly set out as a precondition for counting towards the renewable energy consumption and fuel GHG emission reduction targets. However, the sustainability requirement can be merely implied from the European Commission's implementing measures for the demand-related measures of preferential energy taxes and aviation emission trading. There is no sustainability requirement mentioned for the transfer of CDM credits of biofuel projects. As the binding national renewable energy target and fuel GHG emission reduction target will not be continued after 2020, the sustainability criteria will remain valid but the effect of sustainability criteria on the demand for biofuels will largely depend on how the future legislative framework revises or establishes demand-relevant regulatory measures and support schemes.[289]

activities, ensure that relevant international criteria and guidelines, including those contained in the World Commission on Dams November 2000 Report 'Dams and Development – A New Framework for Decision-Making', will be respected during the development of such project activities. At the request of Member States and market participants, the Commission has endeavoured to develop uniform guidelines on the application of Art. 11b(6) of the Directive. Directive 2004/101/EC, Art. 1(2); European Commission, Climate Action, International Carbon Market, available at: <http://ec.europa.eu/clima/policies/ets/-linking/index_en.htm> (last accessed 25 April 2015).

289 The author posed the question on the sustainability criteria after 2020 to the European Commission via email. In the reply email, the Commission answered that 'As described above the sustainability criteria for biofuels apply not only for biofuels that are counted →

3. EU Regulation on the Supply of Biofuels

Biofuels consumed in the EU are to a great extent produced from first-generation raw materials/feedstock – rapeseed, soybean, palm oil for biodiesel, and wheat, maize, sugar beet and sugar cane for bioethanol, among which rapeseed originating in the EU makes up the most important feedstock source for EU biodiesel.[290] While the second-generation biofuels with raw materials from waste, residues, non-food cellulosic materials (e.g. straws) and ligno-cellulosic materials (e.g. woody biomass) were estimated to comprise a minimal part (about 1.4 per cent) of the biofuels consumed in the EU, their contribution is considered to be twice that made by other biofuels.[291] The sustainability criteria in effect set forth requirements about the production of the raw materials for biofuels through influencing the demand-related measures (i.e. counting towards targets and eligibility for support schemes),[292] but how the raw materials for biofuels could be sustainably produced and managed ultimately hinges on compliance with the regulations for *inter alia* agriculture. As agricultural crops (e.g. rapeseed)[293] constitute the major source of biofuel feedstock in the EU, this section focuses primarily on the EU regulation of biofuel feedstocks from the EU's agricultural sector, with the question of how such regulation could guarantee sustainable production.

towards the renewable energy targets but also generally for all biofuels that are supported via support schemes. Therefore, they remain valid also after 2020 – unless a change in the legislative framework provides otherwise. The way the sustainability criteria for biofuels will remain relevant in the context of Fuel Quality Directive after 2020 will depend on the future legislative framework.' EUROPE DIRECT Contact Centre (9 February 2015).

290 Other feedstocks include barley, rye, triticale, and wine for bioethanol, and sunflower, tallow and used cooking oil for biodiesel, which constitute a less significant source in the overall biofuels feedstock profile. Van de Staaij *et al.* 2012, p. 4-5; Commission Staff Working Document SWD (2013) 102 final, p. 13.

291 According to Art. 21(2) of Directive 2009/28/EC, the contribution made by biofuels produced from wastes, residues, non-food cellulosic material, and ligno-cellulosic material shall be considered to be twice that made by other biofuels. In this sense, the actual share of the second-generation biofuels is only 0.06 per cent of all energy use in transport, and these fuels count as 0.11 per cent points (after double counting) of the 4.7 per cent of 2010 transport energy. However, it is noted by the Commission that the data may underestimate biofuels from waste cooking oil because of different national definitions and reporting methods used in the reporting to Eurostat. Commission Staff Working Document SWD (2013) 102 final, p. 12.

292 If breach of the CAP requirements is shown, the Member State will need to ensure that this is taken into account for the purpose of biofuel sustainability criteria. Communication (2010/C 160/02), section 2.2; Communication (2010/C 160/01), section 2.

293 Agricultural biofuel feedstocks include vegetable oils and oilseeds, cereals and sugar beet, as well as by-products and residues (such as straw) and dedicated energy crops (perennial grasses and short rotation forestry and coppice) which provide non-food cellulosic and ligno-cellulosic biomass. European Commission, Agriculture and Rural Development, Agriculture as supplier of biomass for energy, available at: <http://ec.europa.eu/agriculture/-bioenergy/supply/index_en.htm> (last accessed 26 April 2015).

3.1. *Regulation of Biofuels Feedstock from Agriculture*

The EU has established and implemented a Common Agriculture Policy (hereinafter as CAP) for all the Member States, which is managed and funded at the European level from the resources of the EU annual budget.[294] With the original objectives of improving agricultural productivity and farmers' living standards, the CAP has undergone a series of reforms to decouple subsidies and production and address more recent challenges of *inter alia* climate change, agricultural environment, and rural economy.[295] It is declared that the CAP helps agriculture to provide biomass for energy and encourages the use of bioenergy in rural areas.[296] Although the particular direct payment schemes for growing annual energy crops were abolished,[297] the production of agricultural crops[298] for energy use, including

[294] European Commission, Agriculture, The European Union Explained, The EU's Common Agricultural Policy (CAP): for our food, for our countryside, for our environment, A partnership between Europe and Farmers, p. 3, available at: <http://ec.europa.eu/-agriculture/cap-overview/2014_en.pdf> (last accessed 26 April 2015).

[295] The Common Agricultural Policy was born in 1962, with the objectives of providing affordable food for EU citizens and a fair standard of living for farmers. In 1992, the CAP shifted from market support to producer support, since price support was scaled down and was replaced with direct aid payments to farmers and was encouraged to be more environmentally friendly. In 2003, a new CAP reform cut the link between subsidies and production, and income support payment was conditioned on the fulfilment of environmental, animal welfare, and food safety standards. European Commission, Agriculture, The European Union Explained, The EU's Common Agricultural Policy (CAP): for our food, for our countryside, for our environment, A partnership between Europe and Farmers, p. 3 and 5, available at: <http://ec.europa.eu/agriculture/cap-overview/-2014_en.pdf> (last accessed 26 April 2015).

[296] European Commission, Agriculture and Rural Development, Agriculture and Bioenergy, available at: <http://ec.europa.eu/agriculture/bioenergy/index_en.htm> (last accessed 28 April 2015).

[297] Under Regulation (EC) No 1782/2003, a direct payment scheme was established to grant aid of EUR 45 per hectare per year for areas sown under energy crops which refer to crops supplied essentially for the production of biofuels and electric and thermal energy from biomass. Moreover, lands used to produce materials not directly intended for human or animal consumption, including for growing biofuel feedstocks, were eligible for the set-aside aid provided in the same Regulation. The reason for this abolishment was explained as 'due to recent developments in the bio-energy sector and, in particular, to the strong demand for such products on international markets and the introduction of binding targets for the share of bio-energy in total fuel by 2020.' The set-aside entitlement was abolished because 'market developments in the arable crops sector together with the introduction of decoupled aids no longer justify the need for maintaining this instrument.' Regulation (EC) No 1782/2003, Arts. 88 and 107; Regulation (EC) No 73/2009, paras. (30) and (42) of the preamble.

[298] The scope of agricultural products is provided in Annex I of EU Treaties as a list referred to Art. 38 of the Treaty on the Functioning of the European Union. Particularly, ethyl alcohol or neutral spirits, whether or not denatured, of any strength, obtained from agricultural products listed in Annex I, falls within the scope of Annex I. Consolidated Versions of the Treaty on European Union and the Treaty on the Functioning of the European Union (2012/C 326/01), Annex I; Regulation (EU) No 1307/2013, Art. 4(1)(d).

for use as biofuel feedstock, may still be able to be supported by direct payment[299] and rural development aid under CAP. However, such payments are conditional on compliance with the general requirements regarding *inter alia* environment and climate change provided under CAP. Agricultural requirements only apply to EU farmers and the verification of compliance with the requirements is provided in CAP legislation, such as the control system stipulated in Article 58-66 of Regulation (EU) No 1306/2013, which means that Member States can rely on their existing control systems to ensure that farmers fulfil these requirements.[300] The reform in 2013 for CAP 2014-2020 is declared as continuing 'along this reform path' to strengthen the competitiveness, promote sustainable farming and innovation, and support jobs and growth in rural areas.[301] In pursuit of sustainable agriculture, the CAP 2014-2020 continues the 'cross-compliance' mechanism and adds the 'greening' requirement which also applies to the agricultural activities for biofuels.

3.1.1. Cross-compliance

The requirement of 'cross-compliance' with agro-environment protection provisions under the CAP is set out in the sustainability criteria. Under the CAP, cross-compliance is a mechanism that links direct payments with farmers' compliance with the 'statutory management requirements' at the EU level and the 'good agricultural and environmental condition' at the national level,[302] which in effect creates synergies between CAP payments and compliance with environmental or

299 Although a specific direct payment scheme for energy crops was abolished, agricultural crops may still be eligible for direct payment if conditions provided in Regulation (EU) 1306/2013 and Regulation (EU) 1307/2013 are met. A study on EU subsidies for biofuel feedstocks revealed that producers of feedstock for biofuels within the EU are supported indirectly through the Single Payment Scheme of CAP. The scheme provides payments to farmers based on the land used and have been decoupled from production since 2003, and the areas cultivated for biofuel feedstocks are, like those for food crops, eligible for payments under the Single Payment Scheme. Charles *et al.* 2013, p. 41-43.

300 Communication (2010/C 160/02), section 2.2; Communication (2010/C 160/01), section 2.

301 Agricultural Policy Perspective Brief, Overview of CAP Reform 2014-2020, No. 5/December 2013, p. 2, available at: <http://ec.europa.eu/agriculture/policy-perspectives/policy-briefs/-05_en.pdf> (last accessed 26 April 2015); European Commission, Agriculture, The European Union Explained, The EU's Common Agricultural Policy (CAP): for our food, for our countryside, for our environment, A partnership between Europe and Farmers, p. 3 and 5, available at: <http://ec.europa.eu/agriculture/cap-overview/2014_en.pdf> (last accessed 26 April 2015).

302 Since 2005, all farmers receiving direct payments are subject to compulsory cross-compliance, which has been provided in the legislation of Council Regulation 73/2009 and Commission Regulation 1122/2009. It was provided in Regulation 73/2009 that '[A] farmer receiving direct payments shall respect the statutory management requirements listed in Annex II and the good agricultural and environmental condition referred to in Article 6.' European Commission, Agriculture and Rural Development, Cross-Compliance, available at: <http://ec.europa.eu/agriculture/envir/cross-compliance/index_en.htm> (last accessed 26 April 2015); Regulation 73/2009, Arts. 4-6; Commission Regualtion 1122/2009, Arts. 8-9.

the other sustainability standards.[303] The legislation for CAP 2014-2020 repeals Regulation 73/2009 as provided in Article 17(6) of Directive 2009/28/EC, but it maintains the basic framework of cross-compliance and makes some supplementary changes.

Regulation (EU) No 1306/2013 sets forth the rules on cross-compliance for CAP 2014-2020. Above all, the cross-compliance requirement applies to all the direct payments,[304] certain rural development payments,[305] and certain vine payments,[306] and non-compliance will result in administrative penalties being imposed, such as reduction or exclusion of the total amount of relevant payments, on the beneficiary farmers.[307] The cross-compliance requirement covers an updated and simplified range of the 'statutory management requirements' and the 'good agricultural and environmental conditions' in the areas of environment, climate change, land, health, and animal welfare.[308] For example, in the areas of environment, climate change, and land, the 'statutory management requirements' include the EU's legislation concerning water pollution by nitrates,[309] wild bird conservation,[310] and the

303 Cross-compliance represents the 'baseline' or 'reference level' for agricultural environment measures. For all requirements falling under cross-compliance, the costs have to be borne by farmers due to the 'polluter-pays-principle', and the voluntary activities going beyond the 'baseline' will be rewarded according to the 'provider-gets-principle.' European Commission, Agriculture and Rural Development, Cross-Compliance, available at: <http://ec.europa.eu/-agriculture/envir/cross-compliance/index_en.htm> (last accessed 26 April 2015).

304 The direct payments refer to payments granted directly to farmers provided in Regulation (EU) No 1307/2013, including *inter alia* the Basic Payment Scheme, Single area payment scheme, Payment for agricultural practices beneficial for the climate and the environment. Regulation (EU) No 1306/2013, Art. 92; Regulation (EU) No 1307/2013, Art. 1 and Annex I.

305 For the programming period 2007-2013, cross-compliance began to apply to certain environmental payments forming part of Rural Development policy which covers the voluntary commitments of farmers going beyond the relevant mandatory standards imposed on them. According to Regulation (EU) No 1306/2013, cross-compliance applies to the annual premia under certain provisions of Regulation (EU) No 1305/2013, including payments for afforestation and the creation of woodlands, the establishment of agroforestry systems, agri-environment-climate, organic farming, Natura 2000 and Water Framework Directive payments, areas facing natural constraints, animal welfare, and forest-environmental and climate services and forest conservation. Regulation (EU) No 1306/2013, Art. 92; Regulation (EU) No 1305/2013, Art. 21(1)(a) and (b), Arts. 28-31, 33, 34.

306 They refer to payments for the restructuring and conversion of vineyards and green harvesting under Arts. 46 and 47 of Regulation (EU) No 1308/2013. Regulation (EU) No 1306/2013, Art. 92; Regulation (EU) No 1308/2013, Arts. 46 and 47.

307 The administrative penalty shall only apply where the non-compliance is the result of an act or omission directly attributable to the beneficiary concerned. Member States are mainly responsible for using their existing administration and control systems to ensure compliance with the rules, including administrative checks and on-the-spot checks. The Commission will conduct verifications on compliance with the rules. Regulation (EU) No 1306/2013, Arts. 91, 96, 97, 99.

308 Regulation (EU) No 1306/2013, Art. 93; European Commission, MEMO 13/937, 'CAP Reform – an explanation of the main elements', 25 October 2013, available at: <http://europa.eu/-rapid/press-release_MEMO-13-937_en.htm> (last accessed 27 April 2015); Regulation (EU) No 1306/2013, Art. 93 and Annex II.

309 Directive 91/676/EEC, Arts. 4 and 5.

310 Directive 2009/147/EC, Art. 3(1), Art. 3(2)(b), Art. 4(1), (2) and (4).

conservation of natural habitats and of wild flora and fauna.[311] Compared to the 'statutory management requirements' mentioned earlier,[312] the legislation regarding groundwater protection[313] and the use of sewage sludge[314] is excluded from the list, since these rules do not provide 'clear and controllable obligations for farmers.'[315] The proposal to incorporate the legislation on the water framework[316] and the sustainable use of pesticides[317] into the 'statutory management requirements' was finally overturned,[318] and the deal is that such legislation will be incorporated once it is properly applied in all Member States and provides clearly identified obligations for farmers.[319]

As for the 'good agricultural and environmental conditions', it is supplemented in Regulation (EU) No 1306/2013 that Member States shall ensure that all agricultural areas, including lands which are *no longer* used for production purposes, are maintained in good agricultural and environmental conditions. In addition, Member States shall define minimum standards on the main issues within areas designated by the regulation, taking account of environmental and farming diversities.[320] The list of the minimum standards for the main issues is also simplified and streamlined in the areas of the environment and climate change. For example, on the issue of soil and carbon stock maintenance, Member States shall define the minimum standards for soil cover, land management to limit erosion, and the maintenance of the soil organic matter level.[321]

311 Directive 92/43/EEC.

312 Regulation (EC) No 73/2009, Annex II Part A Point 9.

313 According to Directive 2000/60/EC, Directive 80/68/EEC was repealed on 23 December 2013. Regulation (EU) No 1306/2013, para. (56) of the preamble.

314 Directive 86/278/EEC.

315 It is also stated in Regulation (EU) No 1306/2013 that experience has shown that a number of requirements within the scope of cross-compliance are not sufficiently relevant to the farming activity or the area of the holding or concern national authorities rather than beneficiaries. European Commission, MEMO 13/937, p. 8, 'CAP Reform – an explanation of the main elements', 25 October 2013, available at: <http://europa.eu/rapid/press-release_MEMO-13-937_en.htm> (last accessed 27 April 2015); Regulation (EU) No 1306/2013, para. (54) of the preamble.

316 Directive 2000/60/EC.

317 Directive 2009/128/EC.

318 The proposed extension of cross-compliance to comply with the water and pesticides directive has been overturned by the Parliament. 'Commission Urges MEPs to Reverse Changes to CAP' (ENDS Europe, 28 February 2013), available at: <http://-ends.simplicityweb.com/30832/commission-urges-meps-to-reverse-changes-to-cap> (last accessed 27 April 2015).

319 Regulation (EU) No 1306/2013, Joint statement by the European Parliament and the Council on cross-compliance, para. (11) of the preamble.

320 Member States still shall not define minimum requirements which are not established in Annex II. Regulation (EU) No 1306/2013, Art. 94.

321 In comparison, Regulation (EC) No 73/2009 provided soil matters including soil erosion, soil organic matter, soil structure through machinery use, permanent pasture, and the avoidance of encroachment on unwanted vegetation. The issues of landscape and permanent pastures are addressed separately in Regulation (EU) No 1306/2013. Regulation (EC) No 73/2009, Annex III; Regulation (EU) No 1306/2013, Annex II, Art. 93 (3)-(5).

Thus, in accordance with the changes, to obtain a direct payment, cultivating agricultural crops for the use as biofuel feedstock needs to comply with the EU law on water pollution by nitrates, wild bird conservation, and the conservation of natural habitats and of wild flora and fauna, as well as the national standards on *inter alia* soil and carbon stock maintenance.

3.1.2. Greening Requirement

The 'greening requirement' is for the first time established under the CAP in Regulation (EU) No 1307/2013. According to the requirement, Member States shall use as much as 30 per cent of their national ceilings,[322] in addition to the Basic Payment Scheme,[323] for the holdings (per hectare)[324] declared for the purpose of respecting 'greening' agricultural practices beneficial to the environment and climate change mitigation. The greening agricultural practices take the form of simple, generalised, non-contractual, and annual actions that *go beyond* the mandatory requirements in cross-compliance, including practices that meet the three requirements addressing crop diversification, permanent grassland, and ecological focus areas,[325] as well as the identified equivalent practices.[326] Among the three requirements, the maintenance of permanent grassland[327] and the

322 Under the CAP, the national ceiling refers to the total value of all allocated payment entitlements of the national reserve or the regional reserves each year for a Member State. Regulation (EU) No 1307/2013, Art. 6; European Commission, MEMO 13/937, 'CAP Reform - an explanation of the main elements', 25 October 2013, p. 3, available at: <http://europa.eu/rapid/press-release_MEMO-13-937_en.htm> (last accessed 27 April 2015).

323 Under CAP 2014-2020, Member States will dedicate up to 70 per cent of their Direct Payments national envelope to the new Basic Payment Scheme - minus any amounts committed for additional payments (Young Farmer top-ups, and other options such as Less Favoured Area top-ups, the Redistributive Payment) and 'coupled' payments. 'CAP Reform - an explanation of the main elements', 25 October 2013, p. 1, available at: <http://europa.eu/rapid/press-release_MEMO-13-937_en.htm> (last accessed 27 April 2015); Regulation (EU) No 1307/2013, Arts. 21, 22.

324 The term 'holding' means all the units used for agricultural activities and managed by a farmer situated within the territory of the same Member State. Regulation (EU) No 1307/2013, Art. 4(1)(b).

325 Regulation (EU) No 1307/2013, para. (37) of the preamble.

326 The equivalent practices shall be those which include similar practices that yield an equivalent or higher level of benefit for the climate and the environment compared to one or several of the practices. The equivalent practices shall be the practices listed in Annex IX of Regulation (EU) No 1307/2013 and covered either by Rural Development support for agricultural practices that make a positive contribution to the environment and climate, or by national environmental certification schemes going beyond the relevant mandatory standards. Regulation (EU) No 1307/2013, Art. 43(3).

327 'Permanent grassland' defined in Regulation (EU) No 1307/2013 includes both permanent grassland and permanent pasture and refers to land used to grow grasses or other herbaceous forage naturally (self-seeded) or through cultivation (sown) and that has not been included in the crop rotation of the holding for five years or more. It may include other species such as shrubs and/or trees which can be grazed, provided that the grasses and other herbaceous forage remain predominant as well as land which can be grazed and forms part →

establishment of ecological focus areas are particularly noteworthy for their relevance for biodiversity and carbon conservation.

It is provided that, for the environmental and carbon sequestration benefits of grasslands, Member States shall designate permanent grasslands *within* 'Natura 2000' sites including peatland and wetlands located therein and ban conversion or ploughing in areas and may designate further permanent grasslands *outside* of the 'Natura 2000' sites and ban, where applicable, conversion and ploughing activities.[328] Furthermore, Member States shall establish a reference ratio of permanent grassland to the total agricultural area and ensure that the ratio is roughly maintained.[329] If the ratio has decreased by more than 5 per cent at a national or sub-national level, Member States shall impose obligations on farmers who have land converted from permanent grassland to reconvert such land into permanent grassland or pasture,[330] and this provision also applies when the decrease is caused by *growing trees for energy production*. In this sense, once identified by Member States as permanent grassland, such land will be conserved and protected from conversion and ploughing activities for biofuel feedstock and, on a general level, the land for growing feedstock for biofuel production cannot encroach on the overall size of land conserved as permanent grassland. Meanwhile, it is recalled that the sustainability criteria in Directive 2009/28/EC contain a criterion regarding 'highly biodiverse grassland' which includes both natural and non-natural grassland and biofuel raw material production can be allowed if evidence is provided that the harvesting of the raw material is necessary to preserve the status of the grassland.[331] Thus, more specification about the synergy and difference between the 'permanent grassland' and 'highly biodiverse grassland' is called for.

of established local practices where grasses and other herbaceous forage are traditionally not predominant in grazing areas. Regulation (EU) No 1307/2013, Art. 4(1)(h).

328 In order to assure an efficient protection of permanent grassland, the Commission is empowered to adopt certain acts for the purpose of defining the framework for the designation by Member States of permanent grasslands not covered by the 'Natura 2000', namely the areas covered under Directive 92/43/EEC and 2009/147/EC. Regulation (EU) No 1307/2013, Art. 45(1) and (6), para. (42) of the preamble.

329 Member States shall ensure that the ratio of areas of permanent grassland to the total agricultural area declared by the farmers does not decrease by more than 5 per cent compared to a reference ratio to be established by Member States in 2015 by dividing areas of permanent grassland by the total agricultural area as defined in the EU legislation. The ratio of permanent grassland shall be established each year on the basis of the areas declared by the farmers subject to the obligations for that year. Member States may decide to apply obligation of maintaining the ratio at which level, and shall notify the European Commission their reference ratio and decisions relating to the ratio. Regulation (EU) No 1307/2013, Art. 45(2).

330 This provision shall not apply where the decrease is the result of afforestation that is compatible with the environment, excluding afforestation by plantations of short rotation coppice, Christmas trees or fast growing trees for energy production. Regulation (EU) No 1307/2013, Art. 45(3) and (4).

331 Directive 2009/28/EC, Art. 17(3)(c)(ii); Commission Regulation (EU) No 1307/2014, Art. 3.

The other greening requirement is about ecological focus areas established in particular to safeguard and improve biodiversity on farms.[332] It is provided that where the arable land of a holding[333] covers more than 15 hectares, the farmer shall ensure that at least 5 per cent of the arable land declared for direct payment is an ecological focus area[334] if the area falls within the categories of field margins, hedges, trees, fallow land, landscape features, biotopes, buffer strips, afforested area.[335] In this vein, the farmland for cultivating biofuel feedstock has to comply with the biodiversity requirement if it meets all the conditions.[336]

Thus, to obtain a payment under CAP, other than cross-compliance, agricultural activities that produce biofuel feedstocks have to comply with the greening requirements which exclude using permanent grasslands and ecological focus areas. However, there is not enough data with regard to the amount of EU budget used for supporting agricultural crops dedicated to biofuel production,[337] which may bring difficulties in estimating to what extent the agricultural support affects land use and food availability within or outside of the EU, although the rural development payment[338] under CAP emphasises the use of *non-food* raw materials

332 Regulation (EU) No 1307/2013, para. (44) of the preamble.

333 An ecological focus area is generally located on the arable land or the holding that the farmer declared in accordance with point (a) of the first subparagraph of Art. 72(1) of Regulation (EU) No 1306/2013 or may be adjacent to the arable land of the holding. However, an exception may take place in the cases of areas with short rotation coppice with no use of mineral fertiliser and/or plant protection products, or afforested areas. Regulation (EU) No 1307/2013, Art. 46(2).

334 This figure may rise to 7 per cent after a Commission report in 2017 and subject to a legislative proposal. Regulation (EU) No 1307/2013, Art. 46(1).

335 Member States shall decide that one or more of the enlisted categories of land are to be considered to be ecological focus areas, and only if the land falls within the categories decided by Member States can it be identified by farmers as ecological focus areas. These categories of land can directly or indirectly affect biodiversity: land lying fallow, landscape features, terraces, buffer strips, afforested areas and agro-forestry areas may directly affect biodiversity; while biodiversity may be indirectly affected through a reduced use of inputs on the farm, such as areas covered by catch crops and winter green cover. Regulation (EU) No 1307/2013, Art. 46(1) and (2), para. (44) of the preamble.

336 Nevertheless, exceptions can be provided for farms that already fulfil the objectives of ecological focus areas by being covered to a significant extent by grassland or fallow land, or in the case of predominantly forested Member States or for farmers that pursue an agricultural activity in areas facing natural constraints in certain predominantly forested areas where there is a significant risk of land abandonment. The European Commission is delegated to lay down further criteria for identifying ecological focus areas or even add other categories of land. Regulation (EU) No 1307/2013, para. (44) of the preamble, Art. 46(7) and (9).

337 It was estimated that 3.6 million hectares of EU land has been used for growing biofuel feedstocks, with approximately EUR 1 billion annually provided to farmers producing the feedstocks. Meanwhile, there is controversy over whether to include direct payment to land used for biofuel feedstock production as subsidies for biofuels. Charles *et al.* 2013, p. 41-43; Hamelinck *et al.* 2011.

338 Together with direct payment schemes (Pillar 1), the rural development policy represents the second Pillar of EU CAP, and the rural development policy is funded partly from the central EU budget and partly from individual Member States' national or regional budgets. →

(e.g. agricultural wastes and residues) for bio-economy (including biofuels) as one focus area in shifting towards low-carbon and climate-resilient economy.[339]

3.2. *Regulation on Biofuels Feedstock from Other Sources*

Apart from agricultural crops, wastes, residues, and by-products generated from non-agricultural pathways (e.g. used cooking oil) provide for the other sources of biofuel feedstock.[340] Given the minimal effects on food and land and for the sake of feedstock diversification and recycling, wastes and residues are generally encouraged to be used as biofuel feedstock for bringing 'additional benefits.'[341] Firstly, as mentioned, biofuels produced from wastes and residues, other than agricultural, aquaculture, fisheries and forestry residues, need only fulfil the GHG emission saving criterion.[342] Secondly, in calculating the GHG emission savings of biofuel, no emissions are allocated to wastes or residues, since they are considered to have zero emissions until the point of collection[343] and the calculation can use default values which usually exceed the highest 60 per cent GHG emission saving threshold.[344] Thirdly, for the purpose of demonstrating compliance with the 10 per cent transport renewable energy target, the contribution made by biofuels produced from wastes, residues, non-food cellulosic material, and lingo-cellulosic material shall be considered to be twice that made by other biofuels.[345] The proposed amendment to Directive 2009/28/EC is to strengthen such a favourable treatment by providing double or four times the favourable treatment to the designated waste or residue feedstocks,[346] and a non-binding sub-target is to be set for advanced biofuels produced from *inter alia* wastes or residues.[347] Meanwhile, there is a

Regulation (EU) No 1307/2013, Art. 4(1)(k); European Commission, Agriculture and Rural Development, Rural Development policy 2007-2013, available at: <http://ec.europa.eu/-agriculture/rurdev/index_en.htm> (last accessed 28 April 2015).

339 Regulation (EU) No 1305/2013, Art. 5(5)(c).

340 Whether the feedstock is from agriculture has implications for what law applies. Communication (2010/C 160/02), section 2.2; Communication (2010/C 160/01), section 2.

341 It is stated in the preamble of Directive 2009/28/EC that, when designing their support systems, Member States may encourage the use of biofuels which give additional benefits, including the benefits of diversification offered by biofuels made from waste, residues, non-food cellulosic material, lingo-cellulosic material and algae, as well as non-irrigated plants grown in arid areas to fight desertification. Directive 2009/28/EC, para. (89) of the preamble.

342 Directive 2009/28/EC, Art. 17(1).

343 Directive 2009/28/EC, Annex V Part C Point (18); Communication (2010/C 160/02), section 5.2 and Annex II.

344 Directive 2009/28, Art. 19(3)(c) and Annex V Part B.

345 Raw materials that have been intentionally modified to count as waste (e.g. by adding waste material to a material that was not waste) should not be considered as qualifying. Directive 2009/28, Art. 21(2); Communication (2010/C 160/02), section 5.2.

346 Biofuels from used cooking oil and animal fats are likely to be counted double for their energy content, while biofuels produced from algae and bacteria may be counted up to four times for their energy content. Commission Proposal COM(2012) 595 final, Art. 2 and Annex II.

347 The European Council has included in its position *mandatory* targets for Member States to achieve gradually, at least 0.5 per cent of energy consumption to be met with advanced →

difference between 'waste', 'residues', and 'by-products': while 'waste' is 'any substance or object which the holder discards or intends or is required to discard',[348] a residue is defined under Directive 2009/28/EC as a substance which is not the end product that a production process directly seeks to produce,[349] and a 'by-product' is defined as '[A] substance or object, resulting from a production process, the primary aim of which is not the production of that item.'[350] Due to the potential hazards of waste, the EU has established a legal framework to regulate waste, so it is critical to look into how to safeguard the sustainability of biofuels when waste is used as a feedstock for biofuels.

The definition of 'waste' is a core element in the regulation of waste in the EU. The EU has established a legal framework that sets out general principles in waste management[351] and specific rules for different waste streams and relevant activities.[352] Therefore, the definition of waste is important to decide whether the concerned substance shall comply with the legal framework on waste at the EU level. Directive 2008/98/EC defines 'waste' as 'any substance or object which the holder discards or intends or is required to discard',[353] and excludes a number of categories of substances from the scope of waste.[354] However, the definition of

biofuels in 2016 and at least 2.5 per cent in 2020. However, the Parliament agreed on merely an *optional* sub-target for advanced biofuels. Position of the Council 2012/0288 (COD), p. 8; Recommendation for Second Reading A8-0025/2015, p. 8-9.

348 Directive 2008/98/EC, Art. 3(1).

349 Residue is not the primary aim of a production process and the process has not been deliberately modified to produce it. Examples of residues include crude glycerine and tall oil pitch and manure. Communication (2010/C 160/02), section 5.2.

350 To be identified as a by-product instead of waste, certain conditions have to be fulfilled: *a)* further use of the substance or object is certain; *b)* the substance or object can be used directly without any further processing other than normal industrial practice; *c)* the substance or object is produced as an integral part of a production process; and *d)* further use is lawful, i.e. the substance or object fulfils all relevant product, environmental and health protection requirements for the specific use and will not lead to overall adverse environmental or human health impacts. Directive 2008/98/EC, Art. 5(1).

351 Directive 2008/98/EC.

352 The specific legislation may include rules regarding hazardous waste, packaging waste, electrical or electronic waste, vehicle waste, batteries and accumulators, and law about specific activities such as industrial emissions, extractive industries, ship dismantling practices, offshore oil and gas installations, sewage sludge in agriculture, ship and cargo. There is also specific legislation on activities related to waste such as the disposal, recycling and shipment of waste. European Commission, Environment, EU Waste Legislation, available at: <http://ec.europa.eu/environment/waste/legislation/index.htm> (last accessed 28 April 2015).

353 Directive 2008/98/EC, Art. 3(1).

354 Art. 2 excludes certain substances such as gaseous effluents emitted into the atmosphere and also excludes substances covered by other EU legislation. Accordingly, agricultural waste, such as faecal matter, straw, agricultural or forestry material in farming for energy production, is excluded from the scope of the Directive; and animal by-products or carcasses (unless being slaughtered) are excluded from the Directive to the extent that they are covered by other EU legislation. Directive 2008/98/EC, Art. 2(1) and (2), Annex I.

'waste' has in practice given rise to much complexity.[355] As for biofuels, the complexity can be reflected in Case C-176/05 where a dispute arose regarding a shipment of meat-and-bone meal cargo intended for energy use, and the question is whether this shipped cargo should be identified as 'waste' and thus subject the EU's regulation of waste.[356] The Court approached the question by examining whether the meat-and-bone meal at issue contained 'specified risk material' that was 'required to be discarded' by law[357] so that it could be considered as 'waste' for energy recovery,[358] but it is for the national court to decide whether the meat-and-bone meal contains 'specified risk material' or can be excluded from related public health provisions due to its energy recovery purpose.[359] This case may have implications for the potential problem that will arise when any substance is used to produce biofuels: whether the substance is 'waste' and its transportation or management should comply with waste regulation in the EU and Member States. To align the use of 'waste', Directive (EU) 2015/1513 which amends Directive 2009/28/EC provides that: 'waste' shall be defined as in Article 3(1) of Directive 2008/98/EC, excluding substances intentionally modified or contaminated to meet that definition.[360] However, according to Directive 2008/98/EC, 'waste' may cease to be waste and become a secondary raw material if it meets the end-of-waste criteria provided in Article 6, so there might be ambiguity in distinguishing between waste and secondary raw materials[361] for biofuels. Moreover, differences exist among Member States in how to categorise and report on the used cooking oil (from households, restaurants, or food processing factories) when it is used for producing biofuels.[362]

355 It has been commented that the European waste law is 'strangely engaging', and the Court of Justice of the EU has addressed the concept of waste in numerous cases. Jans & Vedder 2012, p. 473.

356 In the former Waste Framework Directive 75/442, animal carcases and animal by-products used in farming are generally excluded. In the current Waste Framework Directive 2008/98/EC, it is specified that carcasses of animals other than being slaughtered, including for the purpose of eradicating epizootic diseases and being disposed of in accordance with Regulation (EC) No 1774/2002 (laying down heath rules on animal by-products not intended for human consumption), are excluded from the Waste Directive for public health reasons. Animal by-products destined for incineration, landfilling or use in a biogas or composting plant is still defined as waste. Thus, according to Directive 2008/98/EC, the carcasses of animals being slaughtered that are not disposed of according to Regulation No 1774/2002 should be classified as 'waste.' However, the problem in this case is that the meat-and-bone meal could be seen as 'products of animal origin' instead of 'animal carcasses' which refers to the 'entire bodies or parts of animals' according to Art. 2(1)(a) of Regulation No 1774/2002. Case C-176/05, para. 37; Directive 2008/98/EC, Annex I.

357 Case C-176/05, paras. 53-57.

358 Case C-176/05, paras. 50-57, 76.

359 Case C-176/05, paras. 29, 35-38, 53, 64-70.

360 Directive (EU) 2015/1513, Art. 2(1).

361 Jans & Vedder 2012, p. 476-478.

362 For instance, the European Commission has recognised that the difference in national definitions of waste and reporting methods about used cooking oil could be one reason for underestimating biofuels produced from recycled vegetable oil. It is estimated that the EU biodiesel consumption actually included 1276 Ktoe of recycled vegetable oil, but this part of →

If biofuel feedstocks fall within the scope of 'waste', the principles and rules in the EU's waste legislation shall apply. Directive 2008/98/EC establishes the basic principles relating to waste management. On the whole, waste shall be managed without endangering human health and the environment, in particular without risk to water, air, soil, plants or animal.[363] Moreover, the Directive prescribes a waste management hierarchy which shall apply as a priority in Member States' waste legislation and policy:[364] waste 'prevention' is of top priority, which is followed in order by 'preparing for re-use', 'recycling', 'recovery', and 'disposal.'[365] Using waste for biofuel production falls under the definition of waste 'recovery'[366] which, according to the waste hierarchy, has a priority lower than 'prevention', 're-use' and 'recycling' but higher than 'disposal.' Nevertheless, Member States may require that specific waste streams depart from the hierarchy based on the overall life-cycle impact of the waste on environmental protection as well as on the economy, society, and public health.[367] In the case of waste recovery for biofuels, the question is whether the benefit of having high GHG emissions saving or contributing to energy security can constitute justification for a departure from the waste management hierarchy. There is concern about the conflict between the increased use of waste for energy (including for biofuels) and the principle of waste prevention.[368] This may be largely left for national law to decide. For example, French law does not support climate change as an admissible argument for undermining the waste priorities system: energy production from waste has a lower priority than waste recovery for material uses.[369] In Directive (EU) 2015/1513, it is provided that Member States shall have due regard to the waste hierarchy established in Directive 2008/98/EC, including life-cycle thinking on the overall impacts of the generation and management of different waste streams, in designating their sub-target for advanced biofuels.[370]

Furthermore, Directive 2008/98/EC sets forth the principles of self-sufficiency and proximity, meaning that Member States shall take appropriate measures to

the consumption has not been included in Member States' submissions to Eurostat for statistics. Commission Staff Working Document SWD (2013) 102 final, p. 12.

363 Directive 2008/98/EC, Arts. 1 and 13.

364 Member States shall establish waste management plans and waste prevention programmes. The waste management plans shall set out an analysis of the current waste management situation as well as the measures to be taken to improve environmentally sound preparing for re-use, recycling, recovery and disposal. Directive 2008/98/EC, Art. 28.

365 Directive 2008/98/EC, Art. 4.

366 It is explicitly stated that the reprocessing of organic material for fuel use is not 'recycling', and waste 're-use' means waste being used again for the same purpose for which it was conceived. Meanwhile, it is stated in Annex II that the use 'principally as a fuel or other means to generate energy' is among the list of recovery operations. Directive 2008/98, Arts. 13, 15, and 17.

367 Directive 2008/98/EC, Art. 4(2).

368 Langlais 2012, p. 292-293.

369 *Ibid.*, p. 289.

370 The consideration of the waste management hierarchy is also stated in relation to application of the multiple counting for advanced biofuels. Directive (EU) 2015/1513, Arts. 2(2)(d), 2(9), and 2(10)(c).

establish a network for waste disposal or municipal waste recovery in the 'nearest appropriate installations' and by the most appropriate methods and technologies.[371] Thus, the question is whether the support for producing biofuels from waste under Directive 2009/28/EC may incentivise the movement of waste across countries,[372] which may contravenes the self-sufficiency and proximity principles that require waste disposal or recovery to take place in the 'nearest appropriate installations.'[373]

3.3. *Interim Conclusion*

In section 3, the regulation on the production of raw materials from agricultural crops and wastes for biofuels was examined. Agricultural crops still constitute the major source of biofuel feedstocks and their production is subject to the established legal framework of CAP, which sets forth the 'cross-compliance' and 'greening' requirements related to *inter alia* water, soil, and biodiversity for payments to farmers. In this vein, in order to obtain payments, agricultural activities for biofuel feedstock production need to comply with not only the designated mandatory standards at the EU level but also national minimum standards, including the newly added requirements about grassland conservation and ecological focus. However, there is not enough data showing the amount of the EU budget used to support agricultural crops dedicated to biofuel production, which may pose a problem in estimating the effect of CAP on agricultural activities that produce biofuel feedstocks or their consequences for land use and food price changes. While feedstocks from agricultural crops may raise concerns about agro-environment and land or food availability, non-food raw materials such as wastes and residues provide for promising sources of biofuel feedstocks, and their use as biofuel raw materials is supported under the EU legislation on renewable energy and agriculture. Waste streams in the EU need to comply with the EU's waste legislation even when they are used to produce biofuels, but challenges and differences may lie in how to identify 'waste' for use of biofuels and how to reconcile such use with the EU waste management hierarchy and the principles of self-sufficiency and proximity. Thus, the coherence between the EU legislation on agriculture/waste and the sustainability of biofuels need to be further strengthened and specified.

371 Directive 2008/98/EC, Art. 16(1) and (3).

372 Moreover, the transboundary movement of non-hazardous waste for energy recovery is among the 'green list' with an information procedure requirement, so that the shipment of non-hazardous waste intended for recovery can be subject only to information requirements instead of notification. Nevertheless, Member States may limit incoming shipments of waste destined for incinerators that are classified as recovery, and may also limit outgoing shipments of waste on environmental grounds. Regulation (EC) No 1013/2006, Arts. 16(1) and 18 and Annex III; Europa, Shipment of Waste, available at: <http://europa.eu/-legislation_summaries/environment/waste_management/l11022_en.htm#amendingact> (last accessed 28 April 2015).

373 Nonetheless, the principles of proximity and self-sufficiency do not mean that each Member State has to limit the full range of waste recovery to its geographical territory. Directive 2008/98/EC, Art. 16(1), (2), and (4).

4. The EU Regulation of International Trade of Biofuels

Although the major part of biofuels consumed in the EU is domestically produced,[374] the EU has meanwhile increasingly imported biofuels from third countries.[375] The trade on biofuels and biofuel feedstocks between the EU and third countries is subject to the EU's internal legislation on biofuels as well as its external trade policies. Meanwhile, the EU measures affecting the trade of biofuels must comply with its obligation under the WTO agreements. In this sense, this section examines the implications of the EU's legislation on the trade of biofuels, so as to see whether the biofuels imported from non-EU countries need to meet any sustainability requirements, while at the same time taking into consideration the requirements of international trade law.

4.1. *Sustainability Requirement for Trade of Biofuels*

According to Directive 2009/28/EC, the sustainability criteria apply 'irrespective of whether the raw materials were cultivated inside or outside' the territory of the EU.[376] This means that, for biofuels or biofuel feedstocks imported from third countries, they shall comply with the sustainability criteria in order to be eligible to be taken into account in target meeting and support schemes, which thus affects the trade of biofuels between the EU and third countries and has given rise to trade disputes. As mentioned in Chapter 2, a complaint has been brought before the WTO settlement body and one focus of this complaint is on the application of the sustainability criteria.[377] Furthermore, certain biofuels exporting countries have voiced their concerns[378] in relation to the trade distorting effect of the EU's

[374] It is estimated that 83 per cent of biodiesel and 80 per cent of bioethanol consumed in the EU was produced within the EU in 2010, and the Member States such as Germany, France, Italy and Spain are the major producers of biofuels. The biomass resources for biodiesel and bioethanol in the EU mainly come from rapeseeds, cereal and sugarbeet, among which rapeseeds originating from the EU make up the most important feedstock source for EU biodiesel. Van de Staaij 2012, p. 5; Commission Staff Working Document SWD (2013) 102 final, p. 13.

[375] The major biofuel exporters include Argentina, Brazil, the US, Indonesia, and Malaysia. Argentina is the largest biodiesel exporter for the EU, while Brazil and the US are the largest bioethanol exporters, making up shares of 10 per cent, 8 per cent and 4 per cent respectively. Argentina soy comprises the second most important feedstock source both in the biodiesel imported from Argentina and in the EU produced biodiesel. Indonesian and Malaysian palm oil, Brazilian and American soybeans as well as rapeseed from Canada and Ukraine constitute another source of biofuel feedstock consumed in the EU. Van de Staaij *et al.* 2012, p. 5; Commission Staff Working Document SWD (2013) 102 final, p. 13.

[376] Directive 2009/28/EC, Art. 17(1).

[377] This refers to the case DS459 in which Argentina filed a complaint against the EU. See Chapter 2, section 4.

[378] Moreover, exporters or government from major exporting countries, such as Brazil, Argentina, Indonesia and Malaysia, protested against the EU sustainability criteria and argued that the EU's unilateral measure would limit the demand for foreign biofuels on the EU market and may constitute an 'environmental barrier', running afoul of the WTO →

unilateral measure on the sustainability of biofuels.[379] The compatibility of the EU's sustainability criteria with WTO law, particularly regarding the 'national treatment', has been discussed extensively by scholars.[380] According to the discussions, there are two essential questions in arguing for or against consistency with the 'national treatment' obligation: *a)* whether the sustainability criteria constitute 'less favourable treatment' for imported biofuels, and *b)* whether the application of the sustainability criteria has an extraterritorial effect that may exclude the application of Article XX GATT.

4.1.1. Less Favourable Treatment

The EU sustainability scheme has been accused of resulting in less favourable treatment for biofuels produced outside of the EU, which *de facto* disadvantages non-EU biofuels in comparison to the EU's like products, largely due to the GHG emission saving and land use requirements. For example, Argentina argued in its complaint that the 35 per cent GHG emissions saving threshold is 'arbitrary' and appears not to be 'scientifically justified' or 'based on a recognized international norm or standard.'[381] Argentina also complained about the setting of default values, especially the default value of 31 per cent assigned to soybean biodiesel: Argentina

obligation. Yans Felippe Gechler, 'Brazilian biofuels run into EU obstacles' (Inter Press Service (IPS), 22 August 2010), available at: <http://www.ipsnews.net/2010/08/energy-brazilian-biofuels-run-into-eu-obstacles/> (last accessed 6 August 2013); Fredrik Erixon, Indonesian Biofuel Dispute May Burn EU (Jakarta Globe, 29 June 2010), available at: <www.thejakartaglobe.com/archive/indonesian-biofuel-dispute-may-burn-eu/> (last accessed 6 August 2013); Perrihan *et al.* 2010, p. 110.

379 Protests are also based on unfair subsidies to domestic renewable energy sources. It is claimed that subsidies or environment-based trade restrictions provide for protectionism or a 'green blanket' and Argentina and Indonesia are among the most vocal critics. Moreover, trade disputes have increased on the issue of subsidies and protectionism in relation to renewable energy products. For example, in 2009 the EU adopted anti-dumping and countervailing duties on imports of biodiesel from the US via Regulation 598/2009 and 599/2009. The WTO Appellate Body adopted its ruling against Canada in a complaint filed by the EU and Japan regarding the content requirements in the Canadian province of Ontario's feed-in tariff programme. International Centre for Trade and Sustainable Development (ICTSD), Bridges Weekly Trade News Digest (Volume 16, No. 30, 12 September 2012); International Centre for Trade and Sustainable Development (ICTSD), Bridges Weekly Trade News Digest (Volume 17, No. 17, 16 May 2013).

380 The discussions may also include the compatibility of the EU's support for biofuels with WTO subsidy rules or the consistency between use of certifications and the WTO's TBT Agreement. Deshpande 2006; Dufey 2006; Van den Bossche, Schrijver & Faber 2007; de Vera 2007, p. 661-679; Rancourt 2009; Mitchell & Tran 2010; Lydgate 2013; Kent 2013, p. 344; Douma 2010; Puddle 2007, p. 99-140; Payne 2008, p. 891-918; Switzer 2007; Switzer & McMahon 2011, p. 713-736; Motaal 2008, p. 61-86; Swinbank & Daugbjerg 2013, p. 813-834; Machado-Filho 2008, p. 67-77; Taylor 2013; Swinbank 2009; Ackrill & Kay 2011; Lendle & Schaus 2010; Regan 2009, p. 97-123.

381 WTO, Dispute Settlement: Dispute DS459 European Union and Certain Member States – Certain Measures on the Importation and Marketing of Biodiesel and Measures Supporting the Biodiesel Industry, available at: <https://www.wto.org/english/tratop_e/dispu_e/-cases_e/ds459_e.htm> (last accessed 12 April 2015).

is a major soybean biodiesel exporter and the assigned default value cannot meet the 35 per cent threshold, so exporters from Argentina have to either calculate the actual value of GHG emission saving or use voluntary schemes recognised by the EU, both of which will increase the cost and burden on Argentinian suppliers.[382] There has also been a complaint that the criterion on land use change such as the conservation of primary forests will result in the less favourable treatment of forest abundant countries like Brazil, Malaysia and Indonesia, while the EU in fact only has very limited primary forest resources.[383]

A text analysis of the EU sustainability scheme reveals that there is some differential treatment of biofuels originating in third countries, but they might even give more favourable treatment to imported biofuels. Firstly, the agro-environment requirements as provided in Article 17(6) of Directive 2009/28/EC and CAP apply only to biofuels with raw materials from the EU, so biofuel imports do not have to show compliance with the statutory standards on water, soil, habitats or good agricultural conditions to be counted towards the renewable energy targets or support schemes.[384] Secondly, according to Article 19(3) of Directive 2009/28/EC, default values can be used by biofuels with raw materials cultivated outside of the EU, while in comparison only EU raw materials cultivated in the designated areas[385] may use default values. Since the use of default values can relieve the administrative burden and thus save on economic costs, this difference might favour the biofuel raw materials from third countries.[386] Nevertheless, the right to

382 It is argued that actual value calculation represents an additional requirement that must be fulfilled by the biodiesel producers for each transaction. Regarding voluntary schemes, an Argentina voluntary scheme - Cámara Argentina de Biocombustibles (CARBIO), which is an entity comprised of exporters of biofuels in Argentina - has not yet been approved by the European Commission, although it had already presented a voluntary scheme to the Commission and applied for its recognition in December 2010. WTO, Dispute Settlement: Dispute DS459 European Union and Certain Member States — Certain Measures on the Importation and Marketing of Biodiesel and Measures Supporting the Biodiesel Industry, available at: <https://www.wto.org/english/tratop_e/dispu_e/cases_e/ds459_e.htm> (last accessed 12 April 2015).

383 Douma 2010, p. 399.

384 The 'voluntary schemes' for biofuels are not expected to cover the criterion related to agricultural and environmental requirements and standards for EU farmers. Communication (2010/C 160/02), section 2.2; Communication (2010/C 160/01), section 2.

385 The areas classified as level 2 in the nomenclature of territorial units for statistics (NUTS) or as a more disaggregated NUTS level in accordance with Regulation No 1059/2003. These areas are regarded as places where the typical GHG emissions from the cultivation of agricultural raw materials can be expected to be lower than or equal to the emissions reported under the disaggregated default values for cultivation provided in Part D of Annex V of the Directive. Directive 2009/28/EC, Art. 19(2).

386 It has been noted in Directive (EU) 2015/1513 that, while third countries are allowed to use default values, EU producers are required to use actual values, so 'it is appropriate to align the rules for using default values to ensure equal treatment for producers regardless of where the production takes place.' Thus, it is stated that 'current rules should be simplified so that the use of default values is not limited to areas within the Union included in the lists referred to in Article 19(2) of Directive 2009/28/EC.' Directive (EU) 2015/1513, para. (28) of the preamble.

use default values does not absolutely result in compliance with the GHG emission-saving threshold, such as in the cases of soybean biodiesel where a default value is lower than the threshold and the actual value has to be calculated. However, the less favourable allocation of default values is not primarily based on the foreign origin of biofuels,[387] as in some cases biofuels largely imported from non-EU countries are attributed higher default values than the biofuels produced within the EU. For instance, sugarcane ethanol, which is largely imported from Brazil, is allocated a default value of 71 per cent, a value exceeding the highest 60 per cent threshold,[388] while the rapeseed oil (which is largely from the EU) has a default value of 38 per cent, which barely passes the 35 per cent threshold.[389] Thus, it can be argued that the differentiated default values are primarily based on the production process rather than origin: for example, palm oil biodiesel processed with methane capture at an oil mill has a default value of 56 per cent, while palm oil biodiesel with no specified production process is allocated a default value of 19 per cent.[390] Notwithstanding, for more consistency with WTO law, the EU needs to explain well how the thresholds of 35 per cent/50 per cent/60 per cent are established and justified by the best available science,[391] and why the default value (49 per cent) for corn ethanol is only applied to that produced within the EU.[392]

4.1.2. Extraterritorial Effects

The extraterritorial effect or the extraterritoriality of EU's biofuel sustainability criteria is under discussion because the EU sets out criteria relating to land use beyond the territory of EU. For example, before EU's sustainability criteria are adopted, Malaysia argued that sustainability should be defined according to the laws and regulation of the exporting countries, and the requirements about 'land with high biodiversity value' and 'environmentally harmful systems for biofuel production' can only be set out when definitions are agreed internationally.[393] Furthermore, as analysed in Chapter 2, if violation of Article I or III GATT 1994 is found, a preliminary question for invoking the general exception provided in

387 According to case law, less favourable treatment occurs only when it is 'explained by the foreign origin' of the products at stake. See discussion in Chapter 2, section 4.1.

388 Douma 2010, p. 399.

389 *Ibid.*

390 Directive 2009/28/EC, Annex V Part A.

391 For example, Swinbank observed that there could be little real difference between a biofuel resulting in a 34 per cent emission reduction and one that has a 35 per cent emission saving. Swinbank 2009, p. 499; Mitchell & Tran 2010, p. 36.

392 Such a decisive 49 per cent default value is only available for 'community produced' corn ethanol while the pathway for corn produced in, e.g. the United States, would also exceed the 35 per cent threshold. Thus, some may argue that this designation contains 'traces of possible arbitrariness and favoritism in the way that the sustainability thresholds resonate with the domestic GHG emission saving default values.' Nevertheless, this difference will no longer be a problem once the threshold is raised to 50 per cent. Directive 2009/28/EC, Annex V Part A; Jansson & Kalimo 2014, p. 131.

393 Douma 2010, p. 381-382.

Article XX GATT 1994 is whether the general exception can justify a regulatory measure applied outside of the territorial jurisdiction of its issuing state.[394]

Extraterritoriality is characterised by the application of domestic law to facts (e.g. persons, property, acts) occurring outside a state's territory.[395] Due to the transboundary nature of environmental issues, environmental regulation has become a subject of extraterritorial application and disputes.[396] The question is how to prove the legitimacy of the extraterritorial application of unilateral environmental measures to a fact in a third country?[397] Apart from possible conflict with state sovereignty and territorial jurisdiction,[398] consistency with international law is also at issue for the unilateral environmental measures with extraterritorial effect.[399] In the case of the sustainability criteria for biofuels, the question under discussion is: whether the application of sustainability criteria is of an extraterritorial nature and whether the extraterritoriality can be justified under international trade law.

As for the first question, Scott's arguments regarding 'territorial extension'[400] could be considered. Scott uses the concept of 'territorial extension' to explain and analyse the external impact of the EU's internal policies and draws a distinction between 'territorial extension' and 'extraterritoriality.' According to Scott, 'extraterritoriality' implies that the national measure in question is applied to an event with no territorial connection with the regulating state.[401] In contrast, 'territorial extension' refers to the case where a territorial connection exists between the regulatory measure and the regulating state, but the regulatory measure is shaped by conduct or circumstances abroad.[402] Thus, the major difference lies in whether there is a territorial connection on either the levels of transaction, firm, country or globe[403] which 'triggers' the application of EU law.[404] Based on this

394 See the discussion in Chapter 2, section 4.2.

395 Parrish 2009, p. 842; Gevurtz 2014, p. 347.

396 Franconi 1996, p. 122-124.

397 It is generally agreed that there must be sufficient linkage between the fact and the jurisdiction and various principles have been developed to prove such a linkage, such as effects doctrine, universal jurisdiction, and the protective principle. UN International Law Commission (2006) Report of the International Law Commission, U.N. Doc. A/61/10, Annex E, p. 512-522 and 534, available at: <http://untreaty.un.org/ilc/reports/2006/-2006report.htm> (last accessed 30 April 2015); Lewis 2004, p. 2143, 2175-2179.

398 Exercise of extra-territorial jurisdiction would result in the application of two or more national/regional laws to the same fact and thus raises the question of which law should be applied. Zerk 2010, p. 5.

399 There is no controversy that the exercise of extraterritorial criminal jurisdiction is uncontroversial, but whether international law also regulates the ability of states to exercise extraterritorial civil jurisdiction is open to debate. Zerk 2010, p. 8-9.

400 Scott 2014.

401 Scott 2014, p. 90.

402 *Ibid.*

403 Scott classified four spheres of regulation intervention: transaction, firm, country, and globe. Scott 2014, p. 107.

404 For example, the application of an EU aviation ETS measure is triggered by the conduct of landing or taking off or presence of aircraft in EU's airport and the jurisdiction is extended to carbon emissions during the whole fight. Similarly, the application of the EU ETS Directive is triggered by using CDM credits for surrendering allowances or trading allowances and the →

reasoning, it could be argued that the application of EU sustainability criteria to imported biofuels is a 'territorial extension' on the level of the transaction,[405] because the application of the sustainability criteria is triggered by placing the biofuels on the EU market (a transaction of biofuel product), although the application may extend an effect of regulatory intervention to production activities outside of the EU.[406]

Moreover, even if extraterritoriality exists in the case of sustainability criteria, it can be justified by the consideration of carbon emissions and biodiversity which are regarded as 'global concerns' of the humankind. As analysed in Chapter 2, the WTO Appellate Body in its case law has looked into whether a 'sufficient nexus' exists between the protected values and the territoriality of the regulating state, even when the protected substance is not located within the territorial jurisdiction of the regulating state.[407] Following this reasoning, it is possible to argue for the 'nexus' between the objective of avoiding carbon emissions and biodiversity loss due to biofuel feedstock producing activities, since increased carbon emissions in another country might contribute to climate change that causes sea level rise or extreme weather in the EU,[408] and the biodiversity value of the primary forests, protected areas, and highly biodiverse should be the 'common concern of humankind.'[409] Furthermore, the EU does not apply the agro-environment requirement to biofuels of foreign origin and does not include socio-economic criteria,[410] which could be partly due to the difficulty in proving a territorial connection between such concerns and the EU.

jurisdiction is extended to whether there is bilateral or international agreement. The most similar case with biofuels is when the application of Regulation 995/2010 is triggered by placing timber products on the EU market for the first time and the jurisdiction is extended to production (logging) activities abroad and whether there is bilateral agreement. Scott 2014, p. 98-99.

405 In the transaction sphere, the EU regulator is required to take into account the conduct or circumstances taking place or prevailing outside the EU in so far as these pertain to a specific transaction. Scott 2014, p. 107.

406 The regulatory intervention is at the level of 'transaction' sphere - the EU regulator is required to consider the conduct or circumstances outside the EU in so far as it pertains to a specific transaction. Scott 2014, p. 107.

407 See Chapter 2, section 4.2.

408 See discussions about climate change regulation and territoriality in Sampson 2005, p. 102; Van den Bossche, Schrijver & Faber 2007, p. 96-97; Rancourt 2009, p. 97-98; Cottier & Matteotti-Berkutova 2009.

409 It is declared in the CBD that biological diversity is a common concern of humankind. However, in WTO case law, the 'nexus' was identified for the migratory population of sea turtles. As for biofuel sustainability criteria, the protected value is forests and grasslands, so whether there is a sufficient implication of global effect may be open to discussion. Rancourt 2009, p. 97.

410 See Chapter 2, section 4.2.

4.2. *External Trade Policy and Biofuels*

While sustainability criteria for biofuels are internal requirements affecting the internal sale or use of biofuels, trade measures such as levying of customs duties and multilateral/bilateral agreement negotiations belong to the EU's external actions[411] and may have a direct impact on the international trade of biofuels. The EU is a customs union and has a common commercial policy on setting the tariff rates of customs duties and for the conclusion of trade agreements.[412] The EU is meanwhile an international actor and has competence to enter into bilateral or multilateral agreements according to EU law.[413]

The policy regarding the tariff rates of customs duties provides for an immediate way to affect trade of products.[414] The tariff rates in the EU are usually determined by the category to which the goods are assigned in the 'Combined Nomenclature' (CN) system which is largely based on the Harmonized System (HS) nomenclature governed by the World Customs Organisation.[415] However, there is no particular nomenclature code for biofuels, and biofuel products fall under different categories of goods and thus have different tariff rates.[416] In this context,

411 The EU external action primarily includes its common commercial policy, development cooperation and aid, entering into international agreements, and external relations with third countries or international organisations. The external relations of the EU encompasses both its *external policies* and the *external dimension of internal policies*. TFEU, Arts. 205-222; Duran & Morgera 2012, p. 2.

412 Art. 207 TFEU provides a legal basis for the exclusive competence of the EU in its common commercial policy. In areas where the EU has exclusive competence, Member States are refrained from acting individually unless there is a common position or authorisation. Cremona 2011, p. 251.

413 TFEU, Art. 216.

414 Other similar trade measures include *charges* of any kind imposed on or in connection with importation or exportation, the international transfer of payments for imports or exports, the method of levying such duties and charges, and other rules and formalities in connection with importation and exportation. The environmental standards or quality requirements, such as the biofuel sustainability criteria, could be regarded as non-tariff measures in trade.

415 When declared to customs in the EU, goods must generally be classified according to the CN system. Imported and exported goods have to be declared with regard to the subheading of the nomenclature under which they fall. The nomenclature determines which rate of customs duty applies. The CN is comprised of the Harmonized System (HS) nomenclature, which is governed by the Convention on the Harmonized Commodity Description and Coding System under World Customs Organisation (WCO), with further EU subdivisions. Each CN subdivision has an eight digit code number - the CN code - and is followed by a description. European Commission, Taxation and Customs Union, The Combined Nomenclature, available at: <http://ec.europa.eu/taxation_customs/customs/customs_duties/-tariff_aspects/combined_nomenclature/index_en.htm> (last accessed 1 May 2015); Explanatory Notes (2015/C 076/01).

416 For instance, bioethanol may fall under the general category of 'beverages, spirits and vinegar' with the code of 2207, while biodiesel may fall within the category of miscellaneous chemical products with code of 3824. The code 2207 includes undenatured ethyl alcohol of an alcoholic strength by volume of 80 per cent or higher (2207 10 00) and ethyl alcohol and other spirits, denatured, of any strength (2207 20 00). Code 3824 includes chemical products and →

additional or reduced/exempted tariffs on biofuel products from a particular country may be used as tool to implement biofuel policies.[417] For example, consistency with the EU's sustainability criteria for biofuels is taken into account in affording reduced or exempted customs duties to products from a third country when bilateral free trade agreement is negotiated. In the sustainability impact assessment[418] on trade agreements with ASEAN, ANDEAN countries, Central America, and MERCOSUR, it is stated that attention should be paid to compliance with the EU's sustainability criteria for biofuels.[419]

To reach specific bilateral or multilateral agreements on the sustainability of biofuels serves as the other tool for promoting compliance with the EU's sustainability criteria on biofuels. Directive 2009/28/EC provides that the EU 'shall endeavour to conclude bilateral or multilateral agreements with third countries containing provisions on sustainability criteria that correspond to those of this Directive',[420] and it encourages the development of multilateral and bilateral agreements that cover key environmental and social considerations.[421] As yet, there is no such bilateral or multilateral agreement developed for biofuels, but the EU has conducted a series of bilateral and multilateral actions to promote its sustainability criteria for biofuels. The EU has committed to *bilateral* cooperation with certain

preparations of the chemical or allied industries (including those consisting of mixtures of natural products), not elsewhere specified. Explanatory Notes (2015/C 076/01).

417 One instance is that the EU's GSP (Generalised System of Preferences) scheme gave a 15 per cent reduction of the customs duty to bioethanol imported from all GSP beneficiary countries unless they have 'graduated' and the scheme lasted until the end of 2005. A new scheme known as the GSP+ was drafted giving a 100 per cent reduction to bioethanol imports from 'vulnerable' countries having in place a sustainable development model and good governance. Ahamat & Rahman 2014, p. 75-76.

418 A trade sustainability impact assessment is a policy tool for assessing economic, social and environmental implications during trade negotiations and prior to the conclusion of trade agreements, in order to inform negotiators of the possible social, environmental and economic consequences of trade agreements. The assessment is conducted by external consultants, and the European Commission sets out its own views in the 'Position Papers' based on the assessment report. These assessments were first developed in 1999 for the WTO Doha Development Agenda negotiations. Since then they have been applied to all the EU's major multilateral, regional or bilateral trade negotiations. Duran & Morgera 2012, p. 234-249.

419 For example, in the Position Paper on Free Trade Agreement between EU and ASEAN, it is stated that '[d]uring negotiations with these countries the possibility of the using the FTA to ensure better compliance with these criteria will be examined' and '[a]ttention will also be paid to compliance with the EU's sustainability criteria developed for biofuels.' Commission Services Position Paper on The Trade Sustainability Impact Assessment Of The Free Trade Agreement Between The EU and ASEAN, p. 9; Commission Services Position Paper 1 On The Trade Sustainability Impact Assessment (SIA) of The Multiparty Trade Agreement With Andean Countries, p. 6; Commission Services Position Paper on the Trade Sustainability Impact Assessment of the Association Agreement between the EU and Central America, p.6; Position Paper Trade Sustainability Impact Assessment (SIA) of the Association Agreement Under Negotiation Between the European Union and Mercosur, p. 5. European Commission, Trade, Sustainability Impact Assessments, available at: <http://ec.europa.eu/trade/policy/-policy-making/analysis/sustainability-impact-assessments/> (last accessed 1 May 2015).

420 Directive 2009/28/EC, Art. 18(4).

421 *Ibid.*, paras. (74) and (79) of the preamble.

biofuel exporting countries to develop mutually recognised standards on sustainable biofuels or to produce advanced biofuels. For instance, in a Parliament Resolution with Latin American countries in 2010, it was stated that the Bi-regional Strategic Partnership should work to create mutual standards and quality certificates for biofuel production.[422] Biofuel has been on the agenda of the EU-Brazil Summits since 2007,[423] and cooperation to establish an equivalence agreement that recognises the compatibility of Brazilian legislation and European biofuel sustainability criteria was stressed in the Joint Statement of the Sixth Summit.[424] In view of the bilateral efforts, there has been discussion on the possibility of using bilateral negotiations to promote compliance with the EU biofuel sustainability criteria, which could also contribute to building greater consensus on international biofuels criteria in a 'bottom-up' approach.[425] Moreover, the EU has also engaged in *multilateral* efforts regarding the sustainability of biofuels. The EU expressed support for developing international standards under the Convention on Biological Diversity (CBD) to avoid negative impacts on biodiversity and local communities[426] and meanwhile highlighted its sustainability criteria as an example.[427] Moreover, the EU has participated in multilateral cooperation and information exchanges between the major biofuel producing countries, such as the International Biofuels Forum.[428] The reasons underlying the EU's efforts in promoting multilateral

422 Euro-Latin American Parliamentary Assembly Resolution, 'Tackling climate change challenges together: for an EU-LAC coordinated strategy in the framework of the UNFCCC negotiations based on the Resolution of the Committee on Social Affairs, Human Exchanges, Environment, Education and Culture' EP Co-rapporteur: Peter Liese (PPE) LAC Co-rapporteur: Carlos Baraibar (Parlatino) Saturday, 15 May 2010 – Seville (Spain), paras. 36 and 37.

423 The EU-Brazil Summits provide a forum for exchanging respective negotiating positions on topics of climate change and biofuels with commitments to hold bilateral high-level meetings to prepare for key multilateral negotiating sessions. Duran & Morgera 2012, p. 221-222.

424 Council of the European Union, Presse 29, VI Brazil-EU Summit Joint Statement5715/13, Brasilia, 24 January 2013, para. 18, available at: <http://europa.eu/rapid/press-release_PRES-13-29_en.htm> (last accessed 1 May 2015).

425 Duran & Morgera 2012, p. 246 and 270; Al-Riffai, Dimaranan & Laborde 2010, p. 110.

426 IISD 2008, p. 8 and 16.

427 It has been commented that the EU adopted the biofuels sustainability criteria with a view to showing leadership or best-practice in this internationally controversial issue not only for the development of an international standard but also to inspire national actions in partner countries. Kulovesi, Morgera & Muñoz 2011, p. 879.

428 For example, the EU participates in the International Biofuels Forum, a governmental initiative launched in 2007 among Brazil, China, European Commission, India, South Africa, and the US, in a bid to promote an international market for sustainable biofuels, exchange experiences, and build collaboration on biofuel standards and codes. The Forum would initially be established for one year and would meet regularly to discuss ways to promote the sustained use and production of biofuels around the globe. The Forum is not a new international organisation, but rather a mechanism for closer coordination among the field's major players to establish common standards and work towards the commoditisation of biofuels. Commission Decision 2007/446/EC; Press Conference Launching International Biofuels Forum, UN, 2 March 2007, available at: <www.un.org/News/briefings/-docs/2007/070302_Biofuels.doc.htm> (last accessed 1 May 2015).

negotiation or cooperation on sustainable biofuels could be attributed to the motive of increasing consistency with WTO law[429] and 'globalising'[430] or 'exporting'[431] its biofuel sustainability criteria. In fact, such an external effect has been demonstrated in the fact that some biofuel exporting countries, such as Argentina, Brazil, Indonesia, and Malaysia, adopted new regulatory measures in 2009 and 2010 to improve their environmental practices in nature protection or biofuel feedstock cultivation.[432]

4.3. *Interim Conclusion*

In this section, it was found that, by applying the sustainability criteria to biofuels originating from third countries, the EU's internal measure on biofuels thus has an influence on the trade of biofuels between the EU and third countries, but the question is whether such a sustainability requirement is in line with international trade law. An analysis of the text of the sustainability criteria suggests that biofuels from third countries are subject to the primarily similar sustainability requirements, except for some differential treatment in terms of agro-environment criterion and the use of default values, but such differences do not necessarily lead to less favourable treatment based on their foreign origin. The arguments for 'territorial extension' could be used to justify the extraterritorial effect of sustainability criteria on the production of biofuels in the third countries, and the character of 'global concern' may provide for a 'nexus' between the criteria on GHG emission saving/biodiversity and the EU. After all, the consistency between EU sustainability criteria and WTO law is still open to debate and dependent on further case law, but the consideration of WTO law may constitute one challenge for integrating the agro-environmental and socio-economic sustainability requirements into the binding sustainability criteria due to their effect on external trade. Nevertheless, the external policies, such as reducing/exempting of tariff rates and negotiation for bilateral or multilateral agreements on biofuels, may serve as tools for promoting compliance with the EU's sustainability criteria in the third countries and could even go beyond

429 As mentioned in Chapter 2, the WTO jurisprudence has highlighted the importance of coordination and cooperation at the international level. See Chapter 2, section 4.2.

430 Globalising the EU's standards also has the purpose of enhancing the competitiveness of the EU's companies. Kelemen 2010, p. 335.

431 The external implications of the biofuel sustainability criteria have been discussed in literature. It has been commented that the sustainability criteria could be seen as an example to leverage the EU's market size to 'guide' the international community towards more effective action against the global issues of sustainable development and climate change. The sustainability criteria may serve as the means by which the EU attempts to influence environmental policy outside its territory and influence land-use policies in third countries. In this sense, the criteria may provide a tool for the EU to use its market size and economic power to 'export' its regulatory norms, particularly environmental standards, or to influence rule making in multilateral fora. Kulovesi 2012, p. 118 and 141; Duran & Morgera 2012, p. 46 and 48-49; Scott 2011b, p. 27-28.

432 Commission Staff Working Document SWD (2013) 102 final, p. 17 and 32.

the current sustainability criteria by covering other environmental and socio-economic concerns.

5. Conclusion

This chapter examined the regulatory approach to the sustainability of biofuels in the EU and analysed whether it addresses the inclusive sustainability concerns of biofuels. It was found that the sustainability criteria have been established in EU legislation and have played a central role in constraining demand, supply, and trade of 'unsustainable' biofuels with potential negative impacts. However, a reflection concentrating on the integration of sustainability concerns reveals that the EU sustainability criteria are still limited in addressing potential adverse effects in both environmental and socio-economic dimensions, particularly with regard to the concerns about GHG emissions from indirect land use change and socio-economic rights in third countries, and moreover the exhaustive harmonisation of the sustainability criteria makes it difficult for complementary or stricter measures to be adopted at a lower regulatory level.

5.1. *Summary of the Main Findings*

Section 2 analysed the regulatory measures having an effect on the demand for biofuels, including the sustainability criteria and other demand-related measures. The sustainability criteria include requirements about life-cycle GHG emission saving, biodiversity, carbon stock, and agro-environment protection, while the other sustainability concerns, particularly the socio-economic concerns, are addressed with monitoring and reporting and voluntary schemes. Moreover, by requiring compliance with the sustainability criteria, renewable energy targets and fuel GHG emission reduction targets serve as the major drives for the consumption of 'sustainable' biofuels as defined by the sustainability criteria. The other measures, such as preferential energy tax, aviation emission trading, and the acceptance of CDM credits, could incentivise the use of and demand for biofuels, but the link and coherence between these measures and the sustainability criteria need to be strengthened, particularly considering that the targets are not to be continued after 2020.

Section 3 examined the regulation of production of raw materials for biofuels such as agricultural crops and wastes. The production of agricultural crops is subject to the established legal framework of CAP, including the 'cross-compliance' and 'greening' requirements related to *inter alia* water, soil, carbon stock, and biodiversity. In a similar vein, waste streams used for generating biofuels need to comply with the EU's legal framework on waste. However, there is not enough data to calculate the amount of the EU budget used to support agricultural crops dedicated to biofuel production, which may pose a problem to estimate the effect of CAP on agricultural activities to produce biofuel feedstocks, particularly with regard to their consequences for land use and food price changes. Moreover, a challenge may lie in how to identify 'waste' and comply with the waste management hierarchy and the principles of self-sufficiency and proximity.

In section 4, it was found that, by applying the sustainability criteria to biofuels originating in third countries, the EU's internal measures on biofuels thus have an influence on the trade of biofuels between the EU and non-EU countries, but the consistency between the EU sustainability criteria and WTO law is still open to debate and dependent on further case law. However, the consideration of WTO law may constitute one challenge for integrating the agro-environmental and socio-economic sustainability requirements into the binding sustainability criteria. Nevertheless, the external policies such as reducing/exempting biofuels from the application of tariff rates and negotiating bilateral or multilateral agreements on biofuels may provide tools for promoting compliance with the EU's sustainability requirements in the third countries, and may assist in addressing the sustainability concerns which are not currently included in the EU's sustainability criteria.

5.2. *Reflection*

The previous chapters have examined regulation relating to the sustainability of biofuels at the EU level, and the observation is that the sustainability criteria play a central role in constraining demand, production, and trade of biofuels which may have unsustainable impacts. The reflection hereby concentrates on the questions whether such a regulatory approach takes into consideration the inclusive sustainability impacts in the environmental and socio-economic dimensions; and, if not, whether the regulatory approach allows for complementary or stricter sustainability requirements at the national level of the Member States.

5.2.1. The Deficiency of the EU's Sustainability Criteria

As already observed in section 2.1 of this chapter, the EU's sustainability criteria purely consist of requirements relating to environmental aspect, while socio-economic concerns are not included in the binding criteria. The absence is not due to the lack of negative socio-economic impacts caused by the EU's biofuel production and consumption,[433] but it is largely owing to the concern about compatibility with WTO law and the difficulty in defining commonly applicable social standards.[434] Furthermore, the sustainability criteria do not apply to solid or gaseous biomass for electricity, heating and cooling, but biofuels may be produced from the same biomass raw materials, so there is reason to worry that the

433 For example, the current and former UN Special Rapporteurs on the right to food have both addressed the impact of EU pro-biofuel policy on food, land, and other socio-economic rights in developing countries. Elver, H., 'MEPs take note: Biofuels are synonymous with high, volatile food prices', 23 February 2015, available at: <www.euractiv.com/sections/energy/-meps-take-note-biofuels-are-synonymous-high-volatile-food-prices-312347> (last accessed 11 April 2015); Mandate of the Special Rapporteur on the right to food, Note on the Impacts of The EU Biofuels Policy on the Right To Food, 23 April 2013, available at: <www.srfood.org/images/stories/pdf/otherdocuments/20130423_biofuelsstatement_en.-pdf> (last accessed 1 May 2015).

434 See discussion in section 2.1.5 of this chapter.

distinction and exclusion based on usages may essentially diminish the effectiveness and comprehensiveness of the sustainability criteria.[435] The EU intends to fill the gap by obliging the European Commission and Member States to monitor and report and meanwhile allowing voluntary schemes to include the sustainability concerns not covered in the criteria. However, it is questionable whether such supplementary approaches would have a similar effect as the sustainability criteria in addressing the sustainability concerns. Monitoring and reporting are after all a much softer approach than the sustainability criteria which are legally binding, and the Commission has in its report admitted the existence of negative social impacts but largely refuted their causal link with the EU's policy.[436] The effectiveness of the voluntary scheme approach may largely rely on the credibility of the certifying and auditing schemes which shall be examined and recognised by the European Commission, but the current legal framework provides insufficient guidance on the benchmarking standards and procedural requirements to ensure stringency and transparency.[437] Moreover, the 'meta-standard' approach, drawing on privately developed certification schemes, may not be able to avoid the inherent limitation of certification completely, such as the risks of fraud, proliferation, 'forum shopping' 'race-to-the-bottom', disadvantaging small-scale farmers as well as the difficulty in assessing and verifying macro-level impacts (e.g. indirect land use changes).[438] Therefore, although the approaches of monitoring/reporting and voluntary schemes can admittedly provide opportunities for the EU to learn from experience and reduce the risk of violating international law, the approaches meanwhile have certain weaknesses and may not be able to substitute legally binding criteria.[439]

Furthermore, EU's sustainability criteria *per se* may have limits in avoiding or minimising the potential concerns already covered by the criteria. Firstly, the GHG emissions saving criterion may be limited regarding complete accounting of the life-cycle GHG emissions of biofuels. The emissions from biofuels in use is taken as zero, but the emissions from land use change,[440] in particular indirect land use

435 Romppanen 2012a, p. 134-135; Kulovesi, Morgera & Muñoz 2011, p. 881.

436 See discussion in section 2.1.5.1 of this chapter.

437 See discussion in section 2.1.5.2 of this chapter.

438 *Ibid.*

439 There have been arguments to incorporate additional sustainability criteria regarding social aspects. Kaphengst, Wunder & Krista 2012, p. 25.

440 Carbon emissions from land use, land use change and forestry (LULUCF), including those related with agricultural activities are not yet incorporated into the EU's climate change measures such as EU ETS and Effort Sharing Decision. To fill the gap, as well as to come into line with the commitment under Kyoto Protocol, the EU has adopted Decision No 529/2013/EU to harmonise accounting rules for the emissions and removals from the LULUCF sector, which may represent the first step towards integrating the LULUCF sector into the EU-wide rules on GHG emission reductions. However, this decision does not set targets for emission reduction in the LULUCF sector, nor does it lay down any accounting or reporting obligations for private entities, but it just sets out the obligation for Member States to provide information on their LULUCF actions that affect emissions and removals. Nevertheless, the European Commission will propose GHG targets for agriculture and forestry sectors once it has been proven that the accounting systems are robust and effective. Decision No 529/2013/EU, paras. (5) and (9) of the preamble, Art. 1; European Commission, →

change, caused by the production of biofuels is not comprehensively considered.[441] The EU biofuel legislation has been criticised for failing to take into account GHG emissions from indirect land use changes,[442] as consideration of indirect land use changes would greatly affect or even reverse the emission saving performance of biofuels.[443] Although measures have been taken to address indirect land use change mainly by means of capping the share of food crop-based biofuels in meeting the targets,[444] EU legislators still failed to have the indirect land use emissions included in the GHG emission saving criterion accounting[445] in spite of available scientific data.[446] Secondly, the agro-environment criterion does not apply to biofuel feedstocks obtained from third countries, although local environmental pollution in third countries caused by intensified biofuel feedstock cultivation has caused

Climate Action, LULUCF in the EU, available at: <http://ec.europa.eu/clima/policies/-forests/lulucf/index_en.htm> (last accessed 29 April 2015); European Commission, Climate Action, LULUCF in the EU, available at: <http://ec.europa.eu/clima/policies/-forests/lulucf/index_en.htm> (last accessed 29 April 2015).

441 The scientific committee of the European Environmental Agency pointed out the 'accounting error' and 'misapplication' embedded in the way of GHG emissions accounting provided for biofuels in Directive 2009/28/EC, suggesting that either indirect land use change or GHG emissions from the consumption of biofuels (i.e. exhaust pipe emissions) should be accounted for. See discussion in section 2.1.1 of this chapter; EEA Scientific Committee 2011.

442 See discussion regarding EU biofuels emissions accounting and ILUC in Romppanen 2012a, p. 135-136; Di Lucia, Ahlgren & Ericsson 2012; Overmars *et al.* 2011; Fritsche, Sims & Monti 2010; Soimakallio & Koponen 2011; Hellmann & Verburg 2010; Laborde 2011.

443 The European Commission admitted that once the indirect effect (intensification and ILUC) is considered, the original emission saving estimation can be largely reduced, especially as for the food crop-based biofuels which have lower or even no estimated GHG emissions savings. Commission Report COM(2013) 175 final, p. 12.

444 This refers to the 7 per cent limit for the share of food crops-based biofuels in calculating numerator of the 10 per cent transport renewable energy target under discussion with the EU legislators. Other measures are also proposed to be taken, such as raising the GHG emission thresholds, double or four-time counting, and voluntoary sub-targets for advanced biofuels. See discussion in section 2.1.1. of this chapter.

445 This means that the biofuels with indirect land use change effect will still be allowed to count towards renewable targets. Moreover, whether such a measure is the most appropriate way is open to discussion, considering that not all food-crop based biofuels are definitely damaging and not all advanced biofuels are environmentally or socially beneficial. Ecaterina Casinge, 'Parliament rubber stamps EU biofuels reform amid final controversy' (EurActiv.com, 29 April 2015), available at: <http://www.euractiv.com/sections/transport/parliament-rubber-stamps-eu-biofuels-reform-amid-final-controversy-314196> (last accessed 1 May 2015).

446 The EU's Joint Research Centre has published a report on ILUC of biofuels, stating that 'the science has improved significantly, and further investigation of modelling work and sensitivity analysis has allowed uncertainties to be largely reduced' and concluded that 'even with uncertainties, the best estimate of ILUC is not zero.' The JRC scientists recommended that 'it is important that the principle of full GHG accounting is introduced in the proposed amendment by including ILUC' and 'inclusion of the ILUC factors as written, resolving to update them quickly when new data becomes available, before the industry makes investment plans.' Arthur Neslen, 'EU scientists' biofuels warnings were ignored' (EurActiv, 20 February 2014), available at: <www.euractiv.com/energy/eu-scientists-biofuels-warnings-news-533639> (last accessed 3 May 2015).

serious concerns.[447] One reason for the restrained application could be the consideration about consistency with international law and territorial sovereignty.[448] With regard to such environmental impacts in third countries, the European Commission on the one hand admitted the existence of negative impacts in the progress reports;[449] it on the other hand argued that the contribution of the EU's biofuels consumption to the impacts is minimal[450] and has not put forward any substantive suggestions to mitigate the observed environmental risks.[451]

5.2.2. Possibility for Adopting More Stringent Sustainability Criteria

In view of the limits of the EU's sustainability criteria to address inclusive sustainability concerns, the question is whether more stringent criteria, including complementary requirements (e.g. socio-economic criteria) and stricter requirements (e.g. GHG emissions accounting including indirect land use change), can be adopted by Member States for biofuels consumed in their jurisdiction. This question calls for an examination of *a)* to what extent the sustainability criteria harmonise requirements on the sustainability of biofuels in the EU and whether the specific harmonising measure allows for complementary or stricter criteria, and *b)* if not, whether EU law offers the possibility for adopting complementary or stricter criteria.

447 It has been pointed out that the environmental risks are more likely to take place in developing countries. Ottinger & Miller 2010, p. 23; Elbehri, Segerstedt & Liu 2013, p. 1 and 13; Tilman *et al.* 2009, p. 270; Fargione, Plevin & Hill 2010; Hill *et al.* 2009; German, Schoneveld & Pacheco 2011; Gerbens-Leenes *et al.* 2012; Diaz-Chavez *et al.* 2013, p. 184.

448 It has been commented that it is usually questionable whether unilateral environmental actions are consistent with international law and the principle of national sovereignty in relation to natural resources. Romppanen 2012a, p. 133; Shaffer & Bodansky 2012, p. 31.

449 The Commission admitted that biofuels may have an impact on air in every stage: growing feedstock (dust from clearing land, smoke from burnings, nitrogen from fertilisers), transporting feedstock and refined product (vehicle emissions and dust generation), processing (industrial systems emissions), and final use (combustion). Commission Staff Working Document SWD (2013) 102 final, p. 28-32; SWD (2015) 117 final, p. 2-7.

450 For example, the Commission admitted that the widespread burning and semi-accidental forest fires in the vicinity of Indonesian palm oil plantations caused air quality risks, but it argued that these are not directly driven by EU biofuels since those cannot be sourced from areas associated with the clearing of natural forests as this is not permitted under the Renewable Energy Directive. The Commission also argued that the EU biofuel consumption has contributed little to the increase of commodity prices. Commission Staff Working Document SWD (2013) 102 final, p. 28-32. Commission Staff Working Document SWD (2015) 117 final, p. 5-6.

451 There has been a recommendation that, as a minimum basic pre-condition on compliance with relevant legislation, biofuel production is required to be in compliance with national, regional and local soil/air/water protection legislation. Such a requirement is practical, because such a basic compliance criterion is demonstrated by its inclusion in many existing voluntary schemes and compliance with national, regional, and local legislation is also an element of certain EU legislation (e.g. Regulation (EU) No 995/2010 on timber). Diaz-Chavez *et al.* 2013, Summary.

5.2.2.1. *Exhaustive Harmonisation*

In the EU context, harmonisation is 'the creation of rules' by an EU act, such as secondary legislation in the form of a directive or a regulation, with the aim or result of 'changing or supplementing' Member States' national legislation.[452] The purpose of harmonisation is to reduce differentiation in national law and thus increase the degree of uniformity to achieve a common purpose as necessary, such as the functioning of internal market or a high level environmental protection. However, this does not mean that Member States shall adopt the same or uniform national measures on the harmonised matter. Firstly, if the EU legislation is adopted in the form of a directive, Member States need only ensure that the aimed result is achieved while having flexibility in choosing the forms and means of their national measures, in particular with regard to environmental measures.[453] Secondly, the EU law may allow for a derogation from a harmonisation measure to various extents, and Member States can deviate from the harmonised EU measure to accommodate new or specific circumstances. According to extent of harmonisation, EU harmonisation measures can be generally divided into exhaustive (total/full/complete/maximum) harmonisation and minimum harmonisation. Exhaustive harmonisation refers to the situation in which the EU exhaustively regulates the subject-matter by setting both the 'ceiling' and 'floor' in the harmonised provisions, and Member States may not derogate from the 'ceiling' or 'floor' unless permitted to do so by EU law; in contrast, minimum harmonisation sets the 'floor' and Member States may adopt stricter standards above the 'floor.'[454] The exact extent of harmonisation needs to look into relevant provisions of the specific harmonising measure and its legal basis in the EU treaty.[455]

The adoption of sustainability criteria in Directive 2009/28/EC serves as a legal act at the EU level to harmonise national sustainability requirements on biofuels based on Article 114 TFEU. It is stated in the directive that Articles 17 to 19, which contain provisions on biofuel sustainability criteria and their implementing scheme, are adopted having regard to Article 114 TFEU.[456] Article 114 TFEU

[452] Vos 2001, p. 147-148.

[453] Provisions of an EU directive, particularly in an environmental area, may leave room for different levels of discretion. It is concluded that there are two types of environmental directives: a directive that includes the governance approach such as a framework directive and another is the type of directive that contains a classical, detailed regulatory approach. There are three levels of discretion: discretion regarding policy, assessment and interpretation. Van Holten & Van Rijswick 2014.

[454] Vos 2001, p. 148-149.

[455] The provisions in the EU measure at issue should be closely examined to determine extent of harmonisation, as the extent of harmonisation cannot be solely determined by the choice of legal base. The extent of harmonisation should be determined from a close analysis of the harmonised provisions: if an explicit indication of a 'floor' or 'ceiling' is not in the provisions, another important clue is the legal basis according to which the EU harmonisation measure is adopted. Vos 2001, p. 148-149; Maletic 2013, p. 63 and 175-176; Jans & Vedder 2012, p. 112.

[456] It is stated that Directive 2009/28/EC was adopted having regard to 'in particular' Art. 192(1) TFEU (ex Art. 175(1)), and Art. 114 TFEU (ex Art. 95) 'in relation to Arts. 17, 18 and 19.' Directive 2009/28/EC, preamble.

provides the legal basis for the EU to harmonise national laws for the establishment and functioning of the internal market,[457] which has often been invoked for exhaustive harmonisation measures in contrast to minimum harmonisation.[458] The intention of exhaustive harmonisation can be further evidenced by a close examination of the provisions of Directive 2009/28/EC. It is stated that the sustainability criteria have an effect on the functioning of the internal market and aim to facilitate the trade of biofuels between Member States.[459] According to Article 17(8) of Directive 2009/28/EC, Member States shall not refuse to take biofuels that are in compliance with the EU sustainability criteria into account on *other sustainability grounds*, implying that once biofuels satisfy all the criteria provided in the EU directive, it shall be counted towards the targets and eligible for support schemes, and Member States shall not use other sustainability grounds to reject such the biofuels. In fact, the European Commission has explained that Directive 2009/28/EC aims for a *complete* harmonisation of biofuel sustainability criteria,[460] because sustainability criteria have the function of preventing Member States from adopting individual sustainability measures which would create a barrier for internal market of biofuels, and the functioning of internal market serves as the primary objective[461] of the EU sustainability criteria, in addition to the

457 De Búrca 2000, p. 141.

458 It has been pointed that the article itself does not express a predilection for either minimum or exhaustive harmonisation in any categorical manner: Art. 114 does not envisage whether it is possible for Member States to apply more stringent rules, unlike Art. 193 TFEU which provides that 'the protective measures adopted pursuant to Art. 192 shall not prevent any Member State from maintaining or introducing more stringent protective measures.' However, it is observed from practice that there is a general trend or preference for exhaustive over minimum harmonisation under this article. Minimum harmonisation can also take place under Art. 114 TFEU, but there is only one limited type of minimum harmonisation permitted via this Art., which actually turns into full harmonisation. Maletic 2013, p. 63 and 176; Weatherill 2006, p. 46, 47 and 50; Barnard 2010, p. 631-632.

459 Directive 2009/28/EC, para. (94) of the preamble.

460 The European Commission proposed 'dual' legal bases for Directive 2009/28/EC on both Art. 192(1) TFEU and Art. 114 TFEU and the proposal was accepted by the Council without objection: a unanimous vote was achieved with regard to the adoption of the Directive and there was no dispute recorded as for the dual legal base. Commission Proposal COM(2008) 19 final, p. 8; Council of the European Union, Interinstitutional File 2008/0016(COD) 8533/09 Brussels, 6 April 2009; Council of the European Union 2936 meeting of the Council of the European Union on 6 April 2009, 8496/09, Brussels, 27 April 2009.

461 When an EU measure pursues more than one objective and thus can find a bases in multiple Treaty provisions, the question of the legal base choice emerges, since the choice of the correct legal basis is important for 'the participation of other Union institutions', 'the intensity of this participation' as well as the difference in 'residual rights for Member States.' The Court has established the approach of the 'centre of gravity' to distinguish between multiple objectives pursued at the same time by one EU measure. It has been established in case law that it is the objective factors, particularly the content and aim of the EU legislation that determine the legal basis. In case more than one objective is pursued by an EU action, the legal base is dependent on the primary objective – the main and predominant purpose of the EU measure, except when a number of objectives or several components are 'indissociably' linked and it is impossible to distinguish the primary and the incidental ones. However, Kramer once argued that the theory of 'centre of gravity' is not clear enough in specific →

environmental objective based on Article 192 TFEU.[462] In the spirit of exhaustive harmonisation,[463] the sustainability criteria constitute the 'floor' as well as the 'ceiling' for national legislation on the sustainability of biofuels, so Member States in principle shall not provide complementary or stricter sustainability criteria for biofuels to meet targets or for financial schemes, for the sake of the functioning of the internal market.

5.2.2.2. *Complementary or Stricter Criteria through Article 114 TFEU?*

Even with exhaustive harmonisation at the EU level, Member States may still be able to derogate from the harmonised measure in relation to specific issues or for a specific period of time, if thus provided by the harmonisation measure itself or by general EU law.[464] Directive 2009/28/EC does not include any provision that permits a derogation from the sustainability criteria, but Article 114 TFEU, as the legal basis of the sustainability criteria, contains provisions which can be invoked by Member States to derogate from the harmonised sustainability criteria.[465] Among the provisions, paragraphs (4) and (5) of Article 114 TFEU set out conditions under which Member States may respectively *maintain* or *introduce* more stringent measures, which means that the two paragraphs only allow derogation measures which are *not less protective* than the EU harmonisation measure.[466] The general conditions for applying the two paragraphs can be circumscribed into *ratione materiae, personae* and *temporis*.[467] The paragraphs (4) and (5) are different as to the

circumstances and the criterion should be whether 'uniformity is necessary to achieve the Treaty's objective' rather than the wording of preambles and articles of the legislation in question. The legal basis, as Kramer suggested, should be decided according to the types of the measure (such as a ban, restrict on use, or maximum concentration level) and its effect on the internal market, rather than on what is stated in recitals or in the articles. Bradley 2011, p. 92-93; Krämer 2011, p. 6, 72 and 75.

462 Art. 114 TFEU does not pre-empt the pursuit of non-market objectives by internal market regulation. De Witte 2012, p. 26.

463 It is generally agreed that there is exhaustive harmonisation for the sustainability scheme, see e.g. Scott 2011a, p. 831; Romppanen 2012a, p. 127.

464 A derogation may be firstly searched for in the authorisation laid down by harmonisation measures themselves, in the form of allowing one or more Member States to derogate from specific issues and/or for a specific period of time. However, when such provisions are not found, Member States could turn to the Treaty for grounds for differentiation. Vos 2001, p. 146 and 149; De Búrca 2000.

465 According to paras. (4) to (10) of Art. 114 TFEU, Member States may derogate from the harmonisation measures based on Art. 114 TFEU under specific circumstances. The derogation provisions are included in Art. 114 largely as a 'counterweight' to offset the relinquishment of the unanimity requirement and 'a useful tool where an approximation norm does not cater sufficiently for legitimate national interests' or local problems. Maletic 2013, p. 148-149 and 159-160.

466 Commission Decision 1999/835/EC, para. 112.

467 As for *ratione personae*, it has been clarified that Member States can have recourse to the derogation provisions even if they voted for the EU harmonisation measure in the Council. About *ratione materiae*, the provisions apply to an EU harmonisation measure adopted on the basis of Art. 114 TFEU. De Sadeleer 2003, p. 892-894.

ratione temporis: Member States can maintain national measures which already exist at the time of adopting the harmonisation measure but introduce new national measures only thereafter,[468] according to which the two paragraphs provide different substantive conditions to be applied.

a) Maintaining More Stringent Measures

It is provided in paragraph (4) of Article 114 TFEU that, a more stringent existing national measure can be maintained if it is necessary to protect an interest under Article 36 TFEU[469] or is related to the protection of the environment or the working environment. Thus, Member States have to prove that the existing more stringent measure is maintained for the purposes as provided in the EU law and it is necessary to be maintained to achieve the objectives.

Stricter national criteria concerning biodiversity and climate change might be justified under this paragraph, since they are maintained for the protection of the 'environment.'[470] However, complementary measures for the socio-economic sustainability of biofuels may not find grounds in this provision, as the socio-economic concerns such as land rights, food availability, or the labour right are unlikely to fall within the meaning of public morality,[471] public policy,[472] public security[473] of Article 36 TFEU. Nevertheless, protecting the labour rights of workers

468 De Sadeleer 2003, p. 895.

469 The reasons mentioned in Art. 36 TFEU contain the grounds of 'public morality, public policy or public security; the protection of health and life of humans, animals or plants; the protection of national treasures possessing artistic, historic or archaeological value; or the protection of industrial and commercial property', and the grounds listed in the paragraph are exhaustive in nature, precluding any expansion. Maletic 2013, p. 72.

470 Following what is stated in Art. 191(1) and Art. 192(2), it is assumed that the 'environment' should include human health, natural resources (including wild animals), climate, waste, water, land use, town and country planning (man-made environment). Moreover, in from a territorial perspective, the environment mentioned is not limited to the EU's environment, nor is it restricted to trans-frontier environmental problems. Krämer 2011, p. 1.

471 Most of the cases in which the Court has directly admitted the public morality justification have concerned obscene, indecent articles, gambling, or the protection of minors with regard to videos and DVDs. European Commission, Free movement of goods Guide to the application of Treaty provisions governing the free movement of goods, Prepared and drafted by Directorate C, Regulatory Policy, of the Enterprise and Industry DG, 2010, available at: <http://ec.europa.eu/enterprise/policies/single-market-goods/files/goods/-docs/art34-36/new_guide_en.pdf> (last accessed on 18 August 2013), p. 26.

472 Public policy is interpreted very strictly by the Court of Justice and has rarely been accepted alone as justification. This happened in only one exceptional case in which a Member State restricted the import and export of gold-collectors' coins because the national restriction stemmed from the need to protect the right to mint coinage, which is traditionally regarded as involving the fundamental interests of the state. European Commission, Free movement of goods Guide to the application of Treaty provisions governing the free movement of goods, Prepared and drafted by Directorate C, Regulatory Policy, of the Enterprise and Industry DG, 2010, available at: <http://ec.europa.eu/enterprise/policies/single-market-goods/files/-goods/docs/art34-36/new_guide_en.pdf> (last accessed on 18 August 2013), p. 26.

473 The public security justification has been used in the specific area of the EU energy market. In one case a Member State ordered petrol importers to purchase up to 35 per cent of their →

in the production of biofuels might be justified by the protection of the 'working environment' which is defined as non-economic considerations relating to the safety, health and hygiene of workers[474] or the conditions at the workplace.[475] For the measures maintained for the purposes provided in EU law, Member States have to prove the necessity of maintaining such measures to achieve the claimed purpose, which is to be elaborated on later.

b) Introducing More Stringent Measures

On the whole, the substantive conditions set out for introducing more stringent measures are stricter than those for maintaining stricter measure, for the reason that adoption of new national legislation is more likely to jeopardize the EU's harmonisation act.[476]

Firstly, only measures for protecting the environment and working environment can be justified under paragraph (5) of Article 114 TFEU. The question is whether a national measure for protecting food accessibility and thus 'human health' can fall within the meaning of the 'environment', since the interpretation of the 'environment' following Article 191(1) and Article 192(2) may include human health.[477] However, the European Commission once blocked the adoption of stricter national measures for human health under this paragraph, which reflects that the European Commission may not favour a broad interpretation of 'environment' in applying Article 114(5) TFEU.[478]

Secondly, the problem or risk serving as the reason for introducing more stringent measures should be *specific* to the applicant Member State. The intention of this requirement is to avoid derogations of a general character: only particular circumstances such as demography, geography, or epidemiology should render the problem particular, so as to prevent the derogation from solving a problem which is

petrol requirements from a national petrol company at prices fixed by the government. The Court held that the measure was protectionist and violated Art. 34 TFEU but this measure was held to be justified on grounds of public security in order to maintain a viable oil refinery to meet supply in times of crisis. Public security can also serve as a justification in cases about strategically sensitive goods and dual-use goods because 'the risk of serious disturbance in foreign relations or to peaceful coexistence of nations may affect the security of a Member State.' In these cases the Court stated that the scope of Art. 36 TFEU covers both internal security and external security. European Commission, Free movement of goods Guide to the application of Treaty provisions governing the free movement of goods, Prepared and drafted by Directorate C, Regulatory Policy, of the Enterprise and Industry DG, 2010, available at: <http://ec.europa.eu/enterprise/policies/single-market-goods/files/-goods/docs/art34-36/new_guide_en.pdf> (last accessed on 18 August 2013), p. 26-27.

474 De Sadeleer 2003, p. 896; Maletic 2013, p. 71.

475 Krämer 2011, p. 2.

476 De Sadeleer 2003, p. 899.

477 Krämer 2011, p. 1.

478 The reason could be that it is difficult to argue that human health in one Member State is more likely to be endangered than in another Member State. Moreover, Art. 114(8) particularly provides a derogation on the grounds of human health. Maletic 2013, p. 75 and 88; De Sadeleer 2003, p. 900; Sevenster 2000, p. 302.

common to the whole EU.[479] Meanwhile, specificity should not be interpreted too narrowly as 'unique' problems that are present exclusively in a Member State, and it has been confirmed in case law that it is sufficient to demonstrate the problem is 'unusual.'[480] According to the specificity requirement, stricter sustainability criteria with regard to national or local relevance, such as water, soil quality, biodiversity and habitat, or carbon emissions,[481] might be argued to be specific to one Member State.

Thirdly, the specific problem or risk should arise *after* the adoption of the EU harmonisation measure and *before* the end of the prescribed implementation period.[482] The term 'arise' can be interpreted as 'manifest' or 'being discovered', so the problem may have already occurred but only manifest itself or be discovered after adoption of the harmonisation measure.[483] Accordingly, to adopt criteria for the latest biofuel sustainability concerns, Member States have to show that the problem, such as indirect land use, became manifest after the adoption of the sustainability criteria or the relevant evidence needs to be given a different interpretation.[484] However, Directive 2009/28/EC has already taken note of the

479 The requirement of specificity would have been set to prevent the adoption of national measures intended to solve a problem common to the whole of the EU. Maletic 2013, p. 75; De Sadeleer 2003, p. 900.

480 As for the specificity requirement, the Court in the case *Land Oberosterreich and Austria v. Commission* (Joined Cases C-439/05 P and C-454/05 P, paras. 66-67) clarified that the condition of specificity does not presuppose the existence of a 'unique' problem, but Austria still failed to show that this area of Austria had such an unusual or unique ecosystem, compared to other areas of Austria or other similar areas of Europe, that separate risk assessment is needed. This interpretation allows the situation that a similar problem can occur in another Member State. On the whole, specificity 'cannot be ascertained by the presence of characteristics that are absent in all other Member States' and also cannot be rejected because of the appearance of the same problem in two or more Member States. De Sadeleer 2003, p. 901; Doherty 2004, p. 124; Maletic 2013, p. 76.

481 Compared to the other environmental concerns, the emissions risk may be more difficult to argue as being specific to a state, which has been shown in the Dutch Emissions Case concerning measures taken to combat air pollution resulting from motor vehicle emissions. The European Commission rejected the request because health problems caused by diesel particulates were not particularly acute in the Netherlands and existed also in other Member States. The General Court agreed with the Commission and decided that the Netherlands had not proven the existence of a problem particular to its territory that would necessitate derogation. However, the European Commission later approved for the first time a new national derogation measure introduced on basis of Art. 114(5) in which the Commission stated that, although the environmental concern at stake was not specific to the Netherlands, the extent to which the problem existed in the Netherlands has 'specificity' due to the country's abundance of groundwater and surface water. In this vein, it is possible to assume that a common environmental problem such as GHG emissions could still be regarded as 'specific' if the *extent* of the problem is acute or particularly serious in one Member State or if the emission problem can be best or particularly well solved in one Member State. Maletic 2013, p. 133.

482 De Sadeleer 2003, p. 901.

483 *Ibid.*; Maletic 2013, p. 76-77.

484 In the Case *Land Oberosterreich and Austria v. Commission*, Advocate General Sharpston accepted that a problem can be considered as arising after the adoption of EU measures if it →

potential negative effects and required reporting on them, including those that are not addressed by the current sustainability criteria, so it might be difficult to argue for a 'new' problem or risk; nevertheless, there is still a possibility that the reporting might show that a new problem arises.

Fourthly, the introduced stricter measure must be based on 'new scientific evidence.' It means that the scientific evidence as the reason for the derogation shall be novel or 'in progress' in the light of the scientific discoveries made after the adoption of the harmonisation measure.[485] Nonetheless, it is still possible for scientific evidence to be considered 'new' if it already existed but had not been entirely validated or considered during the decision-making process at the time of the adoption of the EU measure.[486] Regarding the quality and certainty of scientific evidence, De Sadeleer emphasises that the serious nature of the evidence and scientific consensus is not necessary, because the use of word 'evidence' instead of 'fact' implies that it is sufficient to show an indication of a *possible* link between the factor in question and the damage which occurred.[487] In case of scientific uncertainty, when the national measure is necessary to protect one of the enumerated interests and is drafted in a way that does not cause a disproportionate effect on the internal market, the measure can probably be approved.[488] In this sense, it is possible to argue that scientific evidence addressing biofuel sustainability appeared after the adoption of Directive 2009/28/EC, including the circumstance that the evidence already existed but was not entirely validated or considered during decision-making process of the Directive or where there is scientific uncertainty. However, the question is which kind of scientific evidence Member States can rely on, since the case law might imply that the European Commission has given preference to risk assessments at the EU level over the national investigations.[489]

Accordingly, it is possible that the afore-mentioned substantive conditions as set out in Article 114(5) TFEU are satisfied cumulatively, as for the environmental concerns of water, soil quality, biodiversity and habitat, or even carbon emissions, but complementary measures regarding socio-economic sustainability concerns can hardly be introduced. However, similar to Article 114(4) TFEU, paragraph (5) also requires that the more stringent measure is necessary for the interest to be protected.

was latent but then discovered only subsequently. However, a mere reassessment of the evidence is not enough; Member States have to provide a detailed explanation as to how the relevant evidence calls for a different interpretation. Opinion of Ms Sharpston – Joined Cases C-439/05P and C-454/05P, Opinion of Advocate General Sharpston, delivered on 15 May 2007, para. 132.

485 Sevenster 2000, p. 304.

486 De Sadeleer 2003, p. 902.

487 *Ibid.*

488 Maletic 2013, p. 171.

489 Maletic 2013, p. 130.

c) Approval by the European Commission

Paragraphs (4) and (5) both require that national derogation measures shall be notified and approved by the European Commission.[490] Apart from the substantive conditions provided in paragraphs (4) and (5), the Commission is also obliged to verify whether the maintenance or introduction of the national measure constitutes 'a means of arbitrary discrimination' or 'a disguised restriction on trade between Member States', or 'an obstacle to the functioning of the internal market.'[491] Such an examination is to focus on the proportionality and suitability of the measure and its purpose, and the proportionality and suitability test is also used to assess the necessity between the maintained or introduced measure and the interest to be protected.[492]

In examining the proportionality, the European Commission tends to evaluate whether the envisaged national measure is 'the least restrictive measure to achieve the stated aim of environmental and health protection' and whether its adoption 'will constitute a disproportionate obstacle to the functioning of the internal market.'[493] In this way, the Commission must analyse whether less restrictive measures could have been taken to meet the stated objectives and whether the national measure is the least harmful to the internal market.[494] Meanwhile, the Commission considers the proportionality of the potential benefits to be brought and the disruption to be caused for the internal market,[495] but the proportionality scrutiny must be difficult to meet in the derogation approval since the proportionality of the EU harmonisation measure will have already been examined in the legislative procedure.[496] Accordingly, to maintain or introduce more stringent biofuel sustainability criteria, Member States need to show that it is suitable to maintain or introduce stricter sustainability criteria to protect the interests as specified in Article 114 TFEU, that there is no other less trade restrictive measure to achieve this objective, and the benefit brought about by the national measure is

490 It is provided that the European Commission shall, within six months of the notifications as referred to in paras. 4 and 5, approve or reject the national provision. Upon receiving the request, the Commission has formed the practice (although not mandatory) of informing the other Member States to give them the chance of expressing opinions on the request; but the applicant Member State is not entitled to a right to be heard in the Commission's decision-making process. The Commission must decide to 'approve' or 'reject' the national application within a time limit of six months. If the Commission fails to make a decision or inform the applicant Member State within the time limit, the derogation request shall be deemed to have been approved. The six months limit could be prolonged to another six months in case of complexity unless there would be danger to human health. TFEU, Art. 114(6); De Sadeleer 2003, p. 904-905; Maletic 2013, p. 81.

491 TFEU, Art. 114(6).

492 De Sadeleer 2003, p. 906-907.

493 The Commission has suggested that the notion of an 'obstacle to the functioning of the internal market' be understood as 'disproportionate effect' in relation to the objective pursue. Commission Decision 2006/372/EC, paras. 46 and 62; Commission Decision 2012/160/EU, para. 89; Maletic 2013, p. 85 and 145.

494 Maletic 2013, p. 85 and 145; Commission Decision 2006/372/EC, para. 61.

495 Maletic 2013, p. 85 and 145; Commission Decision 2006/372/EC, para. 55.

496 Maletic 2013, p. 170.

proportionate with the possible negative effect on internal market of biofuels. These conditions and requirements may increase the difficulty for Member States to deviate from the 'ceiling' of EU sustainability criteria by invoking Article 114 TFEU.

d) Lack of Safeguard Clause

Paragraph (10) of Article 114 TFEU stipulates that harmonisation measures 'shall, in appropriate cases' include a safeguard clause that permits Member States to take temporary measures for non-economic reasons as listed in Article 36 TFEU. The purpose is to allow a provisional derogation for Member States to deal with exceptional situations of limited duration for 'non-economic' purpose.[497] It is noticed that, compared to paragraph (4) and (5) of Article 114 TFEU, the grounds of protecting the environment and the working environment are not included in this provision. To invoke the safeguard clause, Member States must notify the European Commission about the provisional safeguard measure to be taken, so that the Commission can examine its consistency with EU law and the proportionality.[498] However, different from the other derogation provisions provided in Article 114 TFEU, this paragraph is not addressed directly to Member States, which means that Member States cannot apply this provision directly.[499] Moreover, although this paragraph requires harmonisation measures adopted under Article 114 TFEU to include a safeguard clause. This requirement is then softened by the expression of 'in appropriate cases', according to which the safeguard clause is not necessarily included in all the harmonisation measures,[500] such as in the case of Directive 2009/28/EC where no safeguard clause is provided. Thus, in the absence of a safeguard clause provided for the sustainability criteria, Member States cannot invoke Article 114 (10) TFEU to adopt provisional derogation measures.

497 These include 'public morality, public policy or public security; the protection of health and life of humans, animals or plants; the protection of national treasures possessing artistic, historic or archaeological value; or the protection of industrial and commercial property.' According to the case law of CJEU, aims of a purely economic nature cannot provide grounds for national measures to adopt safeguard measures. De Sadeleer 2003, p. 913; Maletic 2013, p. 90.

498 It is provided in Art. 114(10) that the provisional measures of Member States are subject to a Union control procedure. De Sadeleer 2003, p. 914.

499 It has been pointed out that both Art. 191(2) and Art. 114(10) TFEU apply where secondary EU legislation expressly contains a provision which enables Member States to take safeguard measures, so both provisions require a specific provision in secondary EU legislation and do not directly give rights to Member States. De Sadeleer 2003, p. 912; Krämer 2011, p. 130.

500 One study observes that the safeguard clauses in EU harmonisation measures on food issues are absent to some extent, which reflects that, in spite of the provision of Art. 114(10), safeguard clauses are not necessarily included in EU legislation. Vos & Weimer 2016 (forthcoming)

5.2.3. Inclusive Biofuels Sustainability Regulation in View of the EU's Commitment to Sustainable Development?

As set forth, the EU's biofuel sustainability criteria have limits in addressing the inclusive sustainability concerns and, for the sake of the internal market, restrict complementary or stricter criteria to be adopted by Member States. Although Article 114 TFEU provides for the possibility to adopt stricter national measures, the possibility is much limited by the substantive and procedural conditions which are to be met cumulatively. In this sense, inclusive biofuel sustainability regulation is called for, because the deficiency in regulation may result in continuous demand, supply, and trade of biofuels that have potential negative impacts, and moreover the commitment to sustainable development has been set out in EU law and policy guidelines.[501]

Sustainable development has been entrenched as one of the fundamental objectives in the EU Treaty and has been part of the '*raison d'être*' of the EU internal and external actions.[502] The Lisbon Treaty reinforces the status and cross-cutting nature of sustainable development in EU internal and external policy.[503] Sustainable development is enshrined as the objective that the EU shall work for and this objective is to be achieved on the basis of both socio-economic development and environmental improvement.[504] Meanwhile, sustainable development is established as one of the objectives in external relations and actions.[505] These treaty provisions do not clarify the meaning of sustainable development, and sustainable development has been considered to be 'more a guideline to policy action than a normative-legal concept'[506] or 'a source of inspiration for EU policy formation' instead of 'an enforceable obligation.'[507] However, beyond being an objective or

501 Douma has analysed the consistency of the sustainability of EU biofuels with the EU's policy guidelines and principles of sustainable development, see Douma 2014.

502 Pallemaerts 2006, p. 19; Duran & Morgera 2012, p. 35.

503 Vedder 2010, p. 285; Duran & Morgera 2012, p. 35.

504 The article reads as 'It shall *work for* the sustainable development of Europe *based on* balanced economic growth and price stability, a highly competitive social market economy, aiming at full employment and social progress, and a high level of protection and improvement of the quality of the environment. It shall promote scientific and technological advance.' TEU, Art. 3(3).

505 Art. 3(5) reads as '[I]n its relations with the wider world, the Union ... shall contribute to peace, security, the *sustainable development* of the Earth....' Art. 21 provides that '[T]he Union shall define and pursue common policies and actions, and shall work for a high degree of cooperation in all fields of international relations, in order to: ...(d) foster the *sustainable economic, social and environmental development* of developing countries, with the primary aim of eradicating poverty; ...(f) help develop international measures to preserve and improve the quality of the environment and the *sustainable* management of global natural resources, in order to ensure *sustainable development*.' Before the Lisbon Treaty, sustainable development was limited to development cooperation which is one aspect of external actions in the EU, and is interpreted as 'sustainable economic and social development....' TEU, Arts. 3(5) and 21(2).

506 Jans & Vedder 2012, p. 8.

507 Such comments on sustainable development are often elaborated on and implemented by EU policy documents, addressing a political question, but sustainabile development has been →

policy guideline, the following normative implications of sustainable development might be drawn.

Firstly, environmental integration forms a critical part of sustainable development. Environmental integration is set out in Article 11 TFEU, and the word 'must' suggests that it is a legal obligation which demands the substantive integration of environmental protection.[508] It is generally agreed that the environmental integration provision entails both internal and external integration,[509] the latter of which requires the environmental dimension to be incorporated into other legislative instruments outside the field of the environment.[510] The relationship between environmental integration and sustainable development can be implied from the wording of Article 11 TFEU: sustainable development serves as a broader *objective* for environmental integration, and environmental integration is a *means* to achieve sustainable development.[511] The question is how to interpret this relationship: either as an emphasis on the importance of the environmental dimension in achieving sustainable development,[512] or as a constraint on the environmental integration in view of the socio-economic dimension of sustainable development.[513] In spite of the divergent

rarely precisely defined by the law or the Court. Duran & Morgera 2012, p. 36; Gillivray & Holder 2001, p. 150-151.

508 However, the actual binding force of Art. 11 TFEU is doubted by Krämer: EU institutions have a very broad discretion in putting the principle of Art. 11 TFEU into practice, and the wide discretion would not make such an action successful. Plus, the Court of Justice has never discussed whether an EU act would be annulled if it had not taken environmental requirements into consideration. Krämer 2011, p. 20-22; Duran & Morgera 2012, p. 30-31.

509 Internal integration, which requires the EU environmental law to be interpreted broadly in light of the objectives set out in Art. 191 TFEU, and the integration of environmental protection is regarded as one principle of EU environmental law. External integration is a requirement and an interpretative tool for EU law used outside the environmental field. Duran & Morgera 2012, p. 31.

510 It has been commented that Art. 11 TFEU is the primary provision aiming at coherence between environmental policy and other policies. Cremona 2012, p. 36 and 39.

511 Pallemaerts 2006, p. 25-26.

512 Krämer argues that environment integration is a *basis* for realising sustainable development considering the text of Art. 3(3) TEU. It is further set out that the close relationship between environment and sustainable development in the Treaty articles can be interpreted as implying the importance of environment in realising sustainable development: environmental requirements must be integrated into other EU policies in order to achieve sustainable development; or even sustainable development can only be reached when environmental requirements are integrated into other policies. Krämer, L., 'Giving a voice to the environment by challenging the practice of integrating environmental requirements into other EU policies', The European Union Aarhus Centre, ClientEarth, 2012, available at: <www.clientearth.org/aarhus-centre-documents/lk-integrating-requirements.pdf> (last accessed 31 July 2013), p. 4.

513 According to this interpretation, the notion of sustainable development is not limited to environmental aspects or restricted to the development of environmental law. In this sense, the environmental integration requirement is purposely linked to the broader objective of sustainable development and it does not prescribe a priority for environmental protection to the detriment of other EU policy objectives. Duran & Morgera 2012, p. 35-36, 39, 54-55; Pallemaerts 2006, p. 25; Gillivray & Holder 2001, p. 144.

interpretation, it can be understood that, sustainable development is closely related to environmental integration, which serves as one important dimension of sustainable development.

Secondly, sustainable development entails an impact assessment of all the economic, social, and environmental dimensions in the process of law and policy making. The EU implemented the 'Sustainable Development Strategy' in 2001,[514] which aims to establish 'a new approach to policy-making' through examining the economic, social, and environmental effects of EU policies in a coordinated way.[515] Such an impact assessment is different from environmental impact assessments or strategic environmental assessments,[516] because it requires that all major policy decisions are assessed in a balanced way in view of the social, environmental, and economic dimensions of sustainable development.[517] Moreover, the precautionary principle should be applied if appropriate and the impact assessment should be based on the best available science.[518]

Thirdly, sustainable development in the EU context contains an external dimension. It is recalled that Articles 3(5) and 21(2) TEU establish sustainable development as one objective in external relations and actions, and EU external policy concentrates on the global perspective of sustainable development as an 'international construct' particularly in poverty eradication and in environmental protection.[519] In fact, according to the Lisbon Treaty, sustainable development in external policy is not limited to development cooperation but is established as an overall objective for the broader range of external actions.[520] The EU Sustainable

514 In 2001, the Gothenburg European Council agreed on a 'European Union Strategy for Sustainable Development' as a declaration of commitment to implementation of the objective at the highest political level. Pallemaerts 2006, p. 19.

515 For instance, the 2006 renewed Sustainable Development Strategy further requires that all EU institutions should ensure that proposals for major policy decisions undergo an Impact Assessment (IA) that assesses the social, economic and environmental dimensions of sustainable development in a balanced way, and policy making tools such as 'ex-post-assessment of policy impacts' should be widely used not only by the EU institutions but also by Member States. Pallemaerts 2006, p. 29; Krämer 2011, p. 35; Duran & Morgera 2012, p. 234-253.

516 Environmental impact assessments have been set out in law principally in Directive 2011/92/EU (known as EIA Dirctive) for individual projects, and Directive 2001/42/EC (known as SEA Directive) for public plans or programmes. There is another kind of impact assessment - a sustainability impact assessment (SIA) which is to ensure environmental integration in the definition and implementation of EU external relations. Pallemaerts 2006, p. 29; Krämer 2011, p. 35; Duran & Morgera 2012, p. 234-253.

517 Douma 2014, p. 32.

518 According to Art. 191 TFEU, in preparing its environmental policy, the EU shall take account of available scientific and technical data and that, where necessary, it shall be based on the precautionary principle. Furthermore, according to the European Commission's Communication on Impact Assessment, the precautionary principle should be applied when appropriate in accordance with the Commission's guidelines. Douma 2014, p. 30, 31-32.

519 Duran & Morgera 2012, p. 34-35 and 43.

520 The objective of fostering 'the sustainable economic and social development of the developing countries' was established for the EU development cooperation policy in the Maastricht Treaty. The external actions cover a broader range of issues, such as common →

Development Strategy also has an external dimension that stresses the issues of globalisation, trade, poverty eradication, social development, and sustainable management of natural resources beyond the EU.[521] In this sense, the EU law and policy guidelines demand that the EU's policies in the areas of, *inter alia*, environment, energy, and the internal market also contribute to the sustainable development of third countries.[522]

The afore-mentioned implications from commitments to sustainable development in EU law and policy may provide a basis for arguing for an inclusive sustainability regulation on biofuels. The sustainability regulation of biofuels should include requirements not only to address environmental concerns but also to minimise socio-economic impacts, in particular regarding land and food, within and outside of the EU, based on the best available knowledge.[523] Admittedly, there should be consideration for compliance with international law, which in turn urges the EU to take proactive measures at the international level and strive for international consensus on the sustainability of biofuels.

In conclusion, from the perspective of inclusiveness and integration, the EU sustainability criteria do not include requirements that address socio-economic concerns, and the supplementary approaches of monitoring/reporting and voluntary schemes have limits to different extents. Moreover, the sustainability criteria even have limitations in dealing with the sustainability concerns already included in the criteria, such as the life-cycle GHG emissions considering indirect land use changes. Meanwhile, a probe into the extent of harmonisation reveals that the sustainability criteria aim for exhaustive harmonisation for the sake of the internal market and restrict more stringent national biofuel sustainability measures to be adopted by Member States. Although the legal basis of Article 114 TFEU may provide for the possibility to maintain or introduce stricter national measures, this possibility is greatly limited by the substantive and procedural conditions set forth in Article 114 TFEU, and by the lack of safeguard clauses for the sustainability criteria. In this view, an inclusive regulation on the sustainability of biofuels could find grounds in the commitment to sustainable development in the existing EU law and policy.

foreign and security policy, development, trade and cooperation policies, and external aspects of its internal policies such as agricultural, fisheries, transport and energy as well as environmental policies. Duran & Morgera 2012, p. 43.

521 Douma 2014, p. 31-32.

522 *Ibid.*, p. 29.

523 It has been argued that a strict interpretation of the precautionary principle should be adopted for biofuel regulation, which means that the EU should refrain from adopting policy measures when it cannot be ascertained that these measures do not carry the risk of serious negative effects for developing countries. Douma 2014, p. 29.

Chapter 4

CHINESE REGULATION OF THE SUSTAINABILITY OF BIOFUELS

1. Introduction

This Chapter aims to analyse the regulatory approach towards the sustainability of biofuels in China.[1] 'Biomass energy' (*sheng wu zhi neng*) is identified as one of the renewable energies in the Renewable Energy Law of China, which is defined as including solid and gaseous biomass and liquid biofuel (*sheng wu ye ti ran liao*) such as bioethanol and biodiesel.[2] For coherence in terminology, this chapter uses the term 'biofuels' to refer to the biomass-derived fuels for use in transport. China is currently one of the leading biofuel (particularly bioethanol) producing countries,[3] which could be attributed to the biofuel support policy since 2002.[4] In line with the research questions of this thesis, an analysis of the regulatory approaches that address sustainability of biofuels in China will be carried out, as well as an examination of how they can address the inclusive sustainability concerns. Such an analysis, however, has the following characteristics in this chapter.

First, an analysis of the regulation of the sustainability of biofuels will include both the legal and policy measures of the central government for constraining any environmental or socio-economic impact of biofuels. China has a unitary legal

1 This chapter is restricted to the law and policy in mainland China, excluding Hongkong and Macao Special Administration Regions and Taiwan.

2 Art. 16 of the Renewable Energy Law uses the term 'biomass fuel' (*Sheng wu zhi ran liao*) to refer to the energy carrier produced from biomass in various forms for all energy generating purposes, while liquid biofuel refers only to biomass fuel that substitutes oil. In any case, the biomass fuel falling within the scope of 'renewable energy' does not include the traditional way of using biomass, such as directly burning straw, fuel wood, or dung in low-efficient ways. Renewable Energy Law of China, Arts. 2 and 16.

3 European Biofuels Technology Platform, Global Biofuels – An Overview, available at: <www.biofuelstp.eu/global_overview.html> (last accessed 5 May 2015); Statista, Leading countries in biofuel production in 2013 (in 1,000 metric tons of oil equivalent), available at: <www.statista.com/statistics/274168/biofuel-production-in-leading-countries-in-oil-equivalent/> (last accessed 5 May 2015); Scott & Jiang 2013, p. 1.

4 In 2002, several ministries under the State Council jointly launched fuel ethanol pilot schemes that mandated bioethanol blending on the market in several provinces and cities in China.

system with multiple (or two) levels,[5] and there are various legal sources including the Constitution, law (*fa lv*), administrative regulations (*xing zheng fa gui*), local regulations (*di fang xing fa gui*), departmental rules (*bu men gui zhang*), and local government rules (*di fang zheng fu gui zhang*).[6] In addition to the stipulated legal sources, policy documents[7] issued by the central government play an important role in regulating and governing the society. The Chinese government has a tradition of relying on policy documents as the primary tool of governing the society until a shift to the 'rule of law' comes about.[8] As yet, the government has kept the practice of issuing policy documents to implement experimental reforms and then translates

5 Chinese scholars often find it difficult to qualitatively define the Chinese law-making system. Some describe the system as a 'one level unitary system', while others believe it to be a 'two level system' or a 'two level, multi-layered system.' There are also some scholars believe it to be 'multi-type and multi-level system under unified central leadership.' Chen finds that a general agreement among scholars in China and abroad is that laws and regulation issued by the 'unitary' system can be divided into three levels: primary level - national laws, secondary level - national administrative regulation, and tertiary level - local regulation. Chen 2008, p. 181.

6 Under the supremacy of the Constitution, the NPC and its Standing Committee exercise legislative power of the state to make 'basic law' (*ji ben fa lv*) and other laws (*fa lv*). In accordance with the Constitution and laws, the State Council has the power to enact administrative regulation (*xing zheng fa gui*). The ministries and commissions under the State Council can issue rules (*gui zhang*). The People's Congresses and their Standing Committees at the province level can enact local regulation (*di fang xing fa gui*); and the People's Congress of autonomous regions, prefectures, and counties can promulgate autonomous regulation (*zi zhi tiao li*) and separate regulation (*dan xing tiao li*). The Law on Legislation classifies legislative documents into three groups according to their legal status: laws; administrative regulations; and local regulations, autonomous regulations and separate regulations, and rules. Particularly, the Law on Legislation entails the competence and scope of 'rules' including departmental rules (*bu men gui zhang*) and local government rules (*di fang zheng fu gui zhang*), which is not specified in the Constitution. Ministries, commissions, the People's Bank of China, the National Audit Office, and other departments directly under the State Council (*guo wu yuan zhi shu ji gou*) are competent to issue departmental rules for implementing national laws as well as State Council's regulations, decisions and orders. Local administrations of provinces and prefectures promulgate local rules (*di fang zheng fu gui zhang*) to implement national laws, administrative regulations, and local regulations of the same jurisdiction. Constitution of China 1982, Arts. 5, 57, 58, 62(1)-(3), 64, 67(2)-(4), 89(1), 90, 100, and 116; Law on Legislation of China 2000, Arts. 72, 73, 80, and 82; Chen 2008, p. 124-125, 179-181 and 188.

7 Some scholars refer to them as 'policy laws', which may include policy statements, administrative orders and instructions. Lubman 2000, p. 391.

8 Policy was once alone articulated and applied by the Party to direct and guide the entire state, while law was used only formalistically to declare policy and it was imprecise, exhortational, tentative, and subject to unlegislated revision. It is the 1982 Constitution that recognised the promulgated laws rather than the policy as the 'appropriate vehicles both for defining and implementing policy.' Since then, law making has provided the major framework for the organisation and operation of the Chinese government, because law is viewed as a better tool than policy for securing and institutionalising *ad hoc* policies in a more universal manner, providing stability and order through state coercive forces for economic development and defining rights and duties in relation to the state. Lubman 2000, p. 384; Chen 2008, p. 52-53.

them into legislation based on experience gained from the experimentation.[9] Moreover, policy documents are used to set out strategic social and economic development plans: as a legacy from the planning economy, China has continued to formulate 'five-year plans' (*wu nian ji hua*) (renamed as 'five year guidelines' since the 11th Five Year Plan)[10] to set out macro-level social and economic development policies and strategic plans, and this approach remains to be an integral part of China's system of governance.[11] The national 'Five Year Plans' for economic and social development are drafted essentially on the basis of the Central Communist Party's 'recommendations',[12] which is to be adopted by the National People's Congress[13] and implemented by administrative authorities.[14] There are divergent views as to the status of the national Five Year Plans for social and economic development: one extreme view is to classify them as Constitutional legal documents (*xian fa xing fa lv wen jian*), while the other extreme view sees that they

9 Keller observes that the Communist Party of China (CPC) is used to issuing policy documents to initiate political, economic and social changes, while law is used to consolidate the changes when the relevant circumstances are 'ripe.' Keller 1994, p. 730; Chen 2008, p. 59.

10 The term of '*ji hua*' was purportedly changed to '*gui hua*' since the 11th Five Year period in order to reduce the sense of planning economy. There are different translations of the new name '*gui hua*': it is primarily translated as 'guidelines' but it can also be translated as 'programme', while the NPC English website still keeps the word 'plan.' Thus, this thesis uses the term 'plan.' Xu 2012, p. 43-44; The National People's Congress of the People's Republic of China, Full text: Report on China's economic, social development plan, available at: <www.npc.gov.cn/englishnpc/Special_12_3/2015-03/19/content_1930758.htm> (last accessed 7 May 2015).

11 As a legacy from the planned economy, China has continued to formulate such strategic plans every five years to set out national policy goals, to highlight the focuses of government, and to guide market operators, as well as to provide a basis for the government's economic adjustment, market surveillance, social management, and public service. 我国国民经济和社会发展十二五规划纲要(全文) [The 12th Five Year Guideline for National Economic and Social Development (full text)], available at: <http://news.sina.com.cn/c/2011-03-17/055622129864.shtml> (last accessed 7 May 2015).

12 The Communist Party of China is to provide a recommended version of the Five Year Plans before their adoption by the National People's Congress (NPC). For instance, the 12th Five Year Guideline was adopted by the NPC in March 2011 and before that the Central Committee of CPC adopted the recommendation for the 12th Five Year Guideline in October 2010. 授权发布：中共中央关于制定国民经济和社会发展第十二个五年规划的建议 [Authorised publication: Recommendation for the 12th Five Year Guideline of National Economic and Social Development] (News.cn, 27 October 2010), available at: <http://news.xinhuanet.com/politics/2010-10/27/c_12708501.htm> (last accessed 7 May 2015).

13 The Five Year Plan needs to be adopted by the NPC as the powers of the NPC include the 'examination and approval of government budgets and economic and social development plans' as well as the 'supervision of their implementation', and the Standing Committee of the NPC has the power to partially adjust the plans for economic and social development when the NPC is not in session. Constitution of China 1982, Arts. 62(9) and 67(5).

14 The national Five Year Plan only sets forth general guidance, and it is further elaborated on and implemented by local or departmental administrations by formulating their own local or sectoral five year plans. Thus, the sectoral plans might be issued at a time later than the Five Year Plan's starting year, such as in the case of 11th Five Year Plan for renewable energy which was issued in 2008.

are not legal documents and they have no legal force; the opinion in between argues that they are legal documents but only have legal force for the government, or they are the secondary social norms next to the law.[15] The sectoral Five Year Plans for renewable energy relevant for biofuels are purported to implement the general guidelines in the national Five Year Plan for social and economic development and their binding feature is more ambiguous. Nevertheless, policy and law are interlinked in China: as Chen summarised, policy is the foundation of law and law is the mature form of policy.[16] Policy documents have the features of being *ad hoc and non-legally binding*, but in many cases they might be more effectively implemented than the law because they usually contain more specific targets and requirements.[17] Thus, when probing into the regulation on the sustainability of biofuels, both the relevant legal measures from various sources and policy documents will be considered. Second, as China is a unitary state with a centralised governmental structure, the analysis is focused on what regulations are set out by the central government and to what extent they allow complementary or stricter measures by the lower (provincial) level of government. Third, in cases where the term 'sustainability' is not mentioned in relation to biofuels, the analysis will look into any regulation that in effect addresses the potential environmental or socio-economic impacts of biofuels.

This chapter is structured as follows. Section 2 addresses the legal or policy measures relating to the demand for biofuels, in particular with regard to environmental or socio-economic requirements set out in such measures. Section 3 analyses the regulatory approaches to addressing environmental and socio-economic sustainability concerns in the process of producing biofuel feedstock in the agriculture and waste sectors. Section 4 analyses the implications of the China's internal and external measures on biofuels for trade and external action. Section 5 finally comes to a conclusion by reflecting on the observations of the chapter from the perspective of whether sustainability concerns are addressed inclusively in view of China's commitment to sustainable development.

2. Chinese Regulation of the Demand for Biofuels

Similar to other countries, the use of biofuels in China is largely driven by the government's support policy. Biofuels have been promoted by the Chinese government as a renewable energy mainly to mitigate the oil shortage,[18] reduce air pollutants[19] and carbon emissions,[20] and to promote rural development and the

15 Xu 2012, p. 45; Hao 2007, p. 99-100.

16 Chen 2008, p. 53.

17 Policies can be as important as laws or may even be considered more important in practice, because law and the other legal instruments tend to be rather general without details about specific goals and methods of implementation, while plans with target requirements are to be implemented by government agencies and officials by integrating such requirements into the cadre system. Lin 2012b, p. 308-309.

18 Renewable Energy Medium- and Long-Term Plan 2007, p. 19.

19 China has concluded that the use of biofuels would reduce emissions from pollutants such as CO_2 and hydrocarbon. Measures on Expanding Bioethanol Pilot Projects 2004.

economy.[21] In fact, the consumption of biofuels has increased due to the support policy from the central government primarily in the form of blend mandates and consumption targets since 2002.[22] Other measures such as preferential energy taxes and the emerging emission trading rules are also likely to affect the demand for biofuels. However, the central question is whether a sustainability requirement is integrated into these measures. This section will firstly look into the requirement in relation to the environmental or socio-economic impact of biofuels, and then an analysis will be carried out to see whether such a requirement is considered in the biofuel's demand-driving measures.

2.1. *General Environmental and Socio-economic Principles*

General environmental and socio-economic principles have been set out in the biofuel development policy that there shall be no competition for food and arable land and no damage to the ecological environment. These principles are stated in the 2007 Renewable Energy Medium- and Long-Term Plan[23] and are reiterated in the 11th Five Year Plan for Renewable Energy Development[24] as guidelines for adopting development policy or plans in the biofuel sector. It is stated that the development of biofuels (as well as biomass for other energy use) shall properly address its relationship with the environment and shall not illegally occupy arable

20 Liu Yuanyuan, 'China seeks to develop biofuels industry despite production difficulties' (Renewable Energy World, 24 April 2012), available at: <www.renewableenergyworld.com/-rea/news/article/2012/04/china-seeks-to-promote-development-of-biofuels-industry-despite-production-difficulties> (last accessed 5 May 2015).

21 According to Li Shizhong, deputy director at New Energy Research Center of Nuclear Research Institute, Tsinghua University, 'the biofuels industry is an integrated system that can drive the development of a variety of other industries, including agriculture, chemicals, plastics, automotive, power generation, transportation and services while helping boost domestic demand' and 'unlike other industries, biofuels help maintain social stability by addressing a number of agriculture-related issues, including increasing farmers' income and creating jobs for millions of farmers.' Liu Yuanyuan, 'China seeks to develop biofuels industry despite production difficulties' (Renewable Energy World, 24 April 2012), available at: <www.renewableenergyworld.com/rea/news/article/2012/04/china-seeks-to-promote-development-of-biofuels-industry-despite-production-difficulties> (last accessed 5 May 2015).

22 Chang *et al.* 2012, p. 4.

23 The original text could be translated as follows: '[P]articular emphasis shall be placed on the relationship between biomass energy development and the grain and ecological environment; arable land shall not be illegally occupied, the grain shall not be consumed in large quantities, and the ecological environment shall not be damaged.' Renewable Energy Medium- and Long-Term Plan 2007, para. 4(2), p. 16.

24 The original text could be translated as follows: '[M]ake rational development and utilisation of marginal land resources; energy crops/plants', and the 'cultivation of biofuel feedstock shall not compete with people for grains or compete with grains for land, and shall not damage environment; and shall deal well with the relationship between energy use of biomass and other uses.' The 11th Five Year Plan for Renewable Energy Development, para. 3(2), p. 20.

land, delete large quantities of food, or damage the environment.[25] The 12th Five Year Plan for Renewable Energy Development also states that the development of biofuels shall rationally make use of marginal lands (such as saline land, waste grassland, and hillside land) and construct non-grain biofuel feedstock supply bases.[26] In this sense, China has established very general principles with regard to biofuels to reduce their potential negative environmental and socio-economic impacts, which can be summarised as: *a)* feedstock from non-food grains, *b)* the rational use of marginal lands, *c)* no environmental damage. However, the principles are set out in vague and slogan-like wording, and the development plans do not further specify the meaning of the principles, particularly as to what 'non-food grains', 'marginal lands', or 'environmental damage' entail. Moreover, the general principles are set out in the renewable energy development plans which have very ambiguous legal features as discussed earlier in section 1, and they need to be further translated into more specific legal or administrative measures. Nevertheless, the general principles may still have an effect on the governmental authorities of lower levels: according to Article 8 of Renewable Energy Law, the provincial energy authorities must compile renewable energy development plans within their jurisdiction in accordance with national renewable energy development plans.[27]

By all means, the establishment of these principles reflects a major change in China's biofuels policy. The central government initially promoted biofuels in an attempt to deal with the excessive grain reserves after continuous harvests,[28] and the state-owned bioethanol producing plants[29] were constructed to use the old/stale food grains[30] (unconsumed for more than two years)[31] such as rice and corn.[32]

25 Renewable Energy Medium- and Long-Term Plan 2007, p. 16.

26 The 12th Five Year Plan for Renewable Energy Development.

27 Such provincial renewable energy plans shall be approved by the competent provincial government and be recorded by the national energy authorities. Renewable Energy Law of China 2005, Art. 8.

28 China happened to have harvests in the years before 2002 which resulted in increased grain reserves and plunged grain prices. Another result was an increase in the central government's total cost (up to 2.8 billion yuan) to administer and maintain the grain reserve. The low grain prices resulted in farmers losing motivation to grow grains and increased the income gap between urban and rural residents. Thus, to use the stale grains to produce bioethanol was expected to be a 'silver bullet' to deal with the surplus stale grains, raising grain prices and thus increasing farmers' income. Dong 2007, p. 3.

29 To supply the bioethanol needed in the blend mandate pilot project, the central government established four state-owned fuel ethanol plants during 2002 and 2004. The four plants are Henan Tianguan Group (with a production capacity of 300,000 tons), Anhui Fengyuan Bio-chemical Corporation (with a production capacity of 320,000 tons), Heilongjiang Hua Run Ethanol Corporation (with a production capacity of 100,000 tons), and Jilin Fuel Ethanol Corporation (with a production capacity of 300,000 tons). Measures on Expanding Bioethanol Pilot Project 2004.

30 Grain has been defined in China's law as including wheat, rice (paddy), corn, miscellaneous grains (*za liang*) and processed grains (*cheng pin liang*). Administrative Regulation on Grain Circulation 2004, Art. 2.

31 China has a food self-sufficiency strategy for the sake of national security and the government has implemented a national grain reserve (*chu bei liang*) system – a certain →

However, as the old/stale food grains were used up, some producers began to produce bioethanol from new food grains.[33] In particular, the sharp increase in food prices in international market in 2006 and 2007 led to growing concerns about food security in China.[34] The central government expressed serious concern about the impact of biofuels on China's food prices and security[35] and started to reassess and adjust the bioethanol development policy. In 2006, the National Development and Reform Commission (NDRC) announced that any new bioethanol producing plant must be approved by the central government before construction and any new bioethanol plant using food grains as feedstock shall not be supported or subsidised.[36] This instruction was later translated into a 'non-grain' principle for the whole biofuel industry.[37] Hence, it is interesting to note that China's biofuel policy has been closely related to the government's strategy and macro-control of food, as biofuels were initially promoted for the consumption of surplus food grain but were

percentage of food grain is to be kept in national or local governmental stock in case of poor harvest years. It is required that the maximum reserve period is two to five years and the grains reserved longer than the provided period are called 'stale grains' (*chen liang/chen hua liang*) which shall not be sold on market as food. The reserve period is five years for wheat, three years for rice and corn, two years for edible oil/fat and beans. Dong 2007, p. 2-3; Regulation on Central Grain Reserves 2003, Art. 16; Administrative Regulation on Grain Circulation 2004, Art. 9; Administrative Measures on Grain and Oil Storage 2009, Art. 20.

32 For example, storehouses in the Jilin Province in 2001 bulged with corn stocks after several years of good harvests, which caused corn prices to plunge to a ten-year low. The nominal stockpile of cereal in Henan Province totaled 35 million tons, reflecting an annual surplus of maize and wheat of 4-5 million tons. Qiu *et al.* 2012, p. 3100.

33 The inventory of old grain was consumed rapidly and was almost depleted by 2006, but an increasing number of corn-based bioethanol projects were started by local governments, some of which used fresh grain as feedstock. Chang *et al.* 2012, p. 4.

34 International food prices started to spike since 2006, even though the price increase of major domestic grains in China (such as corn, wheat, and rice) were not remarkable, it demonstrated a strong implication for the potential food insecurity. Chang *et al.* 2012, p. 3-4.

35 As corn is the major bioethanol feedstock in the blend mandates, there were concerns that the use of corn for biofuels would cause an unbalance between corn demand and supply, considering that China would face an irreversible decrease in arable land and water and an intensified demand of food in medium- and long-term and food security has always been significant strategic issue of the national government. Plus, corn was already largely used for feed in China, while the increased development of corn processing (for energy or other non-agricultural uses) began to compete with the use for feed. Furthermore, the increasing prices of corn would reduce the production of other grain crops such as wheat and rice, which thus might result in an unbalanced structure of grain production. Urgent Notice on Corn Processing 2006, para. 2(1) and (2).

36 Urgent Notice on Corn Processing 2006; Notice on Bioethanol Project Management 2006.

37 There is less food concern for biodiesel, largely because China is a net importing country of vegetable oil, so the market price of vegetable oil for cooking is higher than that for energy use, which makes it unprofitable to use vegetable oil (such as from rapeseeds) for biodiesel production. Biodiesel produced in China mainly comes from non-food oil crops or used cooking oil, waste animal fact, or residue oil from vegetable oil crushers, instead of from fresh vegetable oil or soybeans. Furthermore, China has been looking to develop biodiesel based on seeds of energy trees (such as Jatropha, Xanthoceras sorbifolia, and Pistacia chinensis), among which Jatropha is considered as one of the most favorable feedstocks in China. Scott & Jiang 2013; Qiu *et al.* 2012, p. 3098.

then restricted due to the food security concern. However, the food security concern embedded in the non-food grain principle is only one of the potential sustainability impacts of biofuels in China, and there has been debate about the impacts on land use and the environment if large-scale cultivation of biofuel feedstocks occurs.[38]

2.2. *The Other Regulatory Measures Addressing the Demand for Biofuels*

The general environmental and socio-economic principles, as set out in the renewable energy development plans, may affect the demand for biofuels if they are indeed adhered to in the regulatory measures that directly drive or restrict use of biofuels. In this vein, it is important to examine the regulatory measures which can exert a major influence on the demand for biofuels in China, focusing on the question of how they can affect demand for biofuels and integrate environmental or socio-economic sustainability requirements.

2.2.1. Blend Mandates

Biofuel blend mandates hereby refer to current obligatory requirements on fuel suppliers to incorporate a certain percentage of biofuels in the fuels to be sold. Pilot projects of an experimental nature have been carried out for mandating blend of bioethanol in several cities and provinces in China since 2002. As mentioned in section 2.1, such a promotion policy with blend mandates was introduced in the context of the food grain reserve surplus and price drop at that time. In the pilot project, the central government[39] mandated a 10 per cent bioethanol blend (E10) in gasoline supplies in the designated cities and provinces. The first phase of the pilot project lasted for one year from 2002 to 2003 in five cities,[40] and the second phase started in 2004 and was extended to more cities and five whole provinces[41] based on the conclusion that the first phase was successful.[42] To implement the blend

38 Chang *et al.* 2012, p. 3; Zhang *et al.* 2009.

39 The governmental agencies involved included the National Economic and Trade Commission, the National Planning Commission (the former National Development and Reform Commission), the Ministry of Finance, the Administration of Taxation, the Ministry of Police, Administration of Industrial and Commercial Management, the Administration of Quality Supervision, the Administration of Environmental Protection (which became the Ministry of Environmental Protection in 2008). Measures on the Bioethanol Pilot Project 2002.

40 The first phase of the pilot project was undertaken in three cities in the Henan province (Zhengzhou, Luoyang and Nanyang) and two cities in the Heilongjiang province (Ha'rbin and Zhaodong) for one year until July 2003. Measures on the Bioethanol Pilot Project 2002.

41 The second phase of pilot project was extended to cover five whole provinces (Heilongjiang, Jilin, Liaoning, Henan, and Anhui), and nine cities in the Hubei province, seven cities in the Shandong province, six cities in the Hebei province, and five cities in the Jiangsu province. These places were chosen due to their abundance of grain feedstocks such as corn and rice. Measures on Expanding the Bioethanol Pilot Project 2004.

42 It was concluded that the use of fuel ethanol had increased stably and was recognised by consumers. Moreover, it was concluded that appropriate use of bioethanol could to a certain extent reduce pollutant emission such as CO_2 and hydrocarbon. Therefore, a general conclusion for the first phase was that bioethanol is suitable for China and its promotion →

mandate, the two largest state-owned oil companies, China Petrochemical Corporation (Sinopec)[43] and China National Petroleum Corporation (CNPC),[44] were required to blend bioethanol produced from the designated state-owned bioethanol plants with gasoline to be sold in their networks.[45] Thus, such a blend mandate approach has the following characteristics: bioethanol is supplied by the designated producing plants and blenders and is mandatorily sold within the provincial or municipal jurisdictions.

The blend mandate approach was later established in the Renewable Energy Law of China in 2005. It is provided in the law that oil-selling companies shall incorporate the biofuels which meet the issued national standards into their fuel selling networks if it is thus required by energy authority of the State Council, primarily referring to the National Development and Reform Commission (NDRC)[46] or the provincial governments.[47] If oil-selling companies fail to do so, they

would contribute to social, economic, and environmental improvements. Measures on Expanding the Bioethanol Pilot Project 2004.

43 The Sinopec (Group) is an extremely large petroleum and petrochemical enterprise group solely invested by the state, functioning as a state-authorised investment organisation in which the state holds the controlling share. Its key business activities include marketing and the comprehensive utilisation of oil and natural gas, oil refining, the wholesale of gasoline (as well as kerosene and diesel), and the production, marketing, storage, transportation of petrochemicals and other chemical products. SINOPEC, About Sinopec Group, available at: <www.sinopecgroup.com/group/en/companyprofile/AboutSinopecGroup/> (last accessed 13 June 2014).

44 The CNPC (Group) was established on the basis of the former Ministry of Petroleum Industry and is a state-owned oil company endowed with certain governmental administrative functions. Its main business activities cover oil and gas upstream and downstream operations as well as oilfield services and engineering construction. CNPC, History, available at: <http://classic.cnpc.com.cn/en/aboutcnpc/companyprofile/history/> (last accessed 13 June 2014).

45 In China, the refining and retailing of oil are principally controlled by the big three state-owned oil companies: CNPC, Sinopec, and CNOOC, and a market division has formed between them. Thus, it was required that Sinopec was responsible for establishing blending centers in Zhengzhou, Luoyang and Nanyang which was charged with blending and distribution of the three cities, while CNPC was responsible for establishing blend centers in Ha'rbin and Zhaodong. For example, all the bioethanol output of the Jilin Fuel Ethanol Co. Ltd, the largest bioethanol producer in China, is shipped to the nearby blenders of CNPC by rail. The CNPC blenders (as many as 20 ethanol blending stations in the Jilin Province) are in charge of all the remaining businesses, including the storage, blending, and delivery of E 10 to the retailers. Nevertheless, an exception is in place for the oil supply to the military and for other special uses. Measures on Expanding the Bioethanol Pilot Project 2004, para. 2(3) and (4); Qiu *et al.* 2012, p. 3098.

46 The NDRC is China's most important economic and social planning department under the State Council and is responsible for drafting economic and social development plans for China. Due to the importance and broad scope of the mandate, it also undertakes certain responsibilities in the fields of energy and climate change as well as food and the environment. The NDRC has a subsidiary body, the National Energy Agency (NEA), in charge of the energy-related issues. The NEA exercises an essential power in setting overall strategies and goals in particular with regard to setting energy prices, while the other departments under the State Council are responsible for issues within their own jurisdictions. It is worth mentioning that in 2010 China established the 'National Energy Committee' above →

shall compensate the biofuel producing plants for the incurred economic losses. If they still refuse to incorporate within a set period, a penalty of maximum twice the economic losses can be imposed by the relevant energy authorities of the central or provincial governments.[48] The provisions do not impose a biofuel mandate nation-wide, but they delegate the power to decide whether and how to start to implement a blend mandate to the energy authorities of the State Council and provincial governments. As yet, biofuel blend mandates have been implemented in 7 provinces (including Guangxi[49] and Hainan[50]) and 27 cities.[51]

However, such biofuel incorporation provisions in the Renewable Energy Law leave much ambiguity and uncertainty. In January 2014, a privately-owned biodiesel producing enterprise in the Yunnan province sued Sinopec for refusing to incorporate its biodiesel into Sinopec's selling network, and requested Sinopec not only to incorporate its biodiesel but also compensate its economic loss by invoking the Renewable Energy Law.[52] The Court of first instance ruled that Sinopec (Yunnan Subsidiary) shall integrate the biodiesel that meets national standards into its fuel

the NDRC and NEA as the superior energy policymaking and coordination body, but it is not the 'energy administration department' provided in law. McElwee 2011, p. 93-94; Wang 2011, p. 206.

47 A similar requirement is also set for the incorporation of biomass-generated gas and heating into urban gas or heating pipelines. Renewable Energy Law of China 2005, Art. 16.

48 A similar provision is also provided for gas and heating supply enterprises regarding their obligation to incorporate biomass-generated gas and heating. Renewable Energy Law of China 2005, Art. 31.

49 Guangxi was included in the bioethanol pilot project since April 2008 with a blend mandate of 10 per cent bioethanol produced from the cassava-based bioethanol plant (with production capacity of 200,000 tonnes) invested by the the biggest state-owned agricultural group COFCO. Guangxi to blend 10 per cent ethanol into gasoline in December (Biopact, 19 September 2007), available at: <http://news.mongabay.com/bioenergy/2007/09/guangxi-to-blend-10-ethanol-into.html> (last accessed 5 May 2015).

50 It was stipulated by the Hainan provincial government that since January 2010 a blend mandate of 5 per cent biodiesel (BD5) shall be implemented in the whole province. Hainan Biodiesel Promotion Work Plan 2009.

51 In 2014 the National Energy Administration (NEA) issued a policy document for biodiesel development that urges oil companies to integrate biodiesel meeting national standards into their selling networks.The NEA stresses that oil companies shall strictly implement the contracts with biodiesel enterprises and integrate biodiesel into their networks. The share of biodiesel purchased and sold by an oil company in the total biodiesel shall be proportional to the market share of that oil company. However, this is a policy document addressed to the energy administrative agencies within the Development and Reform Commissions of the lower levels, which is not equal to the biofuel blend mandate. Biodiesel Industry Development Policy 2014, Arts. 28 and 31.

52 The plaintiff argued that the Sinopec Yunnan Subsidiary violated Art. 16 of the Renewable Energy Law by refusing to incorporate the biodiesel produced in line with national and local standards and regulation, as well as Art. 17 of Anti-Monopoly Law by abusing its predominant status in the market, since Sinopec occupies above 80 per cent of down-stream oil selling market share in Yunnan. Lucy Hornby, 'Lawsuit Puts China Biodiesel Policy in Spotlight' (FinancialTimes, 9 October 2014), available at: <www.ft.com/intl/cms/-s/0/613ee6b0-494e-11e4-8d68-00144feab7de.html#axzz3HvZoAXoA> (last accessed 2 November 2014).

selling network, but it rejected the other claims of the plaintiff.[53] Both Parties appealed in 2015 to the Higher Court of Yunnan Province which rescinded the original judgement and remanded a retrial.[54] One focal point in the case is whether Sinopec (Yunan Subsidiary) has the obligation to purchase and incorporate the biodiesel produced by the plaintiff. The biodiesel producer argued that the Yunnan Provincial Government has the policy of promoting biodiesel and had required the large oil companies like Sinopec and CNPC to cooperate in selling biodiesel.[55] Sinopec, however, argued that there is a lack of rules from the Yunnan Provincial Government about the incorporation of biodiesel and incorporation without going through experimental phases would risk product's quality and consumers' safety.[56] Meanwhile, the Renewable Energy Law is ambiguous about what the 'requirement of the energy authority of the State Council or the provincial governments' refers to. Thus, it is unclear whether the biofuel blend mandate should be set out in formal legal sources of the competent authorities such as departmental rule or local governmental rule,[57] or policy documents which are not regarded as formal legal sources suffice.[58]

53 The Court of first instance decided that the Sinopec Yunnan Subsidiary has a predominant place in the refined oil market of Yunnan with a market share of 50 per cent and violates its obligation under the Renewable Energy Law of purchasing, blending, and selling biodiesel. 危昱萍, 石油垄断第一案 中石化输了 [Sinopec Lost in the First Case about Oil Monopoly] (jiemian.com, 19 December 2014), available at: <http://m.jiemian.com/article/215860.html> (last accessed 5 May 2015).

54 中石化拒售云南民企生物柴油案发回重审 [The Higher Court of Yunnan rescinded the original judgement of the biodiesel case] (Xinhua, 9 September 2015), available at: <www.pkulaw.cn/fulltext_form.aspx?Db=news&Gid=77634&keyword=&EncodingName=&Search_Mode=accurate> (last accessed 4 March 2016).

55 It was argued that the Yunnan Provincial Government Office's 46th Document stated that 'oil company has to establish biodiesel demonstration stations, and the State-owned oil companies should take responsibility in implementing low carbon strategy and set examples.' It also cited this in April 2012 in the official document on the promotion of biodiesel from used cooking oil and planned that the production and utilisation of biodiesel should increase from 50,000 to 100,000 tonnes by 2015. The biodiesel company also argued that the provincial government and the energy authority of the provincial government had coordinated and adopted instructions for integration. 云南省高院二审民企诉中石化"垄断"案聚焦双方八大诉求 [The Higher Court of Yunnan hearing in the second instance of the case on the oil monopoly, Parties focused on eight claims] (Yunnan.cn, 23 April 2015), available at: <http://society.yunnan.cn/html/2015-04/23/content_3702675.htm> (last accessed 5 May 2015); 王力凝,我国石油系统首例反垄断案:民企状告中石化不卖油) [The first anti-monopoly case in China's oil sector: Private company sued Sinopec for not selling its oil] (China Business Journal, 16 August 2014), available at: <http://finance.sina.com.cn/chanjing/-cyxw/20140816/015420029304.shtml> (last accessed 2 November 2014).

56 Sinopec also argued that there is a quality problem related to the biodiesel in question. 云南省高院二审民企诉中石化"垄断"案 聚焦双方八大诉求 [The Higher Court of Yunnan hearing in the second instance of the case on the oil monopoly, Parties focused on eight claims] (Yunnan.cn, 23 April 2015), available at: <http://society.yunnan.cn/html/2015-04/23/content_3702675.htm> (last accessed 5 May 2015).

57 Local government rules and departmental rules are the formal legal sources in China. It is provided in the Law on Legislation of China that ministries, commissions, the People's Bank of China, the National Audit Office, and other departments directly under the State Council →

In addition to the ambiguity, the provisions in the Renewable Energy Law lack a requirement about the sustainability of the biofuels to be incorporated. As provided, oil companies shall incorporate the biofuel meeting 'national standards', which largely refer to the quality specifications issued by the relevant national standardisation agencies,[59] such as GB 18350-2013 for bioethanol and GB 18351-2013 for a bioethanol blend,[60] while there are no available sustainability standards/criteria regarding the environmental and socio-economic impacts of biofuels. How to define the 'national standard' is important for deciding on the incorporation obligation. For example, in the Yunnan biodiesel case, Sinopec argued that the biodiesel produced by the plaintiff may threaten consumer's interests and safety because the quality is not guaranteed, in spite of the fact that the plaintiff provided evidence to show that its biodiesel product meets the relevant national and local standards.[61] It is only stated in the Renewable Energy Law that the state

(*guo wu yuan zhi shu ji gou*) are competent to issue departmental rules (*bu men gui zhang*) for implementing national laws as well as the State Council's regulations, decisions and orders. It is provided that local administrations of provinces and certain prefectures can promulgate local rules (*di fang zheng fu gui zhang*) to implement national laws, administrative regulations, and local regulations of the same jurisdiction. Law on Legislation, 15 March 2000, revised 15 March 2015, Arts. 80 and 82.

58 Different from departmental or local 'rules', which can only be adopted by governments at certain levels, 'normative documents' are created and applied by governments of all levels, including state organs at the county level and below. Since such documents are not provided in the Law on Legislation, and they do not qualify as formal legal sources and they lack binding effect before the courts and thus lack legislative effect. The reason such documents exist is due to their role in bridging the gap between law and reality. Corne 1995, p. 263, 265-266; Corne 2002; Chen 2008, p. 191-192.

59 The General Administration of Quality Supervision, Inspection and Quarantine and the National Standardisation Administration Committee under the State Council are respectively responsible for supervising and setting quality standards of products in China, and the standards issued can be either mandatory or voluntary. Mandatory standards are indicated by 'GB' in their title, while voluntary standards are indicated by 'GB/T.' Report on the Policies for the Biofuel Industry 2014, p. 68-69.

60 These standards were jointly issued by the Administration of Quality Supervision and the Standardisation Administration of China and are effective from 1 January 2014. The current national standards for bioethanol and bioethanol blend E10 are GB 18350-2013 and GB 18351-2013; the national standards for biodiesel BD100 and biodiesel blend B5 are GB/T 20828-2014 and 25199-2014. The current national biodiesel standards are recommended standards, but the NDRC promised to adjust the biodiesel standards into mandatory standards and to upgrade them with the newly implemented gasoline and diesel standards. It is said that the standard on the biodiesel blend B7 is under development. Biodiesel Industry Development Policy 2014, Art. 37; Report on the Policies for Biofuel Industry 2014, p. 68-69.

61 The biodiesel company has adopted a report from the local oil products quality supervision centre. It is noted that in addition to the national standards, the Yunnan province has issued its own mandatory standards on biodiesel blend B10 and B20. Furthermore, the Hainan Province has also issued its B5 biodiesel blend standard DB 46/189-2010, which just needs to be backed up in the national standardisation agencies. 云南省高院二审民企诉中石化"垄断"案聚焦双方八大诉求 [The Higher Court of Yunnan hearing in the second instance of the case on the oil monopoly, Parties focused on eight claims] (Yunnan.cn, 23 April 2015); 吕明合, 生物柴油"最后一战"：政府力推，身份合法；地上"遭拒"，地下"偷卖" [The 'last battle' of biodiesel: Government's promotion, legitimate identity; but being refused on ground and →

encourages 'clean' and 'efficient' development and utilisation of energy from biomass,[62] without specifying what 'clean' and 'efficient' utilisation entails and whether oil companies can refuse to incorporate biofuels because their production/consumption is not clean or efficient. While the general principles of non-food feedstock, using marginal land, and no environmental damage were set out in the development plans in 2007, the Renewable Energy Law as revised in 2009 did not insert such requirements into the provisions on biofuels' incorporation. In fact, the four state-owned bioethanol plants established during 2002 and 2004 for the blend mandate pilot project are still allowed to produce bioethanol from food grains, such as wheat and corn, based on their existing production capacities,[63] which are still used to meet the bioethanol blend mandate. Although biofuel in the blend mandate pilot projects in Guangxi and Hainan are respectively produced from cassava and waste cooking oil, which both comply with the non-food grain feedstock principle, there have been concerns about their socio-economic[64] or environmental impacts.[65]

being secretly sold under ground] (infzm.com, 23 January 2015), available at: <www.infzm.com/content/107359> (last accessed 6 May 2015); 云南启动生物柴油产业发展规划 [Yunnan starts biodiesel industry development plan] (bioon.com, 10 December 2012), available at: <http://news.bioon.com/article/6565869.html> (last accessed 6 May 2015); 海南试用生物柴油 已在12家中石化加油站试销售 [Biodiesel trial in Hainan, being sold in 12 Sinopec gas stations] (news.cn, 21 November 2010), available at: <http://news.xinhuanet.com/2010-11/21/c_12798234_2.htm> (last accessed 6 May 2015).

62 Renewable Energy Law of China, 28 February 2005, revised 26 December 2009, Art. 16.

63 It has been observed that when the stale grains in reserves were gradually consumed, some bioethanol plants began to use fresh grains in production, so that in the production process both new and old grains were used. It was reported that the ratio of fresh grain feedstock has reached 80 per cent in the bioethanol plants. Although the central government banned support for the production of bioethanol based on food grains, the four bioethanol plants are exempted from this, provided that they produce food-based bioethanol based on their existing production capacity. Notice on Bioethanol Project Management 2006; Scott & Jiang 2013; Qiu *et al.* 2012, p. 3100; 高立萍, 中国生物燃料乙醇尴尬:政策转向,企业闪腰 [The awkward situation for Chinese bioethanol as the policy shifts] (Yicai.com, 20 January 2014), available at: <www.yicai.com/news/2014/01/3381443.html> (last accessed 8 May 2015).

64 For example, the cassava-based bioethanol plant in Guangxi might put pressure on local cassava supplies, raising the country's need for imports of the crop from Vietnam and Thailand, because Guangxi can grow cassava only during the three months between November and January. It was reported that more than 45,000 farmers have been contracted to cultivate cassava for the bioethanol pilot project, but, due to the problem in marketing since 2010, the Guangxi bioethanol producing plant reduced and eventually failed to implement the contracts with farmers, which caused economic loss for the farmers. Guangxi to blend 10 per cent ethanol into gasoline in December (Biopact, 19 September 2007), available at: <http://news.mongabay.com/bioenergy/2007/09/guangxi-to-blend-10-ethanol-into.html> (last accessed 5 May 2015); 世界最大非粮燃料乙醇企业被迫停产 [The largest non-grain bioethanol plant is forced to stop production] (Chinanews.com, 20 May 2011), available at: <www.chinanews.com/cj/2011/05-20/3053998.shtml> (last accessed 6 May 2015).

65 From an environmental perspective, cassava processing, especially in areas where the industry is highly concentrated, is regarded as a polluting burden for natural resources. FAO Corporate Document Repository, Impact of Cassava Processing on the Environment, →

2.2.2. Consumption Targets

A consumption target is the proportion of biofuel consumption in the gross nationwide transport fuel mix to be achieved in future years.[66] The Renewable Energy Law does not set out a consumption target for biofuels, but it contains a rule on how to establish a target on renewable energy. It is provided that the central government shall establish a total volume target (*zong liang mu biao*) for the development and use of renewable energy,[67] and local medium- and long-term targets shall be set in accordance with the national target as well as local diversity in economic and energy resources.[68] Renewable energy development and utilisation plans are to be compiled based on the established volume targets at both national and sub-national levels.[69]

According to the Renewable Energy Law, the NDRC formulated and issued the Renewable Energy Medium- and Long-Term Plan in 2007,[70] which sets out the medium- and long-term targets for renewable energy as a whole as well as for each type of renewable energy including biofuels. In this Plan, the share of renewable energy in the total primary energy consumption should reach 10 per cent by 2010 and 15 per cent by 2020.[71] Moreover, based on the estimation that bioethanol and biodiesel production capacity had already respectively reached 1.02 million tons and 50,000 tons,[72] the Plan sets the targets of increasing annual consumption of *non-grain* based bioethanol *by* 2 million tons[73] by 2010 and to 10 million tons by 2020; the targets for the annual consumption of biodiesel is 200,000 tons by 2010 and 2 million tons by 2020. The targets for the consumption of biofuels are also set out in the 12th

available at: <www.fao.org/docrep/007/y2413e/y2413e0d.htm> (last accessed 6 May 2015); Leng 2008, p. 374.

66 To a certain extent, the consumption target is also a kind of blend mandate, but the difference lies in the fact that the consumption target counts as the share of biofuels in the overall consumption volume of transport fuel, while the blend mandate can be understood as the share of biofuels in the transport to be sold to consumers. In some studies, they are regarded as in the same category. For example, it is estimated that over 50 countries have adopted blend targets or mandates. De Gorter & Just 2009, p. 738; Elbehri, Segerstedt & Liu 2013, p. 13.

67 It is the national energy authority of the State Council that sets and issues the national medium- and long-term total volume target after being approved by the State Council. Renewable Energy Law of China 2005, Arts. 4 and 7(1).

68 Renewable Energy Law of China 2005, Art. 7(2).

69 The renewable energy plan at the national level shall be approved by the State Council, while the provincial renewable energy plan shall be approved by the competent provincial government and backed up by national energy authorities. Renewable Energy Law of China 2005, Art. 8.

70 Renewable Energy Medium- and Long-Term Plan 2007.

71 In 2005, the share of renewable energy in primary energy consumption was 7.5 per cent, equal to 166 million metric tons of standard coal. Renewable Energy Medium- and Long-Term Plan 2007.

72 The 2004 pilot project aimed to produce a total volume of fuel ethanol of 1.02 million metric tons. Measures on Expanding the Bioethanol Pilot Project 2004.

73 The 2 million tons of non-grain bioethanol is in addition to the existing 1.02 million tons grain-based bioethanol.

Five Year Plan for Renewable Energy Development from 2011 to 2015:[74] the consumption targets for bioethanol and biodiesel (including bio-kerosene for aviation) in 2015 are set at 4 million tons[75] and 1 million tons.[76] In reviewing the achievement of 2010 targets, it is estimated that the realised bioethanol utilisation in 2010 was 1.86 million tons with merely 0.37 million tons from *non-grain* feedstocks, which did not fulfil the established 2010 target for bioethanol. In contrast, the biodiesel utilisation quantity exceeded the 2010 target by 300,000 tons.[77] Meanwhile, there is no target set for the share of biofuels in the transport energy mix.[78] The failure to achieve the bioethanol target is related to the fact that the central government prohibited the expansion of grain-based bioethanol.[79] Moreover, it is noticed that the 2015 bioethanol target no longer mentions the 'non-grain' condition, largely because of the failure and difficulty to fulfil the 2010 non-grain bioethanol target.

As for the consumption target for biofuels, one question is to what extent the target is binding. The national Five Year Plans for social and economic development conventionally contain both mandatory (*yue shu xing*) and indicative (*yu qi xing*) targets, the former of which is binding on government of various levels[80] particularly through a 'target responsibility' system,[81] such as the energy intensity

74 The 12th Five Year Plan of Renewable Energy Development.

75 In another place of the 12th Five Year Guidelines, the bioethanol consumption target is expressed as 3.5-4 million tons. The 12th Five Year Plan of Renewable Energy Development.

76 The targets for biofuels were correspondingly scaled down from the original ambitious 6 million tons by 2010 and 15 million tons by 2020. Considering the excessive achievements in biodiesel and the gap in bioethanol, the target for biodiesel is largely enhanced (from 0.2 million to 1 million for biodiesel), while the bioethanol target is modestly increased. Qiu *et al.* 2012, p. 3100; 发展受挫 "生物质能"目标调整 [The goals of 'biomass energy' are adjusted due to development setback] (中国联合商报, 27 August 2012), available at: <http://www.newenergy.org.cn/swzn/xydt_15713/201208/t20120827_227353.html> (last accessed 8 May 2015).

77 The 12th Five Year Plan of Renewable Energy Development; Chang *et al.* 2012, p. 2-3.

78 According to IEA's data, in 2012 the biofuel (and waste) consumed for transport in China counted for 1.18 million tons, constituting about 0.5 per cent of the total transport energy consumption in 2012 (estimated as 238 million tons by IEA). IEA, Statistics, China, available at: <www.iea.org/statistics/statisticssearch/report/?year=2012&country=CHINA&product=Balances> (last accessed 6 May 2015).

79 Chang *et al.* 2012, p. 2-3; IISD 2008, p. 14.

80 The mandatory targets/indicators have to be disintegrated into relevant sectors, among which the arable land maintenance ratio, the energy intensity reduction rate, and the major pollutant emission shall be disintegrated and implemented by every provincial government. It is also set forth that such targets are to be disaggregated into provincial targets, taking into consideration the targets proposed by the provinces in their provincial five year plans, and the provincial targets are to be submitted by the National Development Reform Commission (NDRC) to the State Council for approval. Once the State Council approved the NDRC's plan, provincial governments were required to re-allocate the targets to the city and county level and to industrial sectors and major enterprises. In practice, the targets are commonly inflated for lower levels of government to ensure the achievement of the overall provincial targets. Hao 2007, p. 101-102; Teng & Gu 2007, p. 7.

81 The targets are mandatory largely in the sense that they shall be integrated into the 'target responsibility system' (*mu biao ze ren ping jia kao he*) for the government or cadre evaluation →

target and CO_2 emission targets.[82] However, the biofuel consumption targets are not set out as mandatory targets and can hardly be legally binding. Firstly, although a target for non-fossil fuel consumption is mandatory in the national Five Year Plan, the renewable energy penetration targets for hydro, wind, or solar power are set out as anticipatory ones.[83] Secondly, while the national Five Year Plan for Economic and Social Development is adopted by National People's Congress and might thus be argued as constituting a legal document, the Five Year Plan for Renewable Energy Development is to implement the national overall plan in a specific sector and is issued by NDRC, an administrative agency under the State Council, so the targets established therein are binding to a less extent. Thirdly, the biofuel consumption targets *per se* do not specify whether the targets are binding on governments of various levels or who should be accountable for fulfilling the targets.[84] Nevertheless, the National Energy Administration (NEA) under the NDRC has requested their counterparts of the lower levels to establish a biodiesel use target based on production capacity and to include such a target in the 'target responsibility evaluation' for local governments,[85] which might strengthen implementation of the national consumption target for biodiesel.[86]

system (*gan bu kao he*) for officials, which means that provincial and local officials and business leaders are evaluated based on their progress in meeting the mandatory targets. See more about the Chinese target responsibility system or cadre system in: Chen 2008, p. 242-243; McElwee 2011, p. 59; Heimer 2006; Lin 2011b, p. 310; Whiting 2004, p. 101; Edin 2003, p. 35; Teng & Gu 2007, p. 7; Kostka & Hobbs 2012; Lin 2012b, p. 310-312; Wang, A., 'Meeting China's Climate Targets' (Switchboard, 10 December 2009), available at: <http://switchboard.nrdc.org/blogs/awang/meeting_chinas_climate_targets.html> (last accessed 9 July 2014).

82 The State Council has required an annual target responsibility evaluation regarding energy saving and emission reductions at the provincial level, and the results of the evaluation shall be made public. The achievement of target meeting and policy implementation is to be established as an important indicator for evaluating the government/State-owned enterprises' heads and leading teams. The 12th Five Year Work Programme on Energy Saving and Emission Reduction 2011, para. (5).

83 It is set out that the proportion of non-fossil fuel energy in the total primary energy consumption shall be 11.4 per cent by 2015 (compared to 8.3 per cent in 2010) with an annual growth rate of 3.1 per cent. In the 12th Five Year Plan for Energy Development issued by the State Council in 2013, more energy-related targets are set out, including the anticipatory targets for electricity produced from renewable sources (hydro, wind, solar). 我国国民经济和社会发展十二五规划纲要(全文) [The 12th Five Year Guideline for National Economic and Social Development (full text)], available at: <http://news.sina.com.cn/c/2011-03-17/055622129864.shtml> (last accessed 7 May 2015); The 12th Five Year Plan for Energy Development.

84 Nevertheless, it is stated in the 12th Five Year Guideline for Renewable Energy that a system of renewable energy target evaluation is to be established for local areas and the major energy companies, in accordance with national and provincial renewable energy plans. The 12th Five Year Plan for Renewable Energy Development.

85 The NDRC requested that the provincial governments design the annual promotion schemes and targets in accordance with national biodiesel development plans, standards, market entry, and other regulation, and report their schemes and targets to the State Council and then coordinate to implement the schemes and targets. It is further required to establish a binding target responsibility management system (*yue shu xing mu biao ze ren guan li ti zhi*) on →

Another question is whether any sustainability requirement is set out for biofuels used to achieve the consumption targets. The general principles about non-food feedstock, marginal land, and no environmental damage are set out in the same development plans with the consumption targets, but there is no explicit requirement that the biofuel used to count towards the targets has to meet the general principles. Moreover, it is noticed that the 'non-grain' condition for the 2010 bioethanol target is no longer mentioned for the 2015 bioethanol target.

2.2.3. Emissions Standards for Fuels and Vehicles

Biofuels are regarded in China as contributing to carbon emissions reduction in the transport sector[87] and hence the regulation of emissions from fuel and vehicles can have an effect on the consumption of biofuels. However, the current fuel emission standard consists of quality specifications[88] on environmental and health grounds which as yet do not include a requirement about GHG emission intensity, largely because the emphasis of a fuel emission standard has been placed on air pollution control.[89] The current emissions standards for fuels are mandatory.[90] In recent years China has accelerated the pace of upgrading fuel emission standards[91] that gradually filled the gap in stringency. For example, the China V gasoline (GB 17930-2013) standard, which mandates the maximum sulphur content of 10 parts-per-

biodiesel's sale quantity or share. Biodiesel Industry Development Policy 2014, paras. 25 and 28.

86 This request is largely an internal instruction between the National Energy Administration and its counterparts of local levels, which does not have external binding effect. See fn. 49 of the chapter.

87 For example, it is stated in both the 11th and 12th Five Year Guideline for Renewable Energy that energy use of agricultural and forestry biomass will not increase CO_2 emissions in the life-cycle. The 11th Five Year Plan for Renewable Energy Development; The 12th Five Year Plan for Renewable Energy Development.

88 The parameters include *inter alia* sulfur content, manganese, lead, RVP, oxygen for gasoline, and polyaromatics, sulfur, and octane number for diesel. 'China: Light-duty: Fuel Consumption', TransportPolicy.net, available at: <www.transportpolicy.net/index.php?-title=China:_Light-duty:_Fuel_Consumption> (last accessed 4 September 2014).

89 It is estimated that the number of civilian vehicles on China's roads has increased at a compound rate of 15 per cent per annum over the past three decades, which has essentially paralleled nominal GDP growth. With the increased use of vehicles, pollutant emissions such as NOx, CO, hydrocarbons, and fine particulates from fossil fuels become an important source of air pollution. Zhang & Crooks 2012, p. 61.

90 In addition to laws, administrative regulation and rules, standards may be promulgated by the General Administration of Quality Supervision, Inspection and Quarantine or the Standardisation Administration to govern specific, technical, or complex compliance areas and are particularly used to define and help achieve or protect a quantifiable level of environmental quality. McElwee 2011, p. 104-105.

91 In September 2013, the State Council issued the action plan for air pollution control, in which the government is required to accelerate the revision of emission standards and vehicle fuel consumption standards. Air Pollution Prevention Action Plan 2013, Art. 7.

million (ppm), is to be completely implemented nation-wide from January 2017,[92] while some heavily polluting areas (e.g. Beijing) have already phased in the standard before 2017;[93] the more stringent China VI gasoline and diesel standards are under development and will be issued by 2016 and implemented by 2019.[94] The stricter emission standards also call for the alignment of biofuel quality standards such as GB 18350-2013 for bioethanol and GB 18351-2013 for bioethanol blend, since biofuel will be blended in the fuel and might to certain extent affect fuel specifications (e.g. vapour pressure).[95] In fact, the NDRC has mentioned that the China V fuel standard phase-in should also include a standard upgrading of bioethanol blend (E10) and biodiesel blend (B5), and promised to issue mandatory stricter China V standards for E10 and B5 as soon as possible.[96]

Although the fuel quality standard as yet does not include a requirement about GHG emission intensity, China has established fuel economy standards which set a limit for a vehicle's fuel consumption over a fixed distance, so as to require manufacturers to produce vehicles with greater energy efficiency and less

92 The Phase V standard originally planned to be effective as of 1 January 2018. However, in 2015 the NDRC issued a working programme to accelerate oil quality upgrading, according to which the areas that implement the China V gasoline and diesel standards will be expanded to the whole eastern China and the nation-wide implementation of China V will be postponed to 1 January 2017. Hence, before 31 October 2016, oil suppliers have to possess the capacity to produce gasoline and diesel that meet the China V standards, and from 1 January 2017 gasoline and diesel products for vehicles will not be supplied in China. Working Programme for Accelerating Oil Quality Upgrading 2015.

93 The heavily polluted provinces or cities can voluntarily or be required to implement the China V standards before 2018. For example, Beijing has already implemented the China V standards since 1 February 2013. 国五标准2018年执行重污染省份提前实行 [China V standards to be implemented from 2018 but heavily Polluted Provinces to implement prior to that year] (China Association of Automobile Manufacturers, 18 June 2013) available at: <www.caam.org.cn/biaozhuidongtai/20130618/0905093474.html> (last accessed 7 May 2015).

94 Meanwhile, it is stated that mandatory fuel standards for ships will be issued by the end of 2015, and new mandatory diesel standards will be issued by June 2015. Working Programme for Accelerating Oil Quality Upgrading 2015.

95 Biofuels might have other influences on fuel emissions, such as higher nitrogen oxide emission and octane content. The change from China IV to China V fuel standards is not only reflected in the reduction of sulfur content but also in relation to vapour pressure. Gaffney & Marley 2009; Council on Clean Transportation, China V Gasoline And Diesel Fuel Quality Standards, Update, January 2014, available at: <www.theicct.org/sites/-default/files/publications/ICCTupdate_ChinaVfuelquality_jan2014.pdf> (last accessed 3 September 2014); 国五车用汽油标准发布实施 [Issuance and implementation of China V automobile gasoline standards] (People.cn, 19 December 2013), available at: <http://politics.people.com.cn/n/2013/1219/c70731-23881794.html> (last accessed 7 May 2015).

96 The current national biodiesel standards are recommended standards, but the NDRC promised to adjust biodiesel standards into mandatory standards and to be upgraded with the newly implemented gasoline and diesel standards. Biodiesel Industry Development Policy 2014, Art. 37.

carbon emissions.[97] China's first fuel economy standard for passenger cars was adopted in 2004, and the Phase I and Phase II standards were respectively implemented during 2005-2006 and 2008-2009, which required each individual vehicle model to comply with the fuel economy standard before it enters Chinese market.[98] Similar to the fuel emission standard, the fuel economy requirement has been increasingly strengthened. In 2009, China announced the phase-in of Phase III fuel consumption standard from 2012 to 2015, aiming to reach a fuel consumption target of approximately 7 L/100km (equivalent to 167 g CO_2/km) on average, and the Phase IV standard currently under development is to achieve the target of 5 L/100km (equivalent to 116 g CO_2/km) by 2020.[99] Vehicles using a biofuel blend must also meet the increasingly stricter fuel economy standards, which might even disadvantage or disincentivise the use of biofuels because the use of biofuels may theoretically increase fuel consumption, because ethanol contains less heat energy per litre than petrol.[100]

It is thus observed that the existing fuel emission standard emphasises the control of air pollution, and the strengthened standard may impose stricter requirement on air pollutant emissions of biofuels. The fuel economy standard that aims to reduce fuel consumption intensity of vehicles gives no special treatment to vehicles using biofuels and might even disadvantage or disincentivise the use of biofuels.

97 The fuel economy in China is calculated using the unit of 'L/100km' which can be converted into 'g /km CO_2': 1L/100km is equivalent to 23.2 g/km CO_2. Parry *et al.* 2011; Oliver *et al.* 2009.

98 The fuel economy standard is applied to each model of light-duty passenger vehicles less than 3,500 kg irrespective of origin. The fuel consumption limit differentiates between vehicles according to their *weight levels* rather than *fleet average*, as the standard based on vehicle weight can offer enterprises more flexibility in enhancing energy efficiency. This contrasts with policies in the EU, which permit auto manufacturers to meet targets by averaging emissions over their entire fleet of models. The fleet average approach is said not to be suitable in China because the Chinese automobile market is highly fragmented, with more than a hundred manufacturers each producing a handful of models. See 'China: Light-duty: Fuel Consumption', TransportPolicy.net, available at: <www.transportpolicy.net/-index.php?title=China:_Light-duty:_Fuel_Consumption> (last accessed 4 September 2014); Wagner, An & Wang 2009; Lo &Wang 2015.

99 In 2012, China's State Council released the Energy-Saving and New Energy Vehicle Industrialisation Plan, which states an expected fleet average target of 6.9L/100km by 2015 (slightly below that estimated in GB 27999-2011) and 5.0L/100km by 2020. 'China: Light-duty: Fuel Consumption', TransportPolicy.net, available at: <www.transportpolicy.net/-index.php?title=China:_Light-duty:_Fuel_Consumption> (last accessed 4 September 2014); Wang Z. *et al.*, 'New fuel consumption standards for Chinese passenger vehicles and their effects on reductions of oil use and CO_2 emissions of the Chinese passenger vehicle fleet', 38(2010) Energy Policy 5242-50; Lo &Wang 2015.

100 Fontaras *et al.* 2009; NSW Trade & Investment, E10 fuel economy, available at: <www.resourcesandenergy.nsw.gov.au/energy-consumers/sustainable-energy/office-of-biofuels/e10-fuel-economy> (last accessed 7 May 2015).

2.2.4. Preferential Tax Treatment

In China, excise duties constitute the major source of taxation on energy products, which is also the case for biofuels.[101] As biofuels do not constitute a unitary taxation category, taxation rates may differentiate between different types of biofuel (e.g. between bioethanol and biodiesel). Any reduction of taxation rates or exemption from taxation can give preferential treatment to biofuels and thus can incentivise the demand.[102] This sub-section will explore the preferential taxation measures for biofuels and see what sustainability requirements are provided.

Bioethanol can be exempted from excise duties. Bioethanol (before being blended with gasoline) used to be subject to excise duties with a tax rate of 5 per cent levied on all the alcohol (including wine, ethanol, and menthol) products for industrial or food uses. However, the central government removed the 5 per cent excise duty on alcohol for industrial use from 1 December 2014 in order to avoid double taxation.[103] The question is whether excise duty for energy products, such as gasoline, diesel, and aviation kerosene,[104] is levied on biofuels used as fuels for transport. It is noticed that the legal document from 2008 on oil product excise tax reform[105] includes gasoline and diesel with the blends of bioethanol and biodiesel. However, it meanwhile provides that bioethanol is exempt from the oil product

101 Taxes on energy products include value-added (VAT) tax and excise (consumption) tax. Carbon tax imposed on CO_2 emissions from the use of energy products has been under consideration by the Chinese government for the purposes of GHG emission reduction and energy saving, but no concrete measure has been adopted thus far. Carbon tax may become an emission control instrument in addition to an emission trading scheme, especially for the transport sector that is not covered within emission trading. Xu 2011; Zang 2009, p. 214-215; Curnow, Tait & Millar 2010, p. 105; 'China reconsiders carbon tax, citing Australia's planned repeal' (The Sydney Morning Herald, 13 March 2014), available at: <www.smh.com.au/-business/carbon-economy/china-reconsiders-carbon-tax-citing-australias-planned-repeal-20140313-34nfp.html> (last accessed 5 September 2014); 'China to introduce carbon tax: official' (Xinhua, 19 February 2013), available at: <http://news.xinhuanet.com/english/-china/2013-02/19/c_132178898.htm> (last accessed 5 September 2014).

102 A reduction of tax rates or an exemption from taxation for the use of biofuels may provide a financial incentive for the consumption of biofuels. The Chinese government has adopted taxation measures, such as an excise tax, a vehicle acquisition tax, or a vehicle and vessel tax, to promote low emission vehicles and reduce the consumption of fossil fuels. Xu 2011; Lo & Wang 2015.

103 The taxation category of 'wine and alcohol' will be adjusted to 'wine', based on the reason that alcohol (or ethanol and methanol) is usually used for industrial production or transport and the exemption is to avoid double taxation and cut down on costs of relevant enterprises. Notice on Adjustment of Excise Duties 2014; 崔文苑, 财政部:成品油等消费税政策调整取消酒精消费税 [Ministry of Finance removes excise duties on alcohol and adjusts excise duties on oil products] (Xinhua net, 29 November 2014), available at: <http://jiangsu.-china.com.cn/html/finance/sankei/625926_2.html> (last accessed 15 January 2015).

104 It is provided that the levy of excise duties on aviation kerosene is temporarily postponed. Notice on Increasing Excise Duty Rates of Oil Products 2008, Art. 3.

105 China increased the excise duties on oil products around 2008 and, since then, claims that the increase is aimed at reducing fossil fuel consumption and air pollution and the incurred revenues are used for air quality, energy saving, and renewable energy, etc. Notice on Oil Products Taxation Reform 2008; Notice on Increasing Excise Duty Rates of Oil Products 2008.

excise duties, if the bioethanol is blended with gasoline that has already been taxed;[106] if the gasoline has not been taxed, the oil product excise duty is to be calculated based on the consumed gasoline.[107] Such a provision implies that only the gasoline in the bioethanol blend is subject to the oil product excise duty. In view of the fact that the excise duty on alcohol is removed, bioethanol could be exempted from excise duties entirely.

Biodiesel is subject to the excise duty on oil products and no such tax exemption is provided.[108] However, biodiesel is conditionally exempted from the oil product excise duty. It is provided that biodiesel (BD 100) is exempt from excise duties since 1 January 2009 if the proportion of waste animal fat and vegetable oil[109] in the biodiesel feedstock is not lower than 70 per cent.[110] Such a supportive measure attempts to encourage the recycling and reuse of waste or used oil (the so-called 'gutter oil') for energy production rather than for illegal food production.[111] Moreover, the National Energy Administration stated that biodiesel, particularly the biodiesel produced from waste oil, is to be supported by further preferential excise taxes.[112]

Preferential tax treatment has been provided to the designated biofuel enterprises. To implement the bioethanol blend mandate pilot project, four state-owned grain-based bioethanol plants have been established during 2002-2004, and a cassava-based bioethanol plant was established later. These state-owned plants enjoyed preferential tax treatment in the form of excise duty exemptions and value-added tax (VAT) refunds.[113] However, since the central government banned the expansion of grain-based bioethanol, such fiscal supports for grain-based bioethanol plants have been gradually reduced and finally stopped. The excise duty exemption

106 Notice on Increasing Excise Duty Rates of Oil Products 2008, Arts. 4 and 5.

107 *Ibid.*

108 It is provided that, if purchased or re-purchased diesel is used to continuously produce biodiesel blend, it is permissible to deduct the tax that has already been levied, which does not exempt biodiesel from the taxation. Notice on Increasing Excise Duty Rates of Oil Products 2008, Arts. 4 and 5.

109 The government has further specified the scope of 'waste animal fat and vegetable oil' as including: *a)* non-edible animal or vegetable oil produced by catering, food processing, and family; *b)* non-edible oil and fat produced from slaughtering, leather processing, and meat processing; *c)* combination of acid, fat and oil produced from edible oil refinery processing; *d)* non-edible oil produced from oil processing and oil and fat storage. Notice on the Scope of Excise Duty Exemption for Biodiesel (BD100) 2011.

110 Additionally, the biodiesel meets national quality standard about BD100. Notice on the Excise Duty Exemption for Biodiesel (BD100) 2010.

111 China's waste oil has been reused illegally for food due to attractive economic profit, which is a serious social hygiene and health problem. Thus, it has been advised that governments should provide financial subsidies for the legal collection of waste oil, which can make the legal collection more competitive than illegal waste oil collection and sale. Liang *et al.* 2013, p. 144.

112 It is stated that the government is studying further tax preferences for biodiesel. Biodiesel Industry Development Policy 2014, Art. 47.

113 Furthermore, they have been subsidised for purchasing feedstocks of stale corn and wheat from national grain reserves as well as for overcoming the higher cost of producing biofuels. Measures on Expanding the Bioethanol Pilot Project 2004.

and VAT refund for the grain-based biofuel plants stopped from 1 January 2015.[114] In contrast, fiscal support for non-grain biofuels, particularly the second-generation biofuels that use cellulosic biomass feedstock, has been increasing. For example, it has been reported that a company primarily using corncob as bioethanol feedstock has gained excise tax exemption and VAT refund approved by the Ministry of Finance and the State Council.[115] Hence, the preferential tax treatment for biofuel plants are shifting from the state-owned grain-based enterprises to the ones making use of non-food and second-generation feedstocks such as corncobs and crop stalks.[116]

Accordingly, it is observed that the central government has provided preferential excise duties to biofuels, which may have the effect of promoting the use of biofuels. Particularly, the tax exemption for biodiesel is conditional on the use of waste feedstock, and the preferential tax treatment have shifted to non-food and second-generation biofuel enterprises, in order to be in line with the general principle of non-food grain feedstock. In fact, the Ministry of Finance set forth in 2006 that the fiscal support for biofuels shall consider the food security impact as

114 It was decided by the Ministry of Finance and State Administration of Taxation in 2011 that the excise duty exemption and VAT refund would be gradually phased out from 2011 to 2014 and would be completely stopped by 1 January 2015. The VAT refund for grain-based bioethanol production was adjusted as: 80 per cent of VAT would be refunded during 1 October and 31 December 2011; 60 per cent VAT refund during the year 2012; 40 per cent during 2013; and 20 per cent during 2014; and there would be no refund as of 1 January 2015. The excise duty exemption for grain-based bioethanol was adjusted as: the excise duty rate would be 1 per cent for the bioethanol during 1 October and 31 December 2011; and the rate would be gradually increased to 2 per cent, 3 per cent, 4 per cent respectively in 2012, 2013, and 2014, and finally to the normal level of 5 per cent as of 1 January 2015. Furthermore, the subsidy for the grain-based bioethanol plants has kept shrinking until complete its withdrawal from 1 January 2016. 关于调整变性燃料乙醇定点生产企业税收政策的通知 [Notice on adjusting taxation policy for the designated bioethanol plants] 18 November 2011, available at: <www.chinafeedonline.com/news/show_news_detail_p.jsp?id=659795> (last accessed 30 October 2014).

115 The Shandong Longlive Bio-technology Co., Ltd is reported as the first enterprise in China that has realised cellulosic bioethanol's scale production and supplied it as fuel. 龙力生物非粮燃料乙醇实行消费免税政策 [Shandong Longlive Bio-technology Co., Ltd obtains excise duty exemption] (stcn.com, 3 November 2014), available at: <http://kuaixun.stcn.com/2014/1103/11823482.shtml> (last accessed 8 May 2015).

116 For example, Hainan Yedao has invested in cassava-based bioethanol and there are five to six cellulosic biomass projects with corncob and sorghum stalks. The State-owned bioethanol plants, Henan Tianguan and COFCO Biochemical, established for the blend mandate pilot project have also engaged in producing second-generation biofuels. For instance, the Henan Tianguan Group has engaged in bioethanol production technology using cellulose biomass and has built a plant with an annual production capacity at 3,000 tons. 金微, 2016年取消粮食乙醇补贴 上市公司纷纷转型非粮乙醇 [Subsidy for grain-based bioethanol to be removed in 2016, Listed companies turn to non-grain bioethanol] (15 August 2014), available at: <www.nbd.com.cn/articles/2014-08-15/856475.html> (accessed 8 May 2015); APEC, 'China Biofuels Activities', available at: <www.egnret.ewg.apec.org/Archive/-me_china.htm> (last accessed 11 May 2015); 非粮燃料乙醇将获优惠政策扶持 [Non-grain bioethanol will obtain favourable policy's support] (金陵晚报, 22 September 2014), available at: <www.mcshe.com/zcny/35956.html> (last accessed 8 May 2015).

well as factors of natural resources, environmental protection, and energy consumption,[117] and subsidies shall be allocated upon on a case-by-case expert review mechanism that takes into account technology maturity, financial situation, and feedstocks.[118] However, there is no clear legal framework that provides rules on the preferential tax treatment for biofuels, particularly with regard to the preferential tax treatment that can be provided and the conditions for receiving the preferential treatment. While the non-food principle is emphasised, requirements about the wider environmental or socio-economic impacts need to be more specified and clarified for biofuels to be exempted from taxation.

2.2.5. Emissions Trading Rules

Renewable energy policy and law in China have a close relationship with climate change policy: the Chinese government largely frames climate change as predominantly an energy issue to realise energy or economy restructuring.[119] Meanwhile, China's 'low-carbon strategy' emphasises exploring market-based mechanisms for energy saving and GHG emission reduction,[120] such as emission trading and CDM.[121] Due to the linkage between renewable energy and climate change mitigation, the market-based mechanisms of emission trading and CDM may motivate the use of renewable energy including biofuels.

117 It has been recognised that biofuel production may lead to environmental pollution and natural resource depletion, so the support should be restricted to a certain scope and extent. Opinion on Fiscal Support Policy 2006, Art. 3.

118 It is set out that fiscal support is to be allocated to designated biofuel enterprises, and the selection of the designated enterprises is based on an expert review taking into account technology maturity, the financial situation, and the feedstock used - non-food grain biofuel production must have feedstock bases. The selection should be transparent and equitable. Opinion on Fiscal Support Policy 2006, Art. 9.

119 It has been commented that the Chinese authorities have 'reconceptualised' climate change as predominantly an energy issue so as to tackle emission reduction, energy security, and economic growth at the same time. China's mitigation strategy is principally centred on energy policy, particularly energy saving and renewable energy. Zang 2009b, p. 543; Lin 2012b, p. 304-307; Yang 2008, p. 684.

120 In 2011, the State Council specified the main policies regarding low-carbon development in the future, including improving financial instruments for low-carbon development and establishing market-oriented mechanisms for energy saving and carbon reduction. The 12th Five Year Work Programme on Energy Saving and Emission Reduction 2011; Wang & Chang 2014, p. 127; Zhang 2011, p. 134.

121 The 'command and control' approach has been the dominant regulatory approach in China, but after the 11th Five-Year Plan it became clear that the space for the implementation of command-and-control policies was getting smaller. It is expected that themission trading could become a core industrial policy and promote the strategic adjustment of China's economic structure, and the carbon pricing introduced by emission trading can encourage consumers to purchase low-carbon products and guides investment towards low-carbon processes. Li, Zhang & Cai 2012, p. 2-3.

2.2.5.1. *Emissions Trading and Biofuels*

Greenhouse gases emissions trading scheme had not been undertaken until the 12th Five Year period when a national binding GHG emission reduction target was set forth – it declared that China would reduce its CO_2 emissions by 17 per cent per unit of GDP compared to the 2010 levels by 2015.[122] The State Council issued a work programme to achieve the binding national target in GHG emissions reduction,[123] stating that the government is to carry out regional emission allowances trading pilot schemes by issuing relevant law and regulation, working out GHG emission allowance allocation programmes, and gradually forming regional emission trading schemes.[124] The roadmap for a national emission trading scheme is to be firstly experimented with various emission trading models at the regional level in selected cities or provinces,[125] and an integrated national emission trading market is to be gradually built based on the experiences.[126]

Among the regional pilot emission trading schemes, Shanghai's pilot scheme includes the aviation sector.[127] According to Shanghai's local governmental rule,[128]

122 Other climate-related mandatory targets include energy intensity reduction by 16 per cent, a non-fossil fuel consumption proportion of 11.4 per cent, and a forest coverage increase of 21.66 per cent. Comparatively, only an energy intensity target had been set out in the 11th Five Year Guideline. The 12th Five Year Plan for National Economic and Social Development; Liu 2012; Zang 2009a, p. 214-215.

123 The 12th Five Year Work Programme on Energy Saving and Emission Reduction 2011.

124 *Ibid.*, Art. 17.

125 The selected provinces include two provinces and five cities: Guangdong province, Hubei province, Beijing, Shanghai, Tianjin, Chongqing, and Shenzhen. These pilot areas are to carry out emission allowances trading between 2013 and 2015 and issue their own local carbon emission management measures, verification methods, allowances allocation, and other trading rules, which should be firstly approved by the National Development and Reform Commission. Notice on Carbon Emission Allowances Trading Pilot Schemes 2011; Bellevrat 2011, p. 6; Song & Lei 2014.

126 In 2014, the State Council approved the National Plan on Climate Change of China (2014-2020) in which it was declared that the national carbon emission trading market was to be built during this period. It is expected that a national carbon market will begin to operate in 2016 and become mature by the end of the 13th Five Year period (2016-2020). National Plan on Climate Change of China (2014-2020); 刘毅, 我国将建全国碳排放交易市场（在国新办新闻发布会上）到2030年,空气质量将大大改善 [China will establish a nation-wide carbon emission trading market, Air quality will be greatly ameliorated by 2030] (People's Daily, 26 November 2014), available at: <http://paper.people.com.cn/rmrb/html/2014-11/26/nw.D110000renmrb_20141126_3-02.htm> (last accessed 9 May 2015).

127 From 18 June 2013 to 19 June 2014, the pilot schemes have gradually issued rules about *inter alia* emission trading administration, allowance allocation, and verification in the form of local regulation (*di fang xing fa gui*) or local government rules (*di fang zheng fu gui zhang*). According to the accessible rules, the industries with high emissions such as power plants, steel, petro-chemistry, chemistry are most likely to be included, and in some cases the enterprises exceeding a certain carbon emission threshold are also included. Beijing and Shenzhen pilot schemes have begun to include transport sector respectively from 2015. 张博, 分析我国碳排放权交易试点行业范围 [Analysis on the scope of sectors included in China's Emission Trading Pilot Schemes] (易碳家期刊, 21 December 2014), available at: →

the Shanghai pilot carbon emission trading scheme includes both industrial and non-industrial sectors including aviation.[129] Six airlines based in Shanghai are designated to be included in the Shanghai emission trading pilot scheme,[130] and the airlines are required to submit emission allowances for their commercial flights within China.[131] Meanwhile, the Shanghai government issued the pilot GHG emissions accounting and reporting guidelines for the aviation sector,[132] according to which the CO_2 emissions are accounted for by multiplying fuel consumption by an emission factor.[133] It is important to note that biofuels are excluded from the fuel consumption accounting,[134] which means that airlines do not need to surrender allowances for the CO_2 emissions of biofuels. Later, the central government issued

<www.tanpaifang.com/tanguwen/2014/1221/41044.html> (last accessed 9 May 2015); Qi & Cheng 2015, p. 11.

128 Shanghai's pilot emission trading scheme has been provided in the form of a governmental rule, instead of a regulation issued by the local People's Congress. In contrast, Beijing and Shenzhen established their pilot schemes in the form of a local regulation (*di fang fa gui*) which was adopted by local People's Congress. 张博, 解读"五市两省"碳交易试点政策制定情况 [Analysis on the policy making state-of-affairs of carbon emission trading pilot schemes in the five cities and two provinces] (易碳家期刊, 21 December 2014), available at: <www.tanpaifang.com/zhengcefagui/2014/122141038.html> (last accessed 9 May 2015); 张昕,范迪,桑懿,上海碳交易试点进展跟踪调研报告 [Follow-up report on Shanghai Pilot Carbon Emission Trading Scheme] (National Center for Climate Change Strategy and International Cooperation (NCSC), 13 June 2014), available at: <www.ncsc.org.cn/article/yxcg/dybg/201406/20140600000936.shtml> (last accessed 9 May 2015).

129 The Shanghai emissions trading scheme started on 26 November 2013 and the first trading period runs until 2015. The non-industrial sectors include *inter alia* aviation, ports, airports, railways that emit more than 10,000 tons CO_2 per year in 2010-2011. Opinions on the Implementation of the Shanghai Carbon Emissions Trading Pilot Scheme 2012; Measures on Carbon Emissions Management of Shanghai 2013; Shanghai includes aviation emissions in CO_2 market' (Thompson Reuters, 10 January 2013), available at: <www.greeninvestmentservices.com/news/1-latest-news/175-shanghai-includes-aviation-emissions-in-co2-market> (last accessed 24 October 2014).

130 The six airlines are: China Eastern, Shanghai Airlines, China Cargo Airlines, Juneyao Airlines, Spring Airlines and Yangtze River Express. Notice on the First Group of Participating Enterprises for Shanghai Pilot Scheme 2012.

131 The government will allocate each airline emission allowances based on their historical emissions in 2009-2011. The airlines are forced to surrender allowances for each tonne of carbon dioxide they emit during domestic routes when both the departure and arrival are within China, excluding the flights with departure or arrival outside of China (e.g. international, Hongkong, Macao, or Taiwan). Opinions on the Implementation of Shanghai Carbon Emissions Trading Pilot Scheme 2012; Song & Lei 2014.

132 The guidelines were effective since 1 January 2013, but they are only the 'trial' version, meaning that they are subject to further revision depending on implementation and national rule making. Shanghai government issued also CO_2 accounting and reporting guidelines for water transport in 2016. Guidelines on GHG Emission Accounting and Reporting of Shanghai (Trial) 2012; Guidelines on Water Transport GHG Emission Accounting and Reporting of Shanghai (Trial) 2016.

133 An emission factor refers to the quantity of CO_2 emitted for every tonne of aviation fuel during the flight. Guidelines on GHG Emission Accounting and Reporting of Shanghai 2012.

134 Guidelines on GHG Emission Accounting and Reporting of Shanghai 2012.

national accounting and reporting guidelines as references for the aviation sector[135] which require operators to report biofuels' consumption volume and proportion of biomass-derived part, presenting a nuanced approach from accounting guidelines for biomass in power generation.[136] Afterwards, accounting and reporting guidelines for other sectors have been issued by the central government, including the guidelines for road transports which account their energy consumption based on the volume of fossil-based fuels.[137]

In sum, biofuels are exempt from emission allowances surrendering in Shanghai pilot emission trading scheme, while it is still uncertain whether transport sector will be included in the future national emissions trading scheme of China and how consumption and emissions of biofuels will be accounted.[138] Regional and national guidelines have been developed for monitoring and reporting GHG emissions from air and road transports, but their accounting methodologies for biofuels vary. In fact, China's aviation sector has already engaged in piloting flights with biofuels[139] which are largely produced from non-grain feedstock such as

135 It is indicated that the guidelines are a 'trial' version (*shi yong*), and they are issued to be used as references for the relevant work on emissions trading and feedback and suggestions are welcomed. Hence, the national guidelines are issued with an experimental intention, with the purpose of providing *reference* methodologies for the pilot emissions trading schemes. Nevertheless, according to the 2014 Interim Measures on Carbon Emissions Trading Management, enterprises shall report on their GHG emissions in line with the accounting and reporting guidelines issued by national administrations, which means that local accounting guidelines are to be gradually replaced by the national ones with the establishment of national emissions trading schemes. Guidelines for GHG Emission Accounting and Reporting 2013; Interim Measures on Carbon Emission Trading Administration 2014; Song & Lei 2014.

136 In the Guidelines for the power generation industry, it is stated that in calculating CO2 emission from biomass blended fuels, only the emissions from fossil fuel fraction are counted. In answering the question about emissions accounting of biomass on a Q&A platform of NDRC, it is explained that emissions from biomass do not count in the total emissions calculation, except for biofuels in aviation. Guidelines for GHG Emission Accounting and Reporting 2013, Annex 1, p. 3, Annex 10, p. 4 and 7, 6-7; 全国碳市场MRV技术平台 [National carbon market MRV technical platform] (SinoCarbon, 24 March 2016), available at: <www.tanpaifang.com/tanguwen/2016/0324/51636_3.html> (last accessed 25 April 2016).

137 In an explanation manual issued by NDRC, the energy consumption of ethanol-driven cars is accounted according to the proportion of fossil fuel in the blend. 指南使用常见问题解答 [Answers to frequently asked questions in using the Guidelines] (NDRC, 2 March 2016), available at: <www.ccchina.gov.cn/archiver/cchinacn/UpFile/Files/Default/-20160302102636627741.pdf> (last accessed 25 April 2016).

138 It has been commented that at the initial stage of set-up, China's emissions trading scheme might well be limited to those carbon-price sensitive sectors (such as the power sector) considering management costs and environmental effectiveness. On the other hand, apart from Shanghai, pilot schemes of Beijing and Shenzhen have begun to include transport sector. Wu & Huo 2014, p. 38; Song & Lei 2014; Qi & Cheng 2015, p. 11.

139 For instance, it was reported that Boeing announced plans in 2010 to work with Chinese partners to develop aviation biofuel; a one-hour long experimental flight with a biofuel blend was made in October 2011 in Beijing. Kevin Bullis, 'China Ramping Up Biofuels- Western Companies are Partnering with Chinese Ones to Make Renewable Fuel' (MIT Technology →

waste/used cooking oil and jatropha.[140] As China's emission trading scheme is under development and the accounting guidelines are of an experimental nature, how the accounting and reporting rules are framed to incentivise sustainable biofuels in transport sector is a critical question to be further addressed by the government.[141]

2.2.5.2. CDM in China

As mentioned in the previous chapters, CDM is a project-based mechanism established under the Kyoto Protocol that provides for opportunities to achieve both GHG emission reductions and sustainable development goals in non-Annex I Parties to the Kyoto Protocol. China, as a non-Annex I Party, has greatly benefited from this mechanism by hosting the largest number of CDM projects and supplying the greatest number of certified emission reduction (CER) credits.[142] Renewable energy projects, particularly small or medium-sized hydro, wind, and biomass power projects, account for a large part of China's CDM projects. However, compared to the large number of biomass projects,[143] the number of biofuel CDM projects is minimal - according to a search in the CDM Project Database, only one biodiesel project in China is registered with the CDM Executive Board,[144] perhaps

Review, 2 June 2010), available at: <www.technologyreview.com/view/419197/china-ramping-up-biofuels/> (last accessed 9 May 2015).

140 For example, the Boeing project was based on waste cooking oil, and the other is based on biokerosene from jatropha. James T. Areddy, 'China's First Aviation Biofuel Test' (Chinarealtimereport, 1 November 2011), available at: <http://blogs.wsj.com/-chinarealtime/2011/11/01/chinas-first-aviation-biofuel-test/> (last accessed 26 August 2014).

141 It has been reported that the Chinese government has seriously considered biofuels as a way to reduce carbon emissions from aviation as well as to deal with the inclusion of Chinese airlines in the EU aviation ETS. James T. Areddy, 'China's First Aviation Biofuel Test' (Chinarealtimereport, 1 November 2011), available at: <http://blogs.wsj.com/-chinarealtime/2011/11/01/chinas-first-aviation-biofuel-test/> (last accessed 26 August 2014).

142 As of 30 June 2014, China hosts 3,758 registered projects out of the total 7,530 registered project activities, accounting for 49.9 per cent of the total registered projects. UNFCCC, Clean Development Mechanism, Project Activities, 'Distribution of registered projects by Host Party', available at: <http://cdm.unfccc.int/Statistics/Public/CDMinsights/index.html> (last accessed 12 July 2014); Zang 2009a, p. 205.

143 A search within the registered CDM projects in China showed that there are 136 registered biomass projects, with a search title 'biomasss', scope 'renewable energy; fuel substitute', and status 'registered.' CDM Project Database, available at: <http://cdm-en.ccchina.gov.cn/-NewItemList.aspx> (last accessed 5 September 2014).

144 The registered biodiesel project is the Shandong Province Jinjiang Biodiesel Production Project. According to the project design document, this project is located within the Laiyang City, Shandong Province, China. The project is constructed and operated by Shandong Jinjiang Shengwu Group Co., Ltd. (SJSW) and is designated to produce 50,000 tonnes of biodiesel annually based on *waste oil/fat* from oil/fat plants in China. The biodiesel produced will be supplied to *consumers within China* to substitute petrodiesel for use as fuel. After commissioning, the project is expected to reduce greenhouse gas emissions of approximately 122,495 tCO2e annually. The search was carried out using the title 'biodiesel', the scope →

due to the dominant role played by large state-owned enterprises in the biofuel industry in China.[145] Despite the minimal number, the existence of CDM projects with biodiesel[146] demonstrates the need to explore whether China, as a CDM hosting country, has subjected its biofuel CDM projects to sustainability requirements.

China has established a regulatory framework that primarily sets out the governing bodies and approval procedures of CDM.[147] The NDRC is the Designated National Authority (DNA) of China and responsible for approving the applicant projects. Renewable energy development and utilisation is among the areas prioritised by the Chinese government.[148] It is provided that the documents to be submitted by project operators in the application process include the approval of an environmental impact assessment report by relevant authorities.[149] According to the Environment Impact Assessment Law of China,[150] the environmental impact assessment report[151] of construction projects[152] shall include an assessment of

'renewable energy; fuel substitute', and the status 'registered.' CDM Project Data, available at: <http://cdm-en.ccchina.gov.cn/NewItemList.aspx> (last accessed 5 September 2014); more information on this project can be found on CDM Project Cycle Search, Project 8158: Shandong Province Jinjiang Biodiesel Production Project, available at: <http://cdm.unfccc.int/Projects/DB/JCI1352698730.58/view> (last accessed 5 September 2014).

145 The bioethanol industry in China is dominated by the state-owned bioethanol plants established for the blend mandate pilot pojects. While the biodiesel industry is more fragmented, small-scale private enterprises usually face difficulties in investment and technology and cannot compete with the large oil companies which have in recent years begun to invest in biodiesel production. IISD 2008, p. 19-26; Qiu *et al.* 2012, p. 3098; iCET, Policy Recommendations for Supporting the Development of Low Carbon Automotive Fuels in China (Report, G-0907-11265, 2010), p. 10, available at: <www.icet.org.cn/adminis/-uploadfile/201142612552738878.pdf> (last accessed 3 June 2012).

146 It is found that 15 biodiesel projects have been approved by the Chinese CDM authority. See search result with title as 'biodiesel', scope as 'renewable energy; fuel substitute', and status as 'approved.' CDM Project Database, available at: <http://cdm-en.ccchina.-gov.cn/NewItemList.aspx> (last accessed 5 September 2014).

147 Measures on Operation and Administration of CDM Projects 2011.

148 The prioritised areas for CDM projects in China include energy saving, energy efficiency improvement, new and renewable energy development and utilisation, and methane recycling. Measures on Operation and Administration of CDM Projects 2011, Art. 4.

149 Furthermore, the documents include the project feasibility report, a project design document, an environmental impact assessment report, and other documents deemed as necessary by the NDRC. Measures on Operation and Administration of CDM Projects 2011, Art. 15.

150 Environmental Impact Assessment Law of China, Arts. 16-28.

151 The form of the environmental impact assessment of construction projects differentiates according to the extent of the impact. The project with a 'significant' environmental impact has to compile an Environmental Impact Assessment Report; the project with a gentle environmental impact has to fill out the Environmental Impact Assessment Report Form including an analysis of the possible impact; the project with a small environmental impact has to at least fill out the Environment Impact Assessment Registration Form. The contents and forms of Environmental Impact Assessment shall be decided by the environmental departments of the national government, referring to the Ministry of Environmental Protection. The report and report forms are compiled by authorised environmental impact assessment institutes. Environment Impact Assessment Law of China, Arts. 16, 17, and 20.

impacts on the surrounding environment[153] and water and soil conservancy measures if required by law.[154] Upon receiving the application, the NDRC will organise experts to examine and review the application documents, but the specific criteria in the process of the expert review are not specified in this regulation. The result of the expert review is to be further examined by the CDM Supervision Council (*xiang mu shen he li shi hui*) consisting of governmental officers,[155] taking into account the 'check-list'[156] designated in the regulation, including *inter alia* GHG emissions saving of the project, the environmental impact assessment report, and the contribution to sustainable development.

Accordingly, without specific rules for biofuel CDM projects, the sustainability of biofuel CDM projects in China is largely dependent on the implementation of the environmental impact assessment law and the examination of the contribution to sustainable development in the approval process. However, environmental impact assessments in China have been subject to criticism for the limited scope of the impacts to be assessed, construction of projects prior to the impact assessment, and rent-seeking. Improvement of the current environmental impact assessment law is called for to achieve greater compatibility with the newly revised Environment

152 The Environmental Impact Assessment requirements differentiate between governmental 'planning' (*gui hua*) over land use (including water and sea) and 'construction projects' (*jian she xiang mu*).

153 The surrounding environment has been interpreted as the geography, landscape, soil, water and underground water, climate and atmosphere, minerals, forests, grassland, wild fauna and flora, agricultural crops, protected natural areas, etc. Environmental Impact Assessment Law of China, Art. 17; National People's Congress, Question and Answers, Interpretation of Environmental Impact Assessment Law, 21 October 2004, available at: <www.npc.gov.cn/npc/flsyywd/xingzheng/node_2186.htm> (last accessed 10 May 2015).

154 It is especially provided that the projects involving water and soil conservancy shall design conservancy measures approved in advance by the water administration of the government in accordance with the Water and Soil Conservancy Law of China. According to the Water and Soil Conservancy Law of China, water and soil conservancy refers to the preventive and remedy measures on water and soil erosion, and construction projects in the areas identified as prone to have water and soil erosion have to compile water and soil conservancy measures. Environmental Impact Assessment Law of China, Art. 17; Water and Soil Conservancy Law, Arts. 2 and 25.

155 The projects passing the preliminary round of expert review shall be further examined and verified by the CDM Project Verification Council (*xiang mu shen he li shi hui*). The Council consists of officials from the NDRC and the Ministry of Science and Technology (jointly as group leaders), the Ministry of Foreign Affairs (group deputy leader), as well as the Ministry of Finance, the Ministry of Environmental Protection, the Ministry of Agriculture, and the Meteorological Administration. Based on opinions of the Council, the NDRC, together with the Ministry of Science and Technology and the Ministry of Foreign Affairs, will make the final decision on whether to issue the approval letter or not. Measures on Operation and Administration of CDM Projects 2011, Arts. 8, 9, and 12.

156 The items in the check-list include: qualification of project participants, methodology application, required approval of relevant governmental departments, accounting of GHG emission reductions, prices of transferred credits, additionality of purchasing captital, technology transfer, credit transfer period, monitoring plan, and the anticipated contribution to sustainable development. Measures on Operation and Administration of CDM Projects 2011, Art. 20.

Protection Law.[157] Moreover, the *de facto* contribution of CDM projects to China's environment, sustainable development, and climate change mitigation is questionable.[158] For instance, China's leading position in the CDM market was largely established by the HFC-23 projects which can hardly be described as contributing to sustainable development.[159] Moreover, considering that the Chinese government has provided favourable financial policies for renewable energy projects, it is doubtful whether the renewable energy CDM projects (including biofuel projects) in China meet the 'additionality' requirement of the Kyoto Protocol.[160] Actually, the question about the 'additionality' of the Chinese CDM projects was highlighted when the CDM EB refused to grant CDM status to 14 wind power projects in China, based on a debate about their compatibility with the 'additionality' requirement.[161] As for sustainable development, while the absence of

157 China revised the Environment Protection Law in 2014, in which it provided that planning, including but not limited to land use planning, has to go through an Environmental Impact Assessment and pollution prevention facilities shall be installed in line with the approved environmental impact assessment documents. Such revisions may make the relevant provisions in Environmental Impact Assessment Law incompatible with the Environmental Protection Law, because the current impact assessment only covers land use planning and pollution prevention facilities can be installed after impact assessment. Moreover, the Environmental Impact Assessment Law has limitations in dealing with the problems in rent-seeking, construction before assessment, and public participation. 吴青, 环评法修改刻不容缓 [It is urgent to revise Environmental Impact Assessment Law] (China Law Insight, 2 March 2015), available at: <www.chinalawinsight.com/2015/03/articles/corporate/-mergersacquisitions/%E7%8E%AF%E8%AF%84%E6%B3%95%E4%BF%AE%E6%94%B9%E5%88%BB%E4%B8%8D%E5%AE%B9%E7%BC%93/> (last accessed 10 May 2015); 环境影响评价法究竟该怎么改 [How should Environmental Impact Assessment Law be revised?] (中国环境报, 19 March 2015), available at: <http://gongyi.sina.com.cn/gyzx/2015-03-19/095252069.html> (last accessed 10 May 2015).

158 There might be inherent flaws in the CDM, which mainly lie in environmental integrity concerns related to (financial) additionality, ethics, and long-term economic costs. It has even been commented that, taking into account the flaws of CDM, the CDM might be a harmful instrument for China in the long-term. Chang, Belis & Bruyninckx 2012, p. 240.

159 Such CDM projects were controversial because of their questionable environmental merits. Ellis & Kamel 2007, p. 10; Lin 2011b, p. 302; Schneider 2009; Wara 2008.

160 To prove 'additionality' requires a determination that the project's emission reduction is in addition to the reduction that already occurs and requires a baseline projection of emissions that would have occurred without the subsidy provided by the CDM. As it is noted by commentators, it is difficult, if not impossible, to prove the 'counterfactual' that emissions reduction would not have happened but for the CDM's credit transfer. Wara also notes that China's electricity sector projects present 'extremely challenging regulatory decisions to the CDM EB because it must decide which projects would or would not have gone forward without the carbon finance funds', so 'the incentives created by the CDM are likely leading to no change in behavior except for widespread claims for credits.' Lewis 2010; Wara 2008, p. 1790-1798.

161 In mid-2009, the EB's 47th and 48th meetings found and considered uncertainties about power tariffs in the Project Design Documents (PDDs) of the Chinese wind-power projects. At the board's 49th meeting, it rejected applications from 10 Chinese projects, and in 2010 the EB's 52nd meeting it rejected six of the projects. The controversy centered on the Chinese power tariff for wind, which is the most important factor in determining additionality. The Chinese stakeholders objected strongly and described the refusal as a 'wrong signal that will seriously →

international rules in this regard under the Kyoto Protocol[162] gives states significant room to decide whether CDM projects meet their domestic policy objectives,[163] consideration of the economic benefits of CDM projects might prevail over environmental concerns, in particular for certain local governments in China.[164] Thus, Lin calls for strengthened legal rules and criteria on additionality and sustainability in the process of the approval and monitoring of China's CDM projects.[165] Taking the Jinjiang Biodiesel Production Project as an example, the environmental impact assessment was approved by Environmental Protection Bureau of the Yantai municipal government in 2008, covering impact items of water and air quality, solid waste, and noise from the biodiesel industrial production process, through which it was concluded that 'the host Party and the project owner both regard that the proposed project will not bring significant impacts on the environment.'[166] As for sustainable development, it was stated that the project would contribute to national sustainable development with regard to preserving limited fossil fuel resources, reducing other pollutants such as SO_2, NOx and CO_2 from fossil fuel consumption and promoting local economic development and

damage the enthusiasm and confidence of investors.' He Gang and Richard Morse, Lessons from controversy (1), Chinadialogue, 10 June 2010, available at: <www.chinadialogue.net/article/3664-Lessons-from-controversy-1-> (last accessed 10 May 2015); Lin 2011b, p. 303; Karl-Erik Stromsta, CDM Approves 32 Chinese Wind Farms; Rejects Six Others, RECHARGE, 15 February 2010, available at: <www.rechargenews.com/-energy/wind/article206525.ece> (last accessed 10 May 2015).

162 See discussion in Chapter 2, section 2.3.

163 It has been commented that the overall CDM approval process gives states significant room to structure the regulatory field to meet domestic policy objectives, as the process of determining 'additionality' and sustainable development of host countries relies largely on documents demonstrated by project operators and the approval of host countries. Lin 2011b, p. 300; Wara 2008, p. 1790-1798.

164 It is noted that, while the approval of CDM projects remains largely within the ambit of the central government, China's local governments respond positively to the profit-oriented enterprises. The CDM has been an attractive new source of revenue for local governments and has provided incentives for investments in renewable energy projects, particularly regarding hydro and wind power. Qi *et al.* 2008, p. 394; Zhang 2010, p. 6644; World Bank, Clean Development Mechanism in China: Five Years of Experience (2004-09), May 2010, available at: <www.worldbank.org.cn> (last accessed 12 July 2014), p. 33; Lin 2012b, p. 320-321.

165 Lin 2011b, p. 303.

166 As for water quality, it refers to the direct discharges of industrial wastewater and domestic wastewater which may have an impact on the water quality, against the requirement of Integrated Wastewater Discharge Standard (GB8976-1996). Air quality refers to the dust from the boiler house that may have some impacts on the air quality, against the Atmospheric Pollutant Discharge Standard (GB16279-1996). Solid wastes refer to the wastes generated by the project such as boiler ashes from boiler house. The noise mainly derives from excavation and the transportation during the construction period, against the Standard of Noise at the Boundary of Industrial Enterprises (GB12348-1990). See Project Design Document, available at: <http://cdm.unfccc.int/filestorage/2/a/1UBGARLO93E7VZCYFPDMN6TQ0SWH8K.-pdf/PDD%20Shandong%20Biodiesel.pdf?t=OWd8bmUzaHNxfDBbWbut-rPlk21OoQz95I_b> (last accessed 10 December 2015), p. 43-44.

providing potential job opportunities for local residents.[167] Based on such assessment, the environmental impact assessment for the CDM project was approved by the local (municipal) environmental protection bureau five years ahead of the registration date (8 May 2013),[168] focusing on pollution from the industrial process rather than from feedstock production and collection.

2.3. *Interim Conclusion*

Section 2 examined the biofuel demand-related measures in China. Above all, the central government has established general principles related to non-food grain feedstock, marginal land, and no environmental damage as guidance for policy making to promote the demand for and development of biofuels, but the meaning and binding force of the principles need to be further specified. Nevertheless, the establishment of the principles illustrates a major shift in China's biofuel policy, particularly reflecting the emphasis on avoiding adverse effects of the demand for biofuels on China's food security. The essential question is whether the general principles are adhered to or further specified in the other regulatory measures relating to the demand for biofuels. Pilot projects that mandate biofuel blend have been implemented in certain areas of China since 2002, and the Renewable Energy Law sets forth the obligation to incorporate biofuels in fuel selling systems. However, compliance with the general principles or other sustainability requirements is not coherently and explicitly set out for the existing pilot projects or legal provisions. In addition to the blend mandate, the central government has been setting specific volume targets for bioethanol and biodiesel to be consumed annually, but a similar problem is the lack of a corresponding requirement about sustainability, and the only condition of 'non-food grain' once attached to the bioethanol target was withdrawn probably due to the difficulty in achieving such a 'non-food grain' target. Nevertheless, conditions of waste cooking oil feedstock and *ex-ante* environmental impact assessments by public authorities are set out for the economic incentive measures in the form of preferential tax treatment and emission allowance exemptions, and the increasingly strengthened emission standards for

167 It was drawn from comments of local stakeholders that the project would contribute to the local environment and some social benefits in transforming waste oil/fat into useful energy resources, which satisfies with national strategic principle of sustainable development. However, the comments were based on a questionnaire survey conducted on 18 December 2008. There were 30 copies of questionnaire distributed and all recovered, and the participants included local inhabitants, social organisations, and related governmental departments. See Project Design Document, available at: <http://cdm.unfccc.int/-filestorage/2/a/1UBGARLO93E7VZCYFPDMN6TQ0SWH8K.pdf/PDD%20Shandong%20Biodiesel.pdf?t=OWd8bmUzaHNxfDBbWbut-rPlk2lOoQz95I_b> (last accessed 10 December 2015) , p. 2 and 44.

168 The environmental impact assessment law provides that it is the government of provincial level to decide environmental impact assessments are approved by which agency, unless the project at stake is related to nuclear energy or involves more than one province or shall be approved by the State Council as provided by law. Environmental Impact Assessment Law of China, Art. 23 and Art. 24.

fuels and vehicles are likely to advance emission standards for biofuels. However, there is still a lack of a legal framework that provides and applies sustainability requirements, or coherently integrates the general principles of non-food grain, marginal land, and no environmental damage to the demand-related regulatory measures.

3. Chinese Regulation of the Supply of Biofuels

China has employed a biofuel promotion policy also because it is estimated that biomass (such as agricultural crops and residues) is the most abundant source of renewable energy in China[169]– if not used in a modern way to generate energy, these biomass resources would have been burned in the traditional way that has adverse effects on the environment and human health.[170] As for biofuels, agricultural crops and waste cooking oil constitute the two major raw materials. Corn and wheat are the main feedstocks for the four state-owned bioethanol producing plants which make up a large share of the bioethanol supply in China.[171] However, since 2006, non-food crops, or the so-called 'one-and-a-half generation' feedstock such as cassava, sweet sorghum, and jatropha,[172] are increasingly used for producing biofuels,[173] and 'second-generation' biofuels based on cellulosic feedstock such as crop stalks and residues are particularly promoted.[174] Waste cooking oil and

169 Gan & Yu 2008, p. 533.

170 Biomass, such as crop stalks and forest and wood residues, has been traditionally burnt directly for cooking and heating in China, and such low-efficient and unhealthy use may still exist in some rural areas as a way of accessing energy. The modern way of using the biomass for energy purposes is to produce biogas, biomass pellets, biofuels through pathways such as pyrolysis, gasification, carbonisation or to install high efficiency stoves. Gan & Yu 2008, p. 533; Zhang 2011, p. 135; Zhu 2005, p. 156; Verbruggen, Moomaw & Nyboer 2011, p. 955.

171 With stale grains running out in 2005, it is said that all the four plants converted to using fresh maize as the major feedstock since then. Qiu *et al.* 2012, p. 3096.

172 Among the range of oil bearing plants such as jatropha, xanthoceras sorbifolia, and pistacia chinensis, jatropha is considered as one of the most favorable feedstocks in China. Jatropha is mainly planted in the subtropical areas of China including Yunan, Sichuan, and Guizhou provinces. Qiu *et al.* 2012, p. 3098.

173 For example, Siyi Ethanol Company in the Heilongjiang Province has been using sweet sorghum for bioethanol production on a trial basis since 2004 with an annual output of about 50,000 tons. In 2007, CNPC constructed and inaugurated a demonstration unit to produce 3,000 tons of ethanol per year from sweet sorghum stalks in Dongtai City of the Jiangsu Province. While several demonstration scale plants have been constructed in the Sichuan, Chongqing and Shaanxi provinces to produce bioethanol from sweet potato, the construction of a commercial-scale bioethanol plant using sweet potato as feedstock in the Hebei province, which had been approved by NDRC in 2007, was temporarily suspended due to the concerns about the potential competition with cereal production in the region and the logistical constraints of transporting feedstock from remote areas. Additional plants using cassava or sugarcane for bioethanol production in the Guangxi and Fujian provinces are under consideration. Qiu *et al.* 2012, p. 3095 and 3097.

174 Although there is not yet a commercial operation because of high production costs, there were eight pilot and demonstration plants in operation by December 2009, with a total capacity of 280,500 tons of bioethanol per year. One cellulosic based bioethanol plant was built by COFCO in the Heilongjiang Province to use maize stalk as the feedstock and was →

animal fat provide for an important feedstock source for the production of biodiesel in China,[175] largely because there is a persistent shortage of vegetable oil in China and China is a net importer of all the major edible vegetable oils.[176] In view of the variety of feedstock sources, whether the feedstocks are sustainably produced and managed is a critical issue for the sustainability of biofuels. This section aims to look into the regulation of the production of biofuel feedstock, which is generally divided into agricultural and non-agricultural, as well as to see whether any requirement is set out to address sustainability concerns, particularly in relation to food, land, and the environment.

3.1. *Regulation of Biofuels Feedstock from Agriculture*

Biofuel policy in China has from the beginning been closely associated with agriculture: the bioethanol blend mandate was initiated with the purpose of consuming surplus food grains and promoting the rural economy and, until now, agricultural crops and residues provide the major sources of biofuel feedstock. The primary problem with agricultural feedstock in China is the concern about the competition between its use for biofuels and its use for food or feed, considering that China is a developing country with a large population but limited arable land. Another concern is the negative environmental impact in the process of agricultural feedstock cultivation, including the GHG emissions, biodiversity loss, and agro-environment of air, water, or soil. As mentioned earlier, the central government has set out general principles of non-food feedstock, marginal land, and no environmental damage, which respectively address the food, land, and environmental concerns. Although these principles may serve as guidelines for adopting policies on biofuel feedstock production, a further look into the specific rules about agricultural production or land management is still needed. Thus, this sub-section will probe into the general principles in relation to feedstock production in more detail.

3.1.1. Non-food Grain Feedstock

In the Renewable Energy Medium- and Long-Term Plan, it is stated that the relationship between biofuel development and food shall be greatly emphasised

among the first in the world to use a continuous steam explosion process. This project was jointly developed by COFCO and Novozymes of Denmark. Due to the relative success of some of the pilot projects, CNPC has teamed up with COFCO to construct 20 cellulosic ethanol plants across China, with a planned total capacity of 2 million tons. Qiu *et al.* 2012, p. 3097.

175 Biodiesel is also produced from the residue oil from vegetable oil crushers. However, the cost of using vegetable oil residues is higher than that of waste cooking oil. Scott & Jiang 2013.

176 China imported more than 8 million tons of vegetable oil each year in 2008 and 2009, and China is the largest importer in the world. Imported soybeans are mainly used for producing edible vegetable oil and animal feed, and China is still far from being able to rely on imported vegetable oil and soybeans for biodiesel production. Qiu *et al.* 2012, p. 3098.

and biofuel development shall not consume food in large quantities,[177] which was later communicated with the slogan-like principle as 'no competition with people for food' (*bu yu min zheng liang*).[178] In fact, before the principle was set out, the central government had taken measures to centralise the approval of construction licences for bioethanol projects,[179] except that the four state-owned grain-based bioethanol plants could continue to produce food grain based bioethanol within their existing capacity.[180] The requirement of 'non (food) grain' is at the core of this principle, but the principle *per se* does not specify what '(food) grain' (*liang shi*) entails. Nevertheless, the development plans may have implied that non-food grains allowed and promoted as biofuel feedstock could include both the 'one-and-a-half generation' crops (e.g. cassava, sweet sorghum,[181] and oil bearing trees/plants) and 'second-generation' or even more advanced raw materials (e.g. crop stalks and algae).

Since 2006, the 'one-and-a-half generation' crops have been viewed as a *transitional* biofuel feedstock in the short- and medium-term before the 'second-generation' biofuels can be commercially produced, largely because they are not used as *staple food* in China and meanwhile incur less costs than second-generation biofuels.[182] The central government promotes such non-food grain feedstock by requiring the establishment of feedstock bases (*yuan liao ji di*)[183] and providing plans and subsidies for these supply bases. The central government provides that only the

177 Renewable Energy Medium- and Long-Term Plan 2007, para. 4(2).

178 The 11th Five Year Plan for Renewable Energy Development, para. 3(2).

179 It is provided that any bioethanol project, except for the already initiated Guangxi cassava bioethanol project, shall be examined *ex ante* and approved by the central government in order to put in place stricter market entry mechanisms; and, without the approval, a project cannot enjoy any tax preferences or other support from government or even could be fined by the relevant administrations. Notice on Bioethanol Project Management 2006.

180 The four plants shall not expand their production capacity and shall try to diversify their feedstock sources and new corn processing projects, including those for bioethanol production, under construction or to be constructed, shall be temporarily frozen until further specific guidelines are worked out. Urgent Notice on Corn Processing 2006.

181 Although sugar-rich crops such as sugar cane and sugar beets can also be used for producing bioethanol, they are not economically viable, because the demand for edible sugar is so great in China due to its large population and thus sugar production from sugarcane and sugar beets is more profitable than for bioethanol production. Wang, Xiong & Liu 2009, p. 345-346.

182 Moreover, they may have the merit of being tolerant with regard to degraded soil or a harsh environment. For example, sweet sorghum has drought tolerance and high water-use efficiency, so it can be produced in marginal environments where rainfall is limited and irrigation is either too expensive or would deplete water reserves. Compared to the second-generation biofuels, biofuels from these crops are already or are more likely to be produced on a commercial scale. Wang, Xiong & Liu 2009, p. 345-346.

183 Feedstock bases refer to the bases that supply cultivated or dedicated raw materials for the approved biofuel demonstration enterprises, and the feedstock bases shall be constructed with benefits to ecological protection and having no impact on soil erosion, and shall be built in rows or in concentrated in order to satisfy scale processing and production. The 11th Five Year Plan for Renewable Energy Development, para. 3(2); Renewable Energy Medium- and Long-Term Plan, para. 6(4); Interim Measures on Financial Support for Feedstock Bases 2007, Arts. 2 and 7.

biofuel producers having feedstock bases can enjoy fiscal support from the government: the government requires the implementation of an 'enterprises plus bases' approach, which means that the licensed biofuel producing plants are responsible for purchasing the raw materials cultivated in designated farms.[184] The central government has mapped the location of the feedstock bases: according to the Renewable Energy Medium- and Long-Term Plan, sweet sorghum is to be cultivated intensively in North-East China and the Shandong province where degraded land is abundant; while provinces in South-West China, such as Guangxi, Chongqing, and Sichuan, should focus on developing tuber crops such as cassava, according to local land and agricultural characteristics.[185] Furthermore, the non-grain feedstock bases *per se* are eligible to obtain governmental subsidies.[186] However, there is still a question about whether the 'one-and-a-half generation' biofuel feedstock indeed has no impact on food supply, since crops like cassava and sweet sorghum could also be used as food (or food stuff) or fodder.[187] Above all, the State Council has issued an administrative regulation on the circulation of grains, in which (food) grain (*liang shi*) is defined as including wheat, rice, corn, and the other varieties of food grains (*za liang*) and processed grains (*cheng pin liang*).[188] In the draft legislative proposal for the Law on Grains, the definition of '(food) grains' is enlarged to include all cereal (*gu wu*) crops and their processed products as well as beans (*dou lei*) and tubers (*shu lei*).[189] Before the final adoption of the legislative proposal, it is still unclear whether the definition of '(food) grains' covers cassava (as a tuber crop) and sweet sorghum (as a cereal). Moreover, due to the link with food, the socio-economic impact of the promoted cultivation of 'one-and-a-half

184 This requirement not only applies to biofuel producers but also applies to other bioenergy enterprises and biochemical enterprises. Biodiesel producers also need to have compatible oil-bearing plant bases. Opinion on Fiscal Support Policy 2006; Interim Measures on Financial Support for Feedstock Bases 2007, Art. 4; Biodiesel Industry Development Policy 2014.

185 The 11th Five Year Guidelines for Renewable Energy planned that sweet sorghum should be cultivated respectively in the estuary area of the Yellow River in the Shandong province, in the coastal areas along the Yellow River in Inner Mongolia, as well as in the provinces of Heilongjiang, Jilin, and Xinjiang. Cassava and the cultivation of sweet potatoes should take place in the provinces such as Guangxi, Chongqing, Sichuan and Hainan. Sweet sorghum and cassava should be selectively cultivated. The 11th Five Year Plan for Renewable Energy Development, para. 3(2); Renewable Energy Medium- and Long-Term Plan, para. 6(4).

186 The subsidy is on average 200 RMB/acre for forestry biomass raw materials and 180 RMB/acre for agricultural crop raw materials, but the specific amount varies according to different types of land and is decided by the standards and individual support scheme made available and verified by the Ministry of Finance. Interim Measures on Financial Support for Feedstock Bases 2007, Arts. 2 and 7.

187 See the introduction to cassava and sweet sorghum in FAO, Why Cassava?, available at: <www.fao.org/ag/agp/agpc/gcds/index_en.html> (last accessed 11 May 2015); FAO Corporate Document Repository, Breeding and Cultivation of Sweet Sorghum, available at: <www.fao.org/docrep/t4470e/t4470e05.htm> (last accessed 11 May 2015).

188 Administrative Regulation on Grain Circulation 2004, Art. 2.

189 Law on Grains (draft) 2012, Art. 2.

generation' crops on market prices and the supply of other food crops is still debatable.[190]

Furthermore, the central government has implemented the 'non-(food) grain' principle by promoting the utilisation of crop stalks[191] which are regarded as promising feedstocks for producing 'second-generation' biofuels.[192] Crop stalks are recognised by the government as an attractive source for bioethanol feedstock, because they are considered to be abundant in China[193] and do not affect the food supply. More importantly, the use of crop stalks for energy can help to mitigate the phenomena of burning useless crop stalks in the field[194] and may meanwhile increase employment and income of farmers.[195] The law has prohibited the burning of crop stalks in the open air,[196] particularly in certain important areas,[197] and

190 For example, a cassava-based bioethanol plant in Guangxi might put pressure on local cassava supplies, raising the country's need for imports of the crop from Vietnam and Thailand. A study about the impacts of China's bioethanol development on the agricultural economy shows that the impacts differ largely among commodities such as staple food grain and non-staple food grain. There has been concern that even if the production of biofuels may improve farmers' income, the potential price increases of food commodities is likely to result in unchanged or even a decline in the purchasing power for farmers. Guangxi to blend 10 per cent ethanol into gasoline in December (Biopact, 19 September 2007), available at: <http://news.mongabay.com/bioenergy/2007/09/guangxi-to-blend-10-ethanol-into.html> (last accessed 5 May 2015); Qiu *et al.* 2010, p. 83; Wang 2011, p. 208; Huang *et al.* 2009; Yang, Zhou & Liu 2009; Zhang *et al.* 2003; Leng 2008; Jansson *et al.* 2009; Chang *et al.* 2012, p. 7-8, 9; Li & Chan-Halbrendt 2009, p. S162-S169.

191 The scope of 'crop stalks' is defined as including the stalks of crops such as rice, wheat, corn, bean, oil crops, cotton, and tubers, and the residues resulting from the primary processing of agricultural products. Interim Measures on Financial Support for Energy Use of Crop Stalks 2008.

192 The feedstocks for 'second-generation' biofuels include not only agricultural residues but also forest residues and energy grasses such as Miscanthus and Panicum. Wang, Xiong & Liu 2009, p. 344.

193 The Ministry of Agriculture completed the first specific investigation on crop stalk utilisation in China in 2010, estimating that the quantity of available crop stalks is 687 million tons, including stalks of corn (265 million tons), wheat (205 million tons), and rice (150 million tons). The stalk utilisation rate has reached 69 per cent, including for feed, fertiliser, paper production, and energy. 农业部组织完成首次全国农作物秸秆资源专项调查 [The Ministry of Agriculture Organised and Finished the First Specific Investigation of China's Crop Stalk Resources] (gov.cn, 16 December 2010), available at: <www.gov.cn/gzdt/2010-12/16/content_1767577.htm> (last accessed 11 May 2015).

194 Chinese farmers have the habit of burning crop stalks in the field or using them as traditional biomass energy for cooking, thus causing both ambient and indoor air pollution. Notice on Strengthening Synthesised Use of Agricultural Crop Stalks 2013.

195 The government also estimated that the use for biofuels would not affect other uses of crop stalks in paper manufacturing and livestock feeding. The government also expected social benefits from using crop stalks, such as rural economic development and employment. Opinion on Accelerating the Synthesised Utilisation of Agricultural Crop Stalks 2008; Renewble Energy Medium- and Long-Term Plan 2007, para. 7(3).

196 Burning crop stalks in the open air has been recognised as one culprit of severe smog in China. Although the government banned stalk burning as early as 1999, this practice continued and contributed, as recognised by the government, to severe smog in big cities. In this context, the NDRC took measures in 2013 to strengthen the responsibility of local →

requires that the crop stalks shall be carefully disposed of in order to prevent environmental pollution and damage.[198] Crop stalks and residues after harvest shall be utilised in a synthesised manner.[199] Support for 'synthesised use'[200] of crop stalks, including the use for biofuel production,[201] is further set forth by the central government: for example, financial support can be given to enterprises that use crop stalks for energy generation.[202] A specific plan is set for the synthesised utilisation of crop stalks in the 12th Five Year period, which places the use for the production of fuel as one of ways for utilization to be developed as a priority[203] and it commits to supporting enterprises that produce crop stalk-based cellulosic biofuels.[204]

3.1.2. Use of Marginal Land

Due to food security concerns, the Chinese central government fears that the conversion of arable land for food into land for biofuel feedstocks would affect food crop supplies, considering the limited arable land resources in China.[205] It is stated

governments in banning crop stalk burning by linking their fulfilment of this work with the evaluation of the local governments' and officials' performance. Notice on Strengthening the Synthesised Use of Agricultural Crop Stalks 2013, para. 5.

197 It is prohibited to burn crop stalks in populated areas, areas around airports, and main transport lines. Law on the Prevention and Control of Environmental Pollution by Solid Waste 2004, Art. 20.

198 Agriculture Law of China, Art. 65.

199 *Ibid.*

200 The synthesised use means industrialised use of crop stalks in a synthesised way, from the establishment of stalk gathering, storage and transport system to the recycled use in various industries such as paper manufacturing, energy generation, and bioethanol production. Opinion on Accelerating the Synthesised Utilisation of Agricultural Crop Stalks 2008.

201 Crop stalks can also be used to produce bioenergy in the form of solid fuel pellets and biogas through gasification or distillation. Opinion on Accelerating Synthesized Utilisation of Agricultural Crop Stalks 2008.

202 The subsidy is channeled to enterprises engaging in producing solid crop stalk pellets, crop stalk gasification and distillation. Interim Measures on Financial Support for Energy Use of Crop Stalks 2008, Art. 4.

203 The Plan identifies that crop stalks have great energy potential, and the energy use (including in the form of biogas, solid biomass pellet) can effectively reduce primary energy consumption. The 12th Five Year Plan on the Synthesised Utilisation of Agricultural Crop Stalks 2011, para. 4(5).

204 The government is to support the enterprises with research and production capacity to carry out demonstration projects and solve technical barriers. The goal is for crop stalk-based clean energy to be used in over 80 per cent of farm households, and the annual energy utilisation should reach 30 million tons, which accounts for 30 per cent of the crop stalks used in the pilot area. The 12th Five Year Plan on the Synthesised Utilisation of Agricultural Crop Stalks 2011, para. 5(3).

205 Although China has a vast area of land, arable lands account for only 12.7 per cent of China's territory and 6.8 per cent of global arable land, and the per capita area of arable land is only one-fourth of the world average, while the population in China constitutes 20 per cent of the global population. Moreover, the area of arable lands is shrinking due to the conversion of high-quality agricultural land to residential use, and the problem remains despite the central government's restrictive order and clampdowns. In addition to the shrinking area of arable land, land quality is also deteriorating. The over-exploitation in the past two decades has →

in the biofuel development plans that the production of energy crops or plants[206] shall not illegally occupy arable lands[207] and shall make appropriate use of 'marginal land' which is not suitable for growing food grains, so that the cultivation of biofuel feedstock crops and plants does not compete with food crops for lands[208] and may even bring environmental benefits.[209] Accordingly, to use marginal land for biofuel feedstock production constitutes the primary way of reducing land use pressure from biofuels. An essential question is how to define 'marginal land' and how to implement such a principle. The development plans do not provide a formal definition, but they cite waste lands and hills (*huang shan huang di*), waste grasslands (*huang cao di*), sandy lands (*sha di*), and saline-alkali soil lands (*yan jian di*) as examples of marginal lands.[210]

However, 'marginal land' is not a category of land that is defined in China's land law. China has established a basic legal framework for land administration and agricultural land protection.[211] According to the law, land in China is divided into three types based on its purpose of use: agricultural land (*nong yong di*), construction land (*jian she yong di*), and unused land (*wei li yong di*). Agricultural land refers to all the lands used for agricultural production, including arable land, agro-forestry land, grazing land, and irrigation land,[212] while agriculture is defined as including crop-planting, forestry, husbandry, and fishery as well as the services directly related to agricultural production processes.[213] Such a categorisation of lands provides for the basis of making land use plans: governments of various levels adopt and implement land use plans to identify areas as basic agricultural

largely degraded the land quality. In this sense, China indeed has limited arable land resources to increase its domestic biofuel supply. Zhang & Crooks 2012, p. 19; Wang & Tian 2011, p. 163; OECD (2006) Environmental Compliance and Enforcement in China: An Assessment of Current Practices and Ways Forward, p. 12, available at: <www.oecd.org/env/outreach/37867511.pdf> (last accessed 28 July 2014).

206 According to the definition in the Renewable Energy Law of China, 'energy crop' refers to the herbaceous and woody plants especially cultivated as energy feedstocks. Renewable Energy Law of China 2005, Art. 32(3).

207 In fact, retaining a certain amount of arable land has been established as a mandatory target in the Nationl Five Year Plans for Economic and Social Development and the government has set a 'bottom line' of keeping 1.8 billion acres arable land. Renewable Energy Medium- and Long-Term Plan 2007, section. 4(2); The 12th Five Year Plan for National Economic and Social Development, Chapter 3; 'China arable land area above food security red-line' (Xinhua, 30 December 2013), available at: <http://news.xinhuanet.com/english/china/2013-12/30/c_133007338.htm> (last accessed 13 May 2015).

208 The 11th Five Year Plan for Renewable Energy Development, section 3(2).

209 It has been stated that marginal lands can be used appropriately for growing energy crops or plants which can be combined with the efforts of 'greening' waste lands or hills. Renewable Energy Medium- and Long-Term Plan 2007, section. 6(2); The 11th Five Year Plan for Renewable Energy Development, section. 4; The 12th Five Year Plan for Renewable Energy Development, section 3(4).

210 The 12th Five Year Plan for Renewable Energy Development, section 3(4).

211 Land Administration Law of China 1986; Regulation on Basic Agricultural Land Protection 1998.

212 Land Administration Law of China 1986, Art. 4.

213 Agriculture Law of China, Art. 2.

land,[214] arable land (*geng di*), construction land, and land prohibited from reclamation.[215] Any change between the designated uses, particularly from agricultural land to construction land, shall be *ex ante* approved by the competent authorities, depending on the size and type of the land at stake; and the use of arable land for non-agricultural purposes is strictly restrained.[216] However, the land law does not mention or define 'marginal land' and the only similar term is 'unused land', which is defined as including land that is neither arable land nor construction land.[217] The law encourages[218] the exploitation of unused land, with a priority for agricultural use,[219] but such exploitation must protect the environment, prevent soil erosion and desertification, comply with land use planning, and go through scientific evaluation.[220] In view of the land law, questions about the scope and implementation of plans on 'marginal land' remain, particularly regarding *a)* whether 'marginal land' is equivalent to 'unused land' and thus excludes 'arable land' and, *b)* whether biofuel crop cultivation falls within the scope of 'agricultural

214 The basic agricultural land is identified according to land use planning and relevant legal procedures and is particularly protected and shall not be converted for other land uses. There is both a quantity and quality requirement for identifying basic agricultural land: the size of the basic agricultural land shall account for above 80 per cent of the total size of arable land in one provincial jurisdiction and the lands used for grains, cotton, oil, and vegetable production bases and land with good soil and water conservancy shall be identified as basic agricultural land. Once identified as basic agricultural land, occupation of the land or the conversion of the land for another purpose is prohibited, unless approved by the State Council. Basic agricultural land is prohibited from being used to develop fruit forestry or aquaculture and the basic agricultural land shall be compensated with an equivalent size of arable land. Land Administration Law of China, Art. 34; Regulation on Basic Agricultural Land Protection 1998, Arts. 2, 9, 10, 15, and 16.

215 Land use planning shall be approved by the government of a superior level and shall be strictly implemented once the designated uses are approved. For example, the land use planning of provincial governments and big cities shall be approved by the State Council, while the land use planning of other levels of government shall be approved by the provincial government. Land Administration Law of China, Art. 45.

216 The conversion of arable land for non-agriculture uses shall be approved by the competent governmental authority: for example, the county government could only approve land use change from arable land to construction land under 3 hectares and for a larger area the change should be approved by superior governments. The occupied arable land shall be compensated by an equivalent plot of land. The provincial government shall make plans to reclaim arable land and ensure the quantity and quality of arable land within its jurisdiction. Arable land shall not be left idle or wasted. Land Administration Law of China, Arts. 31, 36-37.

217 Construction land refers to the land used for constructing buildings, private and public facilities, in urban and rural areas, factory and mines, transport and water facilities, tourism, and military facilities, and unused land includes the land falling outside of both agricultural land and construction land. Land Administration Law of China, Art. 4.

218 For example, the government can lease waste land to individuals or groups for cultivation, forestry, fishery, etc, and the local government shall take measures to renovate low-quality land and idled or wasted land. Land Administration Law of China, Arts. 40 and 41.

219 Land Administration Law of China, Art. 38.

220 It is particularly provided that reclamation shall not destroy forestry, grassland, or drain lakes. Land Administration Law of China, Art. 39.

use' and can thus be exempted from the restrictions on the conversion of arable land.

Nevertheless, the Ministry of Agriculture once specified the scope of 'marginal land.' The Ministry of Agriculture conducted an investigation on the availability of marginal land for energy crops and plants as preparation for feedstock supply bases as required by the renewable energy development plans.[221] In the programme for the investigation, the scope of marginal land for energy crops was defined as including 'fallow fields in winter' (*dong xian tian*) and 'waste land suitable for energy crops' (*yi neng huang di*). The 'fallow fields in winter' refer to the idled *arable land* without any crops growing (including grazing grass) between the autumn harvest and the next spring.[222] 'Waste lands suitable for energy crops' refer to the natural grassland, thin stocked forest land (*shu lin di*), shrubbery forest land (*guan mu lin di*), and unused land (*wei li yong di*)[223] which are suitable to be reclaimed for growing energy crops,[224] excluding the land within protected areas[225] and flood prevention areas.[226] Such an interpretation suggests that marginal lands used for biofuel feedstock cultivation may include both 'arable land' (as long as it is idled from food production) and 'unused land' as referred to in the land law but exclude the areas for special purposes such as nature conservation and flood prevention. One method that has been used to implement the 'marginal land' principle is to establish designated feedstock bases and to give financial support to the bases on the condition that they make use of unused land (e.g. waste mountain, hill, saline-alkali soil land) and fallow fields in winter.[227] However, when arable land can be used to grow energy crops/plants, questions may arise as to whether the conditions provided in law on conversion of arable land to non-agricultural use (e.g. approval by the competent authorities) are complied with, and how to ensure only *idled* arable land is used for biofuel production since some crops can be used both as food and biofuel feedstock. Moreover, it has been pointed out that, although China has a large area of waste lands, there is no unitary standard at the national level to make

221 Programme of Investigating and Evaluating Marginal Land Resources 2007.

222 *Ibid.*, section 1(1).

223 Unused land in the programme refers to the waste grassland, saline-alkali soil land, waste sandy land, barren earth, and tide-flat areas which can be used for agriculture but have not yet been used. *Ibid.*, section 1(2)(1).

224 Land which does not fall within the scope of these types of land may also be identified as marginal land for energy crops as long as it meets certain criteria regarding *inter alia* soil quality and thickness, water content, and temperature. *Ibid.*, section 1(2)(2).

225 The protected areas include the protected areas for natural forests, nature, wild fauna and flora, water conservancy forest, water and soil conservancy, forest for erosion control, and wetland. *Ibid.*, section 1(2)(2)(1).

226 *Ibid.*, section 1(2).

227 It is particularly stressed that the use of fallow fields in winter shall ensure 'no competition with food for land.' The NDRC in 2014 has confirmed that biodiesel feedstock bases shall not occupy arable land or forestry with ecological or public service values and shall comply with the overall land use planning. Interim Measures on Financial Support for Feedstock Bases 2007, Art. 2(1); Biodiesel Industry Development Policy 2014, para. 41.

clear how these lands are suitable for energy crops/plants cultivation and how to evaluate the impact on the environment and livelihood of local communities.[228]

3.1.3. No Damage to the Environment

Concern about the impact of biofuels on the environment is expressed in China's biofuel development plans. The Renewable Energy Medium- and Long-Term Plan states that the relationship between biofuel development and the environment shall be noted and the development of biofuels shall not damage the environment,[229] but it does not further specify what this requirement means for biofuel feedstocks production. Thus, there is a need to probe into the specific environmental regulation relating to biofuel feedstock production, particularly with regard to the concerns about biodiversity, carbon stock, and agricultural pollution such as air, soil, and water pollution.

3.1.3.1. Biodiversity

China undertakes a commitment to protect biodiversity and wetlands under the CBD[230] and the Ramsar Convention,[231] and has established and implemented a series of law to protect specific ecological systems (e.g. wetlands) or species (e.g. endangered wild animals and plants)[232] and has meanwhile set out national plans and programmes for biodiversity conservation.[233] To designate and establish nature

228 Report on the Policies for Biofuel Industry 2014, p. 116.

229 Renewable Energy Medium- and Long-Term Plan 2007, section 4(2).

230 China ratified and became a Party to the Convention on Biological Diversity on 29 December 1993. Convention on Biological Diversity, China – Overview, available at: <www.cbd.int/countries/?country=cn> (last accessed 13 May 2015).

231 The Convention on Wetlands came into force for China on 31 July 1992. China presently has 45 sites designated as Wetlands of International Importance, with a surface area of 3,858,240 hectares. The Ramsar Convention on Wetlands, The Annotated Ramsar List: China, available at: <http://ramsar.rgis.ch/cda/en/ramsar-documents-list-anno-china/main/ramsar/1-31-218%5E16477_4000_0__> (last accessed 13 May 2015).

232 China adopted a series of laws, administrative regulation, and departmental and local governmental rules concerning biodiversity conservation, including Wild Animal Protection Law, Forest Law, Grassland Law, Animal Husbandry Law, Seed Law and Law on the Quarantine of Import and Export of Animals and Plants, Regulation on Nature Reserves, Regulation on Protection of Wild Plants, Regulation on Biosafety Management of Agricultural Genetically Modified Organisms, Regulation on the Management of Trade in Endangered Wild Animals and Plants, and Regulation on the Protection of Wild Medicine Resources. Biodiversity Conservation Strategy and Action Plan of China (2011-2030), section 2(2).

233 China adopted the China Biodiversity Conservation Action Plan in 1994 and has issued a series of biodiversity plans for a certain period such as the China Nature Reserves Program (1996-2010), China Master Plan for Ecological Conservation, China Program for Ecological Environment Conservation and China Program for Conservation and Use of Biological Resources (2006-2020). These national plans are supposed to be implemented and integrated into plans for the protection of nature reserves of wetlands, aquatic species, and genetic →

reserves (*zi ran bao hu qu*) and protected areas (e.g. forest/wetlands/desert/-geological parks, community-based conservation areas, protected sites of wild plants, special marine protected areas) has become a primary way of fulfilling biodiversity *in situ* conservation.[234]

The biodiversity conservation requirement has been stated in the regulation of biofuel production in the following ways. Firstly, in defining the scope of 'marginal land' for biofuel crop/plant cultivation, the Ministry of Agriculture states that whole areas of nature reserves and other protected sites (for natural forests or wild fauna and flora), including the thin stocked forest land (*shu lin di*) and shrubbery forest land (*guan mu lin di*) therein, are excluded.[235] Secondly, the State Forestry Administration has issued guidelines[236] on the cultivation of energy forestry (*neng yuan lin*),[237] in which it is stated that energy forestry cultivation shall avoid the biodiversity-rich areas[238] and 'sustainable cultivation' is to achieve biodiversity of energy forestry.[239] Specifically, if energy forestry cultivation takes place in areas with frequent wildlife activities or of ecological importance, the local forestry administrations shall identify the areas and their buffer zones with 'high protection value' (*gao bao hu jia zhi qu yu*), taking into account the habitats of wildlife and typical forest ecological systems where energy forestry cultivation shall not take place.[240] Moreover, measures are required to be taken to prevent the invasion of alien species in the process of energy forestry cultivation.[241] Thirdly, in setting out a

resources. Biodiversity Conservation Strategy and Action Plan of China (2011-2030), section 2(2).

234 China declares that a system of *in-situ* conservation has been established which is composed primarily of nature reserves and complemented by scenic spots, forest parks, community-based conservation areas, protected sites of wild plants, wetland parks, desert parks, geological parks, special marine protected areas and germplasm conservation farms. Biodiversity Conservation Strategy and Action Plan of China (2011-2030), section 3(3).

235 It is recalled that the Ministry of Agriculture has defined 'marginal land' as including 'waste land suitable for energy crops' which contains natural grassland, thin stocked forest land (*shu lin di*), shrubbery forest land (*guan mu lin di*), and unused land. Programme of Investigating and Evaluating Marginal Land Resources 2007, section 1(2).

236 The Guidelines were issued for the purpose of guiding and regulating the processes of carrying out energy forestry cultivation activities by forestry administrations of various levels and can also be used as a reference for evaluating and managing private entities' activities in the cultivation of energy forestry. Guidelines for Sustainable Cultivation of Energy Forestry 2011, Annex I, Art. 3.

237 Energy forestry is defined by the State Forestry Administration as forestry which is used primarily for producing solid, liquid, or gaseous bioenergy, including trees producing oil, lignin, or starch-rich feedstocks. Guidelines for Sustainable Cultivation of Energy Forestry 2011, Annex I, Art. 2.

238 Guidelines for Sustainable Cultivation of Energy Forestry 2011, Annex I, Art. 4(4).

239 It is defined that 'sustainable cultivation' refers toensuring that energy forestry has biodiversity, productivity and vitality as core issues currently and in the future through scientific and reasonable cultivation methods and that it should achieve a triple win in ecology, economy, and society. Guidelines for Sustainable Cultivation of Energy Forestry 2011, Annex I, Art. 2.

240 Guidelines for Sustainable Cultivation of Energy Forestry 2011, Annex I, Art. 33.

241 For example, inspection and quarantine measures are to be taken and a supervision and alert system is to be established. The guidelines also contain specifications on how to select the →

policy for the development of the biodiesel industry, the National Energy Administration emphasises that the establishment of biodiesel feedstock bases shall avoid nature reserves and other ecologically sensitive areas, and the cultivation of energy forestry/plant or algae for biodiesel production shall give attention to biodiversity and avoid alien species.[242]

The attempt to protect certain biodiverse areas from biofuel feedstock production can also be based on relevant law. For instance, the Regulation on Nature Reserves provides that the activities of *inter alia* reclamation and logging are prohibited in nature reserves, unless it is otherwise provided.[243] Forest Law requires that typical forest ecological areas (*dian xing sen lin sheng tai di qu*), forests with valuable wildlife, natural tropical rainforest, and other natural forests with special protection values shall be designated as forest nature reserves,[244] and once established as nature reserves they are protected from exploitation according to Regulation on Nature Reserves. Agriculture Law forbids reclamation in the way of destroying forests and grasslands or draining lakes and wetlands,[245] and declares to protect biodiversity related to agriculture, in particular the rare, endangered, and precious species and their habitats.[246] The Administrative Rule on Wetland Protection generally prohibits reclamation, drainage, destroying habitat, or introducing alien species into wetlands,[247] and nature reserves shall be established for the wetlands that can meet the relevant conditions.[248]

However, a regulation setting out binding and integrated requirements for biodiversity protection in the production of biofuels still lacks. The scope of marginal lands is set out in the ministerial document just for investigating land resources. The guidelines on energy forest are limited to a specific type of forestry. The legal rules about biodiversity are largely fragmented and provided in general

types of trees for cultivation, and how to manage the forestry. Guidelines for Sustainable Cultivation of Energy Forestry 2011, Annex I, Art. 36.

242 Biodiesel Industry Development Policy 2014, para. 41.

243 The nature reserves include both national nature reserves and local nature reserves, the former shall be approved by the State Council and the latter shall be approved by provincial governments. This prohibition applies to the whole area of nature reserves, including the core zone, buffer zone, and experiment zone. Particularly, anyone is prohibited from entering into the core zone unless otherwise provided, and production or business activities are prohibited in the buffer zone. Regulation on Nature Reserves of China 1994, Arts. 26-29.

244 Forest Law of China 1998, Art. 24.

245 If reclamation by destroying forests, grass, lakes, or wetlands has been carried out, the operators shall gradually restore the reclaimed land to its original status. Agriculture Law of China 2002, Art. 62.

246 Moreover, species introduced from third countries shall be registered or approved and due safety control measures shall be taken. Agriculture Law of China 2002, Art. 64.

247 Wetlands are defined as including both natural wetlands and artificial wetlands. Administrative Rule on Wetland Protection 2013, Arts. 2 and 31.

248 The areas that can be identified as nature reserves shall have typical national ecological systems, concentration areas of rare or endangered wildlife, natural heritages with special significance. The wetlands that do not meet such conditions can be established as wetland parks or community-based conservation areas, etc. Administrative Rule on Wetland Protection 2013, Arts. 11, 19, 20; Regulation on Nature Reserves of China 1994, Art. 2.

terms. As China's Fifth National Report to CBD admits, China has made 'some progress' in undertaking a study on the impacts of biofuel plantations on biodiversity, but a system is yet to be established to govern the environmental safety of biofuel production.[249]

3.1.3.2. Carbon Stock

The impact of land use change on carbon stock changes has not been reflected in the regulation of or policy on biofuels, largely due to the fact that the concept of 'carbon stock' (*tan hui*) or the related terms[250] is relatively new in China and has not been set out in law. Nevertheless, China has been emphasising afforestation and reforestation for ecological and social purposes.[251] First of all, China defines 'forest resources' as including forest (*sen lin*), woods (*lin mu*), and forest land (*lin di*).[252] Particularly, forest land is defined broadly to include (arbour) forest with a canopy density of at least 20 per cent and the other land appropriate for forestation.[253] With China's commitment to climate change mitigation, afforestation and reforestation are beginning to be highlighted in climate change policies. During the 12th Five Year Period, a domestically mandatory target is set for increasing forest coverage and volume,[254] and increasing carbon stocks through afforestation and reforestation is established as one major means to limit GHG emissions.[255] Moreover, the State Council has committed to building a system of collecting and monitoring the GHG emission inventory that covers energy, industrial production, agriculture, land use change and forest, and waste disposal.[256] Corresponding to this commitment, the State Forestry Administration has engaged in the work of compiling technical standards and methodologies for assessing, monitoring, and reporting on carbon

[249] China's Fifth National Report to CBD 2014, p. 7 and 150.

[250] The related terms include carbon sink, carbon pool, and carbon sequestration. FAO Corporate Documant Repositary, Forests and climate change Carbon and the greenhouse effect, available at: <www.fao.org/docrep/005/ac836e/AC836E03.htm> (last accessed 13 May 2015).

[251] Forest Law divides forests into five categories according to uses: *a)* protective forest (*fang hu lin*) for conserving water and soil, shielding wind and fixing sand, or protecting road, coast, farmland; *b)* raw material forest for manufacture or production; *c)* economic forest for producing fruits, oil, drinks, seasoning, medicine, etc.; *d)* fuel forest for supplying fuels; and *e)* forest with special uses. Forest Law of China 1998, Art. 4.

[252] Forest resources include forests, woods and forest lands, as well as wild animals, plants and microbes whose life and survival depends on the forest, wood and forest land. Forests include arbor forests and bamboo forests. Woods include trees and bamboos. Regulation on Implementation of Forest Law of China 2000, Art. 2.

[253] Forest lands include arbor forest lands with a canopy density of 0.2 degrees or more as well as bamboo forest lands, bush shrub forest lands, open forest lands, stump lands, burned areas, non-mature forestation lands, nursery lands, and land appropriate for the forestation planned by the people's government at or above the county level. Regulation on Implementation of Forest Law of China 2000, Art. 2.

[254] The goal is to increase forest coverage to 21.66 per cent and increase forest volume by 600 million cubic meters. The 12th Five Year Plan for Economic and Social Development, section 3.

[255] The 12th Five Year Work Programme on GHG Emission Control 2011, para. 6.

[256] *Ibid.*, para. 14.

sequestration and stocks related to forest and land use change, and has proposed to revise the Forest Law by supplementing rules about forest carbon monitoring and trading for carbon emission trading mechanisms.[257]

3.1.3.3. *Agricultural Pollution*

Pollution from agriculture has become a severe environmental problem in China. Non-point source (*mian yuan*) water pollution is largely caused by fertiliser and pesticide runoff or discharges from livestock and aquaculture.[258] An investigation into soil quality was conducted by the central government in 2014, which revealed that 16.1 per cent of the investigated soil (including all the arable lands, part of forest lands, grasslands, unused lands, and construction lands) is polluted above what the national standards permit,[259] and the use of fertiliser and pesticides in agriculture may constitute an important cause of arable land pollution.[260]

China has adopted a law to tackle environmental pollution in agriculture. Above all, the revised Environment Protection Law provides that government of various levels shall strengthen environmental protection in agriculture, reinforce agricultural pollution monitoring and implement an alarm system, and prevent *inter alia* soil pollution and erosion or the extinction of species.[261] It is also provided that government of all levels and its agricultural departments shall issue guidance on the scientific use of fertilisers and pesticides to prevent non-point pollution, and preventive measures shall be taken to prevent pollution by metallic and toxic elements when pesticides and fertilisers are used.[262] Agriculture Law provides that farmers shall maintain the productivity of arable land and make reasonable use of fertilisers, pesticides, increase the use of organic fertilisers, and prevent the

257 Carbon stock trading has been carried out in China through the CDM afforestation and reforestation projects. The 2013 White Paper on Climate Change Policy and Action in Forestry 2014, para. 6 and 7; Guiding Opinions on Promoting Forest Carbon Trading, 29 April 2014.

258 The nonpoint source water pollution includes nutrient pollution, organic pollution (from organic discharges from the animal husbandry sector), nutrient pollution (from diffuse runoff of fertiliser from agricultural land), and pesticides. It is noted that China is now the biggest fertiliser consuming country in the world, and most farmers over-apply cheap, low-quality, single-ingredient fertilisers which increases the possibility and amount of fertiliser runoff. China has also become the world's biggest producer, consumer, and exporter of pesticides: About 30 per cent of the total pesticides used in China are highly toxic. Zhang & Crooks 2012, p. 47-51.

259 Report on National Soil Pollution 2014.

260 全国土壤污染报告:总超标率16.1 南方比北方严重 [Report on National Soil Pollution: overall pollution rate 16.1 per cent, the south is more severe than the north] (大河网-大河报(郑州), 18 April 2014), available at: <http://news.163.com/14/0418/02/9Q32PPOR00014Q4P.html> (last accessed 14 May 2015; Lucy Hornby, China admits widespread soil pollution in 'state secret' report (Financial Times, 18 April 2014), available at: <www.ft.com/cms/s/0/c250bd4c-c6b4-11e3-9839-00144feabdc0.html#axzz3a5hYHsqu> (last accessed 14 May 2015).

261 Environment Protection Law of China 2014, Art. 33.

262 *Ibid.*, Art. 49.

contamination, destruction, and degradation of farmland;[263] agricultural departments of all levels shall guide farmers in adopting biological methods or using high-efficiency pesticides with low toxic substance.[264] Specifically, the Water Pollution Prevention Law contains a section that addresses water pollution in different agricultural sectors, providing that the pesticides in use shall comply with national law and standards and transport, storage, and the disposal of pesticides shall be strictly managed.[265] The Land Administration Law provides that governments of various levels shall take measures to improve soil quality and productivity so as to prevent desertification, salinisation, soil erosion and pollution, and to improve the quality and productivity of soil.[266] The Water and Soil Conservancy Law has generally prohibited reclaiming slope land (with slope degree above 25) to grow agricultural crops for the sake of soil and water conservancy.[267]

However, one problem is that the existing provisions on the agricultural environment are too vague to be enforced and leave the major responsibility to the local authorities.[268] For example, the law provides that local governments are responsible for guiding the use of fertilisers and pesticides and for supervising the quality of arable land, but the practical problem is that most county-level governmental staff who provide advice and assistance to farmers are not sufficiently trained and there is frequently a shortage of staff and finances.[269] The law does not specify the responsibilities or liabilities of the farmers in using fertilisers and pesticides and this ambiguity may lead to problematic enforcement.[270] Moreover, soil pollution prevention law has been absent for years and the current national soil environment standard was issued almost ten years ago.[271] It has been

263 Agriculture Law of China 1993, Art. 58.

264 The measures that may reduce the use of chemical fertilisers and pesticides include the promotion of organic agriculture, balanced fertiliser application, use of compound fertilisers, and integrated pest management. Agriculture Law of China 1993, Art. 65; Zhang & Crooks 2012, p. 54.

265 The Water Pollution Prevention Law addresses 'water pollution prevention in agriculture and rural areas' in a separate section. Provisions are respectively adopted for preventing pollution from agricultural planting, livestock breeding, aquaculture, and irrigation. Water Pollution Prevention Law of China 1984, Art. 47.

266 Land Administration Law of China 1986, Arts. 35 and 41.

267 The law has also provided that the work on water and soil conservancy shall be established as one indicator to evaluate the government's responsibility in relation to target fulfilment. Water and Soil Conservancy Law of China 1991, Art. 20.

268 McElwee has identified the main obstacles to the implementation of the Chinese environmental law, which includes the delegation of for environmental protection responsibility to local authorities. McElwee 2011, p. 3-9; see more about implementation problems of Chinese environmental law in Van Rooij 2006; Lieberthal 1997; Lan *et al.* 2006; Economy 2004; Bao 2006, p. 49; OECD, *Environmental Compliance and Enforcement in China: An Assessment of Current Practices and Ways Forward*, 2006, available at: <www.oecd.org/env/outreach/37867511.pdf> (last accessed 28 July 2014), p. 20.

269 Dai 2014, p. 141.

270 Drenguis 2014, p. 186.

271 As yet, the soil environment standard being applied is the version issued in 1995; an institute has updated the standard, but the update has not been approved by the central government. Soil Pollution Prevention Law has been prepared to be adopted in recent years, and the draft →

commented that China's pollution control has been concentrating on industrial sectors over the years by applying the traditional command-and-control environment protection instruments, such as the 'three simultaneous process', discharge standards and permits, and total pollutant discharge, which are difficult to apply effectively in agriculture sectors.[272] With these institutional problems, although the central government has emphasised water and soil conservancy to avoid soil erosion in building biofuel feedstock bases,[273] there is still serious concern that biofuel crop cultivation may intensify water and land use and increase the use of fertilisers and pesticides.[274] Thus, the environmental sustainability concerns related to biofuels reveal the urgent need for China to improve the rules on agricultural pollution prevention and land use-related biodiversity protection and carbon stock conservation/monitoring.

3.2. *Regulation of Biofuels Feedstock from Other Sources*

Apart from agricultural crops and residues, the government encourages biofuels to be produced from non-grain feedstocks such as used cooking oil and algae. For example, the 12th Five Year Plan on Renewable Energy states that the government will further guide and regulate the development of the biodiesel industry based on feedstocks from used/waste cooking oil and animal fat and will carry out research and development on algae-based biodiesel.[275] To use these feedstocks for biofuel production can avoid competition for food and land, but it may involve other sustainability questions such as related to the regulation of waste.

In China, waste/used cooking oil or animal fat is a highlighted source of feedstock for producing biodiesel.[276] Similar to the case of crop stalks, to use waste oil or fat for biofuel production is supposed to have a two-fold benefit: solving biofuel feedstock shortages and meanwhile reducing illegal 'gutter oil' (*di gou you*). The 'gutter oil' generally refers to discarded waste cooking/kitchen oil or fat dredged from sewers of restaurants or related businesses. Such oil has been illegally

law is to establish soil management systems according to different types and zones. But it has been reported that there are still disputes regarding agricultural non-point pollution and setting soil pollution limit standards. 邓昕睿, 首部土壤法年内上报人大:将对土壤分区分类管理 (中国联合商报, 8 December 2014), available at: <http://finance.sina.com.cn/china/-20141208/114021021791.shtml> (last accessed 10 May 2015).

272 Zhang & Crooks 2012, p. 54.

273 For example, biofuel feedstock bases shall be built in a way that is beneficial to ecological protection and that will prevent soil erosion. It requires an environmental impact assessment and 'three simultaneous processes' in building the construction projects of biodiesel plants and feedstock bases. Interim Measures on Financial Support for Bioenergy Feedstock Bases 2007, Art. 2(3); Biodiesel Industry Development Policies 2014, Art. 40.

274 There is a possibility that increased biofuels production would inherently compete for the same water needed for growing food crops. IISD 2008, p. 42-48.

275 The 12th Five Year Plan for Renewable Energy Development, section 3(4).

276 Biodiesel producers are forbidden from using edible fat or oil materials as biodiesel feedstocks or from producing or selling the waste/used oil for food. Biodiesel Industry Development Policy 2014, para. 53.

used as food or foodstuff, which has posed a serious threat to food safety in China. Thus, using 'gutter oil' for biofuel production may provide for an opportunity to mitigate the illegal use.[277] The State Council on the one hand suggests cracking down on the illegal collection, production, sale, and use of 'gutter oil' for food, and on the other hand encourages the licensed collection, storage, and reuse of waste/used kitchen oil.[278] In line with the instructions of the State Council, the National Energy Administration requires provincial counterparts to support the 'gutter oil'-based biodiesel enterprises in order to strengthen the management of the collection and disposal of gutter oil to prevent illegal re-sale for food.[279] Corresponding to the requirement and support from the central government, local projects with governmental (financial) support to use waste cooking oil for biodiesel have emerged,[280] although many problems remain with regard to technology, cost, marketing, and dependence on governmental support.[281] Meanwhile, the waste

277 There is a study revealing that eight categories of activities are direct sources that produce waste oil: grain mill products producing rice bran oil; vegetable oil and forage producing acid oil; and food manufacturing, eating and drinking services and residents producing gutter oil; and seven categories of activities were main underlying drivers of waste oil generation throughout the supply chain: knitting mills, wearing apparel, electronic computers, educational services, health services, public administration and residents. Given the Chinese culinary habits, the Chinese are heavily dependent on oil for cooking, and it is reported that about 10 per cent of China's cooking oil is from illegally produced and used gutter oil which harms human health. Thus, gutter oil and other waste fat or oil has shown a significant potential of being main feedstocks for biodiesel production. Liang *et al.* 2013, p. 139-140; Ramzy, A., 'China cracks down on "gutter oil", a substance even worse than its name' (TIME, 13 September 2011) available at: <http://world.time.com/2011/09/13/china-cracks-down-on-gutter-oil-a-substance-even-worse-than-its-name/> (last accessed 14 May 2015).

278 The State Council requires mayors of cities to take the leading responsibility for cracking down on gutter oil and managing kitchen wastes and to strengthen coordinated organisation and administration. Opinion on Strengthening Crack-down on Gutter Oil 2010.

279 Biodiesel Industry Development Policies 2014, Arts. 11, 13, 24, and 53.

280 For example, Shanghai has promulgated local governmental rules on the management of waste/used cooking oil and has experimented with producing biodiesel for use in buses since August 2013. In the project, waste cooking oil is collected and transformed into biodiesel by licensed enterprises. It supplies as many as 104 buses in Shanghai, with B5 used in 84 buses and B10 used in 20 buses. Moreover, it is reported that China's largest private carrier Hainan Airlines has completed the country's first commercial flight powered by biofuel made from waste cooking oil. This flight was propelled by 50 percent biofuel and 50 percent conventional jet fuel. The biofuel is supplied by a pilot facility jointly established by Boeing and state-owned manufacturer Commercial Airplane Corp. of China that converts about 650 liters of cooking oil into biofuel each day. 马肃平, 上海公交低调试喝地沟油喝的是油吐的是苦水 [Shanghai buses try to 'drink' 'gutter oil'] (Nanfaang Weekend, 17 April 2015), available at: <www.infzm.com/content/108920> (last accessed 14 May 2015); Flying fryer: China completes first flight powered by 'gutter oil' (RT, 21 March 2015), available at: <http://rt.com/news/242913-china-gutter-oil-flight/> (last accessed 14 May 2015).

281 Technology innovation and transfer are key issues. For instance, the economic profit of biodiesel has few advantages over that of fossil-based diesel, which may largely reduce the competitiveness of waste oil derived biodiesel. In addition, nearly half of China's waste oil is gutter oil, and the composition of gutter oil is more complex than foreign waste cooking oil; so in order to use gutter oil as raw material for aviation biokerosene production, many pre- →

cooking oil/fat for producing biodiesel has to comply with the law on preventing and controlling environmental pollution from solid wastes, as solid waste in the law is defined as including gaseous waste (if stored within a container) and liquid waste.[282] For example, enterprises that collect, store, transport, utilise, or treat solid waste must prevent any leakage,[283] and movement of solid waste between provincial jurisdictions shall firstly be approved by the environment protection authority of the exporting province after a consultation with its counterpart in the importing province.[284]

3.3. *Interim Conclusion*

In section 3, it was found that the general principles of non-food grain feedstock, marginal land, and no environmental damage are fleshed out by various legal and administrative measures on the supply of biofuel feedstocks, but an integrated legal framework to regulate and constrain the negative environmental and socio-economic impacts of biofuels still lacks. Firstly, to avoid 'no competition with people for food', the central government on the one hand restricts food grain-based biofuel production, and on the other hand stresses the use of non-food grain biofuel feedstock which includes both the 'one-and-a-half generation' crops (e.g. cassava and sweet sorghum) and the 'second-generation' materials (e.g. crop stalks), in

treatment processes such as filtration and extraction are needed to remove the impurities in the gutter oil, which increases the economic costs of gutter oil derived biokerosene. Liang *et al.* 2013, p. 144.

282 The definition of solid waste constitutes the following elements: *a)* any item or substance in a solid, semi-solid, gaseous (if stored within a container), or liquid waste, and *b)* generated during production, daily life, and other activities, that has *i)* lost its original value and can no longer be used for its original purpose, *ii)* has been discarded or abandoned even if it has not lost its utility value, or *iii)* has been defined as a solid waste by law or administrative regulation. The central government has also issued guidelines that provide a non-exhaustive list of 13 categories of substances that are characterised as solid waste, including wastes generated in production processes and scrap products, laboratory waste materials, sludge from wastewater treatment facilities, and residual slag. Law on the Prevention and Control of Environmental Pollution by Solid Waste 2004, Arts. 88 and 89.

283 They need to obtain a license from the government if they intend to pile up, discard, or dump the waste as defined. Law on the Prevention and Control of Environmental Pollution by Solid Waste 2004, Art. 17.

284 Moreover, the law includes provisions regarding the *transnational* import and export of solid wastes. It is prohibited to import solid wastes into China that cannot be used as raw materials or in some other hazardless way; although some solid wastes fall under an exception of the prohibition, the importation of the exempted products is subject to certain restrictions. The Ministry of Environmental Protection, in conjunction with other national agencies, is required to prepare a catalogue that lists solid wastes that it are forbidden to import, for which import is restricted, and that are automatically licensed to be imported. Furthermore, China has adopted the Basel Convention standards for exports of hazardous wastes – companies that wish to export hazardous waste from China must seek approval from the Ministry of Environmental Protection; and the trans-shipment of hazardous waste via the Chinese territory is forbidden. Law on the Prevention and Control of Environmental Pollution by Solid Waste 2004, Arts. 23-25.

principle through planning, establishing feedstock bases, and financial subsidies. However, the government has not clearly defined the meaning of 'non-food grain' feedstocks, which is particularly a question for the 'one-and-a-half generation' crops that can be used as food. Secondly, to avoid biofuel's 'competition with food for land', it is emphasised that the crops and plants cultivated for energy use shall make appropriate use of 'marginal land' which is not suitable for growing food grains. However, in view of the current land administration law of China, there is ambiguity about the scope of 'marginal land' with regard to 'arable land' and 'unused land' and whether land conversion for biofuel feedstock production can comply with the conditions set out in the land law. Thirdly, the production of biofuel feedstocks must comply with the current environmental regulation particularly with regard to biodiversity and pollution prevention and control. The prevention of exploiting nature reserves and soil erosion are specifically set out for defining 'marginal land' and building feedstock bases. However, considering the lack of a biodiversity regulatory framework, rules about carbon stock changes and institutional flaws in the current regulation of agricultural pollution, there should be great concern about the impact of biofuels on biodiversity and carbon stock change particularly for the areas not identified as protected nature reserves, and the impact of intensified use of water, soil, fertilisers, and pesticides.[285]

4. Chinese Regulation of Trade of Biofuels

While the initial intention of promoting biofuels was to resolve surplus domestic grain reserves, the depletion of the grain reserves and the principles of the 'non-food grain' results in a shortage of biofuel feedstock, which prompts the import of or direct investments in biofuels and biofuel feedstock from third countries. For example, the neighbouring countries in South-East Asia are suitable to grow cassava and palm oil trees, and imported cassava is at times more competitive in price than domestically cultivated feedstocks.[286] In this context, cassava imports from South-

285 It has been pointed out that, due to a lack of binding technical guidlines on how to grow biofuel crops/plants, large-scale plantations of biofuel crops/plants will cause carbon stock loss from soil, grassland, and woods, and may also cause soil erosion, alien species, and monoculture. Moreover, cultivation on waste lands may increase the use of fertiliser and pesticides due to the low quality of these lands. Report on the Policies for Biofuel Industry 2014, p. 115-116; Xie 2012.

286 For example, it has been analysed that, although the Guangxi Province is a world leading cassava production base, the cassava-based biofuel may put pressure on local supplies and raise the need for imports from Vietnam and Thailand or locating ethanol plants in those countries - cassava ethanol can be produced profitably since the transportation costs of cassava from nearby Thailand or Vietnam are marginal. 潘海涛 顾文娜, 中国生物能源产业初具规模 原料短缺是主要瓶颈 [China's bioenergy industry takes shape, but feedstock shortage is the major bottleneck] (China news, 18 June 2007), available at: <www.heb.chinanews.com/news/kjws/2007-06-18/15977.shtml> (last accessed 15 May 2015); Guangxi to blend 10 per cent ethanol into gasoline in December (Biopact, 19 September 2007) available at: <http://news.mongabay.com/bioenergy/2007/09/guangxi-to-blend-10-ethanol-into.html> (last accessed 15 May 2015).

East Asia have increased greatly, particularly from Laos, Vietnam, and Thailand.[287] It has been reported that China's 2012 cassava imports reached a record high of 7 million tons, representing an increase of 42 per cent from the previous year, and among the imported cassava about 20 per cent were used for ethanol production.[288] Another report shows that the Sinopec began to import bioethanol from the US in 2014,[289] and China has been a major importer of biodiesel from Indonesia due to low palm oil prices.[290] Furthermore, some Chinese enterprises have invested in South East Asia and Africa to grow cassava and other energy crops, and then they produce biofuels from the feedstock on site and import the biofuel back into China. This practice is accompanied by activities such as leasing land in the host countries for feedstock cultivation and biofuel production.[291] For example, Henan Tian Guan Group – one of the largest state-owned bioethanol plants – entered into a contract with the Laos government to lease land for producing cassava-based bioethanol which is to be imported to China.[292] However, the government has no intention of relying on biofuel imports for implementing its pilot projects, because imported biofuels are not seen as a secure supply.[293]

The question is whether any principle or requirement about food, land, and the environment is applied to the biofuel (feedstock) imported from third countries. The afore-mentioned general principles and regulation on the demand and supply of biofuels do not explicitly set forth whether they are applied to biofuels or biofuel feedstocks produced outside of China. The 'non-food grain' principle may have an influence on the types of feedstock that will be imported – Chinese enterprises are mainly importing the 'non-food grain' feedstock biofuels such as cassava. However, the 'marginal land' and 'no environmental damage' principles are not likely to exert an influence on the imported biofuels, because such requirements are provided for the purpose of safeguarding national food security and the national environment. Moreover, China has been upholding the 'non-intervention' principle in external

287 It has been reported that China continues to import large quantities of cassava from Thailand as non-grain feedstock for fuel ethanol production. 'The rising tigers of China bioenergy trade', 5 July 2011, available at: <www.biofuelsdigest.com/bdigest/2011/07/05/the-rising-tigers-of-the-china-bioenergy-trade/> (last accessed 12 May 2013).

288 Scott & Jiang 2013.

289 The quantity of bioethanol imports was 10,500 tons. Report on the Policies for Biofuel Industry 2014, p. 6.

290 Attempts were also made to import palm oil from Indonesia and Malaysia. Scott & Jiang 2013; APEC, 'China Biofuels Activities', available at: <www.egnret.ewg.apec.org/Archive/-me_china.html> (last accessed 15 May 2013).

291 There has been a report that China leased land for food and agricultural production in countries such as Cuba and Mexico for many years. Joachim von Braun and Ruth Meinzen-Dick, 'Land grabbing' by foreign investors in developing countries: risks and opportunities, (IFPRI Policy Brief, 13 April 2009) available at: <www.ifpri.org/sites/default/files/-publications/bp013all.pdf> (last accessed 27 October 2012).

292 Latner, O'Kray & Jiang 2006, p. 29.

293 It has been reported that, by taking such a chance, some state companies or coastal provinces have called on the central government to conduct a pilot programme of importing bioethanol. Industry sources noted that unstable international prices could potentially discourage the government from considering the imported ethanol as a secure supply. Scott & Jiang 2013.

relations[294] and interfering with the requirements on land use or environment in third countries would contravene this principle.

It is observed that tariff rates have been the major instruments for China to implement trade policy. There is no unitary tariff code for biofuels. The import tariff on denatured ethanol was 30 per cent in 2009. Although the tariff was lowered to 5 per cent in 2011 for imported ethanol used as chemicals, the 30 per cent tariff remained for ethanol for fuel use largely due to the protection of domestic bioethanol producers.[295] In 2012, China implemented a zero import tariff rate for ethanol imported from the ten ASEAN countries and Chile, Singapore, and Pakistan based on free trade agreements.[296] In this sense, the setting of tariff rates is largely based on considerations of domestic biofuel industry and bilateral trade agreements, rather than on the sustainability of biofuels.

Bilateral and multilateral cooperation serves as another instrument to implement external policy. As for biofuels, China has actively undertaken bilateral or multilateral cooperation primarily to make use of foreign investment and technology transfer. China has bilaterally cooperated with developed countries such as the EU and the United States. For example, the EU has selected Nanyang, a city in Henan province, as a model city for bioethanol promotion and has provided the city with 380,000 Euro as financial support under the BEST (Bio-Ethanol for Sustainable Transport) scheme demonstration project.[297] The Nanyang municipal government spent the funds on the development of flex-fuel cars, bioethanol fuelled public transport, and a modification of gasoline stations.[298] In December 2007, the NDRC signed a Memorandum of Understanding with the US Department of Agriculture and the Department of Energy to cooperate in the development of biofuels in scientific, technical, and policy aspects.[299] Multilaterally, China

294 For instance, in a White Paper, China reiterates the principle of no intervention in the internal affairs of other countries and does not impose its own initiatives on others, and also does not allow interference in China's internal affairs. China's Peaceful Development 2011, section 3.

295 Scott & Jiang 2013.

296 *Ibid.*

297 Bio-Ethanol for Sustainable Transport was a four-year demonstration project in several European regions and cities, with support from the European Commission from the beginning of 2006 to the end of 2009. Despite the strong debate on the sustainability of biofuels, the project introduced over 70,000 bioethanol-powered cars and buses to verify the impact of the utilization of bioethanol on the environment and society, and the EU selected cities in third countries also for the investigation of the use of bioethanol. It is unknown whether the EU's sustainability criteria shall be applied in the project, but it is noticed that the project had ended already by 2009 when Directive 2009/28/EC was just adopted. Transport Research & Innovation Portal, BEST Bioethanol for Sustainable Transport, available at: <www.transport-research.info/web/projects/project_details.cfm?id=10967> (last accessed 15 May 2015).

298 Siang 2007, p. 10.

299 In 2009, the National Energy Administration developed a series of seminars to shape mutual biofuels research and development projects in cellulose ethanol, algae biodiesel, non-grain based ethanol production and thermo chemical conversion of biomass, etc. In May 2010, the first Sino-US Advanced Biofuels Forum was held in Beijing which would take place annually. It was reported that 2011 Sino-US Advanced Biofuels Forum has been successfully held, covering the topics such as sustainable development of feedstock for crop and forest residues, →

participated in the International Bioenergy Forum in 2007 and is a partner of the Global Bioenergy Partnership. These bilateral and multilateral cooperations mostly aim to promote research and development in the supply of non-food grain biofuel,[300] but it may also promote China's awareness and participation in international processes dealing with the sustainability of biofuels.

5. Conclusion

This chapter examined the regulation relating to the demand, supply, and trade of biofuels in China, particularly with regard to the existing principles and requirements addressing the environmental and socio-economic concerns related to biofuels. It was observed that the central government has set out general principles and specific requirements on certain environmental and socio-economic concerns about biofuels, but a legal framework is not yet established for regulating the sustainability of biofuels at the national level. A further reflection from the perspective of integrating environmental and socio-economic dimensions shows the limit of the existing general principles and requirements in inclusively address sustainability concerns. Moreover, although it is possible for lower authorities to adopt more stringent requirements for the sustainability of biofuels, there is a lack of willingness or awareness of local authorities to do so, particularly in face of potential economic benefits that might be brought by biofuel industry.

5.1. *Summary of the Main Findings*

Section 2 addressed the regulatory measures that may affect the demand for biofuels. Above all, the central government has established general principles of non-food grain feedstock, marginal land, and no environmental damage as guidelines for policy making and planning for the demand for biofuels, without explicitly setting out the meaning of the principles. An investigation into the other demand-related measures reveals that these principles are not yet further specified or coherently adhered to. This is a problem particularly with regard to the biofuel blend mandate pilot projects and consumption targets which serve as the major

herbaceous, woody non-grain energy crops and microalgae. During the forum the two sides signed several MOUs on biofuel cooperation: developing sustainable biofuels for aviation, building a cellulosic ethanol demonstration project of 10,000 metric tons per year in 2011, a patent and technology license agreement for a yeast-based cellulosic ethanol project, and collaborative research on sustainable algae biodiesel for aviation. Beckman & Rasmussen 2010, p. 2.

300 For example, in 2011 the NEA and US Department of Agriculture and the Department of Energy agreed to conduct technical exchanges and site visits on establishing efficient, coordinated and sustainable supply system of non-grain biomass feedstock. China and the United States have cooperated in the energy area by initiating the Sino-US Energy Cooperation Program which donates money to support businesses in clean energy and energy efficiency; and the US-China Framework for the Ten Year Cooperation on Energy and Environment was established since 2008 during the US-China Strategic and Economic Dialogue (SED) which also puts clean energy in priority. Beckman & Rasmussen 2010, p. 2.

driver for the demand for biofuels. Although differential environmental or socio-economic conditions are attached to the economic incentive measures, such as preferential tax treatment and emission allowance exemptions, there is still a lack of a legal framework that provides sustainability requirements or specifies the general principles for the major demand-side regulatory measures.

Section 3 found that the general principles of non-food grain feedstock, marginal land, and no damage to the environment are fleshed out by various legal and administrative measures on the production of biofuel feedstocks. The central government restricts food grain-based biofuel production and the use of arable land, and it excludes protected nature reserve areas from being used for biofuel feedstock production. Meanwhile, it encourages the 'one-and-a-half generation' crop and the 'second-generation' material, and use of marginal land for building feedstock bases, in particular through land planning, establishing feedstock bases, and financial subsidies. However, considering the ambiguous scope of 'non-food grain feedstock', 'marginal land' as well as the lack of an integrated environmental regulation for addressing biodiversity, carbon stock changes, and agricultural pollution in biofuel feedstock production, there is great concern about the socio-economic and environmental impacts of biofuels during their production processes.

In section 4, it was found that, although trade in biofuels or biofuel feedstocks exists between China and the third countries, the general principles or specific requirements on the consumption and production of biofuels are not explicitly set out for imports from third countries. The trade measures relating to biofuels are mainly reflected in the adjustment of relevant tariff rates, which primarily aim at protecting the domestic biofuel industry and implementing bilateral free trade agreement, rather than on the sustainability of biofuels imported from third countries. Nevertheless, bilateral and multilateral negotiations and cooperation might provide an opportunity for China to make progress in developing efficient and sustainable biofuel production and to be involved in international processes on ensuring the sustainability of biofuels.

5.2. *Reflection*

Based on the observations from the previous sections, the questions to be further reflected upon are: whether the current principles and requirements have already inclusively addressed environmental and socio-economic concerns related to biofuels; and if not, whether complementary or stricter measures can be taken by lower (provincial) authorities?

5.2.1. The Deficiency of the Current Regulation

Above all, it is perceived that the central government has undertaken a major shift in its policy on biofuels, with the focused concern about the potential negative impacts of biofuels on domestic food security, which is largely reflected in the principles of using non-food grain-based feedstocks and marginal lands. Furthermore, the central government has taken specific administrative measures to limit food grain-based biofuel, such as licensing the construction of bioethanol plants, shifting financial support from food grain-based biofuel to the non-food grain-based biofuel, and planning and supporting non-food grain feedstock bases.

However, the 'non-food grain' and 'marginal land' requirements may hardly address the food sustainability concern. Firstly, bioethanol is still produced for a large part from corn and wheat in China,[301] as bioethanol production from food grains is still allowed in the four state-owned bioethanol enterprises within their existing production capacity. Secondly, the meaning and scope of 'non-food grain' is not clear, particularly considering that the 'one-and-a-half generation' crops which can be used as food or food stuff are allowed and supported as biofuel feedstocks. Thirdly, while the principle of using marginal land is purported to avoid converting arable land for food crops, there is ambiguity in defining 'marginal land' in relation to 'arable land' and 'unused land', and there is concern over whether marginal land is available or suitable to be used for the production of biofuel feedstocks.[302]

Moreover, even if the food sustainability concern is fully tackled, there are many other environmental and socio-economic impacts to be considered. Firstly, the life-cycle GHG emissions of biofuels, particularly the emissions due to carbon stock or land use changes, are not yet taken into account in the current biofuel regulation, largely because it is assumed that the use of biofuels does not increase carbon emissions.[303] The lack of a regulation addressing accounting for a biofuel's life-cycle GHG emissions would ignore the GHG emissions from land use and land use changes, including those from the use of 'marginal lands', while they may refer to grasslands and other unused lands.[304] Secondly, although the central government sets out the principles of 'no damage to the environment' for biofuels and the requirements about nature reserves and soil erosion in building biofuel feedstock bases, such principles and requirements are largely dependent on the stipulation and implementation of existing environmental law in the specific areas which are

301 A report in 2014 demonstrated that most ethanol in China is still produced from grains (76 per cent from corn and 14 per cent from wheat). Among the state-owned fuel ethanol enterprises, 64 per cent of their production was sourced from corn in 2012, 30 per cent from wheat, and 6 per cent from cassava. The cassava ethanol plant was built in Guangxi in 2007 with an annual production capacity of 200,000 tons. A sweet sorghum ethanol plant was completed in Inner Mongolia in 2012 with capacity of 50,000 tons. Scott & Jiang 2013; Anderson-Sprecher & Jiang 2014.

302 One might also argue that the land and related resources can be used for the production of food instead of biofuels if food security is the primary concern. Given the fact that land has been the scarcest resource in China's agriculture, it is questionable whether China will still have new marginal land and water resources available to support the production of feedstocks demanded by the expansion of biofuel production. Even if marginal lands are used, the question is whether they should all be utilised to produce biofuel feedstocks when they can also be used for other environmental or food purposes. Chang *et al.* 2012, p. 4, 18, and 78; Qiu *et al.* 2012, p. 3096.

303 In the section that analyses the environmental and socio-economic impacts of the renewable energy development plan, it is stated that agricultural and forestry biomass does not increase CO_2 emissions in life cycles. The 12th Five Year Plan on Renewable Energy Development, section 5.

304 Report on the Policies for Biofuel Industry 2014, p. 115-116; Xie 2012; Qiu *et al.* 2010, p. 83.

still saddled with problems particularly regarding the prevention of agricultural pollution.[305] Moreover, there is no sufficient consideration of the environment when using marginal land for biofuel feedstock production. Since marginal lands are largely unused lands with low productivity, to exploit such lands usually calls for more investment and technology and an intensified use of water, fertilisers, and pesticides,[306] which may have environmental impacts such as *inter alia* the introduction of alien species, monoculture, fertiliser and pesticide runoff, and water use efficiency.[307] Thirdly, there is little mention about local rural dwellers depending on the marginal lands, as well as the possible impacts on their access to water, land, and food if such lands are used to grow biofuel feedstocks.[308] Moreover, it has been pointed out that even if the production of biofuels may improve farmers' income levels, the potential price increases of food commodities is likely to result in unchanged or even declined purchasing power of farmers.[309]

5.2.2. Possibility for More Stringent Rules

With limits in inclusively regulating the sustainability impacts of biofuels at the national level, a further question is whether the authorities of the lower levels can provide more stringent environmental or socio-economic requirements in relation to biofuels? The principles and requirements on biofuels *per se* do not explicitly set out whether stricter measures can be taken by local governments, but the following legal bases could be argued for adopting local measures with more stringent sustainability requirements.

Firstly, according to the Renewable Energy Law of China, oil-selling companies undertake an obligation to incorporate biofuels into their oil-selling networks in accordance with the blend mandates of the energy authority of the central government (i.e. NDRC) or provincial governments. This provision means that provincial governments have the competence to decide whether to initiate blend mandate projects and how to implement the projects within their jurisdiction. For example, the Hainan provincial government has mandated a blend of 5 per cent biodiesel (BD5) in oil selling systems within the whole province, which is to be produced from waste oil and perhaps oil-bearing plants.[310] In this vein, provincial

305 McElwee 2011, p. 3-9.

306 Qiu *et al*. 2010, p. 83.

307 It has been commented that the spreading of biofuel plants over the rural areas would bring challenges for environmental sustainability: there would be challenges of controlling pollution from the production process, as China has experienced in the development process of township and village industrial enterprises; and special attention must be paid to water pollution. Qiu *et al*. 2012, p. 3103; Report on the Policies for Biofuel Industry 2014, p. 115-116.

308 Ribeiro 2013, p. 361.

309 It has been pointed out that expanding the production of biofuels may conflict with growth in non-agricultural sectors, as these other sectors are also competing for limited resources in terms of labour, arable land and water, and whether biofuels can improve farmers' livelihood remains questionable. Wang 2011, p. 208; Huang *et al*. 2009; Yang, Zhou & Liu 2009.

310 Hainan Biodiesel Promotion Work Plan 2009; 海南试用生物柴油12家中石化加油站试点销售[12 SNOPEC gas stations experiment selling biodiesel in Hainan →

governments can set out requirements on the types of biofuels and possibly other environmental and socio-economic sustainability considerations in blend mandate projects.

Secondly, it is explicitly provided in Environment Protection Law of China that provincial governments may introduce new environmental quality standards or pollutant charges when relevant national standards are absent, and they may adopt stricter standards than the existing national standards as long as the Ministry of Environmental Protection is informed.[311] There are many cases that the central government has allowed sub-national rule-makers to set stricter standards or introduce new standards in order to address environmental priorities in the local area.[312]

Thirdly, according to the Constitution and Law on Legislation, local governments have certain flexibility in adopting local legislation complementary to or stricter than national law. As introduced at the beginning of this chapter, China's legal system is depicted as a unitary system with multiple levels: the unity aspect is reflected in the centralisation of state power at the national level,[313] and the hierarchy between the legal sources – generally the legislation of a lower level shall be stipulated for implementing the superior law and shall not contravene the superior law.[314] However, People's Congresses of the provincial level and

Province] (南海网, 21 November 2010), available at: <www.boxun.com/news/gb/china/-2010/11/201011210832.shtml> (last accessed 1 May 2015).

311 Environment Protection Law of China 1989, Arts. 15 and 16.

312 For instance, the introduction by local authorities of pollution charges, bans and waste minimisation and recycling measures addressing the control of one-time-use products such as plastic bags. Provinces can also promulgate ambient standards for pollutants that are not covered in the national law. The central government generally favors such independent local law-making initiatives, especially when they are considered premature to be promoted nationwide. OECD, 'Environmental Compliance and Enforcement in China: An Assessment of Current Practices and Ways Forward', available at: <www.oecd.org/env/outreach/-37867511.pdf> (last accessed 28 July 2014), p. 21.

313 The Constitution emphasises integrity and unity of the state and centralisation of the state power at the national level. The Constitution declares that it has the supreme authority which shall not be contravened by other sources of law and reserves the power of amendment to the National People's Congress (NPC) where the highest state power resides. Constitution of China, Art. 5; Chen 2008, p. 124-125, 181 and 188.

314 The Law on Legislation specifies a hierarchical system among the legal sources to uphold unity and consistency. In principle, the legal effect of legislative documents depends upon the competence and status of their issuing authorities: laws have higher authority than administrative regulations; administrative regulations have a greater importance than other regulations and rules; local regulations have a higher status than the local rules of the same jurisdiction. The hierarchy is to be guaranteed by the following mechanisms. Firstly, the Law on Legislation provides the content to be governed by each legal source so as to delimit the scope of legislative competences. For example, it enumerates the issues which can only be governed by laws adopted by NPC or its Standing Committee; administrative or local regulations and rules shall be enacted only for implementing laws and regulation of superior levels. Secondly, the content of legislative documents shall be made according to (*gen ju*) legislation of a higher authority or shall not contravene (*bu di chu*) legislation of a higher authority. Thirdly, legislation at certain levels has to go through an approval (*pi zhun*) procedure in order to enter into effect, or it has to file for record (*bei'an*) after promulgation. →

prefectural level can enact local regulation (*di fang xing fa gui*) on matters that are not yet regulated by national law based on the local circumstances and practical needs.[315] The People's Congresses of the autonomous regions can deviate from (*bian tong*) national law so as to adapt it to the local characteristics.[316] Provincial or prefectural governments can adopt local government rules (*di fang zheng fu gui zhang*) on matters that shall be regulated by local regulation if the conditions are not in place to adopt national regulations or there is an urgent need.[317] Accordingly, it could be argued that local regulation can include legal measures on sustainability concerns related to biofuels which are not yet addressed by the central government, and local governments can address the sustainability of biofuels if the conditions are not in place to adopt local regulation, particularly considering that the current national principles and requirements are ambiguous.[318]

Based on the afore-mentioned grounds, the general principles and requirements about non-food grain feedstock, marginal land, and no environmental damage may form the *minimum* requirements for biofuels, and more stringent requirements are normally allowed to be taken. While local authorities are generally restrained from regulating matters that are not yet affected by national law, introducing new sustainability requirements for biofuels is possible based on the authorisation for initiating local blend mandates in the Renewable Energy Law. It can be furthermore based on the provisions in the Environment Protection Law and the Law on Legislation which allow for the introduction of local regulations or local

Finally, the legislative authority of a higher status has supervision powers to amend (*gai bian*) or revoke (*che xiao*) norms of lower legal status. Law on Legislation of China 2000, Arts. 87-97; Chen 2008, p. 190.

315 However, these matters shall exclude the issues that can only be provided by law (*fa lv*) adopted by National People's Congress or its Standing Committee, such as *inter alia* levying taxation and criminal penalties. Moreover, once the matter is later regulated by national law, the local legislator shall revise or annul the local regulation. Law on Legislation of China 2000, Art. 73.

316 However, the modification cannot violate the basic principles of laws or administrative regulations, and the law or administrative regulation specially made for the autonomous regions shall not be modified. Law on Legislation of China 2000, Art. 75.

317 The particularly significant matters within the provincial or prefectural jurisdictions shall be regulated by local regulations adopted by their People's Congresses. Governments of provincial and prefectural levels can adopt local government rules to implement national laws, administrative regulations, and local regulations on matters within their jurisdictions. Local government rules of the prefectural level can only regulate matters on urban/rural construction and management, environmental protection, historical and cultural site protection, but local government rules that have already been adopted can continue to be effective. The local governments shall make a proposal to local National People's Congress to adopt local regulations on the matter after two years of implementation. Law on Legislation of China 2000, Art. 82.

318 In fact, it has been commented that the vagueness in legal drafting allows the regime to maintain the appearance of a viable legal system while retaining changeability and adaptation to local structures and conditions. The process of specifying vague and principle-like terms is to address the reality of the diversity, localism, and uneven development in China, which might be deliberately designed to overcome the inherent limitations of a unitary system of law in a large and diverse country. Corne 1995, p. 262-263 and 286.

government laws according to the local circumstances and practical needs. Nevertheless, local measures on biofuel blend mandates and the construction of new bioethanol feedstock bases are still subject to the supervision of the central government, because the building of bioethanol plants should be licensed by the central government, particularly when the biofuel enterprise or feedstock base intends to acquire financial support from the central government.

In China, however, the problem perhaps is whether local governments can adequately implement national law on environment protection. Compared to economic development, tackling pollution is generally a lower priority on the agenda of local governments, especially when pollution control is in conflict with economic growth. Even if there is local will to act, an 'implementation gap' exists for local environmental protection bureaus which are commonly under-resourced and relatively powerless vis-à-vis their economic counterparts.[319] It is not only the general and vague wording of law that poses challenges for implementation, but also the tension between economic development and environmental progress makes stricter enforcement of environmental law difficult.[320] The central government has taken various steps to strengthen the enforcement of national environmental law, such as information campaigns, economic incentives, and legal sanctions.[321] Moreover, the fulfilment of environment and climate goals has been integrated into binding evaluation indicators for the performance of local governments and their cadres, such as energy intensity targets, CO_2 emission intensity targets, and pollutant (e.g. SO_2 and Chemical Oxygen Demand) control targets.[322] Administrative reform of the organisation of governmental authorities in charge of environment protection has been put on agenda of the central government.[323] In this

319 Van Rooij's observation in the Yunnan Province reveals that the non-implementation of environmental law at the local level is attributed to the increasing conflict between state/national law and local interests. He argues that local cadres and officials take shares in polluting enterprises that have not only contributed to the local revenues but have also improved the locals' life by creating new employment and providing more income, and the complex combination of multiple interests gains support among the locals. See discussion on the 'implementation gap' in: Van Rooij 2006, p. 57-74; Lieberthal 1997, p. 3; Lan *et al.* 2006; Economy 2004.

320 It has been argued that general language may exist in other systems of law, but in China the difficulty lies perhaps in the nature of China's economy (lack of governmental control over the growing private economy, and difficulty of enforcing Chinese environmental statutes against state-owned enterprises), reliance on alternative methods of dispute resolution in China, obstacles to private enforcement of Chinese environmental statutes, lack of economic incentives to challenge administrative interpretations of Chinese environmental statutes, etc. Nagle 1996, p. 536-555.

321 Nagle 1996, p. 536.

322 McElwee 2011, p. 73-74.

323 For example, the Ministry of Environment Protection is to establish three departments respectively in charge of the air, the soil, and the water to replace the former departments on pollution prevention and emission control. Moreover, the local environment protection bureaus under provincial level are no longer governed by the local government, but are directly accountable for the environment protection bureaus of the higher levels. 环保部拟推进省以下垂直管理改革 [The Ministry of Environment Protection will promote administrative reform to realise 'vertical' governance under provincial levels] (中国青年报, 22 →

vein, while experimental local pilot approaches to biofuels may provide experience for the development of the biofuel industry, the strengthening of sustainability requirements for biofuels should primarily depend on the central government's political determination and law-making for tackling environmental issues.

5.2.3. Inclusive Sustainability Requirements in View of Sustainable Development?

As analysed, China's main sustainability concern about biofuels is the impact on national food security, but there is ambiguity regarding the 'non-grain' and 'marginal land' requirements and the other sustainability impacts are stressed to a lesser extent. Although it is possible for local governments to introduce more stringent sustainability requirements, they probably lack the willingness or awareness to do so. However, China may need to establish inclusive sustainability requirements for biofuels in law. Firstly, the current basic principles lack clear definitions and consistent implementation, which may cause the continued consumption and production of biofuels having potential unsustainable impacts. Thus, there is a need to further include the general principles for biofuels and set forth clearly-defined requirements related to both environmental and socio-economic sustainability.[324] Secondly, China's regulation of biofuels as yet has primarily been set out by the NDRC and the other relevant ministerial agencies under the State Council, while the departments are 'horizontally' fragmented and the issued documents regarding biofuels have various binding effects,[325] which may reduce consistency and effectiveness in the regulation of biofuels.[326] Thus, setting out comprehensive sustainability requirements in law, perhaps together with biofuel incorporation provisions in the Renewable Energy Law of China,[327] could be

January 2016), available at: <http://gongyi.sina.com.cn/2016-01-22/doc-ifxnuvxc1583820.shtml> (last accessed 6 March 2016); 环保部回应机构调整:设置"水气土"三司 [The Ministry of Environment Protection responded to organisational restructuring: three departments on water, air, and soil are to be established] (People.cn, 3 March 2016), available at: <http://env.people.com.cn/n1/2016/0303/c1010-28169448.html> (last accessed 6 March 2016).

324 It has also been argued by Wang that an integrated approach is required when building regulatory regimes and comprehensive legislation specifically on biofuels, since biofuels are connected with oil security, transportation, rural development, and environmental problems. Wang 2011, p. 209.

325 It has been observed that China regulates biofuels largely by policy initiatives or by administrative regulations or rules, or by administrative plans that are not legally binding, rather than by law. *Ibid.*, p. 203-205.

326 The problem of horizontal fragmentation among administrative bureaus of the same level has been pointed out as problematic in China, and it may be reflected in the regulation of biofuels by multiple administrative departments regarding energy, agriculture, forestry, finance, land resources, and quality supervision. McElwee 2011, p. 92-94, 99-103.

327 It has been argued by Wang that comprehensive legislation specifically on biofuels should be jointly enacted by the NDRC and the relevant administration departments in the form of joint departmental rules and become the third layer of national energy legal system (with Energy Law and Renewable Energy Law as the first and second layer). Wang 2011, p. 209-210.

a way of inclusively addressing and integrating the sustainability concerns related to biofuels. Thirdly, the domestic demand for biofuels prompts China to import biofuels and biofuel feedstocks from third countries, and incorporating the sustainability consideration into trade policies, such as tariff setting and agreement negotiations, would be a possible way of coherently implementing the existing biofuel policies in China. Therefore, in view of the current limits and the need for inclusive regulation, the central government may need to provide and include the requirements on the environmental and socio-economic sustainability of biofuels, based on China's commitment to sustainable development and the existing efforts in relation to the sustainable cultivation of energy forests.

5.2.3.1. *Sustainable Development in China*

China has established 'sustainable development' as a national strategy. 'Sustainable development' was not introduced into China until the late 1980s, although it had been discussed for several years.[328] In 1994, China's Agenda 21 was endorsed by the State Council[329] and in 2003 the Chinese government launched the 'Programme for Action for Sustainable Development in China in the early 21st Century' which lists the objectives, principles, priority areas, and proposed measures for ecological conservation, environmental protection, and the 'rational exploitation of resources.'[330] China claims that it was one of the first developing countries to propose and implement sustainable development as a national strategy, and it has summarised its sustainability efforts in the official document prepared for the 2012 UN Conference on Sustainable Development (Rio+20).[331]

China accepts 'sustainable development' partly for the reason that 'sustainable development' itself was proposed as a compromise between the North and the South and thus reflects China's pledge to pursue economic and social development.

328 In the early 1980s, Qu Geping, the then Minister of the State Environmental Protection Bureau discussed the relationship between environmental protection and economic development, but a formal discussion of the principle of sustainable development was not raised until the late 1980s. Lin 1997, p. 30.

329 The White Paper on China's Population, Environment and Development in the 21st Century, available at: <www.acca21.org.cn/cca21pa.html> (last accessed 16 September 2014).

330 UN Department of Social and Economic Affairs, Division for Sustainable Development, 'China: National Information/National Reports', available at: <www.un.org/esa/agenda21/-natlinfo/countr/china/index.htm> (last accessed 15 May 2013).

331 In the report, China claimed that, after the 1992 United Nations Conference on Environment and Development, the Chinese government issued 'China's Agenda 21 – White Paper on China's Population, Environment and Development in the 21st Century' in March 1994 and then in 1996, incorporated sustainable development into national strategies and began its full implementation. Some of the major achievements include: poverty reduction, population control, economic restructuring, economic restructuring, transforming development patterns, incorporating environmental protection into national economic and social development planning, and implementing environmental and resource legislation and regulation. China's National Report on Sustainable Development (8 June 2012), available at: <www.chinaun.org/eng/zt/sdreng/P020120608816288649663.pdf> (last accessed 17 September 2014), p. 2.

China has been emphasising 'development' in the international arena[332] and has declared that economic development is the right of developing countries,[333] and solutions to environmental problems can only be found in the progress of development.[334] It is claimed in China's Agenda 21 that the primary importance of sustainable development should be given to the development of its national economy, and the precondition for sustainable development in a developing country like China is growth.[335]

Nevertheless, it has been observed that China begins to seriously treat the environment as an integral pillar of sustainable development and environmental protection has gradually gained more importance, although this has been a gradual process.[336] Moreover, sustainable development has increasingly been set out in the law and China starts to take a more serious stance in turning the past rhetoric into affirmative legal action.[337] China's Agenda 21 states that 'establishing laws related to sustainable development is a mechanism for formalising and codifying strategies and policies for sustainable development...'[338] In this sense, the implementation of the China Agenda 21 may serve as the main catalyst for enshrining sustainable development as an objective of law[339] - for instance, to realise sustainable development has been set as one objective of the Renewable Energy Law[340] and the Environment Protection Law of China.[341]

332 Lin 1997, p. 26.

333 As preparation for the 1992 Rio Summit, the Beijing Declaration was delivered after the Conference on Environment and Development in Developing Countries, and the attitude of the gathered 41 developing countries was stated in the second and third declarations. It was emphasised in the Declaration that environmental problems should be solved during the process of economic development and the developing countries' right to grow should be fully recognised and environmental protection and sustainable development are two separate but inter-related concepts. Lin 1997, p. 37.

334 In China's Report on Environment and Development submitted to the UN Conference on Environment and Development (1992 Earth Summit), China stated that environmental protection itself is not the objective; instead, the objective should be for all people to enjoy life in a pleasant environment. It may not be right to abandon development because environmental problems are encountered; instead, a proper way should be identified for each nation to achieve a sustainable development of the environment and economy. Lin 1997, p. 37.

335 China's 21st Century Agenda, para. 2.1.

336 China has been most driven in relation to economic growth and, while at a local or provincial level, this is sometimes still the case, there is a move towards requiring that economic development should not be at the expense of the environment. An example of this is the inclusion of the improvement of environmental performance as a criterion in assessments to promote government officials. Tung 2009, p. 493.

337 Yeh 2007, p. 439.

338 China's 21st Century Agenda, para. 3.1; Tung 2013, p. 598.

339 Tung 2009, p. 491.

340 Renewable Energy Law of China 2005, Art. 1.

341 Environment Protection Law of China, Art. 1.

5.2.3.2. *'Sustainable Cultivation' of Energy Forests*

Although 'sustainable development' remains a general and vague strategy and objective in policy and in law, the Guidelines for the Sustainable Cultivation of Energy Forestry issued by the State Forestry Administration[342] shows an example, and perhaps the potential, of applying 'sustainable development' to the regulation of biofuels in China.

Since using edible vegetable oil (e.g. rapeseed oil) is not a viable method to produce biodiesel in China,[343] a variety of non-edible oil bearing plant seeds,[344] such as Jatropha (or tung trees)[345] and Pistacia,[346] can provide an important source of raw materials. With this background, developing 'energy forestry' (*neng yuan lin*)[347] is deemed to provide an opportunity for the forestry sector in China. In 2011, the State Forestry Administration issued the Guidelines for Sustainable Cultivation of Energy Forestry to provide guidance for energy forestry cultivation and management with respect to designing, planning, selective cultivation, site choice, base construction, environment protection, and local community. The Guidelines are addressed to the forestry departments of lower levels for guiding and standardising their acts in developing energy forestry, and they can also be used as reference by private entities in practical operations.[348] For example, local forestry administrations must compile energy forestry development plans,[349] and individual energy forestry operators must submit cultivation measures and plans to local forest administrations in reference to the Guidelines.[350] Most importantly, the Guidelines

342 Guidelines for Sustainable Cultivation of Energy Forestry, 17 February 2011.

343 China is a net importer of vetegable oil and the National Energy Administration has emphasised the use of non-edible (herbaceous or woody) plant and waste oils as biodiesel feedstock. Biodiesel Industry Development Plan 2014.

344 In China, over 1,000 species of native woody oil plants are available to be used as feedstock for biodiesel production, such as Pistacia chinensis, Jartropha curcas, Xanthoceras sorbifolia. Some woody oil plants are fast-growing trees have the characteristics of having a short growth cycle and a long harvest period, being easy to harvest, low-cost management, and highly adaptable to different kinds of environment. Wang, Xiong & Liu 2009, p. 345.

345 Producing biodiesel from Jatropha curcas has particular appeal for China, because they can be grown on dry and barren land, so the marginal lands can be used. Planting Jatropha trees as biodiesel feedstock has been initiated by the CNPC, and nearly 200,000 hectares of jatropha trees have been planted in various locations around the country. James T. Areddy, 'China's First Aviation Biofuel Test' (Chinarealtimereport, 1 November 2011) available at: <http://blogs.wsj.com/chinarealtime/2011/11/01/chinas-first-aviation-biofuel-test/> (last accessed 26 August 2014).

346 Pistacia also has the trait of being drought and salinity-resistant, and it can be grown in various types of soil. Yuannan, Shaanxi, and Guizhou Provinces have great potential marginal land sources for planting Pistacia. Lu *et al.* 2014, p. 259 and 264.

347 'Energy forestry' is defined as including trees which can provide oil or starch-rich material for producing (gaseous, liquid, or solid) fuel. Guidelines for Sustainable Cultivation of Energy Forestry 2011, Art. 2.

348 Guidelines for Sustainable Cultivation of Energy Forestry 2011, Art. 3.

349 *Ibid.*, Arts. 3 and 5.

350 *Ibid.*, Arts. 3 and 6.

demonstrate the attempt to integrate environmental, social, and economic sustainability concerns in energy forest cultivation.

According to the Guidelines, 'sustainable cultivation' means realising ecological, economic, and social benefits at the same time through scientific and reasonable cultivation to ensure biodiversity, productivity, and the vitality of forests at the present and in the future.[351] Hence, energy forestry cultivation needs to take into account the natural, economic, and social circumstances, and environmental impact assessments and economic feasibility analyses must be conducted. From the environmental perspective, it is required that the designated high-value areas and their buffer zones shall be excluded from planting energy forestry, and shall take measures to conserve soil and water, biodiversity, prevent alien species, and undertake regular monitoring and impact assessments.[352] From the socio-economic perspective, the Guidelines specifically address the rights and interests of local communities.[353] It is required that energy forest operators shall establish negotiation mechanisms with local communities, the outcome of which shall be sealed in formal contracts or agreements. The operators should respect the local customs, protect their legal rights, and actively coordinate with local residents to ensure that they benefit from the economic and social effects of the energy forestry cultivation, particularly in the multi-ethnic residential areas. Furthermore, operators should endeavour to provide employment, training, and other social services to local residents and to safeguard their labour rights. Thus, it is perceived that the Guidelines define the sustainable cultivation of energy forestry in a way that integrates economic, environmental, and social dimensions. This feature could be attributed to the sustainable utilisation principle that has been established in the Forest Law of China.[354] Although the Guidelines still do not mention GHG emissions accounting and need to be evaluated how they have worked out in practice, they may nonetheless serve as an example for integrating environmental and socio-economic considerations in the regulation of biofuels in China.

In conclusion, the analysis carried out in this chapter demonstrates the limits of the current principles and requirements in integrating inclusive sustainability concerns: the ambiguity in defining 'non-food grain' feedstock and 'marginal land' may mitigate the effect of tackling the food security concern, and many other sustainability impacts such as life-cycle GHG emissions are not yet inclusively considered. A further examination of Chinese law shows the possibility of local authorities to adopt more stringent requirements for the sustainability of biofuels, but the problem is a lack of willingness or awareness of local authorities to enact sustainability requirements for biofuels, particularly in view of the potential economic benefits that might be brought by this industry. Thus, there is a need to provide and include sustainability requirements in the law at the national level, which could find grounds in China's political and legal commitment to sustainable

351 Guidelines for Sustainable Cultivation of Energy Forestry 2011, Art. 2.

352 *Ibid.*, Arts. 33-37.

353 *Ibid.*, Arts. 38-40.

354 Art. 5 of the Law states that forestry development shall promote forest conservation and afforestation in order to realise sustainable utilisation. Forest Law of China, Art. 5.

development and the existing example of promoting the sustainable cultivation of energy forestry.

Chapter 5

COMPARATIVE ANALYSIS AND THE IMPLICATIONS FOR A GLOBAL INCLUSIVE APPROACH

1. Introduction

This chapter aims to conduct a comparative study of the regulation of biofuels based on the observations from the previous chapters. Following the conventional steps of comparative study,[1] this chapter will discuss the similarities and differences between the regulatory approaches to biofuels at the international, EU, and Chinese levels, and it will further analyse the possible reasons for the similarities and differences. Meanwhile, the comparative study adheres to the 'demand-supply-trade' analytical framework and is 'problem-oriented' following from the research questions: what requirements are set out for the sustainability of biofuels, whether the sustainability requirements integrate inclusive sustainability concerns, and are there potential for a global inclusive approach to regulating the sustainability of biofuels?

This chapter is structured as follows: sections 2, 3, and 4 respectively compare the regulation of the demand, supply, and trade of biofuels in order to identify the similarities and differences, as well as to analyse the underlying reasons for and possible implications of the similarities and differences. Section 5 furthermore compares the systems through the lens of inclusiveness, which finally leads to arguments for a global inclusive approach based on the observations from the comparative study and concept of 'global environmental law.'

2. Comparative Analysis of the Demand for Biofuels

As demonstrated in the previous chapters, the demand for biofuels may be affected by measures such as sustainability requirements for biofuels, blend mandates, consumption targets, emission standards for fuel and vehicles, preferential tax

1 It is recalled that the conventional steps of a comparative study include: *a)* determining the similarities and differences; *b)* explaining the causes of the similarities and differences, and *c)* evaluating the solutions. De Cruz 2010, p. 36.

treatment, and emission trading rules. Some of the measures are not established at every regulatory level. For example, while the EU has established sustainability criteria for biofuels, China does not have similar criteria; conversely, the EU does not employ a 10 per cent or 5 per cent biofuel blend mandate in oil selling systems as China does. In particular, many of the demand-side measures like blend mandate and preferential tax treatment are not provided for at the international level. Hence, a comparison in this section will concentrate on the sustainability requirements that address environmental or socio-economic concerns, regardless of whether they are referred to as 'sustainability' requirements or not.

2.1. *Sustainability Requirements*

Sustainability requirements could be general principles or specific rules to avoid or minimise the potential negative environmental or socio-economic impacts of biofuels. Such requirements may influence the demand for biofuels and may also provide guidelines for the supply of biofuels, but this sub-section will focus more on the sustainability requirements relating to the demand side.

2.1.1. Differences

At the international level, the demand for biofuels is closely related to the commitments about GHG emission reductions under the UNFCCC, particularly under the Kyoto Protocol, since biofuels are identified as a strategy to reduce GHG emissions in transport.[2] The primary sustainability concern under the international climate regime is the life-cycle GHG emissions from biofuels. The IPCC Guidelines, as adopted under the UNFCCC and the Kyoto Protocol, provide a method for estimating the life-cycle GHG emissions,[3] but the complete estimation of the life-cycle emissions of biofuels could not be guaranteed due to divergent GHG inventory reporting obligations and methods under the Kyoto Protocol.[4] Although the IPCC has denied that biomass used for energy is automatically considered as 'carbon neutral' even if 'biomass is thought to be produced sustainably',[5] divergent applications of the current method at the state level might still be based on a 'carbon-neutral' assumption and overlook emissions related to indirect land use changes.[6] While sustainable development is set as one principle of the UNFCCC, specific sustainability concerns, such as biodiversity and food security, are not sufficiently specified or even might be compromised in pursuit of climate goals, which could be reflected in inadequate CDM rules on the criterion of the contribution to sustainable development.[7]

2 See Chapter 2, section 2.1.
3 See Chapter 2, section 2.2.1 and 2.2.2.
4 See Chapter 2, section 2.2.3.
5 *Ibid.*
6 *Ibid.*
7 See Chapter 2, section 2.3.

The EU has established its own sustainability criteria in law and has set them as a precondition for the major demand-incentivising measures for biofuels, namely renewable energy consumption targets, financial support, and national biofuel blend obligations.[8] Comparatively, instead of establishing sustainability criteria for biofuels in law, China has set forth general principles, which involve the sustainability concerns related to food, land, and environment, to guide policy making and planning in the biofuel sector.[9] The differences with the EU can be reflected in 'what' and 'how' sustainability concerns are addressed.

Firstly, the EU and China differ in *what* sustainability concerns are emphasised. The EU's sustainability criteria consist of the requirements on the environment and climate change issues, such as life-cycle GHG emissions, biodiversity, carbon stock conservation, and agro-environment,[10] while China prioritises concerns related to food security - two of the three principles (i.e. non-food grain feedstock and marginal land) are aimed at avoiding impacts on food.[11] The difference in the sustainability concerns that are emphasised can be primarily attributed to the broad meaning of 'sustainability' or 'sustainable development', and the divergence in which dimension of sustainable development is prioritised by the government. There is no uniform definition of 'sustainability' or 'sustainable development.'[12] Although 'sustainable development' has been accepted as a reconciliation of the social, economic, and environmental dimensions[13] and both the EU and China admit that there is a need to balance the three dimensions,[14] the EU has legally established the obligation of environmental integration as an integral part of achieving sustainable development,[15] whereas China has largely established sustainable development as a strategy for economic restructuring and social stability.[16] Such a difference is grounded in their divergent national interests and circumstances. In the specific case of biofuels, this difference is exemplified by the EU's emphasis on the environment and climate-related criteria, while China prioritises the 'non-grain' requirement for national food security. The divergence may also be attributed to the lack of an internationally harmonised definition of 'sustainable biofuels', or an agreed international regulatory framework on how to apply sustainability criteria for biofuels.[17]

8 See Chapter 3, section 2.1.
9 See Chapter 4, section 2.1.
10 See Chapter 3, section 2.1.
11 See Chapter 4, section 2.1.
12 See Chapter 1, section 2.2.
13 *Ibid.*
14 See Chapter 3, section 5.3; Chapter 4, section 5.3.
15 According to Art. 11 TFEU, environmental protection requirements must be integrated into the definition and implementation of the Union's policies and activities, in particular with a view to promoting sustainable development. See Chapter 3, section 5.3.
16 See Chapter 4, section 5.3.
17 The CBD Executive Secretary once pointed out that the lack of an international regulatory framework to apply sustainability criteria is one gap in the reglation of the sustainability of biofuels. CBD/SBSTTA/16/14 2012; See Chapter 2, section 3.1.3.

Secondly, the EU and China differ in *how* the sustainability concerns are addressed. The EU's sustainability criteria are adopted by legislative act (in the form of a 'directive') and thus have binding legal force: Member States' failure to transpose the criteria into their national law may result in an infringement procedure initiated by the European Commission.[18] With regard to how the criteria are transposed into national law, Member States have flexibility in choosing the 'form' and 'method' to achieve the 'result' of sustainability criteria but they shall refrain from introducing additional criteria.[19] Furthermore, the EU sets out the methods of how to calculate life-cycle GHG emissions[20] and verify compliance with the sustainability criteria.[21] Moreover, the sustainability criteria are stated as a precondition for meeting renewable energy targets and the other support schemes. In contrast, China sets out the non-grain requirement in renewable energy development plans which have an ambiguous status and need to be further translated into specific legal or administrative measures,[22] but consideration must be given to the special role of policy documents in China[23] and the fact that local governments need to set out their local renewable energy plans in accordance with the national plans as provided in the Renewable Energy Law of China.[24] Another difference is that Chinese local governments may adopt more stringent requirements with regard to the feedstocks or other sustainability concerns.[25]

2.1.2. Similarities

Despite the divergence, evolving convergence can be observed from the recognition of the negative environmental and socio-economic sustainability impacts of biofuels and the attempt to minimise such impacts. For example, some efforts have been made to strengthen the 'sustainability' part of CDM projects under the Kyoto Protocol.[26] The European Commission is required to monitor and report on the impact of biofuels on food availability and prices, and the proposed amendment to

18 The European Commission has initiated infringement procedures against several Member States (e.g. Austria and Ireland) for failing to transpose the requirements in Renewable Energy Directive including the sustainability criteria. Commission Report COM(2013) 175 final, p. 11 and 13; Commission Staff Accompanying Document SWD (2013) 102 final, p. 17.

19 Art. 17(8) of Directive 2009/28/EC precludes Member States from refusing biofuels based on other sustainability grounds, and Art. 114 TFEU provides Member States with a limited possibility of adopting more stringent sustainability criteria. Chapter 3, section 5.2.

20 See Chapter 3, section 2.1.1.

21 See Chapter 3, section 2.1.5.2.

22 There are divergent views with regard to the status of the national Five Year Plans for social and economic development. The sectoral Five Year Plans for renewable energy are purported to implement the general guidelines in the national Five Year Plan for social and economic development, so their binding feature is more ambiguous and their binding force largely hinges on whether and how they are translated into more specific administrative targets or requriements. See Chapter 4, sections 1 and 2.1.

23 See Chapter 4, section 1.

24 See Chapter 4, section 2.1.

25 See Chapter 4, section 5.2.

26 See Chapter 2, section 2.3.

Directive 2009/28/EC will set a 'cap' on the share of food crop-based biofuels.[27] In China, avoidance of negative environmental impacts is also set as a basic principle for biofuel feedstock production, and nature reserve protection and soil and water conservancy in the production of biofuels have been emphasised in the regulation of biofuel feedstocks.[28] Moreover, in setting out the development policy for the biodiesel industry, the National Energy Administration stressed environmental protection measures in the production process of biodiesel and in building feedstock bases, including carrying out environmental impact assessments, avoiding nature reserve areas and ecologically sensitive areas,[29] respecting forest and regional land utilisation planning, and protecting biodiversity and preventing aggressive alien species.[30] The convergence could be attributed to a common awareness of the potential negative impacts of biofuels, as well as to the existing international efforts made in relation to the sustainability impacts of biofuels, such as the discussions on biofuels under the CBD and the Ramsar Convention and within the International Biofuel Forum where both EU and China are parties.[31]

2.2. *Blend Mandates and Consumption Targets*

Biofuel blend mandates and consumption targets may provide a guaranteed market for biofuels and thus give a strong incentive for producing biofuels and stimulating the demand for biofuels.[32] Hence, it is necessary to strengthen sustainability requirements for biofuels used to meet the blend mandates or consumption targets.

27 However, the EU's definition of 'food crop' includes all cereal and other starch rich crops as well as sugars and oil crops, which is different from China's identification of the 'grain'-based feedstocks as including the staple food crops such as corn and wheat. See Chapter 3, section 2.1.5; Chapter 4, section 2.1 and section 3.1.

28 It has been stated that the building of biofuel feedstock supply bases shall carry out environmental impact assessments to avoid negative environmental impacts. Interim Measures on Financial Support for Feedstock Bases 2007; See Chapter 4, sections 3.1.3.1 and 3.1.3.3.

29 The Environment Protection Law of China, as revised in 2014, for the first time stipulates in the law the mechanism of an 'ecological red line' (*sheng tai bao hu hong xian*), which is to be applied in areas with significant ecological functions, *ecologically sensitive areas*, and ecological vulnerable areas, and strict protection measures will be implemented in these areas. The 'ecological red line' was firstly put forward in the State Council's document on strengthening environmental protection in 2011, and the Ministry of Environment has issued technical guidelines on how to designate the ecological red lines for ecological function areas, ecologically sensitive areas and vulnerable areas, and biodiversity conservation areas. Environment Protection Law of China, Art. 29; Liu Qin, China's 'ecological red lines' unlikely to protect, 9 January 2014, available at: <https://www.chinadialogue.net/article/-show/single/en/6641-China-s-ecological-red-lines-unlikely-to-protect> (last accessed 10 May 2015).

30 Biodiesel Industry Development Policy 2014, paras. 40-43.

31 See Chapter 2, section 3.1.2.

32 The blend mandate herein refers to a certain percent of bioethanol or biodiesel in transport fuels to be sold, and the consumption target refers to a certain quantity of biofuels to be consumed among the total transport fuels. An FAO study concludes that 'virtually all
→

2.2.1. Differences

At the international level, although the Kyoto Protocol sets forth GHG emission reduction targets (as amended by Doha COP18)[33] that bind Annex I countries, there is not any blend mandate or consumption target for the use of biofuels, largely because the use of biofuels is regarded as one way of achieving GHG emission reductions and the way in which the climate goal is achieved is decided by the individual state, depending on its resources and socio-economic circumstances. In this context, the EU provides for an example of establishing binding renewable energy targets, while China opts for an approach of imposing blend mandates and biofuel consumption targets.

The EU and China differ in how the targets are set out and whether sustainability requirements are attached. The EU's renewable energy targets are set for the targeted proportion of renewable energy, which includes but is not limited to biofuels, in the total transport fuel consumption.[34] This approach thus leaves flexibility for the Member States to achieve the target by using forms of renewable energy other than biofuels, such as electric cars or hydrogen and fuel cells. Meanwhile, the targets are established in law as legally binding for each Member State, and Member States have to submit national renewable energy action plans explaining how they will achieve the targets.[35] Furthermore, the EU requires that only biofuels meeting sustainability criteria can be counted for achieving the renewable energy targets.[36] In comparison, China opts to follow an approach of requiring biofuel blends in gasoline to be sold to gas stations and setting targets for the volume of bioethanol and biodiesel to be consumed annually, which is more concentrated on production and consumption of biofuels. China does not establish sustainability requirements or compliance with the 'non-grain', 'marginal land', or 'no environmental damage' principles as a precondition for the biofuels to be eligible to be used in the blend mandates or to meet the consumption targets.[37] In fact bioethanol consumed for the blend mandates and consumption targets up to now has been primarily produced from food grains like corn and wheat.[38] Nevertheless, it is noticed that the new blend mandate pilot projects (in Guangxi

existing laws to promote ... biofuels set blending requirements, meaning the percentages of biofuels that should be mixed with conventional fuels.' Charles 2013, p. 1; De Gorter & Just 2009, p. 738; Jull 2007, p. 21.

33 Decision 1/CMP.8 2012.

34 See Chapter 3, section 2.2.1.

35 Moreover, the EU's target in percentage form allows the targets to be achieved through improving energy efficiency without increasing renewable energy deployment. See Chapter 3, section 2.2.1.

36 See Chapter 3, section 2.2.1.

37 See Chapter 4, section 2.2.1.

38 A report in 2014 demonstrates that most ethanol in China is still produced from grains (76 per cent from corn and 14 per cent from wheat). Among the five state-owned fuel ethanol enterprises, 64 per cent of their production was sourced from corn in 2012, 30 per cent from wheat, and 6 per cent from cassava. Scott & Jiang 2013; Anderson-Sprecher & Jiang 2014; see Chapter 4, section 5.1.

and Hainan), initiated after the establishment of 'non-food grain feedstock' principle, are all based on non-food feedstock, such as cassava or non-edible oil bearing seeds.[39]

There are many reasons for the differences in how targets relating to biofuels are set in the EU and China. Firstly, the EU's preference for a binding target is largely because an indicative biofuel consumption target (5.75 per cent) for 2010 failed to be achieved and the European Commission owed the failure to a lack of sufficient incentives for Member States to implement the biofuel target. The Commission subsequently proposed to impose a mandatory target of renewable energy in transport energy for 2020.[40] In contrast, China's biofuel consumption targets have been adopted in the context of bioethanol blend mandate pilot schemes, which are characterised by strict governmental planning and control over the quantity of biofuels to be consumed and produced.[41] Moreover, it has been observed that, to introduce economic and opening up reform, China has maintained the practice of introducing policies or local pilot projects for experimentation purposes and then translating them into law based on the experiences gained from the experimental pilot projects,[42] which means that, until the institutions involved have reached a consensus on the rules applied nation-wide, interim regulatory measures or policy documents are normally applied to implement the experimental reforms before the circumstances are judged to be ripe for enacting a law.[43] Such an observation can also be used to explain the pilot project approach of the biofuel blend mandates and the planning of consumption targets for biofuels: as biofuels are novel in China, blend mandate of biofuels is to be experimented with by gradually expanding local pilot schemes and consumption targets are set out in plans which are adjusted every five years based on the experiences in the previous five years and new circumstances. Thus, the EU adopts mandatory targets for each Member State in law, while China inclines to implementing sub-national blend mandate projects and non-legally binding consumption targets.

Secondly, the EU's mandatory target-setting for renewable energy, rather than for biofuels, could be attributed to constitutional reasons: the EU's competence is conferred from Member States[44] and exercised according to the principles of

39 Guangxi was included in the bioethanol pilot project since April 2008 with a blend mandate of 10 per cent bioethanol produced from cassava-based bioethanol plant (with a production capacity of 200,000 tonnes). It was stipulated by the Hainan provincial government that as of January 2010 a blend mandate of 5 per cent biodiesel (BD5) would be implemented in the whole province. See Chapter 4, section 2.2.1.

40 See Chapter 3, section 2.2.1.

41 China's blend mandate pilot schemes are characterised by the approach of 'end-close' designated production and sales, meaning that the bioethanol in the blend mandates shall be produced and distributed by designated bioethanol producers and blenders, and all the gas stations in the pilot areas have to sell the bioethanol thus supplied. See Chapter 4, section 2.1.

42 Chen 2008, p. 59.

43 Keller 1994, p. 730.

44 Arts. 2-4 TFEU define the division of competences between the Union and the Member States, and Art. 5 TEU and Art. 7 TFEU refer to the principle of conferral according to which the Union has only those competences that are conferred upon it by the Treaties. TFEU, Arts. 2-4, and 7.

proportionality and subsidiarity.[45] The 10 per cent transport renewable energy target was adopted based on Article 192 TFEU, according to which 'measures significantly affecting Member State's choice between different energy sources and the general structure of its energy supply' shall go through a special legislative procedure.[46] Thus, the EU's target for renewable energy which leaves much flexibility for Member States could avoid 'significantly affecting Member State's choice between different energy sources'[47] and comply with the proportionality and subsidiarity principle.[48] In contrast, China is a unitary state with a centralised and top-down administration, particularly with regard to energy matters for which the NDRC is responsible, and thus the Chinese central government has the competence and the possibility to mandate blending of biofuels and set volume-based biofuels consumption targets.

Thirdly, the EU requires sustainability criteria to be met for biofuels being counted as being in compliance with the consumption target, largely because the EU's renewable energy target is binding and forms a major incentive for the demand of biofuels. Hence, the establishment of sustainability requirement for biofuels that are counted towards the national targets is a necessary way for the EU to preclude the inclusion of unsustainable biofuels. In China, the existing biofuel blend mandate pilot projects were mostly initiated by the central government between 2002 and 2004 and were originally intended to deal with surplus food grains,[49] and the consumption targets are largely set on the basis of the existing production capacity of biofuel producers, particularly the four state-owned bioethanol enterprises.[50] The failure of meeting the 2010 non-food grain bioethanol target has illustrated the difficulty to shift the production capacity from biofuel feedstocks based on food grains towards other feedstocks. Probably due to the practical consideration for the state-of-the-art, the general principles regarding food, land, and the environment are not explicitly established for the blend mandate projects and consumption targets. Moreover, given the biofuel pilot projects with controlled supply and consumption and *ex ante* licensing of bioethanol enterprises, the Chinese government may be more accustomed to regulating the supply of biofuel in order to implement the non-food, marginal land, and no environmental damage principles.

45 See Art. 5 of Treaty of European Union, and Europa, Summaries of EU Legislation, The Principle of Subsidiarity, available at: <http://europa.eu/legislation_summaries/-institutional_affairs/treaties/lisbon_treaty/ai0017_en.htm> (last accessed 10 January 2015).

46 TFEU, Art. 192(2).

47 Commission Communication COM(2012) 271 final.

48 A statement on compliance with the subsidiary principle and proportionality principle was set out by the Commission in the Proposal for Directive on Renewable Energy Sources. Commission Proposal COM(2008) 19 final, p. 8 and 9.

49 See Chapter 4, section 2.2.1.

50 See Chapter 4, section 2.2.2.

2.2.2. Similarities

Despite the differences, the EU and China both established targets that promote a certain amount of biofuels to be consumed in a certain period. Although the EU's target can be achieved by other kinds of renewable energy, it has actually led to a continuous increase in the consumption of biofuels in the transport sector.[51] In China, the use of biofuels has also experienced a continuous increase since the implementation of blend mandate pilot projects, which makes China one of the leading bioethanol consumption countries in the world.[52] The convergence in employing consumption targets to promote the demand for biofuels is largely because biofuels are recognised in both the EU and China as a kind of renewable energy that can bring in environmental and socio-economic benefits[53] and the target-setting provides for a major incentive from the government to promote the consumption of biofuels within a fixed period.

Meanwhile, there might be more convergence between the EU and China in the way of setting post-2020 renewable energy targets, as the EU will not continue to set binding targets for individual Member States.[54] A non-binding goal setting approach[55] seems to have become the preferable way of promoting biofuels. For example, in 2011 the European Commission issued a White Paper on the 'Single European Transport Area', which includes the goals of 'no more conventionally-fuelled cars in cities' and '40 per cent use of sustainable low carbon fuels in aviation.'[56] In 2013 the European Commission set out the alternative fuels strategy which contains the aim of building a comprehensive mix of alternative fuels including biofuels.[57] This convergence might be attributed to the uncertain contribution of biofuels to decarbonising transport in science and the diversification of low-carbon fuel options in technology, so setting goals and plans in policy documents rather than in law might provide for a more flexible and adaptable approach in the face of scientific and technological uncertainties.

51 According to the data of the EurObserv'ER, the consumption of biofuels continued to increase from 2002 until 2012, when the consumption firstly dropped by 6.8 per cent. Biofuels Barometer, EurObserv'ER, Graph 1, p. 5, available at: <www.energies-renouvelables.org/-observ-er/stat_baro/observ/baro222_en.pdf> (last accessed 11 January 2015).

52 European Biofuels Technology Platform, Global Biofuels - An Overview, available at: <www.biofuelstp.eu/global_overview.html> (last accessed 5 May 2015); Statista, Leading countries in biofuel production in 2013 (in 1,000 metric tons of oil equivalent), available at: <www.statista.com/statistics/274168/biofuel-production-in-leading-countries-in-oil-equivalent/> (last accessed 5 May 2015).

53 For instance, see Directive 2009/28/EC, para. (1) of the preamble; Measures on Expanding Pilot Projects in Use of Fuel Ethanol Gasoline for Automobiles, 10 February 2004, para. (1).

54 The European Commission explains that this is largely due to the recognition of the limited role of first-generation biofuels in decarbonising the transport sector. See Chapter 4, section 2.2.

55 Jacobsson 2004; Senden 2004; Snyder 1994; Trubek & Trubek 2005; Falkner 2005; Cini 2001.

56 White Paper COM(2011) 144 final.

57 Commission Communication COM(2013) 17 final, section 2.4, p. 7.

2.3. *Emission Standards for Fuels and Vehicles*

The emissions standards mentioned in the chapters on the EU and China refer to the limitation of GHG or pollutant emissions in transport. Biofuels normally have to comply with such emission limitations for conventional fuels, unless otherwise provided. Thus, what and how the emission standards are set out could affect the demand for biofuels: for example, strengthened carbon emissions reduction standards may increase the use of biofuels when biofuels are identified as reducing carbon emissions.

2.3.1. Differences

Emission standards for road transport (or non-road machinery mobile applications) are normally established in domestic regulations, while international regimes draw attention to emissions from international aviation and shipping, because these sectors are more likely to be influenced by international competition. At the international level, the UNFCCC covers GHG emissions from all sources and does not specifically exclude international aviation and marine emissions, but the emissions from international aviation and maritime fuel are not included in inventory totals of national reporting.[58] The Kyoto Protocol Annex 1 Parties commit to pursuing 'limitation or reduction of GHG emissions from aviation and marine bunker fuels' respectively through the International Civil Aviation Organization (ICAO) and the International Maritime Organization (IMO).[59] Particularly, the ICAO provides for an international regulatory framework for the aviation sector including for aviation fuel emissions: the ICAO council has vowed to limit or reduce the impact of aviation emissions on local air quality and global climate[60] and it has established the Committee on Aviation Environmental Protection (CAEP) to assist in developing engine emissions standards.[61] The emission standards do not specify any special rules for biofuels, but the ICAO has emphasised 'sustainable' alternative fuels[62] and has established the alternative fuels task force (AFTF) under

58 IPCC Guidelines 2006, p. 1.5.

59 Kyoto Protocol, Art. 2(2).

60 The limitation of CO_2 emissions from aircrafts through market-based measures have been discussed within ICAO since the 1990s. The ICAO has asked member states to submit action plans on aviation CO_2 emission reduction activities and has endorsed the development of global market-based measures in the 38th assembly resolutions. Huber 2013, p. 84-85; ICAO 38th Session 2013, Appendix H, p. 93.

61 ICAO, Committee on Aviation Environmental Protection (CAEP), available at: <www.icao.int/environmental-protection/pages/CAEP.aspx> (last accessed 10 May 2015).

62 At the 38th session of the ICAO Assembly, the promotion of 'sustainable alternative fuels for aviation' was included in the Assembly Resolution. It was stated that a coordinated approach in national administrations is to be developed in order to accelerate the appropriate development, deployment and use of sustainable alternative fuels for aviation according to national circumstances. Particularly, the Resolution supports further work and national measures to be taken to assess the sustainability of all alternative fuels based on net life-cycle GHG emissions reduction, respect for areas of high importance for biodiversity, and the →

CAEP to work on life-cycle GHG emission reduction ranges of alternative fuels.[63] IMO has committed itself to taking technical and operational reduction measures and market-based mechanisms to enhance the energy efficiency of ships and thereby control and reduce GHG emissions.[64] Due to the functional overlapping, the secretariats of the three regimes have continuously communicated to each other on their activities and progresses.[65] Overall, emission standards for international aviation and shipping are more likely to be subject to the international processes under the UNFCCC, ICAO, and IMO.[66] Although, until now, no special emission

contribution to local sustainable development. UNFCCC SBSTA 37th Session 2012, para. 2.8; ICAO 38th session Executive Committee Assembly Agenda Item 17 2013.

63 Under the ICAO, the Alternative Fuel Task Force (AFTF) was established to assess the potential range of emission reductions from the use of alternative fuels in aviation up to 2050. The Alternative Fuel Task Force under the Committee on Aviation Environmental Protection (CAEP) plans to review existing production projections of alternative fuels and report on life-cycle assessment methodology in 2015; to produce methodology for alternative fuel production assessment, a final report on a range for emissions reduction, and an alternative fuel emission report in 2016. ICAO, Committee on Aviation Environmental Protection (CAEP), AFTF, available at: <www.icao.int/environmental-protection/Documents/CAEP/Images/AFTF-Large.png> (last accessed 12 January 2015).

64 IMO member states addressed issues related to GHG emissions under its Martine Environment Protection Committee (MEPC). The Committee focused on technical and operational measures relating to energy efficiency for ships. In particular, the Committee addressed the development of technical and operational energy efficiency regulations under the new chapter 4 of the International Convention for the Prevention of Pollution from Ships (MARPOL) Annex VI, which entered into force on 1 January 2013. Mitropoulos, Efthimios E., Secretary-General, International Maritime Organization, 'The strategy for climate change in the maritime sector', keynote speech, 14 October 2010, Seoul International Maritime Forum, available at: <www.imo.org/mediacentre/secretarygeneral/speechesbythesecretary-general/pages/the-strategy-for-climate-change-in-the-maritime-sector.aspx> (last accessed 31 March 2015); Climate Change Policy & Practice, 'March 2014 Update on Emissions from Aviation and Maritime Transport', available at: <http://climate-l.iisd.org/news/march-2014-update-on-emissions-from-aviation-and-maritime-transport/> (last accessed 31 March 2015).

65 The UNFCCC secretariat maintains close relations with the secretariats of ICAO and IMO. Cooperation between the three organisations is based on the reciprocal exchange of information. The secretariats of the ICAO and IMO have been continuously providing reports and information on relevant work to the SBSTA and organised several expert meetings to address methodological issues relating to the estimation, compilation and reporting of GHG emissions data from international aviation and maritime transport. UNFCCC, International bunker fuels under the SBSTA, available at: <http://unfccc.int/methods/-emissions_from_intl_transport/items/6142txt.php> (last accessed 31 March 2015).

66 China is member to both the ICAO and IMO. Although the EU has no membership of the ICAO or IMO, the European Commission in September 2005 established an Office in Montreal and appointed its Representative to ICAO, and the European Union and its Member States have a strong preference for a global approach to reducing GHG emissions from international shipping led by the IMO, in particular the use of global market-based measures (MBMs). Mobility and Transport, Air, available at: <http://ec.europa.eu/transport/modes/air/international_aviation/european_community_icao/> (last accessed 1 May 2015); Climate Action, Reducing Emission From Shipping, available at: <http://ec.europa.eu/clima/policies/transport/shipping/index_en.htm> (last accessed 1 May 2015).

standard has been provided for biofuels, the emphasis on carbon emission limitations for international transport might promote the use of biofuels[67] and the development of relevant standards.

The EU and China have established domestic standards for emissions from road (and non-road mobile machinery) fuel and vehicle fuel, but they differ in whether the goal of limiting the GHG emissions is integrated in the fuel emission regulation. The EU has established GHG emission reduction target to be achieved by fuel suppliers by 2020,[68] which has the effect of incentivising the use of biofuels because it is suggested that using biofuels is one way of achieving the target.[69] The promotion effect is reinforced when biofuels are allowed to conditionally derogate from vapour pressure limits,[70] and there are conditional derogations from carbon emission limitations for passenger cars running on a high blend of biofuels.[71] However, biofuels have to meet sustainability criteria if they are used to meet the GHG emission reduction targets[72] or if they are to derogate from the current emissions limitation standards.[73] Moreover, the amendment to the fuel quality directive sets out an obligation for fuel suppliers and Member States to report annually on the production pathways, volumes, and life-cycle GHG emissions of biofuels by taking into account indirect land use changes based on prescribed factors.[74] In contrast, China has not integrated requirements about GHG emissions in the national fuel emission standards and has not established preferable derogations for biofuels due to their possible contribution to GHG emission reductions.[75] Meanwhile, with the increasingly strengthened fuel emission and economy standards, biofuels in China are likely to face more stringent emission requirements on environmental and health grounds.[76]

The difference in incorporating GHG emission reductions in fuel emission standards could be attributed to the divergent GHG emission reduction commitments at the international level. The EU undertakes a binding target to

67 Air Transport Action Group, Sustainable Aviation Biofuels, available at: <www.atag.org/our-activities/sustainable-aviation-biofuels.html> (last accessed 18 May 2015); Sustainable Aviation Fuel Users Group, Biofuel Use, available at: <www.safug.org/biofuel-use/> (last accessed 18 May 2015).

68 Member States shall require fuel suppliers to reduce life-cycle GHG intensity from fuel supplied by up to 10 per cent by 2020. See Chapter 3, section 2.2.2.1.

69 Fuel suppliers shall be required to take concrete actions to achieve the mandatory 6 per cent GHG emission reduction target by 2020, and the target can be achieved by fuel suppliers through the use of biofuels. See Chapter 3, Section 2.2.2.1.

70 Biofuels in general must meet the same technical specifications on environmental and health grounds as set out for conventional fuels, but vapour pressure limits for bioethanol blends can be increased since bioethanol blending results in a non-linear change of the vapour pressure, if designated substantive and procedural conditions can be met, so as to assist the development of the biofuel market. See Chapter 3, section 2.2.2.1.

71 See Chapter 3, section 2.2.2.2.

72 See Chapter 3, section 2.2.2.1.

73 See Chapter 3, section 2.2.2.2.

74 Directive (EU) 2015/1513, Art. 1(2)(d) and 1(9).

75 See Chapter 4, section 2.2.3.

76 *Ibid.*

reduce GHG emissions - including GHG emissions from transport - under the Kyoto Protocol,[77] and the EU has bound itself to a domestic GHG emission target of 20 per cent by 2020 and has declared an intention to reach a target of 40 per cent by 2030.[78] As carbon emissions from road transport are not included in the EU ETS but alone contribute about one-fifth of the EU's total carbon emissions,[79] they may pose a challenge for the EU to achieve its international or domestic targets. In this context, biofuels are seen as an important source of alternative fuel that can contribute to substantially reducing carbon emissions in road transport, if they can be produced sustainably and do not cause indirect land use change.[80] In contrast, China has not undertaken binding GHG emissions reduction obligations under the Kyoto Protocol. Domestically China has established a carbon intensity target that allows for much flexibility and ambiguity,[81] and therefore China may face less pressure to reduce GHG emissions in transport and does not have an urgent need to establish a particular GHG emissions target for the transport sector, particularly at a time when air pollution presents a more pressing and serious problem for China in the short term.

2.3.2. Similarities

Despite all the differences, convergence can be observed in that the requirements in relation to emissions in the transport sector have been gradually tightened, which may affect the use of biofuels to different extents. For example, measures to limit such GHG emissions in international aviation and shipping have been increasingly urged to be undertaken by the UNFCCC[82] and the ICAO or IMO, which may

77 The 15 EU Member States (the 'EU-15') committed to reduce their collective emissions on average over the first commitment period (2008-2012) by 8 per cent below the base year levels (1990 in most cases). Since Doha, the EU is committed to reducing its emissions by 20 per cent under the Kyoto Protocol's second period (2013 to 2020).

78 Parliament Resolution on a 2030 framework for climate and energy policies; Commission Communication COM(2014) 15 final, section 2.2.

79 See European Commission, Climate Action, Reducing Emissions from Transport, available at: <http://ec.europa.eu/clima/policies/transport/index_en.htm> (last accessed 12 January 2015).

80 Commission Communication COM(2013) 17 final, section 2.4, p. 7.

81 China has established targets regarding 'carbon intensity' and 'energy intensity' which refer to the carbon emissions and energy consumption in view of the economic development, thus the target would also be reached by increasing the use of energy or GHG emissions if the GDP also grows. In the 12th Five Year Period, the energy consumption per GDP (energy intensity) is to be reduced by 16 per cent and by 2015 compared to that in 2010 level; CO_2 emissions per GDP (carbon intensity) is to be reduced by 17 per cent by 2015 compared to the level in 2010 and by 40 per cent-45 per cent by 2020 compared to the level in 2005. The 12th Five Year Guideline of Renewable Energy Development 2012.

82 The issue of international 'bunker' fuel emissions is addressed under the UNFCCC, especially under the Ad Hoc Working Group on Long-term Cooperative Action (AWG-LCA) and the Subsidiary Body for Scientific and Technological Advice (SBSTA). UNFCCC, Emissions from fuel used for international aviation and maritime transport (international bunker fuels), available at: <http://unfccc.int/methods/emissions_from_intl_transport/items/1057.php> (last accessed 1 May 2015).

prompt the use of biofuels and studies on the life-cycle emissions of biofuels. Particularly, it is observed that China is trying to bridge the gap in fuel economy standards with the EU's car emission standards – China's fuel economy target measured in unit of 'litre per 100 kilometres' is related to and can be converted to the EU's car emission target expressed as 'grams of carbon emissions per kilometre.'[83] Such a trend can also be observed in the setting of fuel quality standards: China has largely followed the EU's lead in decreasing the sulphur content, one of the major pollutant parameters in petrol, from 800 ppm to 10 ppm,[84] although a difference lies in whether preferential treatment is provided for biofuels.

The reason for such a convergence could be attributed to the 'transplantation' or 'borrowing' effect in environmental and climate standards. A 'legal transplant' is defined by Watson as 'the moving of a rule or system of law from one country to another'[85] and can be used to categorise the laws in one society that are borrowed from others rather than 'logical outgrowth of its own experience'[86] and, since Watson, 'legal transplant' or 'legal borrowing' has become a 'familiar subject' in comparative law.[87] In the field of environmental law, the practice of 'legal transplant' or 'legal borrowing' is 'nothing new.'[88] Yang and Percival further

83 For example, China's 2015 target of 6.9 liters/100 km (equivalent to 160.08 gCO2/km) is close to the EU's 2015 target of 130g CO2/km; and China's 2020 target of 5.0 liters/100km (equivalent to 116g CO2/km) is close to the EU's 2020 target of 95 gCO2/km. See Chapter 3, section 2.2.2 and Chapter 4, section 2.2.3.

84 Since 1 January 2003, China has mandated a fuel quality standard Phase I with a sulfur content of 800 ppm, and it continued with Phase II at 500 ppm (since 1 July 2005), Phase III at 150 ppm (since 1 January 2010), Phase IV at 50 ppm, until Phase V at 10 ppm, which has generally followed the EU's lead from Phase I at 1000 ppm, Phase II at 500 ppm, Phase III at 150 ppm, Phase IV at 50 ppm, and Phase V at 10 ppm. From the perspective of the timeline, the EU adopted these fuel quality standards earlier than China. The EU has already implemented Phase V standard, while China has applied Phase IV nation-wide since January 2014 and applies the Phase V standard only in several large cities such as Beijing, Shanghai, and major cities in the Jiangsu and Guangdong Provinces. The Phase V standard was issued by the State Standardisation Commission on 18 December 2013, and is to be applied nation-wide as of 2018. 国五汽油标准发布 [Phase V petrol standard has been issued] (National Energy Administration, 19 December 2014), available at: <www.nea.gov.cn/2013-12/19/c_132980563.htm> (last accessed 18 October 2014);- 中石化10月1日起全国置换国四国五汽油 [Sinopec shifts to Phase IV and Phase V petrol nation-wide since 1st October] (Chinanews.com, 28 August 2013) available at: <http://finance.chinanews.com/ny/2013/08-28/5217213.shtml> (last accessed 18 October 2014).

85 Watson 1993, p. 21.

86 Alan Watson's main argument is that a society's laws do not usually develop as a logical outgrowth of its own experience. Instead, he contends, the laws of one society are primarily borrowed from other societies; therefore, most law operates in a society very different from the one for which it was originally created. Watson 1993.

87 Wiener 2001, p. 1298-1299.

88 Watson once quoted Roscoe Pound 'history of system of law is largely a history of borrowings of legal materials from other legal systems and of assimilation of materials from outside the law.' Wiener observed that transplantation does not only include 'horizontal' borrowing between nations but also covers 'vertical' or 'trans-echelon' borrowing between national law and international law due to emerging the recognition of global public goods. →

distinguish between 'transplantation' and 'convergence'[89] and cite, among others, Beijing's adoption of fuel emission standard following the European fuel quality standard as an example of transplantation.[90] The motive for the deliberate borrowing, on the one hand, may follow from China's urgent need to address air pollution by reducing fuel emissions and consumption and, on the other hand, it is due to the strictness and 'diffusion' effect of EU's standard,[91] which is probably deliberately pursued by the EU to play a 'leading' role or to be a 'role model' by adopting stringent environmental and climate change regulations and standards.[92]

Notwithstanding, there is uncertainty with regard to the EU's GHG emission reduction target for fuels after 2020, as the current binding target for the GHG emission reductions will expire after 2020 and the EU is not going to set a similar target for the 2020-2030 period.[93] The European Commission states that it does not think it appropriate to establish new targets for the GHG intensity of fuel used in transport after 2020.[94] The Council invites the Commission to examine further the instruments and measures to reduce GHG emissions in the transport sector, but it also agrees to use a 'technology neutral approach' for reducing GHG emissions in transport.[95] In this vein, the EU might converge with China in not providing preferential treatment to biofuels in regulation over fuel emissions, which could be owed to the diversification of low-carbon transport solutions and the uncertainty with regard to the contribution of biofuels to decarbonising transport fuels.[96]

Watson 1993, p. 22; Wiener 2001, p. 1296, 1298-1299, 1302, 1308, and 1321; Yang & Percival 2009, p. 625.

89 Yang and Percival define 'transplantation' as the 'deliberate copying and adaptation of significant portions of statutes or particular doctrines of law by one country from another', based on the definition given by Watson, while 'convergence' refers to unintended similarities due to common external pressures. Yang & Percival 2009, p. 626; Watson 1993, p. 22-24.

90 Yang & Percival 2009, p. 630.

91 There is rich literature about the effect of the EU's unilateral measures on other parts of the world, such as literature on 'diffusion', 'Europeanisation', and 'The Brussels Effect', see for example Vogel 2012, p. 12-13; Bradford 2012; Scott 2009.

92 From the 1990s (Dublin Declaration) onwards, the EU declared itself to be economically, politically, and morally predestined to play a role in global environmental leadership. The ambition was partly driven by economic motives aiming for a competitive advantage of European enterprises. Brande 2009, p. 161; European Commission, Research & Innovation, available at: <http://ec.europa.eu/research/environment/print.cfm?file=/comm/-research/environment/policy/article_1435_en.htm> (last accessed 13 January 2014).

93 See Chapter 3, section 2.2.2.1.

94 Communication from the Commission, 'A policy framework for climate and energy in the period from 2020 to 2030', COM(2014) 15, section 2.2, para. 6.

95 European Council, Conclusions, 23/24 October 2014, EUCO 169/14, para. 2.1.3, p. 4, available at: <www.consilium.europa.eu/uedocs/cms_data/docs/pressdata/en/ec/145397.pdf> (last accessed 13 January 2015).

96 The Commission has admitted that the assessment of how to minimise indirect land-use change emissions made clear that first generation biofuels have a limited role in decarbonising the transport sector. The Commission has already indicated, for example, that →

2.4. *Preferential Tax Treatment*

Preferential tax treatment, in the form of tax reductions or exemptions, is frequently used as financial support to help overcome the higher costs of biofuels and to reduce their selling prices. As a kind of fuel product, excise duties (taxes) are levied on biofuels,[97] so any reduction or exemption of excise duties on biofuels would drive the demand for biofuels.

2.4.1. Differences

At the international level, there is no special rule about the taxation of biofuels but biofuels used in aviation may enjoy an exemption from energy taxation.[98] In domestic law, the EU and China provide preferential tax treatment for biofuels used in road/non-road transport in different ways and under different conditions.

Firstly, whether and how to implement preferential tax treatment for biofuels are primarily decided by Member States in the EU: the EU allows Member States to provide an exemption or a reduction of the taxation rates (from the EU's harmonised minimum rates) to be applied to biofuels, and it is up to the Member States to decide whether and how to apply any tax reduction or exemption, as long as they inform the European Commission about the preferential tax measures.[99] In comparison, the levy of excise duties in China is centralised – the levels and rates of such taxation as well as its exemptions are generally applied uniformly nation-wide, unless otherwise provided by the central government.[100] The State Council is authorised to decide on adjustments of taxation categories and the rates of excise

food-based biofuels should not receive public support after 2020. Commission Communication COM(2014) 15 final, section 2.2, para. 6.

97 Bioethanol and biodiesel frequently belong to different categories of taxed items and may be subject to different exercise duties. While biodiesel is subject to excise duties on energy products, bioethanol may fall within the category of alcohol and is subject to excise duties on alcohol.

98 According to the Chicago Convention, aviation fuels on board international flights shall be 'temporarily admitted free of duty.' A resolution adopted by the ICAO Council in 1999 further clarifies the provision and states that the expression 'customs and other duties' shall include import, export, excise, sales, consumption and internal duties and taxes of all kinds levied upon the aviation fuels. The tax exemption constitutes the reason for exempted or postponed excise duties on international aviation fuels in the EU and China, even including an exemption for domestic flights: the EU provides that Member States shall exempt energy products for air navigation from energy taxation; China sets a taxation rate for aviation kerosene but decides to temporarily postpone levying the taxation without fixing a time limit. Chicago Convention, Art. 24(a); ICAO, 'ICAO's Policies on Taxation in The Field of International Air Transport, Third Edition 2000, Approved by the Council on 24 February 1999 Published by direction of the Council, 1(d), available at: <www.icao.int/-publications/Documents/8632_3ed_en.pdf> (last accessed 15 January 2015); Directive 2003/96/EC, Art. 14(1) and (2), and para. (23) of the preamble; Notice on Increasing Excise Duty Rates of Oil Products 2008, Art. 3; see Chapter 3, section 2.2.3; see Chapter 4, section 2.2.4.

99 See Chapter 3, section 2.2.3.

100 See Chapter 4, section 2.2.4.

duties[101] and the exemption from excise duties on bioethanol and biodiesel has been issued[102] or approved[103] by the State Council. This difference can be attributed to the differential taxation systems. The EU's competence in harmonising energy taxes shall be exercised to the extent of guaranteeing the functioning and integration of the internal market[104] and is subject to the principles of 'subsidiarity' and 'proportionality.'[105] By contrast, China has a centralised taxation system,[106] with tax categories and rates established by the central government, and local governments have to implement the nationally-unified tax categories and rates as well as the tax exemptions and reductions provided by the central government.[107]

Secondly, the substantive conditions for acquiring the preferential tax treatment are different. In the EU, a tax reduction or exemption, if it falls within the meaning of State aid, must be examined by the European Commission according to

101 Law on the Administration of Tax Collection of China 2001; Interim Regulation on Excise Duties 2008, Art. 2.

102 Notice on Excise Duty Policies 2008; Notice on Oil Products Taxation Reform 2008; see Chapter 4, section 2.2.4.

103 For example, the Notice on the Excise Duty Exemption of Biodiesel (BD100) Produced from Waste Animal Fat and Vegetable Oil was promulgated by the Ministry of Finance but was approved by the State Council. Notice on the Excise Duty Exemption for Biodiesel (BD100) 2010; Notice on Increasing Excise Duty Rates for Oil Products 2008; see Chapter 4, section 2.2.4.

104 Within the EU, direct taxation remains the sole responsibility of the Member States; and indirect taxation, which affects the free movement of goods and the freedom to provide services, is subject to EU harmonisation, to the extent that the harmonisation is necessary to ensure the establishment and the functioning of the internal market and to avoid distortion of competition. TFEU, Art. 113.

105 The limits of Union competences are governed by the principle of conferral. The use of Union competences is governed by the principles of subsidiarity and proportionality. TFEU, Art. 5.

106 In China, the taxation power is tightly governed by the central government. Basic taxation and the fiscal framework shall be adopted by the National People's Congress or its Standing Committee in the form of law (*fa lv*), while the State Council can be authorised to set out detailed regulations to implement the tax law. In the amendment to the Law on Legislation in 2015, it is provided that the basic taxation systems such as establishment of taxation categories (*shui zhong*), taxation rates (*shui lv*), and the administration of tax collection shall be stipulated in the law (*fa lv*) adopted by the National People's Congress or its Standing Committee, which aims to reduce the power of the State Council in establishing taxation categories or rates. In China, taxes can be generally classified into central taxes, local taxes, and shared taxes, according to the power levying and managing the taxes. For example, excise duties belong to the central taxes, while VAT belongs to shared taxes as the revenue is shared proportionally between the central and local governments. However, the law does not clearly specify the power of local government in taxation matters: it is only generally provided that local governments and local taxation departments have to levy taxes to implement national laws or regulations. 刘剑文 [Liu, J.], 应以法律形式合理确定地方的税收立法权和收益权) [Local taxation law-making power and income/usufruct power] (China Taxation News, 20 Nov. 2013), available at: <www.gdsswxh.com/product.asp?id=3444> (last accessed 31 October 2014); Law on Legislation 2000, Art. 8(6).

107 刘剑文 [Liu, J.], 应以法律形式合理确定地方的税收立法权和收益权) [Local taxation law-making power and income/usufruct power] (China Taxation News, 20 November 2013), available at: <www.gdsswxh.com/product.asp?id=3444> (last accessed 31 October 2014).

the EU law on State aid.[108] The tax reductions or exemptions for biofuels may qualify for the 'block exemption', but certain conditions have to be met, including compliance with sustainability criteria, and they should be based on non-food based feedstock.[109] In China, substantive conditions, if there are any, are specifically set out for each individual preferential tax measure, such as the 70 per cent of waste fat and oil in biodiesel feedstock.[110] This difference could be grounded in the different purposes of and backgrounds to the preferential tax measures. In the EU, the Energy Tax Directive 2003/96/EC was adopted in the same year as the former Biofuel Directive 2003/30/EC, when sustainability criteria were not yet adopted and the European Commission was determined to promote biofuels for the 'greatest potential in the short to medium term.'[111] After the adoption of the sustainability criteria, the European Commission proposed to amend 2003 Energy Taxation Directive, and part of the amendment was to connect preferential energy taxes for biofuels with sustainability criteria, although the proposal has not yet been adopted due to disagreement among the Member States about the energy taxation reform based on energy content and carbon emissions.[112] In comparison, China at first provided an exemption from excise duties for bioethanol produced by designated state-owned plants to help them overcome cost barriers and recently provides an exemption from excise duties for industrial alcohol to save costs for the producing enterprises;[113] the exemption from excise duties for waste oil-based biodiesel is primarily to solve the 'gutter oil' problem in China.[114]

2.4.2. Similarities

Despite the differences, both the EU and China have changed the original unconditional financial support for biofuels. For example, the EU's Energy Tax Directive does not set out sustainability requirements for preferential tax treatment for biofuels, but the European Commission later emphasises that State aid for renewable energy, even when falling within the scope of the 'block exemption', shall be given to sustainable biofuels in line with Directive 2009/28/EC and shall

108 See Chapter 3, section 2.2.3.

109 *Ibid.*

110 See Chapter 4, Section 2.2.4.

111 It is stated that 'The Commission considers that the use of fuels (such as vegetable oils) derived from agricultural sources (*i.e.* biofuels) is the technology with the greatest potential in the short to medium term.' The promotion of biofuels has been incorporated mainly within Art. 16 of Directive 2003/96 and provides Member States with the option of applying a reduced rate of excise duty to pure or blended biofuels when used as heating or motor fuel. European Commission, Taxation and Customs Union, Encouraging the use of alternative fuels, available at: <http://ec.europa.eu/taxation_customs/taxation/excise_duties/-energy_products/alternative_energy/index_en.htm> (last accessed 16 January 2015).

112 See Chapter 3, section 2.2.3.

113 See Chapter 4, section 2.2.4.

114 China has prohibited the use of 'gutter oil' for food and meanwhile encourages the use of such waste cooking oil for other uses including for producing biodiesel. See Chapter 4, sections 2.2.4. and 3.2.

exclude food-crop based biofuels.[115] China used to provide an exemption from excise duties or tax refunds for the state-owned bioethanol producing enterprises, but such preferential tax treatment has been withdrawn for food grain-based bioethanol.[116] A particular convergence lies in the withdrawn support for food crop-based biofuel, although the scope of 'food crop' feedstock is defined differently: the EU defines 'food-based biofuels' as biofuels produced from cereal and other starch-rich crops, sugars, and oil crops,[117] while China does not exclude the use of certain crops that can be used as food (stuff) such as cassava.[118] The convergence in withdrawing or conditioning preferential tax treatment for food-based biofuels could be attributed to the common concern about the potential negative impacts of biofuels, as well as the awareness that the production of first-generation biofuels has become relatively technically and commercially mature (compared to advanced biofuels).[119]

2.5. *Emission Trading Rules*

Emission trading is a market-based mechanism to achieve GHG emissions reduction in a more cost-efficient manner, primarily through trading emission allowances or project-based emission reduction credits in the carbon market. Emission trading rules are relevant for biofuels if the transport sector is included in the emission trading scheme and biofuels can be used to save emission allowances or generate emission reduction credits.

2.5.1. Emission Trading System

2.5.1.1. Differences

At the international level, the Kyoto Protocol provides for a legal framework for emissions trading, but it does not include rules that specify which sectors are covered or how emissions shall be calculated,[120] and there are no specific rules about emissions trading modalities for the use of biofuels.

Emission trading schemes of the EU and China differentiate in scale and scope. The EU has launched a binding EU-wide emissions trading scheme since

115 See Chapter 3, section 2.2.3.
116 See Chapter 4, section 2.2.4.
117 Commission Proposal COM(2012) 595 final, p. 3.
118 See Chapter 3, section 3.1.1.
119 For example, the Chinese central government has gradually reduced subsidies to state-owned bioethanol producing plants partly due to the fact that the first-generation biofuel production in the plants became technically mature and financially profitable. Measures on the Administration of Flexible Subsidy 2007; Gongjing (宫靖), Ranliao Yichun Butie You Ding'ezhi Gaiwei Tanxingzhi (燃料乙醇补贴由定额制改为弹性制) [The Bioethanol Subsidy Changes from Fixed Amount to Flexible Mechanism] (Caijing.com.cn, 13 December 2007) available at: <www.caijing.com.cn/2007-12-13/100041709.html> (last accessed 31 October 2014).
120 See Chapter 2, section 2.1.

2005, which is now in the third phase (2013-2020) and includes carbon emissions from commercial flights within and between European countries[121] as well as international flights departing from or arriving in European countries.[122] In contrast, China just began with the establishment of an emission trading scheme since 2011, which is as yet at the 'pilot' stage in regional areas.[123] In terms of scope of the covered sectors, transport sector is thus far included in regional pilot emission trading schemes, and whether it will be covered by national scheme is not clear.[124] The difference in scale and scope can be attributed to the differentiated emission reduction commitments undertaken by the EU and China under the Kyoto Protocol. The EU, as an Annex B Party, undertakes a binding GHG emission reduction commitment and has established an emission trading scheme as a key tool for fulfilling its assigned commitment.[125] In contrast, China does not have a similar international obligation with regard to emission reductions, and it just began to establish a domestically mandatory target to reduce CO_2 emission intensity and highlight emissions trading as a way of achieving the target since the 12th Five Year Period (2011-2015).[126]

Moreover, the EU and China have certain differences in the treatment of carbon emissions from biofuels for aviation. As the EU in fact exempts the obligation of surrendering allowances for the carbon emissions of biofuels,[127] the European Commission requires compliance with the sustainability criteria as the precondition for enjoying such preferential treatment.[128] In comparison, China's accounting and reporting guidelines, issued either by the Shanghai government or by the NDRC, do not specify whether biofuels should comply with any sustainability requirement or with the general principles of non-grain, marginal land, and no environmental damage,[129] which is firstly due to the lack of a binding sustainability requirement in the law, and secondly perhaps because the use of

121 See Chapter 3, section 2.2.4.1.

122 The EU ETS is temporarily suspended for international flights. Chapter 3, section 2.2.4.1.

123 See Chapter 4, section 2.2.5.1.

124 *Ibid.*

125 The EU emission trading scheme is claimed to cover around 45 per cent of the EU's greenhouse gas emissions. It is determined that, in 2020 emissions from the sectors covered by the EU ETS will be 21 per cent lower than in 2005; by 2030, they would be 43 per cent lower. Climate Action, The EU Emission Trading System (ETS), available at: <http://ec.europa.eu/clima/policies/ets/index_en.htm> (last accessed 1 May 2015).

126 Nevertheless, China and the US signed a deal in which China intends to achieve a peak in CO_2 emissions around 2030 and to make best efforts to peak early and it intends to increase the share of non-fossil fuels in primary energy consumption to around 20 per cent by 2030. See Chapter 4, section 2.2.5.1; The White House Office of the Press Secretary, U.S.-China Joint Announcement on Climate Change, 11 November 2014, available at: <https://www.whitehouse.gov/the-press-office/2014/11/11/us-china-joint-announcement-climate-change> (last accessed 19 May 2015).

127 The EU attributes a zero emission factor to the biomass-derived part in aviation fuels. See Chapter 3, section 2.2.4.1.

128 See Chapter 3, section 2.2.4.1.

129 See Chapter 4, section 2.2.5.1.

biofuels in aviation accounts for a still minimal amount which has been experimentally supplied with biofuels produced from 'gutter oil.'[130]

2.5.1.2. *Similarities*

Despite the differences, the EU and China have both included or tried to include the aviation sector in their emissions trading schemes. Although China's emissions trading scheme is still in the phase of regional pilots, China plans to establish a nation-wide emission trading scheme by 2016,[131] and, as a preparatory step for the national emission trading, China has already issued national GHG emissions accounting and reporting guidelines for the first group of sectors which includes the aviation sector.[132] Moreover, both the EU ETS and Shanghai's pilot scheme exempt carbon emissions from biofuels used for aviation from the obligation to surrender emission allowances.

The convergence could be attributed to the effects of international integration and mutual borrowing. The Kyoto Protocol establishes a legal framework for trading assigned emission allowances and credits, thus having an international integration effect. Yang and Percival point out that the Kyoto Protocol's provisions on emission trading have an 'integration' effect by providing how the emission units or credits are generated, traded, and used.[133] Such an integration effect may have inspired State Parties like the EU and China to build carbon emissions trading systems to achieve their domestic climate change mitigation targets. Meanwhile, without internationally harmonised rules on how to treat biofuel emissions, China may borrow from the emissions trading scheme experiences from the other countries, particularly from the EU with regard to its aviation emission trading rules. For example, both the Shanghai and National aviation GHG emissions accounting and reporting guidelines refer to the EU's monitoring and reporting

130 For example, it has been reported that China's largest private carrier Hainan Airlines has completed the country's first commercial flight powered by biofuel made from used cooking oil in 2015 and before that an experimental flight was completed by using biofuels produced from Jatropha seeds in 2011. Flying fryer: China completes first flight powered by 'gutter oil' (RT, 21 March 2015), available at: <http://rt.com/news/242913-china-gutter-oil-flight/> (last accessed 14 May 2015); James T. Areddy, 'China's First Aviation Biofuel Test' (Chinarealtimereport, 01 Nov. 2011), available at: <http://blogs.wsj.com/chinarealtime/-2011/11/01/chinas-first-aviation-biofuel-test/> (last accessed 26 August 2014).

131 See Chapter 4, section 2.2.5.1.

132 See Chapter 4, section 2.5.

133 Yang and Percival observe that the result of such monitoring is a global administrative system that is involved in supervising and monitoring the activities of private entities and businesses. The Kyoto Protocol's international transactions log coordinates with national GHG registries and the EU's Transaction Log, which thus links the business and regulatory decisions within individual nations to Kyoto Protocol requirements and to the regulatory schemes of other nations. In this way, the Kyoto Protocol's international transaction log has facilitated international and transnational regulation by promoting the integration of emission trading regulatory schemes worldwide. Yang & Percival 2009, p. 637-638.

rules and guidelines for aviation emission trading.[134] Apart from the reason that the EU has taken the first step in initiating an aviation emission trading scheme, China may want to extend emission trading to aviation in response to the coverage of international flights (including with Chinese airlines) in the EU's emission trading system.[135] Although China has expressed its opposition to the EU's action of including Chinese airlines[136] and the EU has temporarily suspended the application of its emission trading scheme to international flights, the establishment of China's own aviation emission trading scheme could be a reasonable defensive measure either to the possibly resumed application of the EU aviation emissions trading after 2016[137] or to the potential construction of a global market-based mechanism under the ICAO.[138] In fact, it has been pointed out that the inclusion of the aviation sector in Shanghai's pilot emissions trading scheme came at the time when China was involved in a dispute with the EU about the aviation emission trading.[139] There might be potential for more convergence: China's emission trading accounting and reporting rules and guidelines are still in the pilot phases and subject to further revisions based on the experiences gained in China or from the EU.

134 The National guidelines refer to the 'Guidance Document The Monitoring and Reporting Regulation – General guidance for installations MRR Guidance document No. 1, Version of 16 July 2012.' The Shanghai guidelines refer to the 2007/589/EC (Commission Decision of 18 July 2007 establishing guidelines for the monitoring and reporting of greenhouse gas emissions pursuant to Directive 2003/87/EC of the European Parliament and of the Council) and 2009/339/EC. Guidelines for National GHG Emission Accounting and Reporting 2013; Guidelines on GHG Emission Accounting and Reporting of Shanghai 2012.

135 Lan 2011; Borger 2012.

136 Chinese air transport authorities and airlines have expressed their explicit opposition to the EU's initiative to include international flights into the EU ETS. 我国八家航企联手抵制欧盟碳税体系) [Eight Chinese Airlines Jointly Resist EU ETS] (China.com, 19 May 2012), available at: <http://finance.china.com.cn/industry/-20120519/735365.shtml> (last accessed 24 October 2014).

137 The EU ETS may exempt the flights that are already subject to similar emission trading mechanisms. Directive 2008/101/EC, Art. 25a(1).

138 China has declared its intention to adhere to the principle of common but differentiated responsibility in the potential market-based mechanism under ICAO. ICAO Assembly 38th Session Report of the Executive Committee on Agenda Item 17 (Section on Climate Change), Paragraph 2(b), available at: <http://ec.europa.eu/clima/policies/transport/aviation/-docs/a38_wp_430_en.pdf> (last accessed 24 October 2014); Representative of PRC on the Council of ICAO, Statement of Reservation of China regarding Resolution 17/2 of the 38th Session of the Assembly: Consolidated Statement of Continuing ICAO Policies and Practices Related to Environmental Protection – Climate Change, available at: <www.icao.int/-Meetings/a38/Documents/Resolutions/China_en.pdf> (last accessed 24 October 2014).

139 'Shanghai includes aviation emissions in CO_2 market' (Thompson Reuters, 10 January 2013), available at: <www.greeninvestmentservices.com/news/1-latest-news/175-shanghai-includes-aviation-emissions-in-co2-market> (last accessed 24 October 2014); 'Shanghai to include aviation emissions in its ETS' (Climate Policy Watcher, 10 January 2013), available at: <www.climate-policy-watcher.org/?q=node/455> (last accessed 24 October 2014).

2.5.2. Clean Development Mechanism

2.5.2.1. Differences

At the international level, the Kyoto Protocol establishes a legal framework for governing the registration of CDM projects and the issuance of the Certified Emission Reduction credits, under which project-based estimation and monitoring modalities are adopted to ensure the 'additionality' of GHG emission reductions.[140] However, compared with the emphasis on the 'additionality' of GHG emissions reduction, the determination of the contribution of CDM projects to sustainable development is largely determined by hosting State Parties.[141] As a CDM credit-importing Party, apart from requiring compliance with the international rules on 'additionality' and 'sustainable development', the EU furthermore sets a limit on the type and location of the projects which generate the imported CDM credits.[142] As a CDM project hosting country, China establishes administrative regulation on the approval and management of CDM projects, according to which the contribution to sustainable development is one criterion to be considered by the Chinese central government in the approval the applicant projects.[143] Thus, it is observed that the relevant bodies under the Kyoto Protocol, the EU, and China play different roles in the operation of the CDM, which may determine their differential regulation of the CDM. The CDM Executive Board plays a supervising role in ensuring the 'additionality' of CDM project credits in line with the principles and modalities adopted by the Conference as the Meeting of Parties of the Kyoto Protocol, but it has a limitation in defining and investigating the impact on sustainable development for the consideration of sovereign states. For the EU, CDM credits may provide a means to achieve its GHG emission reduction commitments, but it could meanwhile create surplus allowances that inflate the EU's carbon market and impair its GHG emission reduction goal,[144] so there is a need and motivation to

140 'Additionality' means that the human-induced emissions reduction achieved by CDM project activities would not have occurred in the absence of the project. See Chapter 2, section 2.1.

141 It is the host Party to confirm that the CDM project activity assists it in achieving sustainable development. See Chapter 2, section 2.3.

142 For example, the EU limits CDM projects related to nuclear energy, afforestation or reforestation, the destruction of industrial gases, and hydroelectric power over 20 MW of installed capacity, and refuses CERs from new CDM projects starting after January 2013 in countries other than the Least Developed Countries. See Chapter 3, section 2.2.4.2.

143 CDM projects have to demonstrate Environmental Impact Assessment reports, but the Environmental Impact Assessment system in China has been considered problematic in the assessment scope and actual implementation. See Chapter 4, section 2.2.5.2.

144 The EU deems that many CDM projects lack credibility regarding environmental and emission reduction effects, so the EU wants to see CDM further reformed in order to improve their environmental integrity and efficiency through greater use of standardised baselines and alternative ways of assessing additionality. Furthermore, the EU announced that CDM offsets from advanced developing countries should be replaced over time by the new market mechanism covering broad segments of the economy and incentivising net emission reductions, and CDM would then be focused on least developed countries. The EU changed the policy largely because it has been argued that the CDM does not work in favour of the →

employ stricter requirements on CDM projects. China, on the other hand, has benefited from CDM projects by making use of foreign investments,[145] and the drive for economic benefits may need to be balanced by the requirement and examination of the 'additionality' and contribution to sustainable development.[146]

2.5.2.2. *Similarities*

To a certain extent, the lack of specification on the sustainability of biofuel CDM projects may constitute a convergence amongst the governing approaches at the international, EU, and Chinese level. Compared to the emphasis on 'additionality', the Kyoto Protocol Conference as the Meeting of Parties or the CDM Executive Board does not provide much guidance on how to assess and supervise the contribution to sustainable development.[147] The EU has not specified whether the biofuels used in CDM projects should comply with the EU's sustainability criteria before being accepted as transferable credits in EU's emissions trading scheme,[148] whereas it does so in the case of hydroelectric projects.[149] China also does not specify how to evaluate the contribution to sustainable development by applicant biofuel projects.[150]

Such a convergence may be firstly owing to the minimal number of CDM biofuels projects: there are not many registered CDM projects purporting to produce or using biofuels as defined in the thesis,[151] which could be attributed to the comparatively higher costs and difficulty in proving 'additionality' and GHG emissions reduction of biofuel projects.[152] The small number of biofuel CDM

European industry, the European taxpayers nor the EU at large, except for those EU ETS operators that benefit directly and in the short term from CER revenue. European Commission, Climate Action, International Carbon Market, available at: <http://ec.europa.eu/clima/policies/ets/linking/index_en.htm> (last accessed 25 October 2014); Amatayakul & Berndes 2012, p. 197; Chang, Belis _ Bruyninckx 2012, p. 240.

145 See Chapter 4, section 2.2.5.2.

146 For instance, China's lead position in the CDM market was largely established by the HFC-23 projects which have a disputed impact on the environment and has approved projects that have disputed additionality. See Chapter 4, section 2.2.5.2.

147 See Chapter 2, section 2.3.

148 See Chapter 3, section 2.2.4.2.

149 Acceptance in the EU ETS of credits from JI or CDM hydroelectric projects exceeding 20 MW of installed capacity is subject to certain conditions. See Chapter 3, section 2.2.4.2.

150 The requirement in China consists of the submission of s project design document, an environmental impact assessment report, and the examination of GHG emissions reduction methods and the contribution to sustainable development. Measures on Operation and Management of Clean Development Mechanism Projects, Arts. 15 and 20.

151 In contrast with the large number of biomass projects for electricity or heating and cooking, the biofuel projects for transport are minimal in number. See Chapter 2, section 2.1.

152 Another reason Bakker identified in 2006 was the lack of EB approved biofuel baseline and monitoring methodology, but there have been eight methodologies approved for projects related to biofuels up to 2014. Bakker 2006.

projects may mitigate the necessity of adopting special rules on the sustainability of biofuel projects.[153]

Another reason may lie in the need to consider local circumstances in evaluating the contribution to sustainable development. Unlike the carbon emission reduction, the contribution to sustainable development, particularly the sustainable development of the hosting countries, is difficult to define and estimate uniformly, and thus whether CDM projects contribute to sustainable development should be decided by the authorities of the project host countries. Moreover, there is no agreed international definition or guidelines for assessing sustainability of biofuels as those for the hydroelectric projects. However, China's regulation of CDM shows that complete dependence on national regulation to define and monitor the criterion of the 'contribution to sustainable development' may not be reliable.[154] Therefore, specified requirements regarding a project's sustainability are, to some extent, needed on the part of international governing bodies or credit importing countries; otherwise, there would be a risk that host countries might easily come to the conclusion of sustainability without clarifying the assessment criteria. The efforts could perhaps be made by the CDM Executive Body to accelerate its work on evaluating the use of a 'sustainable development tool'[155] or by the EU to make acceptance of biofuel project credits conditional on their compliance with sustainability criteria, like the limitation for hydroelectric project credits.[156]

2.6. *Interim Conclusion*

According to the comparative analysis in section 2, differences and similarities can be found in the regulation of the demand for biofuels.

2.6.1. Differences

Differences first of all are seen in the *content* of the sustainability requirements. Demand for biofuels is closely associated with the GHG emission reduction

153 For example, in the Jin-Jiang Biodiesel project, accordingly, the environmental impact assessment is based on an EIA approved by local (municipal) environmental protection bureau five years ahead of the registration date (8 May 2013), focusing on pollution from industrial processes rather than from feedstock production and collection, and the conclusion is rather ambiguous with 'no significant impacts on the environment'; meanwhile, the comments regarding the contribution to sustainable development lacks any concrete criteria or proof about how to assess the environmental and socio-economic benefits. See Chapter 4, section 2.2.5.2.

154 For example, Lin points out the risks of rent-seeking activities in CDM certification by designated operational bodies as well as by host country agencies. Lin 2012a, p. 43; see Chapter 4, section 2.2.5.2.

155 See Chapter 2, section 2.3.

156 For example, the EU has already required all credits from CDM hydropower projects larger than 20MW sold in the EU Emissions Trading Scheme to meet World Commission on Dams environmental and social standards. European Commission, Climate Action, International Carbon Market, <http://ec.europa.eu/clima/policies/ets/linking/index_en.htm> (last accessed 25 April 2015).

commitments under the international climate regime, which concentrates on the character of the GHG emissions or the GHG emission saving from the use of biofuels. In the EU, the sustainability criteria for biofuels focus on the environmental and climate impacts such as life-cycle GHG emission saving, biodiversity, carbon stocks, and agro-environment. In comparison, China has not established sustainability criteria for biofuels in law, and the general principles set out in the renewable energy development plans largely emphasise the food security concern. The reasons for divergent substantive sustainability requirements can be attributed to the different focus on sustainability concerns and the lack of an international agreed regulatory framework or definition of 'sustainable biofuels.'

Differences are evident in *how* the sustainability requirement is established for the measures that influence the demand for biofuels. At the international level, Parties undertake an obligation to different extents in reporting on their GHG inventories according to the IPCC Guidelines, which may thus generally provide methods and guidance for calculating the life-cycle GHG emissions of biofuels. The EU requires compliance with its sustainability criteria as a precondition for obtaining favourable treatment, such as being counted towards the renewable energy targets and financial support for consumption. The EU also endeavours to align the sustainability criteria and specific demand-side measures, such as in the fuel quality directive, State aid rules, and emission trading monitoring rules. China establishes the general principles as the overall guideline for policy or rule making for the biofuel sector. The general principles are not explicitly stated as a precondition for the specific demand-side measures such as blend mandate and consumption targets. However, certain conditions, in line with the general principles, have been set out for some specific measures, like in the case of prohibiting further expansion of food grain-based bioethanol production, exempting biodiesel with feedstock from waste cooking oil from excise duties or withdrawing preferential tax treatment for food grain-based biofuel enterprises. The reason can be primarily attributed to different regulatory approaches to promote the use of biofuels: the international climate regime undertakes a commitment to reduce the overall GHG emissions but does not specify the measures to achieve that goal. While the EU has promoted the use of biofuel through target-setting in law, China has incentivised the consumption of biofuels primarily through 'pilot' projects with the characteristics of planned supply by dominant state-owned enterprises and administrative mandates for distribution and consumption.

2.6.2. Similarities

Despite the differences, convergence can be observed in the common concerns about potential the negative impacts of biofuels, particularly regarding food security, which prompts both the EU and China to set requirements such as biofuel feedstocks should be 'non-food'-based and financial support to food-based biofuel is withdrawn. Convergence could be also reflected in that: the EU is not going to set binding renewable energy or fuel GHG emission reduction targets for Member States after 2020, while China has gradually strengthened its fuel emission and economy standards; both the EU and China exempt biofuels from allowance surrendering in aviation emissions trading.

The convergence can be firstly attributed to the influence from the international level. The EU and China are both Parties to the UNFCCC and the Kyoto Protocol, and they undertake the general commitment of taking climate change mitigating measures as provided in Article 4(1) of UNFCCC, which may underlie the convergence between the EU and China in adopting climate change mitigation measures such as promoting biofuels through renewable energy or biofuel targets and setting fuel emissions or economy standards. Apart from the international influence, the convergence may also be attributed to the mutual influence between the EU and China. It is observed that China follows the EU's emissions standards for fuels and vehicles, and aviation is included in Shanghai's pilot emission trading scheme where EU's monitoring and verifying rules and guidelines are taken as references. These facts might exemplify the 'transplantation' effect due to copying or adapting to the EU's high-level standards and advanced experiences. The convergence between the EU and China in their concerns about food-based biofuels and withdrawing financial support for such biofuels might be an unintended similar response to the food pressure possibly caused by biofuel.

3. Comparative Analysis of the Supply of Biofuels

The discussion on the supply of biofuels in this thesis has centred on the regulation of the supply of biofuel feedstocks, because the sustainability of biofuels is primarily dependent on whether the feedstocks are produced sustainably. Biofuels can be sourced from a variety of feedstocks, but the feedstocks until now primarily come from conventional or 'first-generation' raw materials. In view of this, the comparative study on the supply of biofuels focuses on the regulation that addresses the major sustainability impacts of conventional biofuel feedstocks, namely biodiversity, carbon stock, agro-environment, and socio-economic rights.

3.1. Regulation Addressing the Biodiversity Concern

3.1.1. Differences

International agreements on biodiversity provide the general commitment to 'sustainable use' or 'wise use' which does not absolutely exclude human intervention and the exploitation of biological diversity.[157] Specific resolutions and decisions regarding the impact of biofuel's on biodiversity have been respectively adopted under the Ramsar Convention and CBD, which contain some substantive guidance (e.g. impact assessment and national biodiversity inventory and 'go' or 'no go' areas) for states (including Parties and non-Party states) to deal with the relationship between biofuels and biodiversity,[158] but they have ambiguous legal

[157] See Chapter 2, section 3.1.1.
[158] See Chapter 2, section 3.1.2.

statuses and may have weak binding force on Parties according to the analysis in Chapter 2.[159]

As one Party to CBD, the EU protects biodiversity by defining and identifying the lands or areas with 'high biodiversity value' to be protected from biofuel production. The CBD COP decisions invite Parties to identify areas of 'high biodiversity value', 'critical ecosystems', and 'areas important to indigenous and local communities', but they do not define these terms in detail.[160] In contrast, the EU specifies the types of lands identified as having 'high biodiversity value' and attempts to define each type of land basically in line with the available definitions commonly recognised at the international level.[161] For example, the EU defines 'primary forest' based on the meaning adopted by the Global Forest Resource Assessment under the FAO,[162] but additionally it includes those areas where non-wood forest products collection occurs with minimal human impact.[163] It is noteworthy that the highly biodiverse lands and areas precluded from providing biofuel feedstocks include those located outside of the EU, while raw materials from the EU should additionally comply with the EU environmental law on the conservation of wild birds[164] and natural habitats and wild flora and fauna.[165]

China is a Party to both the CBD and Ramsar Convention. Biodiversity could be protected from biofuel production primarily by respecting the current law on nature reserves,[166] which is reflected in excluding nature reserves from the scope of marginal lands (for biofuel feedstock cultivation) or the establishment of feedstock supply bases.[167] Moreover, it is required that energy forestry cultivation shall avoid ecological zones with 'high protection value.'[168] Such a definition and requirement, however, remains to be fragmented and of little effect. For example, the scope of 'marginal land' is defined in an internal ministerial document purported to investigate marginal land resources, and the 'high protection value' requirement is only stated for energy forestry. After all, there is no clear definition of which types

159 There is a discussion on the legal status of the measures adopted by treaty bodies of the mentioned environmental treaties. The analysis is based on criteria addressed by Wiersema: authorisation, voting, language, and the effect of implementation. The analysis based on the criteria reveals that although the resolutions or decisions provide certain substantive guidance or requests for states, they do not intend to legally bind states, judging from the criteria of level of authorisation and lanuage. The major implementation measure - the national reports - remains a limited source of information on biofuels due to inadequate reporting. See Chapter 2, section 3.1.3.

160 See Chapter 2, section 3.1.2.1.(c).

161 See Chapter 3, section 2.1.2.

162 Directive 2009/28/EC, para. (69) of the preamble.

163 However, it is suggested that other types of forests identified by the FAO, such as modified natural forests, semi-natural forests and plantations, should not be considered as primary forests. Directive 2009/28/EC, para. (69) of the preamble.

164 Council Directive 79/409/EEC, Art. 3(1), Art. 3(2)(b), Art. 4(1), (2) and (4) and Art. 5(a), (b) and (d).

165 Council Directive 92/43/EEC, Arts. 6 and 13(1)(a).

166 Chapter 4, section 3.1.3.1.

167 *Ibid.*

168 *Ibid.*

of land shall be precluded from being used for biofuel feedstock production, and the dependence on nature reserve system may dismiss the lands or areas which have certain biodiversity value but are not yet established as nature reserves. Differences also exist in the definition of 'land with high biodiversity value' in the EU and the 'nature reserve areas' of China: the EU identifies the 'lands with high biodiverse value' as including primary forest and wooded lands, protected areas and species, natural or non-natural highly biodiverse grasslands; 'nature reserves' in China are defined as the areas with typical natural ecological systems, concentrated rare or endangered wildlife and species, and natural heritage with special values. A further difference lies in the fact that China prohibits reclamation activities in nature reserve areas, while the EU's biofuel sustainability criteria allow certain exceptions for raw materials obtained from protected areas or non-natural highly biodiverse grassland.[169] Finally, the EU's sustainability criteria apply to lands with high biodiversity values not only within the EU but also in non-EU countries, while China's requirements about nature reserves are limited to the lands/areas within China.[170]

The differences between the EU and China can be attributed to the fact that, although the CBD provides an international legal framework for biodiversity protection, the interpretation and implementation of commitments under the CBD are still largely dependent on states. Moreover, neither the resolution under Ramsar Convention nor the decisions under CBD have defined the key terms, such as 'sustainable agricultural practices', 'high biodiversity or conservation value', or 'critical ecosystem', to identify sustainable biofuels in term of biodiversity.[171] There is also a lack of specified methodology to analyse the direct and indirect impacts on biodiversity in the life cycle of biofuels. The Executive Secretary of CBD has admitted that gaps exist with regard to the definition 'high biodiversity or conservation value' and life-cycle analysis of biodiversity impact.[172]

3.1.2. Similarities

Overall, there is convergence between the EU and China in identifying lands/areas with a high biodiversity value to be protected from biofuel feedstock production. Particularly, although China lacks explicit requirements about biofuels and biodiversity, it provides in its regulation that farming or commercial activities shall be prohibited in all the national or local nature reserve areas, and thus biofuel feedstock cultivation or collection is prohibited in these nature reserve areas.[173]

169 For example, an exception can be made if evidence is provided to show that the production of raw materials for biofuels does not interfere with nature protection and biodiversity values in protected areas. For the non-natural grassland, biofuel raw material production can be allowed if evidence is provided that the harvesting of the raw material is necessary to preserve its grassland status. See Chapter 3, section 2.1.2.

170 Regulation on Nature Reserves of China 1994, Art. 3.

171 See Chapter 2, section 3.1.3.

172 CBD/SBSTTA/16/14 2012.

173 See Chapter 4, section 3.1.3.1.

Furthermore, the nature reserves identified in China may in fact include the lands/areas listed as 'lands with high biodiversity value' according to the EU's sustainability criteria, given that a quantity of nature reserves have been established for conserving wetlands, grassland, critical ecosystems for fauna and flora, and primary forests,[174] and wetlands are protected from being used for agricultural reclamation.[175] Moreover, China has revised the Environmental Protection Law to strengthen the ecological and biodiversity protection provisions, such as protection of biodiversity in the aspects of natural resource utilisation and the introduction of alien species.[176] The reason for the convergence can be attributed to the influence of CBD and the other biodiversity-related treaties, particularly with regard to the 'protected area' approach established under the biodiversity treaties – lands or areas with certain biodiversity values are identified and especially protected for controlled utilisation or preservation.

3.2. *Regulation Addressing the Carbon Stock Concern*

3.2.1. Differences

At the international level, the carbon stock inventory is primarily addressed within the theme of 'agriculture, forestry and other land use' (AFOLU) under the legal framework of the UNFCCC and the Kyoto Protocol, and GHG emissions and removals from these sources are to be reported and monitored according to methodologies set forth in the adopted IPCC Guidelines.[177] Meanwhile, other mechanisms have taken place under the UNFCCC, such as CDM afforestation and reforestation projects and REDD-plus, in order to reduce emissions from deforestation/degradation and enhance forest carbon stocks in developing countries.[178] Nevertheless, the conservation of carbon stocks may overlap with the international processes to protect biodiversity or specific eco-systems such as wetlands.

In the EU's regulation of biofuel, carbon stock concerns are addressed by *a)* counting GHG emissions from direct land use changes in the life-cycle emissions of the biofuels, and *b)* identifying 'land with high carbon stock' as 'no-go' areas for biofuels. As for the GHG emissions of carbon stocks from biofuel, the EU has particularly provided guidelines to calculate soil (mineral and organic soil) and above- and below-ground vegetation carbon stocks,[179] which largely draw upon the

174 Biodiversity Conservation Strategy and Action Plan of China (2011-2030), section 3(3). See Chapter 4, section 3.1.3.1.

175 See Chapter 4, section 3.1.3.1.

176 Environmental Protection Law of China, Art. 30.

177 See Chapter 2, section 2.2.2.

178 UNFCCC, Reducing emissions from deforestation and forest degradation and the role of conservation, sustainable management of forests and enhancement of forest carbon stocks in developing countries (REDD-plus), available at: <https://unfccc.int/-land_use_and_climate_change/redd/items/7377.php> (last accessed 21 April 2015).

179 The EU has also adopted Decision No 529/2013/EU to harmonise the accounting of carbon inventories of LULUCF sectors, and has adopted Regulation (EU) No 525/2013 to implement →

2006 IPCC Guidelines,[180] but indirect land use change emissions are currently not included in the EU's accounting for the life-cycle emissions of biofuels.[181] Meanwhile, similar to land with high biodiversity values, the EU protects lands with high carbon stocks by identifying the types of lands as including wetlands and continuously forested areas and preventing biofuel feedstocks from being obtained on such lands.[182] Compared to the EU, China currently lacks a regulation or requirement on the estimation of GHG emissions from carbon stock changes, particularly for the lands not established as nature reserves or other protected areas,[183] but biofuel feedstock production in China shall comply with the laws on forest management and conservation.[184] Although both the EU and China protect forested land with high carbon stocks, there is a difference between them in defining the scope of such lands. Compared to the available definitions in certain international initiatives (see Table 5), the EU's definition of 'continuously forested land' is relatively narrower regarding the indicators of area, canopy, and height, and it does not include land that is predominantly under agricultural or urban land use,[185] while China defines 'forested land' as including forests/trees with canopy no less than 0.2 (20 per cent) without setting conditions for their area or height.[186] Moreover, in China, not all of the forest lands are protected from biofuel production: only the *natural* forests identified as nature reserves and the forests with important ecological and social functions have been explicitly protected from being used for biofuel feedstock cultivation;[187] and it is unclear whether the forestry dedicated for biofuel production falls within the definition of 'forested land' or not.[188] The difference between the EU and China in setting out requirements on carbon stocks could be attributed to the lack of unified/harmonised definitions or a legal framework on 'forested land' at the international level,[189] as well as the

its obligations regarding the monitoring and reporting of GHG emissions under the UNFCCC and the Kyoto Protocol Commission Decision 2010/335/EU; Decision No 529/2013/EU, Art. 1; Regulation (EU) No 525/2013; European Commission, Climate Action, LULUCF in the EU, available at: <http://ec.europa.eu/clima/policies/-forests/lulucf/index_en.htm> (last accessed 29 April 2015).

180 Commission Decision 2010/335/EU, paras. (2)-(4) of the preamble.

181 See Chapter 3, section 2.1.1 (a) and section 5.1.

182 Such a status should be recognised on January 2008, and such land no longer has such status. See Chapter 3, section 2.1.1.

183 See Chapter 4, section 3.1.3.2.

184 See Chapter 4, section 3.1.3.2.

185 Communication (2010/C 160/02), section 4.2.

186 See Chapter 4, section 3.1.3.2.

187 See Chapter 4, section 3.1.3.1.

188 The State Forestry Administration has issued guidelines regarding the sustainable cultivation of energy forestry, which implies that such energy forestry may also be regarded as forested land if it meets the definition set out in the law.

189 'Forested land' has been respectively defined by the UNFCCC, CBD, and FAO, but there is not yet a uniform definition or legal framework on forests. See FAO, Comparative framework and Options for harmonisation of definitions, available at: <www.fao.org/docrep/005/-y4171e/y4171e10.htm> (last accessed 23 November 2014); The Need for the Review of UNFCCC's Forest Related Terms, Definitions, and Classifications, Civil Society Submission to the SBSTTA meeting of the Convention on Biological Diversity, May 2010, →

differentiated commitments regarding carbon stock changes under the Kyoto Protocol due to the complexity and sensitivity of undertaking obligations regarding land-use matters.[190]

Table 5. Comparison between the definitions of forested area by EU RED, China, UNFCCC, CBD, and FAO

	Areas (*ha*)	Canopy	Height (*m*)	Young Trees	Agro-forest
EU RED	1	30%	5	1 *ha*, 10-30%, 5 *m*	No
China	-	20%	–	–	–
UNFCCC	0.05-0.1	10-30%	2-5	Young natural stands forming part of forest area	–
CBD	0.5	10%	–	Where tree growth is climatically suppressed, the trees reach a canopy of more than 10% and height of 5 *m*	Exclude trees primarily under agricultural land use
FAO	0.5	10%	5	Young trees expected to reach 10% canopy and 5 *m* height	Include forest stands on agricultural lands, but exclude trees primarily for agricultural production or planted for agro-forestry systems

3.2.2. Similarities

Although China does not include the carbon stock impact in biofuel regulation, the Chinese government has already been engaged in compiling technical standards and methodologies for accounting and monitoring forest carbon stocks.[191] Provisions on forestry carbon sinks, carbon sink accounting and monitoring, and forestry carbon trading are proposed to be included in the amendment to Law on

<www.wetlands.org/Portals/0/publications/submission%20or%20policy%20doc/Annex%20on%20the%20Need%20for%20the%20Review%20of%20the%20UNFCCC.pdf> (last accessed 23 November 2014).

190 For instance, the progress is land-related LULUCF and agriculture rule-making under the UNFCCC is slower than other sectors. Firstly, there are greater technical challenges and uncertainties associated with changes in GHG emissions and removals from LULUCF and agriculture if compared with those of fossil fuel emissions. Secondly, LULUCF and agriculture play important but different roles in various economies: for example agriculture is viewed by China as a sovereign right linked with food security and therefore China is reluctant to open up this sector to any perceived control by an international body. Murphy, De Vit & Nolet 2009, p. 15.

191 See Chapter 4, section 3.1.3.2.

Forests.[192] These measures in forest carbon stock accounting are in response to China's commitments to climate change mitigation by increasing carbon sinks[193] and constructing a carbon market that includes emissions trading on carbon stocks from forests and land use changes,[194] from which the integration effect of the international climate law regime on carbon stock monitoring and trading could be observed. Moreover, China converges with the EU in protecting wetlands and natural forest areas from reclamation activities including the cultivation of biofuel feedstocks.

3.3. *Regulation Addressing the Local Environment Concern*

Apart from their impact on biodiversity and carbon stocks, biofuels may result in adverse effects on the local environment in the process of feedstock cultivation. As biofuel feedstocks are currently mainly sourced from agricultural crops, the adverse effect is reflected in the pollution of air, soil, and water in agricultural practices. Moreover, as wastes provide for another important source of biofuel feedstocks, although their use for biofuels may help waste recycling and recovery, the waste *per se* may have adverse effects on the environment and may need to be prevented and well managed. Thus, this sub-section will compare the regulation addressing the potential adverse effects and local environmental impacts caused by biofuel feedstock production.

3.3.1. Differences

Rules on the protection of the local environment are usually provided for in domestic law, unless transboundary pollution occurs[195] and movement of waste may require transboundary regulation. At the international level, conventions have been reached on the transboundary movement of *hazardous* wastes: the Basel

192 *Ibid.*

193 China undertakes the commitments of increasing forested land areas, forest coverage rate, and growing stock (森林蓄积量) by afforestation. The 12th Five Year Work Programme on GHG Emission Control 2011.

194 It is stated that these efforts could be particularly associated with potential benefits brought about by CDM afforestration and reforestation projects, REDD+, and negotiations on LULUCF under the UNFCCC. The 12th Five Year Work Programme on GHG Emission Control 2011.

195 The international legal framework addressing water, air, or soil pollution is to address cross-border pollution. For example, the two international instruments on water – the UNECE 1992 Convention on the Protection and Use of Transboundary Watercourses and International Lakes and the UN 1997 Convention on the Law of the Non-navigational Uses of International Watercourses –both concern the equitable and no harm utilisation of *transboundary* or *international waters*. The inter-state binding instruments on air pollutions are focused on *transfrontier* atmospheric pollutants such as sulfate and nitrogen oxides from fixed or mobile sources, and the instruments are mostly agreed upon at the regional or bilateral level. Soil protection is a cross-cutting issue and has been touched upon in the international binding instruments on desertification, biodiversity, and climate change, such as the UN Convention on Combat Desertification (UNCCD). Kiss & Shelton 2007; Wyatt 2008, p. 166 and 181.

Convention,[196] the Rotterdam Convention,[197] and the Stockholm Convention,[198] but feedstocks from waste for biofuels production, such as waste cooking oil and animal fat, are usually not hazardous wastes and fall outside of the scope of the three international conventions.

The EU's sustainability criteria particularly require 'cross-compliance' with agriculture-related environmental rules adopted by the EU and its Member States.[199] Such agro-environmental rules include provisions at the EU level on the conservation of wild bird, natural habitats and wild flora and fauna, groundwater and surface water, soil quality, plant protection products, and water management and the sustainable use of pesticides, as well as national environmental rules on soil, maintenance and the deterioration of habitat, and water protection and management issues.[200] Moreover, agricultural practices for biofuels have to respect the 'greening' requirements with regard to crop diversification, permanent grassland, and ecological focus areas.[201] While biofuels produced from waste may enjoy favourable treatment,[202] the waste feedstocks have to comply with the EU's established legal framework on waste which prioritises a management hierarchy and the principles of self-sufficiency and proximity.

In comparison, China does not explicitly set out a similar requirement about cross-compliance in the law and policy on biofuels.[203] Although rules are provided on the use of pesticides and fertilisers, soil pollution, habitat conservation, they are

196 The Basel Convention entered into force on 5 May 1992. The Convention controls the transboundary movement and trade in hazardous waste in response to growing concerns of the illegal dumping of hazardous waste in developing countries. The Convention provides categories of waste that are considered hazardous due to reasons of *toxicity, flammability* and others. The principal focus was the elaboration of controls on the 'transboundary' movement of hazardous wastes, that is the movement of such wastes across international frontiers, and the development of criteria for the environmentally sound management of the wastes.

197 The Rotterdam Convention controls the trade of hazardous and harmful chemicals through a list of hazardous and harmful chemicals and through a prior informed consent procedure where states are made aware of potential imports and exports to and from their countries and can decide whether to allow the import or export. The Rotterdam Convention entered into force on 24 February 2004.

198 The Stockholm Convention on Persistent Organic Pollutants (POPs) was adopted in 2001 in response to the urgent need for global action to protect human health and the environment from 'POPs' which are chemicals that are highly toxic, persistent, bioaccumulate and move long distances in the environment. The Convention seeks the circular economy, elimination or restriction of production and use of all intentionally produced POPs (i.e. industrial chemicals and pesticides). It also seeks the continuing minimisation and, where feasible, the ultimate elimination of releases of unintentionally produced POPs such as dioxins and furans. Stockpiles must be managed and disposed of in a safe, efficient and environmentally sound manner.

199 See Chapter 3, section 2.1.4 and section 3.1.

200 See Chapter 3, section 2.1.4 and section 3.1.1.

201 See Chapter 3, section 2.1.4 and section 3.1.2.

202 For example, only the criterion of the GHG emission saving requirement applies to biofuels from waste feedstocks, and in emission saving calculations biofuels from waste feedstocks can use default values instead of actual values. See Chapter 3, section 3.2.

203 See Chapter 4, section 3.1.3.3.

mostly provided in vague language and may have problems in their effective implementation[204] - for example, while the major responsibility to guide farmers in the use of fertilizers and pesticides falls on local authorities, insufficient rules or standards are adopted to oblige and guide local officials to do so.[205] Such a gap in agricultural environment protection would raise serious concern about the environmental impact of the cultivation of crops/plants for biofuels, even for the non-grain crops grown on marginal lands, since the use of low-quality land may intensify the use of fertilisers or pesticides.[206] China also encourages biofuels to be produced from waste, particularly waste cooking oil and animal fat, in order to solve the food safety crisis caused by 'gutter oil.'[207] China has adopted a law on solid waste pollution prevention, and waste feedstocks for biofuel production also need to comply with the rules in the law,[208] but there are no similar provisions on waste management hierarchy or the principles of proximity and self-sufficiency.

The difference between the EU and China should primarily be attributed to China's insufficient emphasis on the prevention of environmental impacts in agricultural practices. China has traditionally stressed preventing pollution from industrial sources, while pollution caused by agricultural activities has for a long time been neglected in China. The current environmental protection mechanisms, such as *inter alia* environmental impact assessments, three simultaneous process (*san tong shi*), discharge standards and permits, have difficulty in addressing non-point pollution from agriculture, and rules or standards on certain environmental problems such as soil pollution prevention are either lacking or need to be upgraded.[209]

3.3.2. Similarities

Although the requirement of 'cross-compliance' with environmental law on agriculture is not explicitly set out in China, biofuel feedstock cultivation still needs to comply with the existing rules in relation with agricultural pollution prevention. Such rules are provided generally in the environmental protection law and agriculture law, or specifically in the law on water pollution prevention, water and soil conservation, and nature conservation.[210] Although China currently has gaps in its rules to address agricultural pollution, such an issue has attracted increasing attention of law makers. For example, the Environmental Protection Law as revised in 2014 includes provisions on agro-environment and strengthens the government's responsibility in agricultural pollution prevention.[211] The principle of 'no

204 *Ibid.*

205 It is provided in both the revised Environmental Protection Law and the Law on Agriculture that local government and responsible departments provide guidance on scientific cultivation, scientific utilisation of fertilisers and pesticides. See Chapter 4, section 3.1.3.

206 See Chapter 4, section 3.1.3.3.

207 *Ibid.*

208 See Chapter 4, section 3.2.

209 See Chapter 4, section 3.1.3.3.

210 *Ibid.*

211 Environmental Protection Law of China, Arts. 33 and 49.

environmental damage' might be interpreted as containing the meaning of compliance with the rules in the Environmental Protection Law with regard to agricultural pollution prevention, water pollution prevention, water and soil conservation, and nature conservation areas. The reason for the increasingly convergent emphasis on agro-environment may lie in the fact that China begins to realise the urgency of addressing agricultural pollution: for example, fertiliser and pesticide runoff has been a major cause of soil pollution and erosion.[212]

3.4. *Regulation Addressing the Socio-economic Sustainability Concerns*

Socio-economic sustainability concerns related to biofuels primarily concentrate on the impact on food availability and land use rights (particularly for indigenous peoples). These rights have been largely protected in human rights law which represents a very different regime from the afore-mentioned environment and climate change law. This sub-section compares the requirements with regard to the impact of biofuels on the socio-economic rights.

3.4.1. Differences

At the international level, treaty provisions on socio-economic rights are usually drafted in general language, but the 'General Comments' adopted by the treaty supervising bodies of the treaties may shed light on interpreting or applying socio-economic rights in the case of biofuels. State Parties of ICESCR may violate their obligations regarding the right to food, if they adopt biofuel legislation and policy that manifestly affects the quantity (availability) and price (economic accessibility) of food supply, or if they fail to regulate biofuel-related activities of individuals or groups that infringe the right to food. State Parties undertake an obligation with immediate effect to prevent the production of biofuels from infringing on the minimum essential level of food, and may even have an obligation to do so when the consequence takes place in third countries. Protection of land tenure and access has been particularly provided for the indigenous peoples in Convention 169 agreed under the International Labour Organisation and can be contended as being the basis for enjoying the right to food and housing provided in the ICESCR.[213]

EU's biofuel sustainability criteria do not include socio-economic requirements, although the European Commission and the Member States are required to regularly report on the impact of biofuels on food, land, labour rights, and the wider socio-economic issues in the EU Member States and in third countries.[214] Furthermore, voluntary schemes used for showing compliance with the EU's sustainability criteria may include requirements on socio-economic issues, which could serve as a way of accumulating experiences or a 'back-door' to incorporate socio-economic sustainability standards.[215] However, the analysis in

212 Zhang & Crooks 2012, p. 47-51.
213 See Chapter 2, section 3.2.
214 See Chapter 3, section 2.1.5.1.
215 See Chapter 2, section 2.1.5.2.

Chapter 3 shows the limitation of using monitoring/reporting and voluntary scheme approaches in preventing negative socio-economic impacts.[216] Moreover, it is noted that the EU requires reports on the ratification and implementation of a series of international labour rights conventions which do not include Convention 169 on indigenous people and do not mention reporting on the ratification and implementation of the ICESCR, although all the 28 EU Member States have ratified the ICESCR and no EU Member State voted against the UN Declaration on the Rights of Indigenous Peoples.[217]

In contrast, the concern about the impact on food availability is prioritised in China's biofuel regulation. It is observed that two of the three general principles for biofuels are purported to prevent biofuels competing with food: non-food grain feedstock and marginal land principles aim to avoid impacts on the staple food supply in China.[218] Moreover, the central government has taken measures to prohibit the new construction of food grain based biofuel installations and to encourage the use of non-food grain raw materials for biofuel production.[219] However, China does not specifically oblige imported biofuels to comply with the principles and requirements, and how the biofuel production with 'one-and-a-half generation' crops and 'marginal land' may affect food and local communities is open to discussion.[220] Nevertheless, it is noted that China has ratified the ICESCR and voted for the UN Declaration on the Rights of Indigenous Peoples, and the State Forestry Administration has required consideration for socio-economic rights of local communities and ethnic groups in the cultivation of energy forestry.[221]

The difference in setting out socio-economic requirements can be firstly attributed to the divergent sustainability focus. For the EU, as a developed economy, the impact of biofuels on food availability is more a consideration 'in particular for people living in developing countries'[222] than a major domestic concern. In contrast, China has a different national socio-economic situation and focus than that of the EU. Food availability is of great importance to China's national security and social stability, considering the large population and limited arable land.[223] Secondly, there are some practical considerations in adopting or

216 See Chapter 2, sections 2.1.5.1 and 2.1.5.2. and 5.1.

217 United Nations Human Rights, Status of Ratification Interactive Dashboard, available at: <http://indicators.ohchr.org/> (last accessed 10 April 2015); United Nations Human Rights, Declaration on the rights of indigenous peoples, <www.ohchr.org/EN/Issues/IPeoples/-Pages/Declaration.aspx> (last accessed 12 April 2015).

218 See Chapter 4, sections 2.1 and 3.1.

219 *Ibid.*

220 Wang 2011, p. 208; Report on the Policies for Biofuel Industry 2014, p. 115-116; see Chapter 4, sections 3.1.1, 3.1.2 and 5.1.

221 See Chapter 4, section 5.3.2.

222 Directive 2009/28/EC, Art. 17(7).

223 According to comments, the Chinese Communist Party has pegged its political future on 'performance legitimacy', which means to justify its hold on power by providing faster growth and higher standards of living than any other form of government. To a certain extent, economic and social stability provides for the Communist Party's 'performance legitimacy' and food price turbulence or food shortage are negative for maintaining social →

applying socio-economic requirements for biofuel production. The EU does not provide socio-economic criteria largely due to considerations about the difficulty in developing social criteria in legal terms and the possibility of being inconsistent with WTO law,[224] so the EU prefers the approach of drawing experiences from monitoring and reporting and voluntary schemes at the moment.[225] China has a principle about non-food grain and marginal land primarily due to the concerns about China's national food security, so there is no similar explicit requirement for biofuel production and supply taking place in third countries.

3.4.2. Similarities

Convergence is observed in the increasing emphasis on the impact of biofuels on food availability. At the international level, the impact of biofuels on food availability has been addressed by the Special Rapporteur on Rights to Adequate Food on many occasions, and international conferences have been convened by the UN Human Right Council.[226] In the EU, the proportion of food crop-based biofuels for target fulfilment and the subsidies for food crop-based biofuels will be restricted,[227] and China has prioritised the food security concerns and use of non-food based biofuel feedstock.[228] The convergence can be attributed to the reason that the potential impact on food availability is a major concern in relation to adopting biofuel policy and law, as it may not only have domestic effects (stressed by China) but also external or global impacts (a concern at the EU and international level).

3.5. *Interim Conclusion*

The regulation of biofuel feedstocks is compared at international, EU, and Chinese levels along the line of different sustainability concerns. The divergence, convergence, and their underlying reasons can be summarised as follows.

stability. Lin 2012b, p. 306-307; Lieberthal 1995; Laliberte & Lanteigne 2008; see Chapter 4, sections 3.1.1 and 3.1.2.

224 See Chapter 3, section 2.2.

225 See interview with Ignacio Vazquez Larruscain, the Biofuels Policy Officer of European Commission DG CLIMA, on 17 December 2013 in Brussels (Interview script is on file with the author).

226 See Chapter 2, section 3.2.1.

227 See Chapter 3, sections 2.2.1 and 2.2.3.

228 It should be noted that there is a difference in the food concern between EU and China, as the EU is to impose a limit on the contribution of food-based biofuels to target meeting: (1) the 'food' in the EU refers to a wider scope than the 'grain' which primarily refers to the Chinese staple food such as wheat, rice, corn; many crops that can be used as food but not staple food (such as sweet potatoes, sugarcanes, rapeseeds, sweet sorghum) seem to be outside the scope of grain-based feedstocks and belong to the 'energy crops'; (2) while the EU's concern is food availability on a global level or in third countries, China's food concern is about the domestic food security; and (3) the EU's limitation on food-based biofuels is mostly for indirect land use change emission prevention.

3.5.1. Differences

It is found that the international regimes, the EU, and China have different approaches to defining and integrating these concerns into the regulation of biofuels.

Regarding biodiversity, the CBD and Ramsar Convention provide for general commitments about 'sustainable use' or 'wise use' and specific guidance as to the relationship between biofuels and biodiversity. The EU goes a further step to explicitly prescribe and define the types of lands/areas with high biodiversity values as 'no-go areas' for obtaining biofuel feedstocks. China does not prescribe or define the types of highly biodiverse areas prevented from providing for biofuel feedstock production, but the relevant law generally prohibits agricultural activities in the established nature reserve areas and wetlands. However, it is unclear whether and how biodiverse lands/areas outside of the scope of nature reserves can be used for biofuel feedstock production.

As for carbon stocks, based on the international commitments and methods used to calculate GHG inventories, the EU accounts for GHG emissions from direct land use change in estimating the life-cycle GHG emissions of biofuels and meanwhile determines the types of land with high carbon stock values, but indirect land use change emissions are currently not accounted for in its sustainability criteria. In contrast, requirements and methodology for monitoring and accounting for carbon stock changes have as yet not been established in Chinese law, and the protection of areas with high carbon stocks may largely depend on the establishment of nature reserves or specific ecosystems such as wetlands and natural forests.

With regard to the other concerns, while the EU explicitly requires 'cross-compliance' with designated environmental provisions at the EU and national levels, it does not include requirements about socio-economic rights in the sustainability criteria. In contrast, China does not clearly set out requirements regarding agricultural pollution for biofuel production and the agricultural environment protection provisions in law are found to be general and vague, while the emphasis has been on avoiding that biofuels impact on national food security.

The differences could be attributed to the following reasons. Firstly, although international treaties and provisions have been established to address biodiversity, carbon stocks, socio-economic rights, a commonly agreed-upon definition, classification, or quantification of the elements important for identifying sustainable production of biofuel is still lacking, such as definition and understanding of the terms of 'high biodiversity or conservation value', 'sustainable agricultural practices', 'critical ecosystem', 'forest' or 'forested land', 'degraded lands', etc.[229] There is also a lack of consistent and complete methodologies for estimating a biofuel's life-cycle impact on biodiversity, carbon stock changes, and displacement effects caused by indirect land (and other resources) use change.[230] As biofuel

229 CBD/SBSTTA/16/14 2012.

230 *Ibid.*

feedstock regulation is closely associated with agriculture and land use, which are frequently regarded as critical to a state's interests and sovereignty,[231] it is more difficult to reach international agreements on sustainability standards for these issues.

Secondly, the regulation on biofuel feedstock supply involves a range of sectors due to the variety of feedstocks, which is more often than not fragmented according to sectors. Feedstocks for biofuels may come from various resources and locations, involving regulation at least on agriculture, environmental, land management, and perhaps on forestry and waste. Furthermore, the range of sustainability concerns is usually governed by different legal frameworks, such as biodiversity, climate change, forestry, agriculture, or human rights. For example, international regulation relating to biofuel feedstock falls within scope of different treaty regimes; in China, biofuel feedstock regulation involves both the Ministry of Agriculture and the State Forestry Administration.

Thirdly, similar to the difference in demand-side regulation, the divergent emphasis in supply-side regulation might be based on the disparate sustainability focus in sustainable development. The EU and China place an emphasis respectively on environmental and socio-economic aspects largely due to their different development status (respectively as developed and developing countries) and divergent national interests and priorities.

3.5.2. Similarities

In addition to the differences, convergence can still be perceived in the regulation of biofuel feedstocks. For example, there is convergence in protecting biodiversity through the establishment of protected areas where agricultural activities are limited or prohibited. The EU and China adopt similar approaches to preventing the impact of biofuels on food by restricting the use of food crops for biofuel production and meanwhile encouraging the use of feedstocks that have minimal impacts on food and land use. China has made efforts to protect wetlands from reclamation and to pay more attention to preventing agricultural pollution in the revised environmental law, and China has also declared the intention to establish carbon stock accounting and monitoring methodology and systems.

The convergence can be above all attributed to the common sustainability pressure caused by the increased production of biofuels and their feedstocks. With the emerging disputes about the potential negative impacts of biofuels, the biofuel issue has gradually gained attention in the international regimes such as the CBD and the Ramsar and both the EU and China have made policy shifts related to biofuels with increasing restrictions on feedstock production.

[231] For example, on the matter of GHG inventory calculation of the agricultural sector, China is reluctant to open up this sector to any perceived control by an international body since agriculture is viewed by China as a sovereign right linked with food security. In discussions under UNFCCC, representatives from China have noted that China does not want international interference in the agricultural sector because of its importance for food security and rural development and livelihoods. Murphy, De Vit & Nolet 2009, p. 15.

The convergence could be moreover a consequence of the inter-influence between the international and domestic regulation. The treaty provisions and COP resolutions or decisions of the CBD and the Ramsar Convention may have exerted a certain influence on Parties or non-Party states on biodiversity-related matters, such as the establishment of protected areas or nature reserves for biodiversity conservation. Particularly, the EU's sustainability criteria define the terms of 'primary forest', 'wetlands', or 'continuously forested land' with reference to the definitions widely recognised at the international level.[232] Nevertheless, compared to the demand-related measures, fewer effect of international harmonisation/-integration on biofuel feedstocks supply regulation is found,[233] and domestic regulation might in turn have an influence on the international level. It is interesting to notice that the European Commission issued its proposal for adopting sustainability criteria on biofuels in January 2008[234] and in May of the same year the CBD COP adopted a decision on 'biofuels and biodiversity' to include the issue of biofuels in CBD's programme of work;[235] and the CBD decision on identifying a national inventory and 'go' or 'no-go' areas was adopted after the promulgation of EU's sustainability criteria.[236] This convergence, if it is not completely coincidental, may illustrate a 'trans-echelon' influence from domestic legal systems to an international regime.[237]

4. Comparative Analysis of the International Trade of Biofuels

Regulation affecting the trade of biofuels is discussed separately in the thesis because of its close relationship with the international trade regime and the potential effects on third countries.

4.1. *Differences*

At the international level, the WTO provides a regulatory framework for negotiating relevant agreements and settling trade disputes. To date, no special

232 References are made to the FAO's definition of 'primary forest', the IUCN's identification of 'rare, threatened or endangered ecosystems or species', and the Ramsar Convention's definition of wetlands. The definition and classification of land with high carbon stocks are generally in line with the methods commonly adopted by the FAO, UNFCCC, and CBD. See Chapter 2, sections 2.1.2 and 2.1.3.

233 As analysed in Chapter 2, compared to the biodiversity treaties, the international climate change law and decisions adopted by its COP may have a stronger influence on Parties particularly the Annex-I Parties. In effect, although the CBD COP adopted decisions to urge the identification of 'no-go' areas for biofuel production and require reporting on such issues, China has not followed suit except for the already established nature reserves. See Chapter 2, section 3.1.3; Chapter 4, section 3.1.3.1.

234 Commission Proposal COM(2008) 19 final.

235 Decisions IX/2 on 'biofuel and biodiversity' was adopted at the ninth meeting of CBD COP on 19 – 30 May 2008 in Bonn, Germany. See Chapter 2, section 3.1.2.1.

236 The RED was adopted in 2009, and decision X/37 that contains the national biodiversity inventory and 'no-go' areas was adopted in 2010 by the CBD COP.

237 See verticle 'trans-echelon' borrowing in Wiener 2001, p. 1298-1299.

trade agreement has been reached regarding biofuels, but trade disputes about domestic regulations on the sustainability of biofuels have arisen for further settlement under the WTO.[238] The essential point of disputes lies in whether the domestic regulation on the sustainability of biofuels is consistent with the articles relating to equal treatment in WTO law.[239] The answer to the question may largely depend on how the relevant WTO rules are interpreted, for instance Article I, III, and XX of GATT 1994, as well as how domestic regulation affecting trade of biofuels is formulated. The challenges for justifying the regulation of the sustainability of biofuels may particularly lie in identifying whether biofuels with different production methods are 'like' products and whether the foreign source of origin results in less favourable treatment. It is as well disputable as for whether the sustainability requirement falls within the scope of a general exception, particularly with regard to jurisdiction limitation and the proportionality between the trade restriction effect and the claimed goals. Socio-economic sustainability requirements, however, are unlikely to be justified by general exceptions. In this way, WTO law may constitute a limit on the application of unilateral sustainability requirements to biofuels from third countries.

Against this backdrop, the EU and China adopt divergent regulatory approaches relating to the trade of biofuels. The EU applies its biofuel sustainability criteria irrespective of the origin of the biofuel, which means that biofuels or biofuel feedstocks produced in third countries shall comply with the sustainability criteria. This measure has particularly given rise to trade disputes and legal discussions regarding its consistency with WTO law.[240] The EU's sustainability criteria may not be in line with WTO law, focal questions of which are concentrated on possibly less favourable treatment due to the origin of the biofuel, the designation of GHG emission saving threshold and default values, and the extraterritorial effect.[241] In contrast, although China is becoming an importer of biofuels or biofuel feedstocks from other countries, China does not apply its domestic non-grain, marginal land, and no environmental damage principles to biofuels produced in or imported from third countries.[242]

The difference between the EU and China in applying their domestic sustainability requirements to imported biofuels can be attributed to the disparate objectives of regulating biofuels. The EU adopted sustainability criteria for biofuels for a two-fold purpose – the functioning of the internal market and a high-level of environmental protection,[243] and this two-fold objective may underlie as well the application of the sustainability criteria to the imported biofuels. As for the market

238 See Chapter 2, section 4.

239 Still, regulations relating to the trade of biofuels need to comply with the other rules formulated under WTO, such as tariff classification (as agricultural, industrial, or environmental goods), subsidy prohibition and restriction (e.g. subsidies and countervailing measures). See Chapter 2, section 4.1.

240 See Chapter 3, section 4.1.

241 *Ibid.*

242 See Chapter 4, section 4.

243 European Commission Proposal COM(2008) 19 final.

concern, it was estimated that the renewable energy targets need to be achieved by a combination of domestic and imported energy products,[244] so the EU's domestic biofuel producers would be at a competitive disadvantage due to the higher sustainability requirement, if the sustainability criteria were not applied to biofuels originating from third countries. In terms of environmental protection, the sustainability criteria would not be able to achieve their purpose if they were not applied to imported biofuels, because the negative impacts are more likely to have negative consequences in developing countries.[245] Moreover, the EU aims to have a high level of environmental protection, which has been written as a principle of EU law,[246] and sustainable development has an external aspect for the EU.[247] In contrast, China sets the basic principles for the development of biofuels primarily due to concerns about domestic food security and environment protection. Moreover, China may not have the same concerns about market competition, bearing in mind that biofuels are mainly promoted and deployed through blend mandate pilot projects characterised by controlled production and consumption through the several designated state-owned enterprises and oil companies,[248] and in fact it is the state-owned enterprises that are the major importers of biofuels and biofuel feedstocks,[249] so market competition may hardly become a major concern for regulators. Moreover, China applied the 'non-intervention' policy in external relations,[250] and application of Chinese socio-economic or environmental principles to producing activities in third countries might be argued as inconsistent with this policy.

4.2. *Similarities*

Although the EU applies the sustainability criteria to imported biofuels, it still endeavours to avoid inconsistencies with WTO law. It is noted that the EU does not adopt binding socio-economic criteria for biofuels and does not apply the criterion regarding the agricultural environment to imported biofuels, and one underlying

244 Directive 2009/28/EC, para. (16) of the preamble.

245 The concern about negative consequence when increased production of biofuels is incentivised worldwide is explicitly stated. Directive 2009/28/EC, paras. (69) and (74) of the preamble.

246 TEU, Art. 3(3).

247 Sustainable development is established as one of objectives in external relations and actions in Arts. 3(5) and 21(2) in the Treaty on European Union. See Chapter 3, section 5.3.

248 Although biodiesel is dominated by private enterprises, the supply of oil as well as diesel and petrol and their blending with biofuels are controlled by the state owned oil companies, which is one reason for a lawsuit brought by a biodiesel production company against Sinopec because the latter refused to purchase the produced biodiesel. See Chapter 4, section 2.2.1.

249 It has been reported that in 2014 Sinopec imported a shipment of bioethanol from the US. Henan Tian Guan Group – one of the largest state-owned bioethanol plant – entered into contract with the Laos government to lease land for producing cassava-based bioethanol which would be then transported to China. See discussion in Chapter 4, section 4.

250 See Chapter 4, section 4.

reason is the attempt to be compatible with WTO law.[251] Furthermore, the EU and China have both taken measures to participate in bilateral or multilateral cooperation on biofuels,[252] which could contribute to the development of international consensus and standards on the sustainability of biofuels and mitigate the possibility of contravening WTO law. The reason for these convergences might be attributed to the harmonising effect of WTO law. It has been commented that the WTO has constructed a stronger enforcement mechanism that contains the imposition of trade sanctions when legal obligations are violated, in contrast with the rather 'aspirational' commitments contained in multilateral environmental agreements.[253] While limiting 'arbitrary' unilateral trade measures, WTO law encourages international negotiation and cooperation to reduce trade disputes.[254] After all, WTO law does not prohibit regulation for the purpose of sustainability, but it requires the regulation to be conducted in a transparent and cooperative way and thus encourages the development of harmonised international standards. For instance, the Agreement on Technical Barriers to Trade prioritises the establishment or application of international standards through multilateral cooperation and negotiation, and WTO Members shall use international standards as a basis for their technical regulations or play a full part in preparing international standards by international standardisation bodies,[255] which may prompt states to agree on and apply international standards for the sustainability of biofuels.[256]

[251] The EU may expect to make use of internationally or locally developed certifications as well as bilateral or multilateral agreements to develop these sustainability standards, which can reduce the possibility of not complying with WTO law. See Chapter 3, section 2.1.5.

[252] For example, the EU and China have joined the International Forum on Biofuels and the Global Bio-Energy Partnership; and the EU and China have bilateral cooperation in a bioethanol promotion project. See Chapter 3, section 4.2; Chapter 4, section 4.

[253] Yang & Percival 2009, p. 657.

[254] The WTO jurisprudence has highlighted some of the circumstances which may assist in demonstrating compatibility with the chapeau, including coordination and cooperation at the international level, Agreement on Technical Barriers to Trade, Arts. 2.4-2.6. See Chapter 2, section 4.2.

[255] Agreement on Technical Barriers to Trade, Arts. 2.4-2.6.

[256] It is noticed that the International Organization for Standardization (ISO), the world's largest non-governmental international standards developer comprised of members from 162 national standards bodies, is working on sustainability criteria for the production, supply chain and application of bioenergy (ISO/CD 13065) within the project committee 248 since 2009. There are 36 participating countries, including standardisation bodies from China, the US, many EU countries (Denmark, Finland, France, Germany, Italy, Netherlands, Spain, Sweden, and UK) and nine observing countries (such as Belgium). The project committee 248 consists of four subcommittees/working groups respectively in charge of cross-cutting issues (including terminology and verification and audit), greenhouse gases, environmental, economic and social aspects, and indirect effects, which are under development and are to be published by 30 April 2015. These ISO standards are of a voluntary nature, depending on their adoption by States or enterprises. The ISO develops international standards, but it is not involved in the certification of any of the standards it develops – certification is performed by external certification bodies, which are largely private. ISO, SO/PC 248 Sustainability criteria for bioenergy, available at: <www.iso.org/iso/home/standards_development/list_of_iso_-
→

5. Conclusion

This chapter compared the regulation at international, EU, and Chinese levels, and identified the differences and similarities as well as their underlying reasons, as summarised below in section 5.1. Section 5.2 reflects on the comparative observations in line with the research questions of this thesis, concluding that the international, EU, and Chinese regulation have limits to various extents in including the environmental and socio-economic sustainability concerns, which may have the consequence that biofuels with adverse environmental or socio-economic effects continue to be produced and consumed and even be labelled as 'sustainable.' Given the inherent complexity of regulating the sustainability of biofuels by either international or domestic law, a global inclusive approach is explored based on implications from the comparative analysis and the conceptual framework of 'global environmental law.'

5.1. *Summary of the Main Findings*

Section 2 concentrates on the differences and similarities in regulation relating to the demand for biofuels. It is particularly observed that there are differences in whether and what sustainability requirements are established for the demand-side regulatory measures. Demand for biofuels is associated with the international commitments under the UNFCCC and the Kyoto Protocol, but there is a lack of corresponding sustainability requirements for the created or incentivised demand of biofuels. Domestically, the EU and China contrast with each other in respectively emphasising the environmental impacts or socio-economic concerns and also in whether their requirements are explicitly set as preconditions for measures relating to the demand for biofuels. The primary reason for divergent sustainability requirements can be attributed to the different focus of the sustainability measures, a lack of commonly agreed-upon definitions of 'sustainable biofuels' or a regulatory framework for it, as well as the conventional regulatory approach for promoting biofuels. Meanwhile, it is observed that convergence occurs in the common concern about the potential negative impacts of biofuels, in particular regarding food availability. The convergence can be attributed to the influence from international level and between the EU and China, due to the similar food pressure that would be caused by increased biofuel consumption.

Section 3 found that there are differences as well as convergences in defining and regulating the biodiversity, carbon stocks, and socio-economic right concerns about biofuels. For instance, the CBD and Ramsar Convention provide general commitments about 'sustainable use' or 'wise use' and specific guidance for biofuels and biodiversity. The EU goes a further step to explicitly prescribe and define the types of lands/areas with high biodiversity values as 'no-go areas' for obtaining biofuel feedstocks. China does not prescribe or define the types of highly biodiverse

technical_committees/iso_technical_committee.htm?commid=598379> (last accessed 10 April 2015).

areas prevented from being used for biofuel feedstock production, but the law generally prohibits agricultural activities in the established nature reserve areas and wetlands. On the other hand, the Chinese government has taken measures to forbid the further expansion of biofuel production based on grains used as staple food in China and called for the cultivation of biofuel feedstocks on marginal lands. The divergences can be attributed to the difference and even deficiency in defining key terms and developing assessment methodologies for evaluating the sustainable production of biofuels, gaps in legislation or the implementation of environmental rules, and different prioritised sustainability concerns. Meanwhile, convergence exists in the approach to protecting certain areas with biodiversity values from biofuel production in all the international, EU, and Chinese cases, which is owing to the influence from the existing international treaty regimes as well as the common response to the potential adverse effects of increasing the production of biofuels.

Section 4 compares the regulation associated with the trade of biofuels, particularly with regard to the sustainability requirement set out for the biofuel (feedstock) imports. In the context that WTO law may limit the application of unilateral sustainability requirements to biofuel (feedstock) of foreign origin, the EU and China have different approaches to treating biofuel (feedstock) from third countries. The EU applies its sustainability criteria irrespective of the origin of the biofuel, which has given rise to trade disputes and legal discussions regarding its consistency with WTO law. In contrast, China does not set out its domestic biofuel development principles as requirements for biofuels from third countries. This difference could be attributed to the disparate objectives of regulating biofuels as well as different stances in external relations policy. Nevertheless, as the EU and China are both Members of the WTO, the obligation to comply with the WTO agreements may underlie the convergence that neither the EU nor China applies binding socio-economic or agro-environment requirements for biofuels from third countries. Moreover, both the EU and China have taken measures to engage in bilateral or multilateral cooperation on biofuels, which could also be influenced by the WTO law in promoting international consultation and cooperation.

5.2. *Reflection*

This section will reflect on the differences and similarities as observed in the regulation of the demand, supply, and trade of biofuels in view of the research questions of this thesis: whether the regulation is inclusive addressing both environmental and socio-economic sustainability concerns. Moreover, whether 'global environmental law' offers the prospect for an inclusive approach to regulating sustainability of biofuels.

5.2.1. Current Differences and Deficiencies in Inclusive Regulation

Sustainability has the essential meaning of integrating environmental, social, and economic considerations, and whether the three dimensions are inclusively taken into account is essential for achieving sustainability and sustainable development.[257] In the same vein, whether regulation on biofuels integrate a consideration of the impacts on the three dimensions is crucial for guaranteeing the sustainability of biofuels. The comparative analysis in the previous sections demonstrates different extents of inclusive environmental and socio-environmental considerations at the international, EU, and Chinese levels.

Above all, a lack of an integrated regulatory framework for developing and applying binding sustainability requirements on biofuels at international level was observed.[258] Although initiatives have been taken to address the sustainability of biofuels under the existing international legal regimes, sustainability concerns are addressed in a fragmented manner under different international regimes which are limited by their objectives and functions. For example, while specific decisions or resolutions on biofuels have been adopted under the CBD and the Ramsar Convention, they can hardly touch upon the issue of GHG emissions; meanwhile, the international climate regime is considered to be single-minded in pursuing climate change mitigation which might conflict with biodiversity or human rights.

In the absence of a harmonised international regulatory framework on the sustainability of biofuels, the EU and China contrast with each other in what sustainability concerns they address and prioritise. It is observed that the EU's sustainability criteria purely consist of requirements in environment and climate change aspects, while socio-economic concerns are not included in the binding criteria.[259] In comparison, China's primary sustainability concern is the impact of biofuels on national food security, which is largely reflected in the principles or requirements about non-food grain feedstocks and marginal lands;[260] while from the environmental perspective, the life-cycle GHG emission saving requirement lacks. There is an observation about the convergence of certain sustainability requirements, such as non-food feedstocks and the protection of nature reserves and wetlands, but differences lie in the scope of the 'food' and protected areas. For instance, the European Commission proposed that food-based crops include cereal and other starch-rich crops, sugars and oil crops,[261] while China still sees certain cereal or sugar crops as promising sources of biofuel feedstocks, such as cassava.[262] Differences also exist in whether the sustainability requirements are applied to biofuels or biofuel feedstocks produced from third countries.

Furthermore, the EU and China allow different degrees of possibilities to adopt complementary or stricter sustainability requirements. The EU's sustainability criteria for biofuels are exhaustively harmonised, meaning that Member States are generally precluded from adopting more stringent criteria unless otherwise provided and EU law leaves only a limited possibility open for Member

257 Gaines 2014, p. 10; see Chapter 1, section 2.2.
258 Chapter 2, section 5.
259 See Chapter 3, section 5.1.
260 See Chapter 4, section 5.1.
261 Commission Proposal COM(2012) 595 final, p. 3.
262 See Chapter 3, section 3.1.1.

States to maintain or introduce more stringent measures.[263] China does not specify whether more stringent requirements could be adopted by local authorities, but an analysis of the relevant provisions in the Renewable Energy Law, the Environment Protection Law, and the Law on Legislation concluded that the competent local authorities may adopt more stringent measures; nevertheless, the problem perhaps is that local authorities may lack the willingness to set more stringent environmental constraints in view of the economic benefits to the local economy.[264]

Despite the differences, it is observed that the limits in inclusive regulation exists on all the international, EU, and Chinese levels. At the international level, an integrated regulatory framework for the sustainability of biofuels is not yet established and the existing treaty regimes are limited to providing inclusive standards beyond their functional scope. The EU's sustainability criteria do not include socio-economic criteria and are limited in taking account of indirect land use change emissions and applying agro-environment requirements to biofuels from third countries, and the approaches of monitoring/reporting and voluntary schemes still have respective limitations in ensuring sustainability.[265] In China, the principle of 'no environment damage' is ambiguous and Chinese environmental legislation has deficiencies in setting out requirements or accounting rules about life-cycle GHG emissions.[266] The divergent and limited sustainability requirements can be attributed to many reasons, as analysed already in this chapter, including the divergent sustainability focus,[267] the lack of an agreed definition of 'sustainable biofuels' and many other key terms,[268] the contrasted historical background and approaches in promoting biofuels,[269] the different extents of the biofuel market integration,[270] the gap in environmental law-making,[271] and the different external policy stances.[272]

263 See Chapter 3, section 5.2.
264 See Chapter 4, section 5.2.
265 See Chapter 3, section 5.1.
266 See Chapter 4, section 5.1.
267 See Chapter 5, section 2.1.
268 See Chapter 5, sections 2.1 and 3.5.
269 For example, the EU uses legislation and target-setting to promote the use of biofuels, while China mandates the blending of biofuels through pilot projects and designated supply. See Chapter 5, section 2.2.
270 The divergent possibility for adopting complementary or stricter sustainability requirements could be associated with the need of market integration for the free movement of biofuels. For example, the EU has an exhaustive level of harmonisation related to the sustainability criteria for biofuels, primarily to ensure the proper functioning of internal market, because different biofuel sustainability criteria would constitute a barrier to the free movement of biofuel products. In order to ensure proper market functioning, the possibility of Member States to adopt additional or stricter sustainability requirements is restricted in order to guarantee a 'level playing field.' In comparison, the market integration for biofuels is not yet evident at the international level or in China. Although biofuels are increasingly internationally traded, the international market may still be minimal. Only roughly 6 per cent of biofuels (ethanol and biodiesel only) were traded internationally in 2009; in China, the proportion of biofuels in the transport fuel mix was still minimal and moreover the biofuel market has been developed in a local pilot approach and thus a national biofuel market has not taken shape. One report states that biofuel production accounts for less than one percent →

Limits in inclusively addressing the sustainability concerns related to biofuels may have the consequence that 'unsustainable' biofuels continue to be demanded, supplied, or traded. In theory, sustainable development calls for an integrated approach to addressing all the social, economic, and environmental impacts, since 'any policy that is not attentive to all three factors will not lead to sustainable development at global scale.'[273] This comment can be applied in the case of regulating the sustainability of biofuels. For example, without including criteria on socio-economic concerns or indirect land use change emissions, it is possible to recognise biofuels that are produced in a way that has adverse socio-economic effects or does not contribute to climate change mitigation as 'sustainable biofuels' in the EU. Once recognised, Member States generally cannot refuse to take such biofuels into account for meeting national renewable energy targets or financial support schemes due to the exhaustive harmonisation. Moreover, given that the EU is a main importer of biofuels,[274] such limited sustainability requirements might have an external effect of allowing or even expanding the production of biofuels with negative environmental or socio-economic impacts.[275] In China, the strong emphasis on the food security concern may overlook the sustainability concerns of biodiversity, carbon stock change, life-cycle GHG emission reductions, and other socio-economic rights.[276] Also, the domestic support for use of biofuels has prompted import or direct investments on biofuels and biofuel feedstocks in the other developing countries.[277] Thus, it is desirable to establish inclusive sustainability requirements to address the environmental and socio-economic impacts based on the best available knowledge.[278]

5.2.2. Challenges for an Inclusive Regulatory Approach

Although inclusive sustainability regulation for biofuels is desirable, a comparative analysis in this chapter revealed the possible challenges to providing inclusive

of China's liquid fuel. Chapter 3, section 5.2; Chum *et al.* 2011, p. 257; IEA 2011, p. 29; Anderson-Sprecher & Jiang 2014.

271 See Chapter 5, section 3.3.

272 See Chapter 5, section 4.1.

273 Gaines 2014, p. 14.

274 Chum *et al.* 2011, p. 252.

275 The exporting countries are mostly developing countries with a high risk of sustainability concerns, such as countries in Latin America and sub-Saharan Africa, while North America, Europe and Asia are expected to be net importers. Chum *et al.* 2011, p. 252; Heinimö & Junginger 2009.

276 For example, the promotion of cassava as a biofuel feedstock, although meeting the current non-grain and marginal land principles, may still cause socio-economic and environmental problems, including affecting the economic rights of farmers, increasing the price of cassava, and agricultural pollution. See Chapter 4, section 5.

277 See Chapter 4, section 4.

278 Sustainable development is in part about adaptability and the flexibility to change policies and institutions to meet changing conditions. This implies that the energy systems of the future should not favor making heavy commitments to non-resilient technologies or projects with long-tailed economic and ecological consequences. Gaines 2014, p. 15.

regulation at the international or domestic level due to the inherent complexities[279] caused by the broad and vague meaning of 'sustainability', integrating cross-sectoral concerns, and regulating multi-scale impacts.

Firstly, the concept of 'sustainability' or 'sustainable development' has a broad and vague meaning, which may result in the inherent complexity in defining the sustainability of biofuels. The notion of the 'sustainability of biofuels' or 'sustainable biofuels' is to apply the meaning, value, and elements of 'sustainability' in the specific case of biofuels. However, 'sustainability' or 'sustainable development' remains broadly and vaguely defined, and its meaning has rarely been clearly set out in either international law or in the domestic law in the EU or China.[280] The commonly cited definition of 'sustainable development' in the Brundtland Report stresses a balance between the 'needs' of humankind and 'limits' of natural resources, which may come about from a compromise between the developed countries' emphasis on environmental and developing countries' call for economic and social progresses.[281] This feature enables the concept of 'sustainable development' to be accepted by both developed and developing countries, but at the same time makes its meaning broad and vague, which may constitute one reason for the complexity of defining the sustainability of biofuels.

Secondly, the sustainability of biofuels is a cross-cutting issue involving multiple sectors and authorities, which increases the difficulty of integrating the sustainability concerns. It is demonstrated in this thesis that a variety of regulatory instruments may affect the demand for, supply, and trade of biofuels, including those about *inter alia* renewable energy, fuel quality, emission trading, taxation, agriculture and forestry, environment, and waste. For example, regulating biofuels may fall within the scope of different international regimes or administrative departments: at the international level, the sustainability of biofuels can be addressed under the CBD, Ramsar Convention, and UNFCCC; and in China, the NDRC and the National Energy Administration are the major governmental authorities for regulating biofuels, but the ministerial departments on agriculture, forestry, finance, taxation, and quality supervision may also take measures if biofuels fall within their administrative scopes. The 'instrument mix' may go beyond one field of law and multiply the embedded complexity of developing or applying regulation on biofuels.[282]

279 The IPCC once pointed out the inherent complexity of biomass resources that makes the assessment of their combined technical potential controversial and difficult to characterise. Chum *et al.* 2011, p. 300.

280 See Chapter 1, section 2.2; Chapter 3, section 5.3; Chapter 4, section 5.3.

281 For example, China has declared that economic development is a right of developing countries and emphasised that solutions to environmental problems can only be found in the progress of development. See Chapter 4, section 5.3.

282 Peeters in one paper pointed out the 'instrument mix' of the administrative complexity of EU climate change legislation and discussed the possibility that the complexity may lead to challenges for governments (primarily Member States) to apply the law which might form an 'instrument mess.' Sorrell and Sijm had also discussed carbon trading in the policy mix and stated that 'a policy mix may easily become a policy mess.' Peeters 2014, p. 173-174; Sorrell & Sijm 2003, p. 434.

Thirdly, the sustainability of biofuels may have impacts of different scales and thus may increase the complexity of regulation at an appropriate level. The feature of involving different scales of impact in the regulation of biofuels has been noted in both the Cramer report[283] and the IPCC report.[284] According to the Cramer report, there are both company-level and macro-level effects of the production of biofuels: some effects of large scale production, such as indirect land use change, can only be assessed well at the macro-level and they primarily fall within the responsibility of the public authorities, while company-level effects can be managed by producers and proved by certification.[285] The IPCC in its 2014 assessment report lists the potential institutional, social, environmental, economic, and technological implications of bioenergy (including biofuels) according to scales of impacts from the local to the global.[286] A comprehensive regulation on biofuel sustainability may involve *global-scale* impacts such as climate change, biodiversity, and food prices, *regional-scale* concerns for environment and trade, *national-scale or subnational-scale* effects of food security, agricultural pollution, and socio-economic rights of local communities, and there can be impacts involving a *hybrid* of the different levels.[287] The impacts of different scales may pose challenges for integrating the sustainability concerns in the regulation of biofuels. The international legal regimes could hardly address the national- or sub-national scale impacts such as agricultural pollution and in many cases still depend on domestic law-making and implementation,[288] while the domestic legal systems of the EU or China may face limits in providing environmental or socio-economic criteria to address global-scale impacts such as GHG emissions from land use changes.[289] The different scales of impact can affect or

283 'Testing framework for sustainable biomass', Final report from the project group 'Sustainable production of biomass', February 2007, available at: <file:///D:/2014-2015%20semester/Thesis/Introduction/070427-Cramer-FinalReport_EN.pdf> (last accessed 13 December 2014).

284 Chum *et al.* 2011, p. 308.

285 The macro-level effects are often effects that cannot be directly attributed to one company but are only visible on a national or regional scale, such as the crowding out of agrarian production or indirect effects due to changes in land use. 'Testing framework for sustainable biomass', Final report from the project group 'Sustainable production of biomass', February 2007, available at: <file:///D:/2014-2015%20semester/Thesis/Introduction/070427-Cramer-FinalReport_EN.pdf> (last accessed 13 December 2014), p. IV.

286 In a 2011 special report on renewable energy, the IPCC also illustrates the impacts and performance of biomass production and use according to region- and site-specific, or region-oriented and global-oriented considerations. Smith *et al.* 2014, p. 884; Chum *et al.* 2011, p. 308.

287 For example, biodiversity includes global and local biodiversity; food security impact may happen on the global level and national level; and food security may also involve implications from the local to global scale. Smith *et al.* 2014, p. 884.

288 For example, the CBD COP decisions and Ramsar COP resolutions lack clear binding effect on state parties and should rely on the establishment of domestic legal or policy frameworks for implementation. See Chapter 2, section 3.1.3.

289 For example, the application of current EU sustainability criteria with regard to GHG emission saving and land use has already given rise to trade disputes under WTO; and the EU does not include indirect land use change emissions in calculating life-cycle GHG emission savings, while indirect land use changes may usually occur somewhere else in the world. See Chapter 3, sections 2.1.1 and 4.1.

even determine the appropriate level of regulation for achieving sustainability.[290] The appropriate level of regulation, in the case of biofuels, may vary according to the scale of impact of the different sustainability concerns. For instance, GHG emissions have a global impact – as emission in any corner of the globe will increase the total emissions; and its accounting method is to a lesser extent susceptible to diverse local environmental or socio-economic circumstances.[291] These features make the harmonised or centralised regulation of GHG emissions possible at a higher level, particularly with regard to the result of emission reductions and accounting methodologies. In comparison, although biodiversity is also regarded as a global common concern,[292] its regulation is more reliant on its translation and implementation by national or sub-national law and on local environmental and socio-economic conditions. Local environmental protection as for the air, soil, and water hinges even more on the diversity of environmental and socio-economic conditions – for example, even in the unitary legal system of China, it is local government that takes the major responsibility of guiding and implementing rules to prevent agricultural pollution.[293] Socio-economic sustainability concerns normally have an impact at the state or sub-state level and are subject to divergent socio-economic circumstances,[294] but the impact on food availability and price stability can vary from the local to the global scale since food grains are traded on the international market and thus become subject to the volatility of the international market.[295]

5.2.3. 'Global Environmental Law' Offers a Prospect?

Implications could be drawn from the observation about the different extents and challenges of integrating inclusive sustainability concerns in the regulation of

290 The need to engage the correct level of governance is important to achieve sustainable development. Gaines 2014, p. 18.

291 Nevertheless, compared to industrial GHG emissions, accounting of land use-related emissions and removals is more difficult and complex and may be more politically sensitive. Murphy, De Vit & Nolet 2009.

292 See the preamble of CBD.

293 See Chapter 4, section 3.1.3.3.

294 The socio-economic rights of local communities may include the right to adequate food, water, housing, land access for indigenous people, and labour rights. Although the food availability impact has a certain global impact, it can also affect the access to food at local levels. According to an analysis by the IPCC, the other socio-economic impacts, such as access to water, land tenure, and labour rights, primarily have implications that vary from local to regional scales. Smith *et al.* 2014, p. 884.

295 In the fifth IPCC Assessment report, it is estimated that the social implications of biofuels include the 'competition with food security including food availability (through reduced food production at the local level), food access (due to price volatility), usage (as food crops can be diverted towards biofuel production), and consequently to food stability' which has implications from the local to the global scale. Moreover, from the economic perspective, biofuels have implications that 'contribute to the changes in prices of feedstock', which also vary from the local to the global scale. Smith *et al.* 2014, p. 884; Ajanovic 2011; Zilberman *et al.* 2012; Mueller, Anderson & Wallington 2011.

biofuels. As current legal systems at the international, EU, and Chinese levels have limits in integrating the inclusive sustainability concerns related to biofuels, a global inclusive approach is needed to overcome the inherent complexities of sustainability regulation. As observed above, international law or unilateral measures alone are currently not able to inclusively integrate sustainability concerns in both the environmental and socio-economic dimensions. Reinventing an international regime is difficult and time-consuming. Domestic sustainability requirements, in pursuit of inclusively addressing socio-economic or local environmental sustainability in the production processes, may meet with the legal constraints of international law (i.e. WTO law and the state sovereignty principle). While it is desirable to build a binding international regulatory framework for the sustainability of biofuels, it is difficult to reconcile divergent national interests between states, particularly between the exporting countries and importing countries,[296] and it is also difficult to address sustainability concerns that go beyond the mandates of an international treaty regime. In contrast, a global approach based on the conceptual framework of 'global environmental law', which combines the conventional domestic or international regulation and draws on the existing convergences,[297] may provide for a way to overcome the inherent complexities and to combine the existing international and domestic efforts.

As mentioned in Chapter 1, the literature has observed the limit of regulating through a binding 'top-down' approach in both international and domestic legal systems, and such a limit is evident in climate law in view of the difficulty in reaching international agreements under the UNFCCC in contrast with the growing 'bottom-up' and 'soft-law' initiatives on a global scale.[298] These observations could also be applied to the case of biofuels: there is difficulty in developing a harmonised legal framework on the sustainability of biofuels, but meanwhile various non-binding initiatives primarily developed by international organisations[299] and non-

296 For instance, the exporting states have argued against applying internal sustainability requirements to biofuels of foreign origin. See Chapter 3, section 4.

297 For instance, setting aside the divergences, convergence can also be observed from the comparison. The convergence lies in the recognising and taking measures to tackle potential impacts on food and the overall environment. Although the EU concentrates on environmental sustainability, food crop-based biofuels will face quantitative limits and the end of subsidy, which converges with China's restrictions on grain-based biofuels. Meanwhile, although China emphasises the food security concern more, it also stipulates 'no environmental destruction' as one principle and nature conservation areas are protected from biofuel production. The convergence might be grounded in either internal or external pressures in the face of potential risks caused by biofuels.

298 See for example Hoffmann 2014; see Chapter 1, section 2.3.

299 'International organisations' in this study refers to intergovernmental organisations founded based on international agreements, with at least one organ with a will of its own and established under international law. According to Klabbers, the primary delimitation lies in the nature of the body of law governing the activities of the organisation, and the three common characteristics are 'created between states'; 'on the basis of a treaty'; 'an organ with a distinct will.' Shermers & Blokker 2011, p. 30, 36-37; Klabbers 2009, p. 7-12.

state actors[300] are emerging,[301] which may have the merit of avoiding negotiating for states' consent and enhancing the participation of various stakeholders. Taking account of observations from comparative analysis as well as the trend in literature, the current thesis uses the conceptual framework of 'global environmental law' as defined by Yang and Percival to explore the potential for a global inclusive approach, because the argument for the blurred boundary between national law and international law may coincide with the cross-sectoral and multi-scale complexity in the regulation of the sustainability of biofuels.[302] Moreover, instead of merely stressing the importance of polycentric non-binding initiatives, global environmental law also notes the importance of harmonisation, integration, transplantation, and convergence conducted by public authorities, the characteristics of which may fit better with the regulation of biofuels, because rules and policies have played an important role in promoting the demand for biofuels,[303]

300 There is no agreed definition of 'non-state actors.' For example, according to the Max Planck Encyclopedia of Public International Law, the term 'non-State actors is a 'superordinate' concept that encompasses all those actors in international relations that are not States, comprising individuals, entities, organisations and institutions on the global, regional, sub-regional and local levels. These 'non-state actors', in contrast with states which possess full legal capacity, have some form of legal capacity under international law to various degrees. According to the report of the International Law Association, the definition of 'non-State actor' contains three elements: the exclusion of any body comprised of and governed or controlled by states; the actors that actually perform functions in the international arena and have real or potential effects on international law; inclusion in the categories of internationally legally recognised and organised entities. Therefore, the non-state actors defined in the ILA report exclude intergovernmental organisations comprised solely of state members or federal components of a state, and include non-governmental organisations with some form of recognised international status that have formal accreditation or informal relations with the UN bodies and other treaty bodies. The definition by the International Law Association is adopted here to exclude inter-governmental organizations from non-state actors. Wagner 2010; International Law Association the Hague Conference (2010) 'Non State Actors' - First Report of the Committee Non-State Actors in International Law: Aims, Approach and Scope of Project and Legal Issues in International Law Association, available at: <file:///D:/Downloads/non_state_actors_draft_report_2010.pdf> (last accessed 15 April 2014).

301 For instance, the Executive Secretary of the Convention on Biological Diversity has compiled a list of criteria, certification, and indicators for the sustainability assessment of biofuels as well as tools and approaches for sustainable land use, socio-economic issues, and policy-making. Another similar compilation of bioenergy sustainability initiatives is made by the Bioenergy and Food Security (BEFS) approach under the Food and Agriculture Organization (FAO), demonstrating the existing 13 voluntary standards and certification schemes for the sustainability of biofuels. Convention on Biological Diversity, 'Tools and Approaches for Sustainable Use of Biofuels', available at: <https://www.cbd.int/agro/biofuels/tools.shtml> (last accessed 15 April 2014); Food and Agricultural Organization of the United Nations, Bioenergy and Food Security, A compilation of bioenergy sustainability initiatives, available at: <www.fao.org/energy/befs/compilation/en/> (last accessed 15 April 2014).

302 For example, the implementation of EU's biofuel sustainability criteria involves not only the EU law but also international law, national law, as well as private certification schemes. Yang & Percival 2009, p. 616, 623, 664; see Chapter 1, section 2.3.

303 Richardson points out that the biofuel market is dependent on government mandates and tax breaks. Thus, if policy makers really wanted to prevent deforestation, biodiversity loss and →

and sustainability risks are usually rooted in the lack of legislation and the proper enforcement in producing countries which cannot be totally replaced by non-binding initiatives.[304] Hence, the major implications from the concept of 'global environmental law' for the potential of a global inclusive approach to the sustainability of biofuels could be: instead of resorting to a sole 'top-down' or 'bottom-up'/'polycentric' approach, inclusive regulation on the sustainability of biofuels could be built upon the existing harmonisation, integration, transplantation, and convergence effects at different levels according to the scales of impact and subjectivity to diverse environmental or socio-economic conditions. Below, a proposal will be given on how this global inclusive approach could be developed.

5.2.3.1. *Counting Life-cycle GHG Emission of Biofuels: Harmonisation and Integration*

As defined by Yang and Percival, 'harmonisation' and 'integration' both refer to the *joint multi-country* efforts or cooperation to achieve some level of uniformity, which can most easily be seen in the formal international treaties and institutions.[305] Harmonisation in this thesis is reflected in that the UNFCCC and the Kyoto Protocol provide obligations for states to reduce GHG emissions,[306] and the treaty bodies adopt methodologies for accounting and reporting on the reductions.[307] Such harmonisation under the international climate regime has a relationship with

land grabbing, they should just pull the plug on their production incentives. Richardson 2014, p. 205 and 216.

304 De Man commented that global conversion to sustainability standards and full implementation of those principles and criteria will only be feasible once sustainability standards have become part of national legislation and enforcement is guaranteed by well functioning public institutions in the producing countries. De Man 2013, p. 50.

305 Precisely, harmonisation is defined as the adjusting and conforming of disparate requirements of different legal systems to an international system or to each other, which is similar to harmonisation used in EU context because they both have the purpose of reducing differentiation and increasing uniformity for a common objective. However, the difference between the harmonisation defined by Yang and Percival and harmonisation in the EU may lie therein. Yang and Percival define harmonisation in a broad sense as including international standard-setting processes that coordinate (rather than mandate) uniform operating standards with respect to pollution, safety, etc. Yang and Percival define integration as the process of linking, instead of adjusting or conforming, separate legal systems. For example, the emissions trading scheme and clean development mechanism as established under the Kyoto Protocol are identified as the 'most visible driver of regulatory integration.' Yang & Percival 2009, p. 627 and 636, 635-636, 638-640; see Chapter 1, section 2.3.

306 It was observed that for most of the past 20 years the much-maligned United Nations (UN) process that produced the 1992 Framework Convention on Climate Change (UNFCCC), the 1997 Kyoto Protocol and a range of recent agreements intended to move the treaty making forward past 2012 has been the only global climate governance process in play. Hoffmann 2014, p. 433; see Chapter 2, section 2.1.

307 Although the reduction obligations and reporting methods differentiate between developing and developed states, the differentiation is the outcome of adherence to the principle 'common but differentiated responsibilities' provided in Art. 4 of the Convention. See Chapter 2, section 2.2.

biofuels, since the obligation of reducing GHG emissions offers incentives for using biofuels, as long as biofuels are considered to contribute to GHG emission reductions based on the adopted accounting methods.[308] Furthermore, the use of biofuels may also be incentivised by the emission trading and offsetting mechanisms arranged under the Kyoto Protocol if the approved monitoring methodologies could show GHG emission reductions by using biofuels. Due to these connections, it is desirable and viable that the treaties and institutions under the UNFCCC provide harmonised and integrated rules or guidelines for the accounting for the life-cycle GHG emissions from biofuels for the following reasons.

Firstly, more attention to specifying or clarifying the accounting for the life-cycle GHG emissions from biofuels is desirable, since the current methodology based on the IPCC Guidelines for the estimation GHG inventories may not guarantee the complete accounting of emissions from biofuels.[309] The methodology under the UNFCCC is meant to provide a complete assessment method covering all emission sources worldwide, but its application based on divergent reporting obligations and methods may dismiss a large portion of GHG emissions,[310] particularly those caused by indirect land use changes.[311] Moreover, the arrangement of excluding the carbon emissions related to the use of biofuels from inventory totals accounting might be misapplied on the regional or state levels, if GHG emission from biofuel's consumption is assumed to be zero without ensuring a complete accounting for the land use change emissions.[312] Although the IPCC has denied the carbon neutral assumption[313] and stated that there is uncertainty in GHG emission reduction benefits from the use of biofuels,[314] states could be more

308 See Chapter 2, section 2.1.

309 See Chapter 2, section 2.3.

310 Zahar observes that a large proportion of global anthropogenic emissions is not reported or independently reviewed, and there is a possibility of in-built biases in national systems that result in greenhouse gas inventories favouring the economic interests of the reporting states. Zahar 2010a, p. 424; Zahar 2010b, p. 322.

311 As it is admitted by the IPCC, growing crops for biofuels can result in land use changes indirectly by driving land use changes elsewhere, and even though the GHG emissions from land-use changes within one country are reported in the AFOLU sector, GHG emissions from the use of biomass and indirect land use changes can occur in another country where the biomass is transported or cultivated across borders. See Chapter 2, section 2.3; IPCC, Task Force on Greenhouse Gases Inventories, FAQs, Q2-10, available at: <www.ipcc-nggip.iges.or.jp/faq/faq.html> (last accessed 16 October 2014).

312 For example, the EU attributes a zero emissions factor to biofuels, but it does not include indirect land use emissions in the accounting of life-cycle GHG emission savings. In China, there is a general assumption that forestry or agricultural bioenergy does not increase carbon emissions in the life cycle of biomass. See Chapter 3, section 2.1.1, and Chapter 4, section 5.1.

313 IPCC, Task Force on Greenhouse Gases Inventories, FAQs, Q2-10, available at: <www.ipcc-nggip.iges.or.jp/faq/faq.html> (last accessed 16 October 2014).

314 The negative impacts of biofuels and the importance of sustainable feedstock supplies and conversion path are increasingly emphasised. For example, the Working Group II on the adaptation to and impact of climate change mentions that land-use changes to produce bioenergy can affect food crops, biodiversity, and ecosystem services; the Working Group III on the mitigation of climate change concludes that bioenergy can play a critical role for mitigation, while there are issues to consider such as the sustainability of practices and the →

informed about such clarifications or assisted by more specific accounting guidelines about biofuels.

Secondly, to clarify life-cycle emissions accounting for biofuels under the international climate regime would be desirable due to the objectives enshrined in the UNFCCC. To stabilise GHG concentrations in the atmosphere is the ultimate objective of the UNFCCC,[315] and the unsustainable production of biofuels may even give rise to increasing GHG emissions[316] which contravenes the ultimate objective. Furthermore, the UNFCCC and the Kyoto Protocol respectively provide that the COP of the UNFCCC and the COP/CMP of the Kyoto Protocol have competence in promoting and guiding the development and periodic refinement of GHG emissions accounting methodologies.[317]

Thirdly, compared to the other sustainability concerns, GHG emissions have a stronger global feature as 'emissions anywhere are essentially emissions everywhere',[318] so an international approach fits. Moreover, the methodology of assessing GHG emissions can be less affected by the diversity of environmental conditions and socio-economic values, and it is thus more viable to provide for a harmonised accounting methodology at the international level.

Fourthly, the integration effect of emission trading and offsetting under the Kyoto Protocol could be used. Although the amount of GHG offsetting credits generated from biofuel CDM projects is currently minimal, there are numerous bioenergy projects which may use the same biomass materials and thus have some similar potential environmental or socio-economic impacts. Meanwhile, as biofuels are increasingly used in the aviation sector, the inclusion of aviation in regional or national emission trading schemes or perhaps a global market-based mechanism[319] would meanwhile pose challenges for ensuring the sustainability of biofuels. In this view, it is desirable that the international legal framework which links or governs emissions trading and offsetting may provide any harmonised life-cycle GHG emissions accounting method for biofuels.

Admittedly, there is much uncertainty as to whether a clarified or specified life-cycle GHG emissions accounting method for biofuels can be worked out under the UNFCCC. Above all, the progress of multilateral negotiation and treaty-making under the UNFCCC has been slow, while a decentralised and bottom-up approach

efficiency of bioenergy and impacts on environmental and socio-economic objectives and concerns other than climate change. Field *et al.* 2014, p. 50 and 91; Smith *et al.* 2014; Edenhofer *et al.* 2014b, p. 77-78 and 89; IPCC, Task Force on Greenhouse Gases Inventories, FAQs, Q2-10, available at: <www.ipcc-nggip.iges.or.jp/faq/faq.html> (last accessed 16 October 2014).

315 UNFCCC, Art. 2; Kyoto Protocol, preamble.

316 Searchinger *et al.* 2008; Fargione *et al.* 2008; Kendall, Chang & Sharpe 2009, p. 7143.

317 UNFCCC, Art. 7(2)(d); Kyoto Protocol, Art. 5(2); see Chapter 2, section 2.2.

318 Hoffmann 2014, p. 434.

319 Within the ICAO, the Environment Advisory Group (EAG) will oversee the work related to the development of the global scheme, which will be principally undertaken by a Global Market-based Measure Technical Task Force (GMTF), comprised of representatives and experts from ICAO member states, industry and NGOs. 'ICAO groups start work on developing a global market-based measure for international aviation emissions', available at: <www.greenaironline.com/news.php?viewStory=1836> (last accessed 17 March 2015).

that engages diverse actors at multiple levels has gained momentum.[320] For instance, Keohane and Victor depict the current climate governance as 'regime complex', and they argue that a climate change regime complex, if it meets specified criteria, has the advantages of allowing for adaptability and flexibility.[321] Cole, drawing on the theoretical framework of polycentric governance, argued for polycentric climate governance which aims to address the classic collective action problem of climate change mitigation at multiple scales and levels, thus allowing for greater experimentation, learning, and cross-influence among different - either independent or interdependent - levels and units of governments.[322] Moreover, biofuels and their GHG emissions have never been a central issue on the agenda of the UNFCCC or the Kyoto Protocol and no COP or COP/CMP decision has been adopted on biofuels. On top of that, although GHG emission assessment has a larger degree of homogeneity compared to the other sustainability concerns, the life-cycle assessment involves emissions from land use and land use changes which is a more complex and sensitive issue to be agreed on among states.

However, the afore-mentioned considerations should not deny the need for a more common understanding/clarification on life-cycle GHG emissions accounting

320 Hoffmann observed that there are now two fundamentally different models of climate governance: the traditional, centralised, top-down UN process and a decentralised and bottom-up mechanism that engages diverse actors at multiple levels. The latter, the so-called 'alternative governance mechanisms', may include arguments for climate governance experiments and transnational climate governance, and focuses on subnational and municipal networks, NGOs, corporate alliance, private sectors, as well as various governmental units). The decentralised and bottom-up mechanisms are eroding the dominance of inter-governmental negotiations and rapidly changing the shape of global climate governance, while climate governance was conceived of as a problem requiring a universal, multilateral response. Hoffmann 2014, p. 433-434.

321 A climate change regime complex is described as 'the elements of this regime complex are linked more or less closely to one another, sometimes conflicting, usually mutually reinforcing.' The regime complex is a continuum between two ends - one end is comprehensive international regulatory institutions focusing on a single integrated legal instrument and another end is a spectrum of highly fragmented arrangements. The criteria are identified as coherence, accountability, determinacy, sustainability, epistemic quality, and fairness. Keohane & Victor 2011, p. 7-23.

322 It was argued that, as the opposite to a thoroughly top-down global regime in which lower levels of government simply carry out the mandates of international negotiators, a polycentric approach provides for opportunities where global institutions and organisations might be improved by learning from various innovative policies instituted by local, state, and regional governments, and any successful governance system for global climate must function as part of a larger, polycentric set of nested institutions and organisations at various governmental levels. A polycentric approach to tackle climate change was also argued for by Elinor Ostrom, who won the 2009 Nobel Prize in Economic Sciencies. Ostrom argued that single policies adopted only on a global scale are unlikely to generate sufficient trust among citizens and firms so that collective action can take place in a comprehensive and transparent manner. Moreover, a single governmental unit to solve global collective action is inherently weak because of free-rider problems. Therefore, a polycentric approach was proposed because it has the main advantage of encouraging experimental efforts at multiple levels with active oversight of local, regional, and national stakeholders. Cole 2011, p. 395; Ostrom 2014, p. 97-134.

for biofuels at the international level, as this thesis already shows the limit of addressing such an issue by the EU's unilateral approach. Furthermore, although international climate governance faces difficulty and the trend of decentralisation, the recent climate negotiations since Cancún might indicate a return to collective emission reduction goals – states have agreed to launch a new commitment period under the Kyoto Protocol in which the important legal and accounting models of the Kyoto Protocol will remain in place.[323] Furthermore, although the bottom-up and decentralised approach has many merits, there is concern about how the scale-up of such polycentric initiatives can achieve an actual effective solution in the absence of centralised and binding governmental measures,[324] and, as already mentioned in this thesis, certification schemes have limits in monitoring and evaluating macro-level impact such as indirect land use changes.[325] Actually, as studies on life-cycle GHG emissions from biofuels in aviation are already undertaken under the ICAO,[326] it could be expected that the rising importance of biofuels in aviation emission mitigation, as well as the existing cooperation and communication between the ICAO and UNFCCC,[327] might provide an opportunity to put the biofuel emissions accounting issue onto agenda at the international level. Moreover, although many voluntary initiatives have addressed the life-cycle GHG emissions of biofuels, they still largely refer to the methodologies provided by the IPCC,[328] since the IPCC is an inter-governmental organisation gaining epistemic

323 At Doha COP 18, a decision was adopted that strengthened the resolve and set out a timetable to adopt a universal climate agreement by 2015, thereby ensuring that this treaty's important legal and accounting models remain in place and underlining the principle that developed countries lead mandated action to cut greenhouse gas emissions. Hoffmann 2014, p. 437-438; UNFCCC, The Doha Climate Gateway, available at: <http://unfccc.int/-key_steps/doha_climate_gateway/items/7389.php> (last accessed 18 March 2015).

324 Hoffmann observed that, on the one hand, climate governance experiments are operating on many levels and addressing multiple aspects of the problem; on the other hand, climate governance experiments are small in scale and there is no central purpose driving their collective activities and no oversight exercised over the collection of experiments. Hoffmann 2014, p. 447.

325 'Testing framework for sustainable biomass', Final report from the project group 'Sustainable production of biomass', February 2007, available at: <file:///D:/2014-2015%20semester/-Thesis/Introduction/070427-Cramer-FinalReport_EN.pdf> (last accessed 13 December 2014), p. IV; see Chapter 3, section 2.1.5; Chapter 5, section 5.2.

326 Under the ICAO, the Alternative Fuel Task Force was established to evaluate the range of potential GHG emission reductions from the use of alternative fuels in aviation. ICAO, Committee on Aviation Environmental Protection (CAEP), AFTF, available at: <www.icao.int/environmental-protection/Documents/CAEP/Images/AFTF-Large.png> (last accessed 12 January 2015); see Chapter 5, section 2.3.1.

327 Cooperation and coordination has been conducted between the UNFCCC and the ICAO. UNFCCC, International bunker fuels under the SBSTA, available at: <http://unfccc.int/-methods/emissions_from_intl_transport/items/6142txt.php> (last accessed 31 March 2015); see Chapter 5, section 2.3.1.

328 For example, the GBEP has established a task force working on bioenergy GHG emissions and has issued a report. In 2009, the Task Force on GHG methodologies produced a version zero report that aims to provide a common methodological framework for a GHG life-cycle analysis of bioenergy. The framework covers considerations for emissions from all phases of bioenergy production and use, from direct and indirect land use changes to transporting, →

authority in 'mediating between science and politics' and 'certifying, gauging, and reckoning' relevant knowledge about climate change.[329] Supposing any effort to harmonise life-cycle emission accounting for biofuels is made, deviations could be allowed for the circumstances where alternative but compatible accounting methods would be better able to reflect the national situation and would be well documented and scientifically based.[330]

5.2.3.2. *Biodiversity and Food: Harmonisation and Integration from a Collaborative Approach*

Biological diversity is a 'common concern of humankind', as is climate change,[331] but it is more subject to the diversity in environmental and socio-economic conditions at the national or sub-national levels. Similarly, changes in food availability and prices may have global impacts on people in poverty in a third country, but it is meanwhile largely a national issue of great importance for national security. These characteristics involving different impact scales poses a challenge to providing harmonised rules on biodiversity and food availability at the international level, and the existing international commitments on biodiversity or food availability primarily hinge upon their translation and implementation by states. For example, Parties of the Ramsar Convention or CBD have great latitude in deciding the areas protected for wetlands and biodiversity conservation and what constitutes 'wise use' or 'sustainable use' of biodiversity,[332] which are primarily monitored under the treaty regimes through national reports and action plans.[333] Regarding food availability, State Parties to the ICESCR undertake an obligation to

processing and fuel use. The framework consists of a 10-step analysis covering the emissions from biomass feedstock production, including land use changes, co-products and by-products, transport of biomass, conversion, transport of fuel, and fuel use, while options are provided for reporting life-cycle analysis of transport fossil fuels and heat and electricity production systems. However, the framework only aims to provide references of pertinent questions for countries/institutions to ask when developing their methodologies, and it is optional for reporting direct land use change or indirect land use change, or a combination of both. The Global Bioenergy Partnership Common Methodological Framework for GHG Lifecycle Analysis of Bioenergy Version Zero (2009) <www.globalbioenergy.org/fileadmin/-user_upload/gbep/docs/2009_events/7th_SC_NY/GBEP_GHG_report_2306.pdf> (last accessed 19 April 2014); GBEP Clearing House on GHG methodologies for lifecycle analysis of bioenergy, available at: <www.globalbioenergy.org/toolkit/clearing-house-on-ghg-methodologies/en/> (last accessed 19 April 2014).

329 Beck *et al.* 2014b; Rached 2014, p. 5.

330 In COP's adopting IPCC Guideline on GHG inventory and emission accounting, it is provided that Annex I Parties shall apply methodologies in accordance with the guidelines, but they have discretion in applying their national methodologies as long as they can better reflect the national situation, are compatible with IPCC Guidelines, and well documented and scientifically based. See Chapter 2, section 2.2.

331 See the preamble of CBD and UNFCCC.

332 See Chapter 2, section 3.1.1.

333 For example, under the CBD, states are required to submit national reports on their implementation and national strategies or action plans on biodiversity. See Chapter 2, section 3.1.1.

protect the right to adequate food, but they may realise the enjoyment of this right 'progressively' unless the 'minimum essential level' of food is threatened.[334]

The afore-mentioned feature may experience the difficulty of providing harmonised binding rules at the international level and of enforcing such international rules if they are established.[335] To overcome the difficulty, one line of thought draws upon experiences from WTO law and calls for the creation of an overarching environmental organisation under the auspices of the International Environmental Organisation or Global Environmental Organisation.[336] Another trend proposes to promote coordination, linkage, and streamlining based on the existing institutions instead of reinventing one. For example, Biermann et al classify fragmentation into different types and suggest that it is crucial to increase synergies within the policy areas, to integrate regulatory processes, to reduce duplications, and to better coordinate between different institutional arrangements.[337] In depicting a global network of public, private, and mixed institutions and norms, Abbott argues that an 'orchestrator' role should be played by inter-governmental organisations, in order to overcome both the inadequacy of traditional state-based governance and the problems resulted from highly decentralised norm-setting institutions.[338] According to Yang and Percival, 'global environmental law', rather

334 See Chapter 2, section 3.2.1.

335 Yang and Percival observed that international environmental agreements are frequently characterised as entirely unenforceable because of the absence or weakness of international enforcement mechanisms in most environmental agreements, the commitments contained in multilateral environmental agreements appear to be more aspirational than legally obligatory. However, the Kyoto Protocol's Non-compliance Mechanism is of course one of the few exceptions. Carlarne has also depicted the institutional fragmentation in environmental law. Yang & Percival 2009, p. 656-657; Carlarne 2008, p. 456-462.

336 See the discussion in Carlarne 2008, p. 459-460; Yang & Percival 2009, p. 657-658; see more in Driesen 2003, p. 365-366; Esty 2000; Biermann & Bauer 2005; Charnovitz 2002; Tarasofsky & Hoare 2004.

337 Biermann et al classified fragmentation into 'synergistic fragmentation', 'cooperative fragmentation', and 'conflictive fragmentation', among which 'synergistic fragmentation' was regarded as the 'second-best option' in the absence of universal governance architecture in reality. Biermann cited the case of climate governance as demonstrating some elements of synergistic fragmentation, strong elements of cooperative fragmentation, and an indication of conflictive fragmentation. In analysing the consequences of different types of fragmentation, it is found that cooperative forms of fragmentation may entail both significant costs and benefits, while conflictive fragmentation appears to bring more harm than positive effects. Biermann, Pattberg & Van Asselt 2009, p. 31-32.

338 Abbott argued for a 'transnational new governance' which is formed by an intricate global network of public, private and mixed institutions and norms partially orchestrated by international organisations. In another paper, Abbott depicted a 'governance triangle' of climate change and argued for a 'transnational regime complex' featured with multi-scalar systems, as a development from the conceptions of 'regime complex' and 'polycentric' governance. In the 'transnational regime complex' of climate change, Abbott also argued for some form of 'orchestration' to remain essential even for multi-scalar systems, because even if small and medium-sized groups could autonomously generate context-specific responses to climate change, and even if their cumulative impact were sufficient to address the problem, it is questionable whether that result could be achieved as rapidly as necessary. Abbott *et al.* 2012b; Abbott & Snidal 2010, p. 315; Hale & Roger 2014, p. 59.

than a single and entirely new governing body for the global environment, would address the weakness of environmental law implementation by creating a closer relationship with national law and strengthening global institutional linkages or expansion.[339] While agreeing that a certain kinds of international environmental organisation is to be established in the long term, Carlarne recommends that climate change can serve as a vehicle for improving coordination between institutionally distinct but substantively interlinked areas of international law.[340]

In view of the feature and the trends addressed above, it is argued that the potential biodiversity and food availability impacts of biofuels could be addressed by international harmonisation and integration developed from a collaborative approach. It has been observed in this thesis that differences and gaps exist in how to define the key terms essential for the sustainability of biofuels, such as 'high biodiversity value and critical ecosystem', 'forest land' or 'land with high carbon stocks',[341] which may cause potential regulatory conflicts[342] or deficient sustainability regulation.[343] In this sense, a common understanding is needed for a definition of the key terms.

Firstly, a common understanding of the key terms could draw on the efforts already made under the relevant international regimes. For instance, the CBD Executive Secretary has undertaken the work of compiling and communicating among Parties the information and progress about biofuels and biodiversity, as well as identifying biodiversity areas being protected from the production processes for biofuels.[344] The UN Human Rights Council has adopted resolutions that call for states and international organisations to review the impact of biofuel policies or projects on food.[345]

Secondly, the effort for a common understanding could be assisted by increasing collaboration between international regimes/organisations, so as to pool regulatory efforts at various levels to promote a common understanding and thus

339 Yang & Percival 2009, p. 656-659.

340 It was suggested that climate change can function not only as a mechanism for highlighting present shortcomings, but it can also provide an opportunity to examine overlapping mandates more closely and to begin structuring new mechanisms for addressing these issues in a comprehensive, joined-up manner. Carlarne 2008, p. 471-474.

341 The Executive Secretary of the CBD identified certain gaps in biofuels and biodiversity conservation, including the gap in the definition of terms such as 'full life cycle', 'direct and indirect land use change', 'direct and indirect water and other resource use change', 'critical ecosystems, areas of high biodiversity value, cultural, religious and heritage interest and important to indigenous and local communities', 'sustainable and unsustainable production and use of biofuels.' See Chapter 2, section 3.1.3; Chapter 5, section 3.

342 For example, in applying domestic sustainability criteria to biofuels produced in third countries, the different definitions of the key terms such as 'high biodiversity value' or 'forest land' may result in divergent conclusions about the sustainability of biofuels in the consuming country and the producing country.

343 For example, the lack of definitions or identification of the areas with high biodiversity value or high carbon stocks may result in deficiencies in the inclusive regulation on the sustainability of biofuels, as in the case of China.

344 Chapter 2, section 3.1.3.

345 Chapter 2, section 3.2.1.

reduce pluralism and conflict.[346] Collaborative relationships between international regimes and initiatives can already be observed from the existing measures for dealing with biofuels. For example, the CBD COP calls for, rather than obliges, Parties as well as non-Party states and other institutions to address the relationship between biofuels and biodiversity. Meanwhile, the CBD Executive Secretary is instructed to note the voluntary certifications and standards. In a report on the definition gaps, the CBD Executive Secretary notes the IUCN's work on listing critical biodiversity ecosystems and areas. As for the impact on food, the High-Level Conference in Rome was attended by Heads of State and Government as well as non-governmental organisations seeking to identify international consensus on sustainable biofuels through reviewing the existing international instruments and policy options for decision makers.[347] A declaration was produced by the High-level Conference which calls upon national governments and other stakeholders to enhance coherent, effective and result-oriented international dialogues on biofuels.[348]

Thus, it is important to further strengthen the collaboration between the relevant international regimes in an effort for harmonisation and integration regarding the biodiversity and food impacts of biofuels. The biodiversity and food concerns involve at least the treaties of the CBD and the Ramsar Convention,[349] the ICESCR, and the UNFCCC and the Kyoto Protocol,[350] among which there exists

346 Abbott proposes a typology of four modes of governance: hierarchy, delegation, collaboration, and orchestration, among which collaboration and orchestration respectively refer to direct and indirect soft approaches to reaching governance objectives. Abbott argued for an orchestration governance for international governmental organisations, because collaboration can be difficult due the lack of direct access to private targets or actors, while orchestration enables the international governmental organisations to manage or bypass states via intermediaries which are generally private actors, NGOs, or other international organisations. Abbott, K.W. *et al.*, 'Orchestration: Global Governance through Intermediaries', 6 August 2012, available at: <http://ssrn.com/abstract=2125452> (last accessed 5 April 2014), p. 8.

347 Bioenergy, food security and sustainability – towards an international framework. Rome, 3-5 June 2008, para. 64; Climate change, bioenergy and food security: Civil society and private sector perspectives, available at: <ftp://ftp.fao.org/docrep/fao/meeting/013/k2470e.pdf> (last accessed 5 April 2014); see Chapter 2, section 3.2.1.

348 Declaration of the High-Level Conference on World Food Security: The challenges of climate change and bioenergy, available at: <www.fao.org/fileadmin/user_upload/-foodclimate/HLCdocs/declaration-E.pdf> (last accessed 5 April 2014), para. 7(f); see Chapter 2, section 3.2.1.

349 Biodiversity may involve a broader scope of treaties, such as the Convention on International Trade in Endangered Species (CITES), the Convention on the Conservation of Migratory Species of Wild Animals (CMS), the World Heritage Convention (WHC), International Treaty on Plant Genetic Resources for Food and Agriculture (ITPGR), and the International Convention on the Regulation of Whaling (ICRW), as well as biodiversity-related socio-economic rights such as the Nagoya Protocol on Access and Benefit-Sharing.

350 For example, the afforestation and reforestation projects for climate change purposes may also have an impact on biodiversity and food availability. See Chapter 2, section 2.3; see more in Trouwborst 2009; Van Asselt 2011a; Van Asselt 2011b; Morgera 2011; Morgera 2013; Roht-Arriaza 2010a; Roht-Arriaza 2010b.

potential synergy[351] as well as conflict.[352] While 'hierarchisation through a new umbrella agency' is not viable in a near future, coordinating and streamlining institutions, organisations, and bureaucracies provides a more realistic way to deal with potential conflicts caused by regime fragmentation.[353] However, there are divergent proposals as to how to operationalise the coordination. For example, Van Asselt[354] argues for 'autonomous interplay management' through states or non-state actors, and Morgera[355] emphasises the positive interaction between the two regimes which can promote a human rights-based approach to climate change regulation and implementation.[356] Based on the proposals, collaboration could at least happen in the following two aspects in the specific case of biofuels. In the first place, liaison and communication between relevant treaty bodies could be reinforced. The COP decision and resolution on biofuels under the CBD and the Ramsar Convention both stress the coordination between institutions of the two treaties.[357] To a certain extent, such a liaison and communication can be extended to include UNFCCC institutions, either based on the existing Joint Liaison Group between the secretariats of the CBD, the UNFCCC, and the UN Convention to Combat Desertification,[358] or in the form of regular information exchange and conference participation as already conducted between the ICAO and the UNFCCC.[359] In the second place, there should be an integrative substantive consideration of biodiversity, climate change, and food availability in the measures to be taken in relation to biofuels. The climate change regime has been criticised for being single-mindedly focused on GHG emissions while marginalising biodiversity and human rights concerns and biofuels are cited as one example illustrating such a conflict.[360] Although the UNFCCC or the Kyoto Protocol or the Paris Agreement has not taken specific measures to promote biofuels, the GHG emission-related commitments and

351 See Chapter 2, section 3.1.3.

352 See Chapter 2, section 2.3.

353 Biermann summarised that there are three ways to deal with fragmentation: clustering institutions, upgrading UNEP, streamlining institutions, organisations, and bureaus, as well as hierarchisation through a new agency. It is found that upgrading of the UN Environment Programme to a UN specialised agency, with additional and increasing streamlining of other institutions and bureaucracies, offers the most potential for environmental policy integration. Biermann, Davies & Van der Grijp 2009, p. 351.

354 Van Asselt 2011a, p. 1268; Oberthür 2009.

355 Morgera 2013.

356 See more discussion about human rights and climate change in Humphreys 2010; Von Doudda, Corkery & Chartres 2007.

357 For example, the CBD Executive Secretary is invited by the Ramsar COP to include considerations and activities related to wetlands, biodiversity and biofuels in the joint work plan between CBD and Ramsar Conventions, drawing upon expertise from regional workshops convened by the CBD subordinary body. Furthermore, the CBD takes the lead in the Biodiversity Liaison Group, which also includes the World Heritage Convention and the Ramsar, CITES and Bonn Convention. Chapter 2, section 3.1.3.

358 See Chapter 2, section 3.1.3.

359 See Chapter 5, section 2.3.1.

360 See Chapter 2, section 2.3.

methodologies may have the effect of driving the demand for biofuels.[361] Thus, the potential impacts of biofuels on biodiversity and food should be taken into account if any harmonisation or integration measure affecting the demand for biofuels happens under the international climate regime, which is also in accordance with the sustainable development enshrined under the framework of UNFCCC.[362]

5.2.3.3. *Agro-environment and Socio-economic Rights: Transplantation and Convergence*

Yang and Percival use the concepts of 'transplantation' and 'convergence' to describe the *uni-directional*, rather than joint multi-country, transfer of legal doctrines:[363] while legal transplantation is usually driven by the need for less developed legal systems to 'catch-up' with more sophisticated systems,[364] convergence is largely a result of a response to similar external pressures such as common functional goals, governance considerations, and ecological and public health constraints.[365] In sum, transplantation and convergence are determined by either the internal willingness to borrow or the external pressure to adapt.

The effects and motivation of transplantation and convergence can be used for regulating the sustainability of biofuels in relation to their environmental and socio-economic impacts on a local scale. As analysed, sustainability concerns with regard to local impacts, such as local soil, water, air quality and socio-economic rights of local communities,[366] are difficult to regulate in detail in a harmonised way due to the fact that they are more easily subject to diverse environmental and socio-economic circumstances. Such impacts are primarily regulated in national or sub-national legal systems, unless transboundary environmental impacts[367] or serious

361 See Chapter 2, section 2.1.

362 It is stated as a principle of the Convention that 'The Parties have a right to, and should, promote sustainable development. Policies and measures to protect the climate system against human-induced change should be appropriate for the specific conditions of each Party and should be integrated with national development programmes, taking into account that economic development is essential for adopting measures to address climate change.' Furthermore, Art. 2 of the UNFCCC mentions the relationship between GHG emission stabilisation and food. UNFCCC, Arts. 2 and 3(4) of the UNFCCC.

363 Yang and Percival follow Watson in defining 'transplantation' as the deliberate copying and adaptation of significant portions of statutes or particular doctrines of law by one country from another, not only including the borrowing of legal principles and norms between states but also from the state level to the international level. Particularly, Yang and Percival refer to 'convergence' as unintended similarities or similar directions in designing, implementation, and the operation of regulatory systems between disparate legal systems. Yang & Percival 2009, p. 625-630; Watson 1993, p. 22-24.

364 Yang & Percival 2009, p. 626.

365 Yang & Percival 2009, p. 631.

366 Smith *et al.* 2014, p. 884.

367 International law has been adopted on transboundary environmental damage, such as the UNECE 1992 Convention on the Protection and Use of Transboundary Watercourses and International Lakes and the UN 1997 Convention on the Law of the Non-navigational Uses of International Watercourses, 1975 Final Act of the Helsinki Conference on Security and Cooperation in Europe and the 1979 Geneva Convention on Long-Range Transboundary Air Pollution.

fundamental rights violations[368] occur. However, total reliance on national or local authorities may result in divergences and gaps in comprehensive regulation, since the national or local authority may have divergent priorities and limited capacity to establish and enforce relevant environmental and socio-economic rules. For example, the Chinese local authorities may give priority to economic development and attach less importance to environmental issues unless serious consequences occur; although progress in environmental legislation has been made, the current legislation and enforcement of environmental law, particularly regarding agricultural pollution prevention, still leaves much room for improvement.[369] Thus, in the absence of international harmonisation, the divergences and gaps could be preferably bridged by driving transplantation and convergence.

In fact, the effects of transplant and convergence are observed from a comparative study in this chapter. China has partially borrowed from the EU's fuel emission standards and accounting methods for biofuel emissions in aviation emission trading; meanwhile, convergences exist in restricting food crop-based biofuels and protecting nature reserves. Based on the driving factors analysed in the literature, the transplantation and convergence can be mainly attributed to the following factors: China's willingness to learn from world-leading standards,[370] the EU's stringency in standard-setting and its ambition to 'export' the standards,[371] and the similar food and environmental constraints caused by biofuels.[372] Based on the existing experiences and trends, efforts could be made in at least three respects for driving further convergence in biofuel sustainability regulation.

Firstly, the effort of strengthening domestic sustainability regulation with regard to the local environment and socio-economic rights may drive transplantation in the other legal systems. In practice, one reason for China's borrowing from the EU is due to the stringency of EU's environmental standards.[373] Moreover, the commitments to maintaining a high level of environmental

368 According to the supervising committee of the ICESCR, there is a distinction between the core and non-core elements of the socio-economic rights, as the core element – such as the minmum essential level of food or prohibition of forced eviction – imposes immediate and non-derogable obligations upon State Parties. See Chapter 2, sections 3.2.1 and 3.2.2.

369 See Chapter 4, section 3.1.3.3.

370 Qin in a presentation mentioned that the interaction between China's domestic environmental law and international environmental law or foreign environmental law has greatly increased, and the influence from the EU is particularly evident. The increasing impact of the EU on China's environmental law can be attributed to the fact that politically the Chinese officials feel safer referring to European models rather than the US models because: *a)* the EU has been the leader in many environmental regimes, *b)* the EU's continental law system is closer to the Chinese system, and *c)* the EU is inclined to persuading rather than commanding China, in contrast with he US manner in environmental cooperation. Qin, T., 'Development of Environmental Law in China: 1978-2014', 10 February 2015, Maastricht (the presentation script is available on file with the author).

371 See Chapter 5, section 2.3.2.

372 See Chapter 5, sections 2.6.2 and 3.5.2.

373 Qin, T., 'Development of Environmental Law in China: 1978-2014', 10 February 2015, Maastricht (the presentation script is available on file with the author).

protection and sustainable development are enshrined in EU law,[374] while a high level of regulation could meanwhile lead to the exportation of the EU's standards[375] and the effect of 'trading-up.'[376] Hence, by continuously improving and updating the domestic rules on the protection of the local environment and socio-economic rights, the EU could exert a 'leading by example' influence on the less developed regions or countries like China. However, this does not mean that the EU (or developed countries) is necessarily the 'leading example' and China (or developing countries) is necessarily the 'learner', as the roles might be reversed if China has relatively better experiences.[377]

Secondly, direct communication and cooperation in both bilateral and multilateral settings can increase the communication of information and incentivise convergence. This thesis has observed bilateral cooperation and project building in the area of biofuels between the EU and China[378] or between the other biofuel consuming and producing countries.[379] Thus, bilateral cooperation at the state or sub-state level could provide a viable approach to communicating the 'leading example' or experiences of local environmental and socio-economic regulation. In the cooperative processes, experiences from developing countries may also be valuable for developed countries, such as China's experimental approach to developing biofuels[380] and the emphasis on the impact on food availability.

374 See discussion in Chapter 3, section 5.1.

375 There is rich literature about the effect of the EU's unilateral measures on other parts of the world, such as literature on 'diffusion', 'Europeanisation', and 'The Brussels Effect', see for example Vogel 2012, p. 12-13; Bradford 2012; Scott 2009.

376 Vogel observed the 'California effect' in environmental regulation: nations are increasingly adopting the standards of their richer and greener trading partners. Vogel 1997, p. 5-6; Vogel 2012.

377 As Yang and Percival exemplify, the experience of China's importation of foreign environmental law could even contribute to the emerging global environmental law. Yang & Percival 2009, p. 664.

378 At the EU-China Summit in 2013, the EU-China Strategic 2020 Agenda for Cooperation was agreed upon and has served as the guiding document of EU-China relations. The Agenda covers cooperation areas in agriculture, transport, energy, climate change, and environmental protection. See Chapter 3, section 4.2 and Chapter 4, section 4.

379 Biofuel is on the agenda of the EU-Brazil Summits which have been held every year since 2007. In the Joint Statement of the third Summit, there was a bilateral political commitment to promote the use of sustainable biofuels and the promotion of sustainable biofuels. In 2009 China and the US decided to establish the US-China Clean Energy Research Center which focuses on building energy efficiency, clean coal including carbon capture and storage, and clean vehicles. In November 2014, the two countries reached a new deal on climate change and clean energy cooperation, pledging to strengthen cooperation on climate and clean energy as well as expand the ongoing programme of cooperation through policy dialogue and technical work on clean energy and low greenhouse gas emissions technologies. See Chapter 3, section 4.2 and Chapter 4, section 4.

380 It is observed that, in the EU, increasing flexibility would be given to Member States based on a new governance framework. Rather than continue with the binding national renewable energy targets, the European Commission attempts to guide Member States on their national renewable energy policies through a 'new governance' system. The Commission has vowed to attain the European renewable energy target for 2030 with a new governance framework based on national plans for competitive, secure and sustainable energy prepared by the →

Moreover, cooperation could take place in a multilateral setting. For example, both the EU and China have participated in the International Biofuels Forum in order to promote an international market for sustainable biofuels, exchange experiences, and build collaboration on biofuel standards and codes.[381]

Thirdly, the meta-standard approach that combines legally binding criteria and private certifications can be used to promote convergences in defining and regulating sustainability of biofuels. It is shown in this thesis that the EU has been inclined to use market forces and trade measures to extend its regulation and standards beyond its borders. By requiring operators who have economic connections with the EU market to comply with EU law, the EU may exert an external effect on the exporters or exporting countries.[382] The application of biofuel sustainability criteria illustrates such an effect as the criteria are applied irrespective of biofuel's origin. Therefore, biofuels produced in a third country have to comply with the EU's requirements, which have indeed compelled some exporting countries to improve their environmental practices in forestry, nature protection, or

Member States. Particularly, the Commission has referred to the 'soft-law' instruments of 'white papers' and 'communications' for future policy in renewable energy transport, in which the Commission states the goal or policy regarding alternative fuels – primarily referring to biofuels since they are 'currently the most important type of alternative fuels' – in non-binding instruments such as 'white papers' and 'communications.' For example, in 2011 the Commission issued the White Paper on the 'Roadmap to a Single European Transport Area', setting out the goals of phasing out 'conventionally-fuelled' cars in urban transport and increasing low-carbon sustainable fuels in aviation. The Commission is about to develop detailed guidance on the operation of the new governance process and the content of national plans in particular, which is followed by the preparation of Member State plans through an iterative process and the assessment of the Member States' plans and commitments, so as to guide the Member States on their national renewable energy policies, including the reform of renewable energy support schemes. Moreover, systematic monitoring with key indicators is to assess progress over time and to inform any future policy intervention. Commission Communication COM(2014) 015 final, section 3; Peeters & Schomerus 2014, p. 283.

381 The International Biofuels Forum is a governmental initiative among Brazil, China, the European Commission, India, South Africa, and the US launched in 2007. Furthermore, both the EU and China attended the International Renewable Energy Agency and the Global Bio-Energy Partnership. The International Renewable Energy Agency (IRENA) is an intergovernmental organisation that supports countries in their transition to a sustainable energy future and serves as the principal platform for international cooperation. The Global Bio-Energy Partnership (GBEP) is an international private-public partnership concentrating on bioenergy and has established task forces on sustainability indicators of bioenergy and life-cycle GHG emissions. See Chapter 3, section 4.2, and Chapter 4, section 4; IRENA, Membership, available at: <www.irena.org/Menu/Index.aspx?mnu=Cat&PriMenuID=-46&CatID=67> (last accessed 30 May 2015); GBEP, Partners and Membership, available at: <www.globalbioenergy.org/aboutgbep/partners-membership/en/> (last accessed 10 May 2015).

382 Resnik has commented that the EU has used market access as a tool to leverage the 'migration' of its frequently used norms abroad. Scott observed the 'territorial extension' effect in applying the EU's measures to actions that trigger a territorial connection such as a market transaction. Resnik 2006; Scott 2014; see also in the discussion in Chapter 3, section 4.1.

feedstock cultivation.[383] However, the mere application of the EU's unilateral requirement may still have limits in addressing the local environmental and socio-economic impacts in third countries. For instance, the EU does not apply the agro-environment criterion to biofuels of foreign origins and does not provide socio-economic criteria at all, partly to avoid a potential conflict with WTO law. The limit in addressing local impacts by domestic binding rules could be made up by the use of 'voluntary schemes' which are largely private certification schemes, as the voluntary schemes can include additional and stricter environmental and socio-economic standards[384] which have to be complied with to be certified.[385] To strengthen credibility and coherence, the EU adopts a meta-standard approach that benchmarks the certification schemes against the EU's sustainability criteria.[386] The meta-standard approach that combines mandatory criteria and private standardisation/verification undertaken by private actors has caused much discussion in literature.[387] While admitting its advantages in preventing reinventing wheels and reducing the risk of contravening WTO law,[388] it is also pointed out that there are potential risks in implementing sustainability criteria with the EU's meta-standard approach, such as the reliance on private certifications to monitor land use change[389] or unclear process-related requirements about benchmarking and verification.[390] In this sense, the meta-standard approach may provide a useful way to regulate sustainability concerns with local impacts,[391] but improvement can be made to clarify and strengthen the role of public authorities in monitoring macro-level impacts and establishing process-related requirements.[392] Moreover, the meta-

383 Some major biofuel (feedstock) exporting countries, such as Argentina, Brazil, Indonesia and Malaysia, adopted new regulatory measures in 2009 and 2010 to improve their environmental practices in forestry, nature protection or feedstock cultivation. Commission Staff Working Document COM(2013) 175 final, p. 17 and 32; see Chapter 3, section 4.1.

384 See Chapter 3, section 2.1.2 and section 5.1.

385 Kaphengst, Wunder & Krista 2012, p. 20.

386 See Chapter 3, section 2.1.5.2.

387 Lin 2011a; Romppanen 2012b; Endres 2010; Mol 2010, p. 76-77; Lee 2011.

388 It should be noticed that the use of certifications cannot completely guarantee compatibility with WTO law. The compatibility of a biofuels-certification measure with international trade rules is determined from several perspectives, including *a)* whether the nondiscrimination rules of Art. I or Art. III of GATT apply and, if there is prohibited discrimination, *b)* whether it is excused by the health or environmental carve-outs of GATT Art. XX (among other possible exceptions), and *c)* whether the detailed substantive and procedural rules of other WTO agreements apply. Echols 2009.

389 Lin 2011a, p. 43.

390 Romppanen 2014, p. 265.

391 It has been pointed out that certifications are not suitable to monitor the macro-level impacts such as land use changes. See Chapter 3, sections 2.1.5.2 and 5.1.

392 It has been argued that standards, if applied voluntarily, have fundamental limits in ensuring more sustainable consumption and production pathways, and additional and accompanying policy measures and institutional frameworks are essential for sustainable development, because they not only regulate voluntary standards, but they also provide for governance structures that set clear rules and incentives for the responsible use of natural resources. Thus, challenges for the future will be the context related and adjusted interplay between →

standard approach might be applied to integrate local environmental and socio-economic concerns in international governance. For example, more specified sustainability criteria/principles could be established by the Executive Board or COP/CMP of the Kyoto Protocol, and voluntary schemes can be used to certify and verify project operators' compliance with the criteria/principles.[393] In this sense, the meta-standard approach could serve as a chance to gain experiences for introducing or *de facto* implementing additional or stricter sustainability standards,[394] which may also increase convergence by implementing benchmarked sustainability requirements in third countries.

In conclusion, from the perspective of whether sustainability concerns in all the environmental, social, and economic dimensions are inclusively addressed, it is found that there are differences and limits in taking all the inclusive sustainability concerns into account at all the international, EU, and Chinese levels, which may allow biofuels with adverse environmental or socio-economic effects to continue to be produced and consumed or even be labelled as 'sustainable.' The limit could to some extent be attributed to the inherent complexity of regulating the sustainability of biofuels, the broad and vague meaning of sustainability as well as the cross-cutting and multi-scale features of biofuels regulation. The inherent complexity poses challenges for having the inclusive sustainability concerns of biofuels regulated by either domestic or international law, and meanwhile it raises the question of whether there is potential for a global inclusive approach to the sustainability of biofuel that could overcome the challenges brought about by the inherent complexity. Finally, based on implications from the comparative analysis in this chapter and the conceptual framework of 'global environmental law', the potential for a global inclusive approach is explored where the evolution process of 'global environmental law' – 'harmonisation', 'integration', 'transplantation', and 'convergence' – can be used to frame the regulation of the sustainability of biofuels in view of the inherent complexity. Instead of relying on merely one model of 'top-down' or 'bottom-up', this global approach makes a combination of the two by drawing on the effects of international harmonisation and integration as well as interactive transplantation and convergence, according to the different scales of impacts and the likelihood of being influenced by diverse environmental or socio-economic circumstances. As suggested above, the accounting of the life-cycle GHG emissions could be harmonised under the auspices of the UNFCCC, due to its global impact and the viability of international harmonisation. Biodiversity and

policy-driven instruments and voluntary standard-schemes. Kaphengst, Ma & Schlegel 2009, p. S101.

393 It is noted that some initiatives have been taken or considered to address the sustainable development of CDM projects, including the use of voluntary sustainable development tools and the development of guidelines for designated national authorities. See Chapter 2, section 2.3.

394 It is argued that voluntary standards are not the end of the process but rather the beginning: market-based initiatives take the lead to gather motivation for building up legal instruments in producing countries. They should not be regarded as an alternatives to regulation but rather as a powerful input into regulation, both in the consuming countries and producing countries exporting commodities to the EU. De Man 2013; Kaphengst, Wunder & Krista 2012.

food concerns could be harmonised in a way which allows for a common understanding to be reached by strengthened collaboration and coordination. Local environmental and socio-economic right concerns might be better regulated on the national or sub-national level with transplantation and convergence fostered through 'leading by example', bilateral and multilateral cooperation, and meta-standard certification. In sum, the essence of an inclusive approach would be that a common set of principles for the sustainability of biofuels can evolve in a dynamic and interactive way based on the concept of 'global environmental law', because the sustainability impacts of biofuels remain complicated and difficult to evaluate, and rule-making for the sustainability of biofuels has not reached a final stage and is hardly ready to be agreed upon internationally. Hence, common understandings could be fostered through the dynamic processes of harmonisation, integration, transplantation, and convergence, and may stem from interactions between different regulating levels and methods. In this way, it is hoped that the regulation of biofuels would be developed from a fragmented, divergent, and gapped 'mess' to an inclusive 'mix'[395] so as to address sustainability concerns ranging from a domestic to a global scale.

395 It has been commented that the issue of biofuels is exceedingly complex in economic and political terms ad involves authority over policy making and associated regulatory rule-making dispersed among different actors and at different levels, which shares some similarities with the description of 'policy mix' or 'instrument mess/mix' once respectively used by Sorrell and Sijm and Peeters. See De Beer 2011, p. 1-2; Sorrell & Sijm 2003; Peeters 2014, p. 173.

Chapter 6

GENERAL CONCLUSION

Biofuels are considered as a potential solution for mitigating problems of energy security, GHG emissions, and rural development, but biofuels have been increasingly criticised for bringing about unsustainable results and sometimes even giving rise to more perils to sustainability. The key issue is how the demand for, supply, and trade of biofuels are regulated to avoid or minimise adverse environmental and socio-economic effects. In this vein, this thesis concentrated on the regulatory approaches at the international, EU, and Chinese levels and examined whether they include sustainability concerns in all the environmental, social, and economic dimensions. By exploring the research questions, the present thesis aimed to provide comparative insights into the regulation of the sustainability of biofuels and furthermore to explore the potential for a global inclusive approach, drawing on the implications from the comparative study. This chapter will summarise the main findings of the respective chapters and provide a final conclusion of this thesis.

1. Main Findings

This thesis has respectively examined the regulation of biofuels in the international, EU, and Chinese legal systems in Chapters 2, 3, and 4 and then compared their regulatory approaches in Chapter 5, in an attempt to answer the question of 'what are the regulatory approaches to the sustainability of biofuels' in the international, EU, and Chinese legal frameworks? The main findings in the respective chapters are summarised as follows.

1.1. *Mapping the International Regulatory Terrain*

In Chapter 2, in the absence of a specific international legal framework for the sustainability of biofuels, the international regime for regulating the sustainability of biofuels was mapped by looking into the different international legal regimes relating to the demand for, supply, and trade of biofuels. It was observed that, although international instruments in relation to biofuels are emerging, an international legal framework that develops and applies binding rules for the sustainability of biofuels still lacks.

In looking into the relationship between the international climate regime and the demand for biofuels, it was found that the commitment of addressing or reducing GHG emissions under the UNFCCC or the Kyoto Protocol may provide incentives for the use of biofuels, especially for Annex I Parties undertaking the GHG emission reduction obligation, if biofuels are regarded as a means to reduce GHG emissions in the transport sector. However, the international climate regime does not specify how to calculate the GHG emissions of biofuels, and the existing GHG inventory accounting methodology provided in the IPCC Guidelines can hardly ensure a complete calculation of the GHG emissions of biofuels due to disparate GHG inventories communication and reporting commitments of states. Furthermore, it is possible that the pursuit of climate change mitigation by using biofuels may ignore the other sustainability impacts such as biodiversity and human rights.

The international instruments in the field of biodiversity and socio-economic rights are related to the supply of biofuels. It was found that the general commitment to conservation and sustainable/wise use of biodiversity does not exclude human activities and natural resource exploitation for the production of biofuels, while the CBD and Ramsar COPs have adopted decisions or resolutions that contain specific guidance for addressing the relationship between biofuels and biodiversity, particularly by means of conducting environmental impact assessments, establishing appropriate national policies, and developing biodiversity inventories and 'go' or 'no-go' areas. Meanwhile, the possible impact on food availability and land grab may violate State Party's obligations with respect to socio-economic rights (e.g. the right to adequate food) provided under the international human rights regime, in the cases where a state adopts laws or policies inflicting the rights or failing to regulate individuals and groups who undertake activities infringing such rights. However, the commitments and guidelines under the international biodiversity and socio-economic rights regimes remain deficient in addressing the sustainability concerns in relation to biofuels in an inclusive and integrated manner: there are identified gaps in defining the key terms for the sustainable production of biofuels, and both the biodiversity and socio-economic rights regimes have limits in addressing the issues beyond their functional mandates.

The investigation into the relationship between the regulation of the sustainability of biofuels and WTO law concentrated on the question of to what extent WTO law may allow or constrain unilateral regulation on the sustainability of biofuels that affects the international trade of biofuels. It was found that the current provisions and case law under the WTO may present a constraint for the application of unilateral sustainability requirements, particularly socio-economic requirements, on imported biofuels, due to the questions about production and process methods and jurisdictional limitation.

1.2. *The Regulatory Approach in the EU*

Chapter 3 examined the regulatory approach to the sustainability of biofuels in the EU, and analysed whether it addresses the inclusive sustainability concerns of biofuels. It was found that the sustainability criteria provided in Directive

2009/28/EC (and in Directive 2009/30/EC) play a central role in the EU's regulatory framework on the demand for, supply, and trade of biofuels.

In analysing the regulatory measures having an effect on the demand for biofuels, it is observed that the sustainability criteria are set out as binding conditions for biofuels to be treated favourably with regard to achieving national renewable energy consumption targets, receiving financial support, and complying with national blend obligations. The EU also sets forth the methodology for assessing GHG emission savings and the schemes for verifying compliance with the sustainability criteria. The sustainability criteria include requirements about life-cycle GHG emission saving, biodiversity, carbon stock, and agro-environment protection, while the other sustainability concerns, particularly the socio-economic concerns, are addressed by monitoring and reporting and voluntary schemes. Moreover, by requiring compliance with the sustainability criteria, renewable energy targets and GHG emission reduction targets for fuels serve as the major drives for the consumption of 'sustainable' biofuels as defined by the sustainability criteria. The other measures, such as preferential energy tax rates, aviation emission trading, and the acceptance of CDM credits, could incentivise the demand for biofuels, but the link and coherence between these measures and the sustainability criteria need to be strengthened, particularly considering that the national renewable energy targets are not to be continued after 2020.

In examining the regulation on the production of raw materials for biofuels, such as agricultural crops and wastes, it was found that the production of agricultural crops is subject to the established legal framework of CAP, including the 'cross-compliance' and 'greening' requirements related to *inter alia* water, soil, carbon stock, and biodiversity. In a similar vein, waste streams used for generating biofuels need to comply with the EU's legal framework on wastes. However, the current CAP system does not provide a way to calculate the amount of the EU budget used for supporting agricultural crops dedicated to biofuel production, which may pose a problem in estimating the effect of CAP on agricultural activities that produce biofuel feedstocks, particularly with regard to their consequences for land use and food price changes. Moreover, a challenge may lie in how to identify 'waste' and comply with the waste management hierarchy and the principles of self-sufficiency and proximity.

By applying the sustainability criteria to biofuels originating in third countries, the EU's sustainability criteria on biofuels thus has an influence on the trade of biofuels between the EU and non-EU countries, but the consistency between the sustainability criteria and WTO law is still open to debate and dependent on further case law. However, the consideration of WTO law may constitute one challenge for integrating the agro-environmental and socio-economic sustainability requirements into the binding sustainability criteria. Nevertheless, the external policies, such as reducing/exempting tariff rates and the negotiation of bilateral or multilateral agreements on biofuels, may provide tools for promoting compliance with the EU's sustainability requirements in the third countries and may assist in addressing the concerns which are not currently included in the EU's sustainability criteria.

1.3. *The Regulatory Approach in China*

Chapter 4 addressed the existing principles and requirements regarding the environmental and socio-economic concerns related to biofuels in China. It was observed that the Chinese central government has set out general principles and specific requirements on certain environmental and socio-economic concerns related to biofuels, but a legal framework is not yet established for regulating the sustainability of biofuels at the national level.

As for the regulatory measures that may affect the demand for biofuel, the Chinese central government has established general guiding principles for policy making on biofuels, including the use of non-food grain feedstocks, using marginal land that is not used for producing food, and no environmental damages. An investigation into the other demand-related measures reveals that these principles are not yet further specified or coherently adhered to. This is a problem particularly with regard to the pilot projects that mandated consumption of biofuels blend and the quantified consumption targets which serve as the major drives for the demand for biofuels. Although differential environmental or socio-economic conditions are attached to the incentive measure of preferential tax treatment, there is still a lack of a legal framework to provide for sustainability requirements or to specify the general principles for the major demand-side regulatory measures.

Meanwhile, the general principles of non-food grain feedstock, marginal land, and no damage to environment are fleshed out by various legal and administrative measures on the production of biofuel feedstocks. The central government restricts the production of food grain-based biofuels, the use of arable land, and excludes protected nature reserve areas from use for biofuel feedstock production. The central government also encourages the cultivation of non-food grain feedstocks on marginal land through planning for development of the biofuel industry, establishing biofuel feedstock bases, and providing financial subsidies. However, considering the ambiguous scope of 'non-food grain feedstock' and 'marginal land' as well as the lack of specific rules for addressing biodiversity, carbon stock changes, and agricultural pollution, there is great concern about the socio-economic and environmental impacts of biofuels in their feedstock's production processes.

Although trade of biofuels exists between China and third countries, the general principles or specific requirements related to the consumption and production of biofuels are not explicitly set out for imports from third countries. Adjustments of tariff rates for biofuels are primarily aimed at protecting the domestic biofuel industry and implementing bilateral free trade agreements, rather than for guaranteeing the sustainability of biofuels from third countries. Nevertheless, bilateral and multilateral negotiation and cooperation might provide opportunity for China to make progress in the development of efficient and sustainable biofuels and to be involved in international processes on the sustainability of biofuels.

1.4. *Differences and Similarities*

Chapter 5 compared the regulatory approaches at the international, EU, and Chinese levels, and analysed the differences and similarities as well as their underlying reasons.

With regard to regulation of the demand for biofuels, it was observed in particular that there is a difference in whether and what sustainability requirements are established for the demand-side regulatory measures. The demand for biofuels is associated with the international commitments under the UNFCCC and the Kyoto Protocol, but there is a lack of corresponding sustainability requirements for the created or incentivised demand for biofuels. Domestically, the EU and China contrast each other in respectively emphasising the environmental impacts or socio-economic concerns, and also differentiate in whether their requirements about sustainability of biofuels are explicitly set out for measures relating to the demand for biofuels. The primary reason for the divergent sustainability requirements can be attributed to different sustainability focus, a lack of a commonly agreed definition of 'sustainable biofuel' or regulatory framework for it, as well as the conventional regulatory approach for promoting biofuels. On the other hand, convergence lies in the common concern about the potential negative impacts of biofuels, particularly on food availability, which can be considered to be of influence from international level and between the EU and China.

Differences as well as convergences are also found in defining and regulating the biodiversity, carbon stocks, and socio-economic rights concerns in the production of biofuels. For instance, the CBD and Ramsar Convention provide general commitments about 'sustainable use' or 'wise use' and specific guidance for biofuels and biodiversity, while the EU has gone further to explicitly prescribe and define the types of lands/areas with high biodiversity values as 'no-go areas' for obtaining biofuel feedstocks. China does not prescribe or define the types of highly biodiverse areas prevented from biofuel feedstock production, but the law generally prohibits agricultural activities in the established nature reserve areas and wetlands. This divergence can be attributed to the difference and even deficiency in defining key terms and assessment methodologies for evaluating the sustainable production of biofuels, the gaps in legislation or the implementation of environmental rules, and the different prioritised sustainability concerns. Nevertheless, convergence exists in the approach to protecting certain areas with biodiversity values from biofuel production in all the international, EU, and Chinese cases, which can be attributed to the influence from the existing international treaty regimes as well as the common response to the potential adverse effects of the increased production of biofuels.

Regarding the regulation of the trade of biofuels, in the context of WTO law, the EU and China have different approaches in treating biofuel (feedstock) from third countries. The EU applies its sustainability criteria irrespective of the origin of the biofuel, which has given rise to trade disputes and legal discussions regarding its consistency with WTO law. In contrast, China does not set out its domestic non-food grain, marginal land, and no environmental damage principles as requirements for biofuels from third countries. This difference could be attributed to the disparate objectives in relation to regulating biofuels as well as different stances in their external relations policy. Nevertheless, both the EU and China have taken measures to engage in bilateral or multilateral cooperation on biofuels.

2. Reflection

Based on the main findings on what regulatory approaches are provided, this section will reflect on the question of whether these regulatory approaches address sustainability concerns in all the environmental, social, and economic dimensions.

2.1. *Fragmented International Regulatory Terrain for the Sustainability of Biofuels*

In Chapter 2, it was found that the sustainability impacts of biofuels are addressed respectively under the different international regimes to various extents, but they are largely fragmented and there is a limit in the ability of each regime to integrate inclusively the environmental and socio-economic concerns. For example, the commitment to reducing GHG emissions under the UNFCCC and the Kyoto Protocol may give incentives for the use of biofuels due to the benefit for GHG emission reductions in the transport sector. However, the international climate regime does not provide specified or clarified rules or methods on life-cycle GHG emissions accounting for biofuels, and the pursuit of the goal of climate change mitigation may ignore other sustainability impacts such as biodiversity and human rights in the case of biofuels. Under the biodiversity treaties of CBD and Ramsar Convention, although specific decisions or resolutions have been adopted to deal with the relationship between biofuels and biodiversity, a gap exists in defining the key terms essential for determining the sustainability of biofuels and the life-cycle GHG emission concern is not addressed. The fragmentation and limits are primarily owing to the differential objectives and functions under the 'self-contained'[1] regimes, which presents a barrier for individual international conventions to inclusively address both environmental and socio-economic concerns.

Apart from fragmentation, the examined international instruments related to biofuels have weak or uncertain legal effects. It was observed that the international commitments and guidelines examined in the previous sections vary in their legal status and binding effect and demonstrate different scales of normativity and bindingness.[2] For instance, the general commitments under biodiversity treaties

[1] It is commented that the phenomenon of fragmentation in international law is mainly caused by the emergence 'self-contained regimes' and geographically or functionally limited treaty-systems. International Law Commission 2006, para. 10.

[2] Boyle and Shelton both argue for differential scales of normativity and bindingness instead of a binary perception of law or no-law. Shelton argues for the 'relative normativity' which is a question of a hierarchy of norms, which concerns the nature, structure and content of international legal system. The relative normativity arises in determining whether a legal rule →

leave room for states to decide on and implement measures, while the Kyoto Protocol provides binding quantity targets for Annex I Parties to reduce their GHG emissions. Meanwhile, based on an analysis of authorisation, voting, and language, COP decisions or resolutions on biofuels under the CBD and Ramsar Convention do not intend to bind State Parties, while COP decisions relating to GHG emissions accounting under the UNFCCC and Kyoto Protocol require Annex I Parties to follow the provided methodologies.

Therefore, an integrated and binding international regulatory framework to develop and apply rules on the sustainability of biofuels still lacks, and the existing instruments relating to biofuels are fragmented to the extent that different sustainability concerns are addressed respectively under various international regimes and usually have weak or uncertain legal status. The fragmentation and weak legal status would create challenges for regulating the sustainability of biofuels in an inclusive and integrated manner at the international level, and would call for further institutional collaboration for promoting common understanding of sustainability of biofuels.

2.2. *Deficiency in Inclusive Regulation of the Sustainability of Biofuels in the EU*

In spite of the fact that the European Commission claims that the sustainability scheme is comprehensive and advanced, it was found that the sustainability criteria are limited in addressing the potential adverse effects in both environmental and socio-economic dimensions. The EU's sustainability criteria purely consist of requirements from an environmental perspective, while the socio-economic concerns are not included in the binding criteria. Furthermore, the sustainability criteria do not apply to solid or gaseous biomass for electricity, heating and cooling, but biofuels may be produced from the same biomass raw materials, so there is reason to worry that the lack of sustainability requirements for biomass used for non-transport purposes may essentially diminish the effectiveness and comprehensiveness of the current sustainability criteria.

The EU sustainability criteria may also have limits in comprehensively addressing the environmental concerns that are already included in the criteria, such as GHG emissions caused by indirect land use changes and agro-environment in third countries. These limits are largely owing to the concern about compatibility with WTO law and the difficulty in defining commonly applicable social standards. The EU intends to fill the gap by obliging the European Commission and Member States to monitor and report and meanwhile allowing voluntary schemes to include the sustainability concerns not covered in the criteria, but such supplementary measures may have drawbacks to various extents[3] and cannot after all completely substitute the binding sustainability requirements.

exists to govern a problem and in deciding whether priority must be given to a specific rule among several. Boyle 2010, p. 122; Shelton 2010, p. 141.

3 For example, the approach of voluntary schemes has benefits of accumulating standard-setting experiences, avoiding conflicts with international trade law, and phasing in →

Meanwhile, an investigation into the extent of harmonisation reveals that the sustainability criteria aim for an exhaustive harmonisation which restricts more stringent national biofuel sustainability measures adopted by the Member States. Although the legal basis of Article 114 TFEU may provide for the possibility to maintain or introduce stricter national measures, this possibility is greatly limited by all the substantive and procedural conditions set forth in Article 114 TFEU as well as the lack of a safeguard clause for the sustainability criteria. However, an inclusive biofuel sustainability regulation is called for, because the deficiency in the regulation may result in a continuous demand for, supply, and trade of biofuels that have potential negative impacts, and moreover the commitment to sustainable development, which not only aims to address environmental concerns but also to minimise socio-economic impacts, has been set out in EU law and policy.

2.3. *Deficiency in Inclusive Regulation of the Sustainability of Biofuels in China*

In Chapter 4, the analysis revealed the deficiency of the current principles and requirements in integrating inclusive sustainability concerns in China. Above all, China has not yet established the concept of the 'sustainability of biofuels' or 'sustainable biofuels' in law or policy, nor has it set out legal criteria for the sustainability of biofuels. Although the central government has established guiding principles of non-food grain feedstock, marginal land, and no environmental damage for the development of biofuels, it does not explicitly specify the meaning of the principles, and these principles are not yet further coherently set out as a precondition for other regulatory measures on biofuels. China has emphasised avoiding the impacts of biofuels on food security, but the scope of 'non-food grain feedstock' and 'marginal land' is ambiguous, and there is deficient regulation related to biodiversity conservation, life-cycle GHG emissions including carbon stock changes, agricultural pollution in biofuel feedstock production, or the other socio-economic rights than food availability. Although there is a possibility for local authorities to adopt more stringent requirements for the sustainability of biofuels, it can be questioned whether local government has the willingness or awareness to do so without binding rules at the national level. These deficiencies in addressing the sustainability of biofuels shows that the potential negative impacts of biofuels are not yet fully clear and regulated with an inclusive approach in China, and the general rules with regard to monitoring and accounting of life-cycle GHG emissions and carbon stocks, biodiversity protection, and agro-environment need to be established or further improved. Thus, the strengthening of inclusive biofuel sustainability regulation in China may mainly depend on the law and policy making by the central government, which could find grounds in China's political and legal commitment to sustainable development and the existing example of promoting the sustainable cultivation of energy forestry.

complementary or stricter sustainability standards, but it may also have the limits of rent-seeking, green-washing, or race-to-the-bottom, and transparency. See Chapter 3, section 2.1.5.

2.4. *Differences and Challenges*

Sustainability has the essential meaning of integrating environmental, social, and economic considerations, and whether the three dimensions are inclusively taken into account is essential to achieve sustainability and sustainable development. In this way, whether the regulation of biofuels integrates the consideration of impacts on the three dimensions is crucial to guarantee the sustainability of biofuels. This thesis has demonstrated the differences, limits, and challenges of providing inclusively integrated regulation on the environmental and socio-environmental sustainability concerns related to biofuels.

While the sustainability concerns related to biofuels are respectively and fragmentally addressed under different international treaty regimes, the EU and China contrast each other in prioritising different sustainability concerns in the regulation of biofuels: generally the EU emphasises sustainability concerns in the environmental dimension, namely biodiversity, carbons stock, GHG emissions, and agro-environment, but China attaches more importance to the impact on domestic food security. Differences also lie in how the sustainability of biofuels is addressed: the EU sets out biofuel sustainability criteria as binding requirements in law applied irrespective of the origins of the biofuels, but China establishes the general principles in development plans and does not explicitly state their application to biofuel (feedstock) imports. Differences also exist in whether the sustainability requirements are applied to biofuels or biofuel feedstocks produced from third countries. Furthermore, the EU and China allow for different degrees of possibilities to adopt complementary or stricter sustainability requirements. Despite the differences, it is observed that limits in inclusive regulation exist at all the international, EU, and Chinese levels. The difference and limits in regulating the sustainability of biofuels can be attributed to many reasons, including the divergent sustainability focus, the lack of an agreed definition of 'sustainable biofuels' and related key terms, the contrasted regulatory background and approaches, the divergent extents of biofuel market integration, the gap in environmental law-making, and the different external policy stances. Limits in inclusively addressing sustainability concerns of biofuels may have the consequence that 'unsustainable' biofuels continue to be demanded, supplied, or traded. Although inclusive sustainability regulation for biofuels is desirable, the comparative analysis in this chapter has revealed the remaining challenges of providing inclusive regulation at the international or domestic level due to the inherent complexities caused by the broad and vague meaning of 'sustainability', integrating cross-sectoral concerns, and regulating multi-scale impacts.

3. Recommendations

It was stated in Chapter 1 that the comparative study is problem-oriented, which means that the observations about the differences and similarities are oriented by the question of how the regulatory approach can be developed to inclusively address sustainability concerns in both the environmental and socio-economic dimensions at different levels. From this perspective, the identified differences and similarities between the international, EU, and Chinese regulation are to reveal the implications for exploring the potential of an inclusive regulatory approach on a global scale. From the comparative analysis, it was found that the differences and limits are partly owing to the inherent complexity of regulating the sustainability of biofuels, and meanwhile it was also observed that there are similarities due to the effects of international harmonisation and integration as well as inter-state transplantation and convergence. These comparative findings imply that, while inclusive regulation is needed to prevent unsustainable biofuels from being produced and consumed somewhere on the globe, a sole dependence on either domestic or international law may not be able to address the complex issue of the sustainability of biofuels, so there is a need to foster a convergent inclusive understanding and regulation of the sustainability of biofuels at each regulatory level. Hence, it is recommended that the evolving process of 'global environmental law' – 'harmonisation', 'integration', 'transplantation', and 'convergence' – can be used to frame the regulation of the sustainability of biofuels according to different scales of impacts and likehood of being influenced by environmental and socio-economic diversity, so that common understandings, or a common set of principles for the sustainability of biofuels, could be formulated in a dynamic and interactive way.

Firstly, it is desirable and viable that rules or guidelines for the accounting of the life-cycle GHG emissions from biofuels could be harmonised and integrated under the UNFCCC. Above all, more attention could be given to specifying or clarifying the accounting of life-cycle GHG emissions from biofuels, since the current methodology based on the IPCC Guidelines for GHG inventories estimation may not guarantee the complete accounting of emissions from biofuels. Next, to specify life-cycle emissions accounting for biofuels under the international climate regime may find grounds in the objectives enshrined by the UNFCCC. Furthermore, compared to the other sustainability concerns, GHG emissions have a stronger global feature and the methodology of GHG emissions calculation can be less affected by the diversity of the environmental conditions and socio-economic values. Finally, the integration effect of GHG emissions trading and offsetting mechanisms under the Kyoto Protocol could be used.

Secondly, a common understanding of the definitions of the essential terms of the impacts of biofuels on biodiversity and food availability could be reached by further communication and collaboration among relevant international regimes/organisations. A common understanding of the key terms could draw on the efforts already made under the relevant international regimes. Moreover, the effort for a common understanding could be facilitated by increasing collaboration between international regimes/organisations, through reinforcing the liaison and communication between relevant treaty bodies and integrating substantive consideration for biodiversity, climate change, and food availability in the measures to be taken in relation to biofuels.

Thirdly, concerns about agro-environmental and socio-economic rights might be better regulated on a national or sub-national level with transplantation and convergence fostered through 'leading by example', bilateral and multilateral cooperation, and meta-standard. Above all, the effort of strengthening domestic sustainability regulation with regard to the local environment and socio-economic rights may drive transplantation to the other legal systems. Meanwhile, direct communication and cooperation in both bilateral and multilateral settings can increase communication of information and incentivise convergence. Furthermore, the meta-standard approach that combines legally binding criteria and private certification schemes could be used to promote convergence amongst domestic regulation of sustainability of biofuels.

In sum, the essence of an inclusive approach based on 'global environmental law' would be that a common set of principles for the sustainability of biofuels evolve in a dynamic and interactive way, with the view that the sustainability of biofuels has not reached a final stage and is hardly ready to be agreed upon internationally. Hence, convergent understandings and regulatory measures that include both environmental and socio-economic sustainability concerns could be developed through the dynamic processes of harmonisation, integration, transplantation, and convergence from different regulating levels and methods. In this way, the regulation of the sustainability of biofuels could be developed from a fragmented, divergent, and gapped 'mess' to a global inclusive 'mix.'

4. Final Conclusion and Outlook

At the end of the thesis, it is necessary to revisit the main research questions raised in Chapter 1 and see what conclusions could be drawn to answer the questions. Firstly, this thesis finds fragmentation, limits, and deficiencies in the regulatory approaches to the sustainability of biofuels respectively in the international, EU, and Chinese legal frameworks: sustainability concerns related to biofuels are addressed in a fragmented way under different international treaty regimes; the EU's sustainability scheme has limits in including socio-economic criteria and applying agro-environment criterion to biofuels of foreign origins; a clarification of China's general principles regarding the impacts of biofuels on food, land, and the environment is lacking. Secondly, this thesis finds divergent extents and limitations in including sustainability concerns in all the environmental, social, and economic dimensions of the regulation of biofuels: international regimes are usually limited to dealing with the sustainability concerns within their functional scopes; the EU and China contrast with each other in providing and including environmental or socio-economic requirements largely due to a different sustainability focus as well as the inherent complexity of regulating biofuel's sustainability. Finally, this thesis argues that the 'global environmental law' offers prospect for an inclusive regulation to deal with the sustainability of biofuels, by making use of the effects of harmonisation, integration, transplantation, and convergence.

The research carried out for this thesis and its conclusions may provide food for thought or be extended for future studies in the following ways. In the first place, this thesis has explored the regulation of the sustainability of biofuels, which can be extended to the other bioenergy sectors (i.e. biomass for electricity, heating, cooling) or even the 'bio-based economy' which may have similar problems of potentially resulting in negative environmental or socio-economic impacts due to the promoted consumption and production by governments, particularly when the case of biofuels shows that a sufficient evaluation of the sustainability impacts on various scales is crucial before any support policy is made. In the second place, this thesis has provided an analysis from the perspective of whether sustainability concerns in the environmental, social, and economic dimensions are included, and further studies could be carried out to analyse the trade-offs between the three dimensions in a more specific circumstance. In the third place, the sustainability of biofuels is a cross-cutting issue which involves regulation in many other relevant sectors such as land use, agriculture, and forestry, and the sustainability of biofuels in effect reflects the necessity of addressing the root causes of land use changes, unsustainable farming, and deforestation. In the fourth place, although this thesis recommends to develop inclusive regulatory approach to the sustainability of biofuels based on the 'global environmental law', further studies on the viability and implementation of such an approach are needed.

SUMMARY

DIFFERENT PATHS TOWARDS SUSTAINABLE BIOFUELS? COMPARATIVE STUDY OF THE INTERNATIONAL, EU, AND CHINESE REGULATION OF THE SUSTAINABILITY OF BIOFUELS

Biofuels are promoted as a type of renewable energy from biomass that replaces fossil fuels in transportation, in an attempt to achieve the three-fold objectives of energy security, rural development, and GHG emission reductions. However, the increased consumption and production of biofuels have been increasingly subject to criticism for their potential negative impacts on environmental and socio-economic sustainability, such as, *inter alia,* ecological and climate change impacts as well as negative social effects on food availability and accessibility. The environmental, social, and economic facets can be interconnected, and the potential implications may range from a local scale to the global scale, thus involving regulation in multiple sectors and at different levels. The concerns over sustainability of biofuels have already given rise to increasing discussions on how biofuels are and can be regulated to avoid adverse consequences, and eventually led to regulatory measures particularly for biofuels in various legal systems. In this context, this thesis is dedicated to a study on the theme of 'regulation of the sustainability of biofuels' consisting of research in response to the following three questions: *1)* what the regulatory approaches to sustainability of biofuel are in the selected legal frameworks, *2)* whether they include sustainability concerns in all the environmental, social, and economic dimensions, and, *3)* does the concept of 'global environmental law' offer a prospect for an inclusive approach for the regulation of the sustainability of biofuels?

For the first research question, this thesis has respectively examined the regulation of biofuels in the international, EU, and Chinese legal systems in Chapters 2, 3, and 4 and then compared their regulatory approaches in Chapter 5. Due to the 'mix' of instruments from various sectors, the analysis in these chapters is framed according to the effects on demand, supply, and trade of biofuels, which respectively focus on whether and how the regulatory measures constrain the demand and supply of unsustainable biofuels as well as their external influences.

In Chapter 2, in the absence of a specific international legal framework for the sustainability of biofuels, the international regime for regulating the sustainability of biofuels was mapped by looking into the different international legal regimes of climate change, biodiversity, socio-economic human rights conventions, and international trade law. In looking into the relationship between the international climate regime and biofuels, it was found that the commitment of addressing or reducing GHG emissions under the UNFCCC or the Kyoto Protocol may provide incentives for the use of biofuels, especially for Annex I Parties undertaking the GHG emission reduction obligation, if biofuels are regarded as a means to reduce GHG emissions in the transport sector. The international instruments in the fields of biodiversity and socio-economic rights are related to the supply of biofuels, such as the general commitment to conservation and sustainable/wise use of biodiversity, the decisions or resolutions on biofuels and biodiversity, and states' obligations with respect to the right to adequate food. The investigation into the relationship between the regulation of the sustainability of biofuels and WTO law concentrates on the question: to what extent WTO law may allow or constrain unilateral regulation on the sustainability of biofuels that affects the international trade of biofuels, particularly regarding the consistency the current provisions and case law under the WTO about production and process methods and jurisdictional limitation.

Chapter 3 examines the regulatory approach to the sustainability of biofuels in the EU, particularly the sustainability criteria provided in Directive 2009/28/EC (amended by Directive (EU) 2015/1513) which play a central role in the EU's regulatory framework on the demand, supply, and trade of biofuels. The sustainability criteria include requirements about life-cycle GHG emission saving, biodiversity, carbon stock, and agro-environment protection, and the EU also set forth the methodology for assessing GHG emission savings and the schemes for verifying compliance. On demand-side, the sustainability criteria are set out as binding conditions for biofuels to be treated favourably with regard to achieving national renewable energy consumption targets, receiving financial support, and complying with national blend obligations within the EU. In the aspect of supply, the sustainability criteria attempt to protect highly biodiverse areas or lands with high carbon stocks from biofuels' production, and the production of agricultural crops for biofuels is subject to the 'cross-compliance' and 'greening' requirements of the EU Common Agricultural Policy. By applying to biofuels originating in third countries, the sustainability criteria thus have an influence on the trade of biofuels between the EU and non-EU countries, and the external policies, such as reducing/exempting tariff rates and the negotiation of bilateral or multilateral agreements on biofuels, may provide tools for promoting compliance with the EU's sustainability requirements in the third countries.

Chapter 4 addresses the existing principles and requirements regarding the environmental and socio-economic concerns related to biofuels in China. The Chinese central government has set out general principles and specific requirements on certain environmental and socio-economic concerns related to biofuels, but a legal framework is not yet established for regulating the sustainability of biofuels at the national level. In the respect of demand, the central government has established general guiding principles for policy making on biofuels, including the use of non-food grain feedstocks, using marginal land that is not used for producing food, and no environmental damages, but these principles are not yet further specified in or coherently adhered to the other regulatory measures having effect on demand of biofuels, such as the biofuels blend mandate pilot projects and consumption targets. As for supply of biofuels, the general principles restrict the production of food grain-based biofuels and use of arable land, but there is lack of clear and binding scope of 'non-food grain feedstock' and 'marginal land.' Although trade of biofuels exists between China and third countries, the general principles or specific requirements related to the consumption and production of biofuels are not explicitly set out for imports from third countries, while bilateral and multilateral cooperation might provide opportunity for China to make progress in the development of efficient and sustainable biofuels and to be involved in international processes on the sustainability of biofuels.

Chapter 5 compares the regulatory approaches at the international, EU, and Chinese levels, and analyses the differences and similarities as well as their underlying reasons. With regard to regulation of the demand for biofuels, it is observed in particular that there is a difference in whether and what sustainability requirements are established for the demand-side regulatory measures, whilst convergence lies in the common concern about the potential negative impacts of biofuels, particularly on food availability. Differences as well as convergences are also found in defining and regulating the biodiversity, carbon stocks, and socio-economic rights concerns in the production of biofuels, but convergence exists in the approach to protecting certain areas with biodiversity values from biofuel production. Regarding the regulation of the trade of biofuels, in the context of WTO law, the EU and China have different approaches in treating biofuel (feedstock) from third countries, whilst both the EU and China have taken measures to engage in bilateral or multilateral cooperation on biofuels.

For the second research question, the regulation as examined in Chapters 2-5 is further reflected on from the perspective of whether both environmental and socio-economic sustainability concerns are included. In Chapter 2, it is found that the sustainability impacts of biofuels are addressed respectively under the different international regimes in a fragmented manner and there is a limit in the ability of each regime to integrate the environmental and socio-economic concerns inclusively, primarily owing to the 'self-contained' objectives and functions which present a barrier for individual regimes to inclusively address both environmental and socio-economic concerns. In Chapter 3, although the European Commission claims that the sustainability scheme is comprehensive and advanced, it is found that the sustainability criteria are limited in addressing the potential adverse effects in both environmental and socio-economic dimensions, largely due to the concern about compatibility with WTO law and the difficulty in defining commonly applicable social standards. Chapter 4 reveals that China has not yet established the concept of the 'sustainability of biofuels' or 'sustainable biofuels' in law or policy, nor has it set out legal criteria for the sustainability of biofuels, and analysis points out the deficiency in the current principles and requirements in integrating inclusive sustainability concerns in China. Chapter 5 further reflects on the differences in the prioritised sustainability concerns, the preferred regulatory approaches, and the degrees of possibilities to adopt complementary or stricter sustainability requirements in the regulation of biofuels. Despite the differences, limits in inclusively addressing the environmental and socio-economic sustainability dimensions exist at all the international, EU, and Chinese levels, which may have the consequence that 'unsustainable' biofuels continue to be demanded, supplied, or traded, and thus an inclusive sustainability regulation for biofuels is desirable.

The third research question is mainly discussed in Chapter 5, drawing on observations from comparative analysis in the previous sections. While inclusive sustainability regulation is desirable, challenges of providing inclusive regulation may exist at the international or domestic level due to the inherent complexities caused by the broad and vague meaning of 'sustainability', integrating cross-sectoral concerns, and addressing multi-scale impacts. Based on the differences and convergences observed from comparative analysis, the conceptual framework of 'global environmental law' as defined by Yang and Percival is used to explore the potential for a global inclusive approach for sustainability of biofuels. The major implications from 'global environmental law' for the potential of a global inclusive approach could be: instead of resorting to a sole 'top-down' or 'bottom-up'/'polycentric' approach, inclusive regulation on the sustainability of biofuels could be built upon the existing harmonisation, integration, transplantation, and convergence effects at different levels according to the scale of sustainability impact of biofuels and its subjectivity to diverse environmental or socio-economic conditions. Accordingly, recommendations are made as for regulation on the major sustainability concerns of biofuels, including accounting of life-cycle GHG emissions, biodiversity, food availability, agro-environment, and socio-economic rights, at different levels to evolve towards a global inclusive 'mix' in a dynamic and interactive way.

VALORISATION ADDENDUM

1. Societal Relevance

The research on the regulation of the sustainability of biofuels has a pivotal societal relevance with a view on a prevalent pursuit for low-emission and sustainable energies. The Paris Agreement adopted at the UNFCCC COP21 in December 2015 galvanized the long-term goal of achieving balance between anthropogenic greenhouse gas emissions in the second half of this century.[1] Declaration of this long-term goal is acclaimed as sending a signal for an irreversible trend of replacing fossil fuels with low-emission and sustainable energies.[2] Such a momentum has also culminated in the series of renewable energy initiatives or alliances launched under auspices of the Lima-Paris Action Agenda (LPAA).[3] In face of the rush towards a fossil-free economy and society, it is high time we reflect upon what 'clean' and 'sustainable' energy should be promoted and to what extent law and policy on climate and energy may ensure environmental and socio-economic sustainability in the production and consumption of energy. In this sense, biofuels provide a valuable case study.

Biofuels have been promoted by governments for decades as an affordable substitute for fossil fuels, in an attempt to reduce dependence on oil import, reduce

1 Paris Agreement, Art. 4(1).

2 See, for example, Kumi Naidoo, 'COP21: shows the end of fossil fuels is near, we must speed its coming', (Green Peace International, 12 December 2015), available at: <www.greenpeace.-org/international/en/news/Blogs/makingwaves/cop21-climate-talks-paris-negotiations-conclusion/blog/55092/> (last accessed 6 February 2016); John D. Sutter, 'This is the end of fossil fuels' (CNN, 14 December 2015), available at: <http://edition.cnn.com/-2015/12/12/opinions/sutter-cop21-climate-reaction/> (last accessed 6 February 2016).

3 The LPAA was launched in Lima at COP 20 which purported to build momentum to support a universal climate agreement at COP 21 in Paris. At COP 21, a quantity of renewable energy initiatives are reached and launched under spotlight of the LPAA thematic forum on energy, including the SIDS Lighthouses Initiative, the Global Geothermal Alliance, the Africa Renewable Energy Initiative, and the RE100 (Companies Leading with 100 Per Cent Renewable Energy). See more information at LPAA, Renewable Energy, available at: <http://newsroom.unfccc.int/lpaa/renewable-energy/> (last accessed 6 February 2016).

greenhouse gas emissions, while boosting rural economy; more recently, reducing GHG emissions in the transport sector has become the key objective. With increasing consumption and production of biofuels, they are increasingly subject to debates over the potential negative impacts on environmental and socio-economic sustainability. From an environmental perspective, the criticism concentrates on the ecological and climate change impacts during the process of producing biofuels, particularly with regard to the effects of land use changes. Apart from environmental concerns, biofuels are also criticised in view of negative social impact, since the feedstocks used for producing biofuels are mostly food sources and their diverted use for biofuel production may give rise to problems in food availability, land grabbing, and land use rights of local and indigenous communities. From an economic perspective, biofuels were assumed to create new economic opportunities for people in rural areas by increasing the demand for and prices of agricultural products, but the economic viability and benefits may have been overestimated, and many studies demonstrate resource efficiency problems with regard to certain types of biofuels when the input of energy and resources for producing biofuels is considered. Furthermore, the environmental, social, and economic facets can be intertwined – the environmental impacts might incur socio-economic problems and vice versa. Therefore, by looking into the regulatory approach to biofuels in response to the sustainability concerns, this research may question if and how 'sustainable' energy needs to be defined.

In fact, sustainability of biofuels has been addressed already in many regulatory frameworks. For example, sustainability of biofuels is considered under a number of domestic regulatory frameworks such as the US, Switzerland, Brazil, Germany, the UK, and the EU.[4] At an international level, instruments have been adopted under auspices of, for example, the Convention on Biological Diversity to specifically address the relation between biofuels and biodiversity. Moreover, sustainability of biofuels is an extremely cross-cutting issue, involving not only climate and energy law but also regulation relating to, inter alia, agriculture, forestry, land use, and biodiversity. Moreover, the long-lasting environmental and societal problems, such as unsustainable farming, biodiversity and carbon stock loss, and social injustice, underlie the negative effects of biofuels. Thus, there is a need to explore and map the regulatory terrain, and analysis of the key measures affecting demand and supply of biofuels may offer an opportunity to reflect on how sustainability can be regulated and safeguarded by law. Fully aware that one dissertation does not have time or space to look into all the existing measures, this research chose to concentrate on regulatory approaches of the EU and China as well as related international treaty regimes, for the following reasons. Firstly, both the EU and China have become the leading consumers and producers of biofuels, and have adopted regulatory measures in different ways to tackle potentially negative impacts. Secondly, as there is not yet a unified definition of the sustainability of

[4] A review of these regulatory measures has been made by under the FAO BEFS approach. Food and Agriculture Organization (FAO), Bioenergy and Food Security (BEFS), A Compilation of Bioenergy Sustainability Initiatives, available at: <www.fao.org/energy/befs/compilation/en/> (last accessed 10 February 2016).

biofuels, the examination and comparison of the EU and Chinese biofuel regulations may provide insight into how the two different legal systems integrate different sustainability concerns related to biofuels. Thirdly, the sustainability of biofuels is not a completely domestic issue and may have international or global ramifications, considering the effect on GHG emissions and biodiversity due to land use changes, so mapping the international setting may reveal what regulation exists for addressing sustainability of biofuels within the current international treaty regimes.

2. Target Group

This research is first and foremost relevant with policy makers in various areas. In the development of biofuels, the governmental policy and support have acted as the main drives for both the demand and supply of biofuels on market through, for instance, mandate of biofuel blending or energy tax exemptions. With mounting evidences of sustainability concerns, the question is if and how such regulatory instruments take into account the sustainability concerns and constrain the demand and supply of unsustainable biofuels. This thesis concentrates on the regulation which is promulgated by government or inter-governmental international regimes. In view that the biofuel issue involves multiple governmental departments - at least energy, agriculture, and climate change - this thesis adopts a 'demand-supply-trade' analytical model for understanding, mapping, and analysing a mix of regulatory measures in a more systematic and coherent manner. Moreover, by comparing biofuels regulation in the EU, China, and international regimes, this research may provide policy makers with a multi-level and global perspective in reflection on biofuel policies. Based on the comparative analysis, this research not only points out the insufficiencies in the current biofuel regulation, but also puts forward recommendations in view of the challenges and complexity of regulating sustainability of biofuels.

This research could be of great interest to organisations and scholars who attach attention to biofuels. Many regional as well as international inter-governmental organisations have noted or produced reports on sustainability issues of biofuels, such as those of International Energy Agency (IEA),[5] United Nations Environmental Programme (UNEP),[6] Food and Agriculture Organization (FAO),[7] and World Bank.[8] Furthermore, many organisations have contributed to developing indicators, criteria, and principles for the sustainability of biofuels, such as the Global Bioenergy Partnership (GBEP)[9] and FAO.[10] Sustainability of biofuels has also

5 International Energy Agency (IEA), 2011.

6 United Nations Environment Programme (UNEP), 2009.

7 Elbehri, Segerstedt & Liu 2013.

8 Rajagopal & Zilberman 2007.

9 Global Bioenergy Partnership (GBEP), 'The Global Bioenergy Partnership Sustainability Indicators for Bioenergy', First Edition, December 2011, available at: <www.globalbioenergy.org/fileadmin/user_upload/gbep/docs/Indicators/The_GBEP_Sustainability_Indicators_for_Bioenergy_FINAL.pdf> (last accessed 9 February 2015).

10 The FAO has developed Bioenergy and Food Security (BEFS) approach to include a multidisciplinary and integrated set of tools and guidance on sustainability of biofuels. Food →

been a topic cared for by a number of non-governmental organisations such as the International Union for Conservation of Nature (IUCN), the World Wide Fund for Nature (WWF), and European Environmental Bureau (EEB). Particularly, the environmental non-governmental organisations such as ClientEarth have lodged lawsuits relating to biofuels against the European Commission. Moreover, there have been increased discussions in legal literature on how biofuels are and can be regulated to avoid unsustainable consequences. Building on the previous studies and resources, this research aims to engage an analysis that looks into the sustainability issue of biofuels from a legal perspective and in a comparative and holistic approach.

Furthermore, this research may interest enterprises or industry associations in businesses relating to biofuels. As governmental support has played a critical role in the market penetration of biofuels, the biofuel industry is highly sensitive to regulatory and policy changes. In turn, how to balance the protection of existing investment and the prevention of environmental or socio-economic hazards is an important factor to be considered in the policy making of governments. Hence, this thesis may provide insight into the state-of-the-affairs of the regulatory progress in the EU, China, and international regimes, and its implications for the demand, supply, and trade of biofuels.

3. Activities/Products

This thesis is primarily based on a desktop research. Academic literature provides the major resources for information and arguments, which was accessed through sorting through the legal databases provided by Maastricht University's library, such as the HeinOnline, SSRN, Westlaw, and Picarta. The research has also referred to resources and publication from reports and studies of international organisations, non-governmental institutes, and scientific bodies. Regulatory documents and instruments were primarily accessed on the websites of the issuing authorities, and international instruments were obtained on the official websites of the issuing international treaty regimes and the United Nations Treaty Collection database. Besides desk research, face-to-face interviews, email correspondence, and news from media provided an additional source of information for this thesis. Interviews were carried out with biofuel policy officers from the European Commission DG Climate Action, Chinese experts on environmental and climate law, and biofuel consultant from ECOFYS. Information from the European Commission has also been sought by sending questions via the EUROPE DIRECT Contact Centre. News from the EurActiv was subscribed in order to keep abreast with the latest EU legislative and policy progresses.

The output of this research is the conclusions and recommendations regarding how sustainability of biofuels is and can be regulated. It is mentioned above that the analysis on the differences and similarities of regulation is problem-oriented, and its

and Agriculture Organization (FAO), Bioenergy and Food Security (BEFS), available at: <www.fao.org/energy/befs/en/> (last accessed 10 February 2015).

conclusion aims to reveal the implications for exploring the potential of an inclusive regulatory approach on a global scale. Therefore, based on the differences and limits partly owing to the inherent complexity of regulating the sustainability of biofuels, it is recommended that inclusive regulation is needed to prevent unsustainable biofuels from being produced and consumed in a global view by fostering a harmonisation, integration, transplantation, and convergence between different regulatory levels.

4. Innovation

A literature review reveals that attention to biofuels has frequently concentrated on one sustainability concern such as biodiversity or GHG emissions, while the other concerns are not equally addressed in the meantime. Moreover, literature has usually focused on the regulation of developed legal systems such as the EU, while regulation in less developed legal systems such as China is less examined. Meanwhile, recent literature focuses more on governance approaches featured with private standard and certification, but the regulation of the government still deserves detailed examination and analysis. Thus, this research examines regulation relating to the major sustainability concerns in both environmental and socio-economic aspects.

Furthermore, in addition to following the traditional comparative law approach of exploring regulation in at least different jurisdictions, determining their similarities and differences, explaining the causes and evaluating the solutions, the comparative study in this thesis is a functional comparison that identifies problems and effects as the yardsticks and views the even doctrinally different institutions as functionally equivalent and comparable. In the end, instead of determining which law is 'better' or arguing for an international legal unification on biofuels, the observations from comparative analysis will provide the basis for answering the question of how the sustainability of biofuels is regulated to avoid or minimise the potential negative impacts in a global inclusive approach, based on the implications from the conceptual framework of 'global environmental law' which depicts a dynamic process of 'convergence', 'transplantation', 'harmonization' and 'integration.'

5. Schedule & Implementation

This research may provide food for thought or be extended for future studies in the following ways. Firstly, exploration of the regulation in the specific case of biofuel's sustainability can be extended to the other bioenergy sectors (i.e. biomass for electricity, heating, cooling) or even the bio-economy as a whole which may have similar problems of negative environmental or socio-economic impacts, particularly when the case of biofuels shows that a sufficient evaluation of the sustainability impacts on various scales is crucial before any support policy is made. Secondly, this research has provided a normative analysis of the regulation on the sustainability of biofuels from the perspective of whether sustainability concerns in the environmental, social, and economic dimensions are inclusively addressed, while further studies could be carried out to analyse how to reach a trade-off or balance between impacts of the three dimensions. Thirdly, the sustainability of biofuels is a cross-cutting issue which involves regulation in many other relevant sectors such as land management, agriculture, and even forestry, and the unsustainable production of biofuels in effect reflects some root problems such as land use changes, unsustainable agriculture, and deforestation in these sectors, so more research is needed to address these root problems. Fourthly, this research suggests and sketches an inclusive approach that fosters the evolution of a common understanding and convergent regulation by harmonisation, integration, transplantation, and convergence, and further study on viability and functioning of such an approach is needed, particularly regarding the role to be played by public-private partnership and co-regulation. Fifthly, valorisation opportunity lies in developing and implementing standardisation and certification frameworks for sustainability of biomass in a trend of 'bio-based economy',[11] and there is opportunity of combining mandatory criteria and voluntary verifying schemes and establishing public-private partnerships[12]in this regard. The schedule for implementing these potential research projects depends largely on further funding opportunities.

11 For example, the OECD has issued a document entitled 'The Bioeconomy to 2030: Designing a Policy Agenda', and in 2012 the EU presented the strategy of 'Innovating for Sustainable Growth: A Bioeconomy for Europe.' Besides, many countries have either published separate strategies and policies relating to bio-based products and industries, or developed strategies that collect all separate topics under the conceptual umbrella of bioeconomy. Staffas, Gustavsson & McCormick 2013, p. 2752.

12 James Crisp, 'Council rubber stamps private-public partnerships in biomass, air, energy, electronics, medicine' (EurActiv, 7 May 2014), available at: <www.euractiv.com/sections/-innovation-enterprise/council-rubber-stamps-private-public-partnerships-biomass-air-energy> (last accessed 7 February 2016).

RELEVANT DOCUMENTS

1. Relevant Documents at the International Level

Addis Ababa Principles and Guidelines 2004
Secretariat of the Convention on Biological Diversity, Addis Ababa Principles and Guidelines for the Sustainable Use of Biodiversity (CBD Guidelines) (Montreal: Secretariat of the Convention on Biological Diversity, 2004).

CBD
Convention on Biological Diversity, United Nations, 1992.

CBD SBSTTA Recommendation XII/7 2007
Twelfth meeting of the Subsidiary Body on Scientific, Technical and Technological Advice, Recommendation XII/7, 'Biodiversity and Biofuel Production', 2007.

CBD/COP/12/23 2014
Conference of the Parties to the Convention on Biological Diversity Twelfth Meeting, 'Biodiversity and Biofuels: Implementation of Decisions', Note by the Executive Secretary, UNEP/CBD/COP/12/23, Pyeongchang, 2014.

CBD/SBSTTA/16/14 2012
'Biofuels and Biodiversity: Report on the Work in Response to Decision X/37', Note by the Executive Secretary, UNEP/CBD/SBSTTA/16/14, 23 February 2012

CBD/SBSTTA/18/15 2014
Subsidiary Body on Scientific, Technical and Technological Advice Eighteenth meeting, 'Biofuels and Biodiversity Information on Relevant Definitions of Relevant Key Terms to Enable Parties to Implement Decisions IX/2 and X/37', Note by the Executive Secretary, UNEP/CBD/SBSTTA/18/15, Montreal, 2014.

Chicago Convention 1944
Convention on International Aviation done at Chicago, 7 December 1944.

Commission Communication COM(2012) 271 final
Communication from the Commission to the European Parliament, the Council, the European Economic and Social Committee and the Committee of the Regions Renewable Energy: A Major Player in the European Energy Market, 06/06/2012.

Convention 169
Convention 169 – Indigenous and Tribal Peoples Convention, 1989 (No. 169), Convention concerning Indigenous and Tribal Peoples in Independent Countries, Adopted at the 76th ILC session, Geneva, 1989.

Decision 3/CMP.1 2005
Kyoto Protocol, Decision 3/CMP.1 'Modalities and Procedures for a Clean Development Mechanism as defined in Article 12 of the Kyoto Protocol', FCCC/KP/CMP/2005/8/Add.1.

Decision 22/CMP.1 2005
Kyoto Protocol, Decision 22/CMP.1, 'Guidelines for Review under Article 8 of the Kyoto Protocol', FCCC/KP/CMP/2005/8/Add.3.

Decision 4/CMP.7 2011
Kyoto Protocol, Decision 4/CMP.7, 'Greenhouse Gases, Sectors and Source Categories, Common Metrics to Calculate the Carbon Dioxide Equivalence of Anthropogenic Emissions by Sources and Removals by Sinks, and other Methodological Issue', FCCC/KP/CMP/2011/10/Add.1.

Decision 8/CMP.7 2011
Kyoto Protocol, Decision 8/CMP.7, 'Further Guidance relating to the Clean Development Mechanism', FCCC/KP/CMP/2011/10/Add.2.

Decision 5/CMP.8 2012
Kyoto Protocol, Decision 5/CMP.8, 'Guidance relating to the Clean Development Mechanism', FCCC/KP/CMP/2012/13/Add.2.

Decision 1/CMP.8 2012
Kyoto Protocol, Decision 1/CMP.8, 'Amendment to the Kyoto Protocol pursuant to its Article 3, paragraph 9 (the Doha Amendment)', FCCC/KP/CMP/2012/13/Add.1.

Decision 3/CMP.9 2013
Kyoto Protocol, Decision 3/CMP.9, 'Guidance relating to the Clean Development Mechanism', FCCC/KP/CMP/2013/9/Add.1.

Decision 6/CMP.9 2013
Kyoto Protocol, Decision 6/CMP.9, 'Guidance for Reporting Information on Activities under Article 3, paragraphs 3 and 4, of the Kyoto Protocol', FCCC/KP/CMP/2013/9/Add.1.

Decision 5/CMP.9 2013
Kyoto Protocol, Decision 5/CMP.9, 'Guidance on the Implementation of Article 6 of the Kyoto Protocol', FCCC/KP/CMP/2013/9/Add.1.

Decision 4/CMP.10 2014
Kyoto Protocol, Decision 4/CMP.10, 'Guidance relating to the Clean Development Mechanism', FCCC/KP/CMP/2014/9/Add.1.

Decision 4/CP.1 1995
UNFCCC, Decision 4/CP.1, 'Methodological Issues', FCCC/CP/1995/7/Add.1.

Decision 2/CP.3 1997
UNFCCC, Decision 2/CP.3, 'Methodological Issues related to the Kyoto Protocol', FCCC/CP/1997/7/Add.1.

Decision 17/CP.8 2002
UNFCCC, Decision 17/CP.8, 'Guidelines for the Preparation of National Communications from Parties not Included in Annex I to the Convention', FCCC/CP/2002/7/Add.2.

Decision 1/CP.17 2011
UNFCCC, Decision 1/CP.17, 'Establishment of an Ad Hoc Working Group on the Durban Platform for Enhanced Action', FCCC/CP/2011/9/Add.1.

Decision 2/CP.17 2011
UNFCCC, Decision 2/CP.17, 'Outcome of the work of the Ad Hoc Working Group on Long-term Cooperative Action under the Convention', FCCC/CP/2011/9/Add.1.

Decision 15/CP.17 2011
UNFCCC, Decision 15/CP.17, 'Revision of the UNFCCC Reporting Guidelines on Annual Inventories for Parties Included in Annex I to the Convention', FCCC/CP/2011/9/Add.2.

Decision 1/CP.18 2012
UNFCCC, Decision 1/CP.18, 'Agreed Outcome pursuant to the Bali Action Plan', FCCC/CP/2012/8/Add.1.

Decision 22/CP.19 2013
UNFCCC, Decision 22/CP.19, 'Sixth National Communications from Parties Included in Annex I to the Convention', FCCC/CP/2013/10/Add.2/Rev.1.

Decision 24/CP.19 2013
UNFCCC, Decision 24/CP.19, 'Revision of the UNFCCC Reporting Guidelines on Annual Inventories for Parties Included in Annex I to the Convention', FCCC/CP/2013/10/Add.3.

Decision IX/2 2008
Conference of the Parties to the Convention on Biological Diversity at its Nineth Meeting IX/2, 'Agricultural biodiversity: Biofuels and Biodiversity', UNEP/CBD/COP/DEC/IX/2, Bonn, 9 Octobre 2008.

Decision X/37 2010
Conference of the Parties to the Convention on Biological Diversity at its Tenth Meeting X/37, 'Biofuels and Biodiversity', UNEP/CBD/COP/DEC/X/37 (2010), Nagoya, 18-29 October 2010.

Decision XI/27 2012
Conference of the Parties to the Convention on Biological Diversity at its Eleventh Meeting XI/27, 'Biofuels and Biodiversity', UNEP/CBD/COP/DEC/XI/27, Hyderabad, India, 5 December 2012.

GATT 1994
General Agreement on Tariffs and Trade, 1994.

General Comment No. 3
CESCR General Comment No. 3, The Nature of States Parties' Obligations (Art. 2, Para. 1, of the Covenant) adopted at the Fifth Session of the Committee on Economic, Social and Cultural Rights, on 14 December 1990 (Contained in Document E/1991/23).

General Comment No. 4
CESCR General Comment No. 4, The Right to Adequate Housing (Art. 11(1) of the Covenant), Adopted at the Sixth Session of the Committee on Economic, Social and Cultural Rights, on 13 December 1991 (Contained in Document E/1992/23).

General Comment No. 7
CESCR General Comment No. 7, The Right to Adequate Housing (Art. 11.1): Forced Evictions, Adopted at the Sixteenth Session, UN Committee on Economic, Social and Cultural Rights, E/1998/22, 1997.

General Comment No. 12
CESCR General Comment No. 12, The Right to Adequate Food (Art. 11) Adopted at the Twentieth Session of the Committee on Economic, Social and Cultural Rights, on 12 May 1999 (Contained in Document E/C.12/1999/5).

General Comment No. 15
CESCR General Comment No. 15, The Right to Water (Arts. 11 and 12 of the International Covenant on Economic, Social and Cultural Rights), Adopted at the 29th session of the Committee on Economic, Social and Cultural Rights on 11-29 November 2002 (Contained in Document E/C.12/2002/11).

ICAO 38th Session 2013 Assembly Resolution
International Civil Aviation Organization, Resolutions Adopted By The Assembly, Assembly - 38th Session Montréal, 24 September - 4 October 2013 93-99 (2013).

ICAO 38th session Executive Committee Assembly Agenda Item 17 2013
Executive Committee Assembly - 38th session, Agenda Item 17: Environmental protection Consolidated Statement of Continuing ICAO Policies and Practices Related to Environmental Protection - Climate Change (Presented by the Council of ICAO) Executive Summary, A38-Wp/34 Ex/29 Appendix, 2013.

ICESCR
International Covenant on Economic, Social and Cultural Rights, Adopted and Opened for Signature, Ratification and Accession by General Assembly resolution 2200A (XXI) of 16 December 1966, entry into force 3 January 1976, in accordance with Article 27.

ICJ Case Gabcikovo-Nagymaros Judgment 1997
ICJ, Reports of Judgments, Advisory Opinions and Orders Case Concerning the Gabcikovo-Nagymaros Project (Hungary/Slovakia) Judgment of 25 September 1997.

ICJ Case Pulp Mills Judgment 2010
ICJ, Report of Judgments, Advisory Opinions and Orders, Case concerning Pulp Mills on the River Uruguay (ARGENTINA v. URUGUAY), Judgment of 20 April 2010.

International Labour Office 2012
International Labour Office, 'General Survey on the Fundamental Conventions concerning Rights at Work in Light of the ILO Declaration on Social Justice for a Fair Globalization 2008', Report III (Part 1B), Report of the Committee of Experts on the Application of Conventions and Recommendations (Articles 19, 22 and 35 of the Constitution), International Labour Conference, 101st Session, ILC.101/III/1B, 2012.

IPCC Guidelines 2006
Intergovernmental Panel on Climate Change, '2006 IPCC Guidelines for National Greenhouse Gas Inventories', Prepared by the National Greenhouse Gas Inventories Programme, edited by Eggleston H.S., *et al.* (Kanagawa, Japan: IGES, 2006).

Kyoto Protocol
Kyoto Protocol to the United Nations Framework Convention on Climate Change, United Nations, 1998.

Maastricht Principles
Maastricht Principles on Extraterritorial Obligations of States in the area of Economic, Social and Cultural Rights, Final version 29 February 2012.

Paris Agreement
UNFCCC, 'Adoption of the Paris Agreement', FCCC/CP/2015/L.9/Rev.1 (2015), Annex 'Paris Agreement', 12 December 2015.

Ramsar Convention
Convention on Wetlands of International Importance especially as Waterfowl Habitat (Ramsar, Iran, 2.2.1971), as amended by the Protocol of 3.12.1982 and the Amendments of 28.5.1987, Director, Office of International Standards and Legal Affairs United Nations Educational, Scientific and Cultural Organization (UNESCO), Paris, 13 July 1994.

Report of the Special Rapporteur on the rigft to food 2007
Report of the Special Rapporteur on the right to food, United Nations General Assembly, 62th Session, Item 72(b), A/62/289 (2007)).

Resolution VII. 16
The 7th Meeting of the Conference of the Contracting Parties to the Convention on Wetlands (Ramsar, Iran, 1971) Resolution VII.16, 'The Ramsar Convention and Impact Assessment: Strategic, Environmental And Social', San José, Costa Rica, 10-18 May 1999.

Resolution X. 17
The 10th Meeting of the Conference of the Parties to the Convention on Wetlands (Ramsar, Iran, 1971) Resolution X.17, 'Environmental Impact Assessment and Strategic Environmental Assessment: Updated Scientific and Technical Guidance', 'Healthy wetlands, healthy people', Changwon, Republic of Korea, 28 October-4 November 2008.

Resolution X. 25
The 10th Meeting of the Conference of the Parties to the Convention on Wetlands (Ramsar, Iran, 1971) Resolution X.25, 'Wetlands and "biofuels"', Changwon, Republic of Korea, 28 October-4 November 2008.

Resolution XI. 10
The 11th Meeting of the Conference of the Parties to the Convention on Wetlands (Ramsar, Iran, 1971) Resolution XI.10, 'Wetlands and Energy Issues', Bucharest, Romania, 6-13 July 2012.

Rio Declaration on Environment and Development 1992
The United Nations Conference on Environment and Development, 'Rio Declaration on Environment and Development', United Nations publication, No. E.73.II.A.14 and corrigendum, Chap. I., Rio de Janeiro, 3 to 14 June 1992.

UNFCCC
United Nations Convention on Climate Change (the consolidated versions of the Convention text, including amendments to Annex I and II), United Nations, 1992.

UNFCCC SBSTA 37th Session 2012
UNFCCC, SBSTA 37th session 2012, Information Relevant to Emissions from Fuel Used for International Aviation and Maritime Transport, Submissions from International Organisations, 2012.

United Nations Declaration on the Rights of Indigenous Peoples
United Nations Declaration on the Rights of Indigenous Peoples, Resolution adopted by the General Assembly [without reference to a Main Committee (A/61/L.67 and Add.1)], 61/295, March 2008.

Voluntary Guidelines on Tenure of Land
Voluntary Guidelines on the Responsible Governance of Tenure of Land, Fisheries and Forests in the Context of National Food Security, Food and Agriculture Organization of the United Nations, Rome, 2012.

World Commission on Environment and Development 'Our Common Future' 1987
Report of the World Commission on Environment and Development 'Our Common Future', Annex to the Report of the World Commission on Environment and Development, Note by the Secretary-General, Records of the General Assembly, Forty-second Session, Supplement No, 25 (A/42/25), 4 August 1987.

2. Relevant Documents in the EU

Biofuels Progress Report COM(2006) 845 final
Biofuels Progress Report – Report on the Progress made in the Use of Biofuels and Other Renewable Fuels in the Member States of the European Union COM(2006) 845 final.

Case C-176/05
Judgment of the Court (First Chamber) of 1 March 2007, *KVZ retec GmbH v Republik Österreich.*

Case C-201/08
Judgment of the Court (Third Chamber) of 10 September 2009, *Plantanol GmbH & Co. KG v Hauptzollamt Darmstadt.*

Case T-278/11
Order of the General Court (Fourth Chamber) of 13 November 2012, *ClientEarth, Friends of the Earth Europe, Stichting FERN and Stichting Corporate Europe Observatory v European Commission.*

Case T-120/10
Order of the General Court (Fifth Chamber) of 9 November 2011, *ClientEarth and Others v Commission.*

Case T-499/10
Order of the General Court (Fifth Chamber) of 9 November 2011, *ClientEarth and Others v Commission.*

Commission Communication COM(97) 196 final
Communication of 'The Energy Dimension of Climate Change', COM(97) 196 final, 14.5.1997.

Commission Communication COM(97) 599 final
Communication from the Commission 'Energy for the Future: Renewable Sources of Energy' White Paper for a Community Strategy and Action Plan, COM(97) 599 final, 26.11.1997.

Commission Decision 1999/835/EC
Commission Decision of 26 October 1999 on the National Provisions notified by the Kingdom of Denmark concerning the Limitation to the Placing on the Market and Use of Creosote (notified under document number C(1999) 3427), *OJ* L329/82.

Commission Communication COM(2005) 628 final
Communication from the Commission COM(2005) 628 final, Biomass Action Plan SEC (2005) 1573 Brussels, 7.12.2005.

Commission Communication COM(2006) 34 final
Commission Communication 'An EU Strategy for Biofuels', COM(2006) 34 final, 8.2.2006.

Commission Communication COM(2006) 848 final
Communication from the Commission to the Council and the European Parliament 'Renewable Energy Road Map – Renewable Energies in the 21st Century: Building a More Sustainable Future', COM(2006) 848 final.

Commission Communication COM(2007) 642 final
Commission Communication 'Progress Report on the Sustainable Development Strategy 2007', COM(2007) 642 final, 22.10.2007.

Commission Communication COM(2009) 400 final
Commission Communication, 'Mainstreaming Sustainable Development into EU Policies: 2009 Review of the European Union Strategy for Sustainable Development' COM(2009) 400 final, 24.7.2009.

Commission Communication (2010/C 160/01)
Communication from the Commission on Voluntary Schemes and Default Values in the EU biofuels and bioliquids sustainability scheme (2010/C 160/01), *OJ* C160/1, 19.6.2010 (2010).

Commission Communication (2010/C 160/02)
Communication from the Commission on the Practical Implementation of the EU Biofuels and Bioliquids Sustainability Scheme and on Counting Rules for Biofuels (2010/C 160/02), *OJ* C160/8, 19.6.2010.

Commission Communication COM(2011) 31 final
Communication from the Commission to the European Parliament and the Council Renewable Energy: Progressing towards the 2020 target COM(2011) 31 final, Brussels, 31.1.2011.

Commission Communication COM(2011) 244 final
Communication from the Commission to the European Parliament, the Council, the Economic and Social Committee and the Committee of the Regions 'Our Life Insurance, Our Natural Capital: An EU Biodiversity Strategy to 2020', COM(2011) 244 final, Brussels, 3.5.2011.

Commission Communication COM(2013) 17 final
Communication from the Commission, 'Clean Power for Transport: A European Alternative Fuels Strategy' COM(2013) 17 final, Brussels, 24.1.2013.

Commission Proposal COM(2014) 617 final
Proposal for a Council Directive on laying down Calculation Methods and Reporting Requirements pursuant to Directive 98/70/EC of the European Parliament and of the Council relating to the Quality of Petrol and Diesel Fuels COM(2014) 617 final, Brussels, 6.10.2014.

Commission Communication (2014/C 200/01)
Communication from the Commission Guidelines on State Aid for Environmental Protection and Energy 2014-2020 (2014/C 200/01).

Commission Communication COM(2014) 015 final
Communication from the Commission to the European Parliament, the Council, the European Economic and Social Committee and the Committee of the Regions, 'A Policy Framework for Climate and Energy in the Period from 2020 to 2030', COM(2014) 015 final.

Commission Decision 2007/446/EC
Commission Decision of 21 June 2007 on the Participation of the Commission of the European Communities in the International Biofuels Forum (2007/446/EC), *OJ* L169/63 29.6.2007.

Commission Decision 2010/335/EU
Commission Decision of 10 June 2010 on Guidelines for the Calculation of Land Carbon Stocks for the Purpose of Annex V to Directive 2009/28/EC (notified under document C(2010) 3751) (2010/335/EU), *OJ* L151/19 17.6.2010.

Commission Implementing Regulation (EU) No 1101/2014
Commission Implementing Regulation (EU) No 1101/2014 of 16 October 2014 amending Annex I to Council Regulation (EEC) No 2658/87 on the Tariff and Statistical Nomenclature and on the Common Custom Tariff.

Commission Proposal COM (97) 88 final
Proposal for a Council Directive on 'Reduction of Sulphur Content in Certain Liquid Fuels' and modification of Directive EC/93/12 COM(97) 88 final, 12/3/1997.

Commission Proposal COM(2008) 19 final
Proposal for a Directive of the European Parliament and of the Council on the Promotion of the Use of Energy from Renewable Sources (presented by the Commission) COM(2008) 19 final, Brussels, 23.1.2008.

Commission Proposal COM(2011) 169/3
Proposal for a Council Directive amending Directive 2003/96/EC Restructuring the Community Framework for the Taxation of Energy Products and Electricity COM(2011) 169/3.

Commission Proposal COM(2012) 595 final
European Commission Proposal for a Directive of the European Parliament and of the Council amending Directive 98/70/EC and amending 2009/28/EC COM(2012) 595 final, 2012/0288 (COD).

Commission Proposal COM(2014) 617 final
Proposal COM(2014) 617 final for a Council Directive on Laying Down Calculation Methods and Reporting Requirements pursuant to Directive 98/70/EC of the European Parliament and of the Council relating to the Quality of Petrol and Diesel Fuels 2014/0286 (NLE) (2014), Brussels, 6.10.2014.

Commission Regulation (EC) No 1122/2009
Commission Regulation (EC) No 1122/2009 of 30 November 2009 Laying Down Detailed Rules for the Implementation of Council Regulation (EC) No 73/2009 as Regards Cross-Compliance, Modulation and the Integrated Administration and Control System, under the Direct Support Schemes for Farmers Provided for that Regulation, as well as for the Implementation of Council Regulation (EC) No 1234/2007 as regards cross-compliance under the support scheme provided for the wine sector, *OJ* L316, 2.12.2009.

Commission Regulation (EU) No 601/2012
Commission Regulation (EU) No 601/2012 of 21 June 2012 on the Monitoring and Reporting of Greenhouse Gas Emissions pursuant to Directive 2003/87/EC of the European Parliament and of the Council Text with EEA relevance, *OJ* L181, 12.7.2012.

Commission Regulation (EU) No 1123/2013
Commission Regulation (EU) No 1123/2013 of 8 November 2013 on Determining International Credit Entitlements pursuant to Directive 2003/87/EC of the European Parliament and of the Council, *OJ* L299/32, 9.11.2013.

Commission Regulation (EU) No 651/2014
Commission Regulation (EU) No 651/2014 of 17 June 2014 Declaring Certain Categories of Aid Compatible with the Internal Market in Application of Articles 107 and 108 of the Treaty, *OJ* L187/1, 26.6.2014.

Commission Regulation (EU) No 1307/2014
Commission Regulation (EU) No 1307/2014 of 8 December 2014 on Defining the Criteria and Geographic Ranges of Highly Biodiverse Grassland for the Purposes of Article 7b(3)(c) of Directive 98/70/EC of the European Parliament and of the Council relating to the Quality of Petrol and Diesel Fuels and Article 17(3)(c) of Directive 2009/28/EC of the European Parliament and of the Council on the promotion of the use of energy from renewable sources Commission Regulation (EU) No 1307/2014, *OJ* L351/3, 9.12.2014.

Commission Report COM (2010) 11 final
Report from the Commission to the Council and the European Parliament on Sustainability Requirements for the Use of Solid and Gaseous Biomass Sources in Electricity, Heating and Cooling COM(2010) 11 final, Brussels, 25.2.2010.

Commission Report COM (2010) 811 final
Report from the Commission on Indirect Land-use Change related to Biofuels and Bioliquids COM(2010) 811 final, Brussels, 22.12.2010.

Commission Report COM (2013) 175 final
Report from the Commission 'Renewable Energy Progress Report' COM(2013) 175 final, Brussels, 27.3.2013.

Commission Report COM (2015) 293 final
Report from the Commission to the European Parliament, the Council, the European Economic and Social Committee and the Committee of the Regions Renewable Energy Progress Report COM(2015) 293 final, 15.6.2015.

Commission Staff Working Document SWD (2012) 343 final
Commission Staff Working Document Impact Assessment Accompanying the document Proposal for a Directive of the European Parliament and of the Council amending Directive 98/70/EC relating to the Quality of Petrol and Diesel Fuels and amending Directive 2009/28/EC on the Promotion of the Use of Energy from Renewable Sources SWD (2012) 343 final, Brussels, 17.10.2012.

Commission Staff Working Document SWD (2012) 344 final
Commission Staff Working Document Executive Summary of the Impact Assessment on Indirect Land-Use Change Related to Biofuels and Bioliquids Accompanying the document Proposal for a Directive of the European Parliament and of the Council amending Directive 98/70/EC relating to the Quality of Petrol and Diesel Fuels and amending Directive 2009/28/EC on the Promotion of the Use of Energy from Renewable Sources SWD (2012) 344 final, Brussels, 17.10.2012.

Commission Staff Working Document SWD (2013) 102 final
Commission Staff Working Document Accompanying the document Report from the Commission 'Renewable Energy Progress Report' SWD (2013) 102 final, Brussels, 27.3.2013.

Commission Staff Working Document SWD (2014) 259 final
Commission Staff Working Document State of play on the Sustainability of Solid and Gaseous Biomass used for Electricity, Heating and Cooling in the EU SWD (2014) 259 final Brussels, 28.7.2014.

Commission Staff Working Document SWD (2015) 117 final
Commission Staff Working Document Technical Assessment of the EU Biofuel Sustainability and Feasibility of 10 per cent Renewable Energy Target in Transport Accompanying the Document Report from the Commission to the European Parliament, the Council, the European Economic and Social Committee and the Committee of the Regions Renewable Energy Progress Report, SWD (2015) 117 final, 15.6.2015.

Council Conclusion (11 December 2009)
European Council, European Council 10/11 December 2009 Conclusions, EUCO 6/09 CO EUR 6 CONCL 4, Brussels, 11.12.2009.

Council Conclusion on 2030 Climate and Energy Policy Framework
European Council (23 and 24 October 2014) Conclusions on 2030 Climate and Energy Policy Framework Brussels, 23 October 2014 (OR. en) SN 79/14.

Council Decision 93/500/EEC
Council Decision 93/500/EEC of 13 September 1993 concerning the Promotion of Renewable Energy Sources in the Community (Altener programme), *OJ* L235, 18.09.1993.

Council Decision 98/352/EC
Council Decision 98/352/EC of 18 May 1998 concerning a Multiannual Programme for the Promotion of Renewable Energy Sources in the Community (Altener II), *OJ* L159, 03.06.1998.

Council Decision 1999/21/EC
Euratom: Council Decision 1999/21/EC of 14 December 1998 adopting a Multiannual Framework Programme for Actions in the Energy Sector (1998-2002) and Connected Measures, *OJ* L007, 13.1.1999.

Council Implementing Regulation (EU) No 443/2011
Council Implementing Regulation (EU) No 443/2011 of 5 May 2011 extending the definitive countervailing duty imposed by Regulation (EC) No 598/2009 on imports of biodiesel originating in the United States of America to imports of biodiesel consigned from Canada, whether declared as originating in Canada or not, and extending the definitive countervailing duty imposed by Regulation (EC) No 598/2009 to imports of biodiesel in a blend containing by weight 20 per cent or less of biodiesel originating in the United States of America, and terminating the investigation in respect of imports consigned from Singapore, *OJ* L122, 11.5.2011.

Council Implementing Regulation (EU) No 444/2011
Council Implementing Regulation (EU) No 444/2011 of 5 May 2011 extending the definitive anti-dumping duty imposed by Regulation (EC) No 599/2009 on imports of biodiesel originating in the United States of America to imports of biodiesel consigned from Canada, whether declared as originating in Canada or not, and extending the definitive anti- dumping duty imposed by Regulation (EC) No 599/2009 to imports of biodiesel in a blend containing by weight 20 per cent or less of biodiesel originating in the United States of America, and terminating the investigation in respect of imports consigned from Singapore, *OJ* L122/12 11. 5. 2011.

Council Implementing Regulation (EU) No 1194/2013
Council Implementing Regulation (EU) No 1194/2013 of 19 November 2013 Imposing a Definitive Anti-dumping Duty and Collecting Definitively the Provisional Duty Imposed on Imports of Biodiesel Originating in Argentina and Indonesia, *OJ* L315/2, 26.11.2013.

Council Position (10 December 2014)
Position of the Council at first reading with a view to the adoption of a Directive of The European Parliament and of the Council amending Directive 98/70/EC relating to the Quality of Petrol and Diesel Fuels and amending Directive 2009/28/EC on the Promotion of the Use of Energy from Renewable Sources (10 December 2014).

Decision No 377/2013/EU
Decision No 377/2013/EU of the European Parliament and of the Council of 24 April 2013 derogating temporarily from Directive 2003/87/EC Establishing a Scheme for Greenhouse Gas Emission Allowance Trading within the Community, *OJ* L113, 25.4.2013.

Decision No 529/2013/EU
Decision No 529/2013/EU of the European Parliament and of the Council of 21 May 2013 on Accounting Rules on Greenhouse Gas Emissions and Removals Resulting from Activities relating to Land use, Land-use Change and Forestry and on Information concerning Actions relating to Those activities, *OJ* L165/80, 18.6.2013.

Directive 79/409/EEC
Council Directive 79/409/EEC of 2 April 1979 on the Conservation of Wild Birds, *OJ* L103, 25.4.1979.

Directive 80/68/EEC
Council Directive 80/68/EEC of 17 December 1979 on the Protection of Groundwater against Pollution caused by Certain Dangerous Substances, *OJ* L20, 26.1.1980.

Directive 85/536/EEC
Council Directive 85/536/EEC of 5 December 1985 on Crude-oil Savings through the Use of Substitute Fuel Components in Petrol, *OJ* L334/20, 12.12.1985.

Directive 86/278/EEC
Council Directive 86/278/EEC of 12 June 1986 on the Protection of the Environment, and in particular of the Soil, When Sewage Sludge is Used in Agriculture, *OJ* L181, 4.7.1986.

Directive 91/414/EEC
Council Directive 91/414/EEC of 15 July 1991 concerning the Placing of Plant Protection Products on the Market, *OJ* L230, 19.8.1991.

Directive 91/676/EEC
Council Directive 91/676/EEC of 12 December 1991 concerning the Protection of Waters against Pollution caused by Nitrates from Agricultural Sources, *OJ* L375, 31.12.1991.

Directive 92/43/EEC
Council Directive 92/43/EEC of 21 May 1992 on the Conservation of Natural Habitats and of Wild Fauna and Flora, *OJ* L206, 22.07.1992.

Directive 92/83/EEC
Council Directive 92/83/EEC of 19 October 1992 on the Harmonization of the Structures of Excise Duties on Alcohol and Alcoholic Beverages, *OJ* L316, 31.10.1992.

Directive 98/70/EC
Directive 98/70/EC of the European Parliament and of the Council of 13 October 1998 relating to the Quality of Petrol and Diesel Fuels and amending Council Directive 93/12/EEC, *OJ* L350, 28.12.1998.

Directive 2000/60/EC
Directive 2000/60/EC of the European Parliament and of the Council of 23 October 2000 Establishing a Framework for Community Action in the Field of Water Policy, *OJ* L327, 22.12.2000.

Directive 2001/18/EC
Directive 2001/18/EC of the European Parliament and of the Council of 12 March 2001 on the Deliberate Release into the Environment of Genetically Modified Organisms and Repealing Council Directive 90/220/EEC - Commission Declaration, *OJ* L106, 17.4.2001.

Directive 2001/77/EC
Directive 2001/77/EC of the European Parliament and of the Council of 27 September 2001 on the promotion of Electricity Produced from Renewable Energy Sources in the Internal Electricity Market, *OJ* L283, 27.10.2001.

Directive 2002/91/EC
Directive 2002/91/EC of the European Parliament and of the Council of 16 December 2002 on the Energy Performance of Buildings, *OJ* L001, 04.01.2003.

Directive 2003/30/EC
Directive 2003/30/EC of the European Parliament and of the Council of 8 May 2003 on the promotion of the Use of Biofuels or Other Renewable Fuels for Transport, *OJ* L123/42, 17.5.2003.

Directive 2003/54/EC
Directive 2003/54/EC of the European Parliament and of the Council of 26 June 2003 concerning Common Rules for the Internal Market in Electricity and Repealing Directive 96/92/EC - Statements Made with Regard to Decommissioning and Waste Management Activities, *OJ* L176, 15.7.2003.

Directive 2003/87/EC
Directive 2003/87/EC of The European Parliament and of the Council of 13 October 2003 Establishing a Scheme for Greenhouse Gas Emission Allowance Trading within the Community and amending Council Directive 96/61/EC, *OJ* L275, 25.10.2003.

Directive 2003/96/EC
Council Directive 2003/96/EC of 27 October 2003 Restructuring the Community Framework for the Taxation of Energy Products and Electricity, *OJ* L283, 31.10.2003.

Directive 2004/8/EC
Directive 2004/8/EC of the European Parliament and of the Council of 11 February 2004 on the promotion of Cogeneration based on a Useful Heat Demand in the Internal Energy Market and amending Directive 92/42/EEC, *OJ* L52, 21.2.2004.

Directive 2004/101/EC
Directive 2004/101/EC of the European Parliament and of the Council of 27 October 2004, *OJ* L338/18, 13.11.2004.

Directive 2008/98/EC
Directive 2008/98/EC of the European Parliament and of the Council of 19 November 2008 on Waste and repealing Certain Directives, *OJ* L312/3, 22.11.2008.

Directive 2008/101/EC
Directive 2008/101/EC of the European Parliament and of the Council of 19 November 2008 amending Directive 2003/87/EC so as to Include Aviation Activities in the Scheme for Greenhouse Gas Emission Allowance Trading within the Community, *OJ* L8/3, 13.1.2009.

Directive 2009/28/EC
Directive 2009/28/EC of the European Parliament and of the Council of 23 April 2009 on the Promotion of the Use Of Energy From Renewable Sources and Amending and Subsequently Repealing Directives 2001/77/EC and 2003/30/EC, *OJ* L140/16, 5.6.2009.

Directive 2009/29/EC
Directive 2009/29/EC of the European Parliament and of the Council of 23 April 2009 amending Directive 2003/87/EC so as to Improve and Extend the Greenhouse Gas Emission Allowance Trading Scheme of the Community, *OJ* L140, 5.6.2009.

Directive 2009/30/EC
Directive 2009/30/EC of the European Parliament and of the Council of 23 April 2009 amending Directive 98/70/EC as regards the Specification of Petrol, Diesel and Gas-oil and Introducing a Mechanism to Monitor and Reduce Greenhouse Gas Emissions and amending Council Directive 1999/32/EC as regards the Specification of Fuel Used By Inland Waterway Vessels and Repealing Directive 93/12/EEC, *OJ* L140/88, 5.6.2009.

Directive 2009/41/EC
Directive 2009/41/EC of the European Parliament and of the Council of 6 May 2009 on the Contained Use of Genetically Modified Micro-organisms, *OJ* L125/75, 21.5. 2009.

Directive 2009/128/EC
Directive 2009/128/EC of the European Parliament and of the Council of 21 October 2009 establishing a framework for Community action to achieve the sustainable use of pesticides, *OJ* L309, 24.11.2009.

Directive 2009/147/EC
Directive 2009/147/EC of the European Parliament and of the Council of 30 November 2009 on the Conservation of Wild Birds, *OJ* L20/7, 26.01.2010.

Directive 2010/30/EU
Directive 2010/30/EU of the European Parliament and of the Council of 19 May 2010 on the Indication By Labelling and Standard Product Information of the Consumption of Energy and Other Resources By Energy-Related Products, *OJ* L153/1, 18.6.2010.

Directive 2012/27/EU
Directive 2012/27/EU of the European Parliament and of the Council of 25 October 2012 on Energy Efficiency, amending Directives 2009/125/EC and 2010/30/EU and repealing Directives 2004/8/EC and 2006/32/EC, *OJ* L315/1, 14.11.2012.

Directive (EU) 2015/412
Directive (EU) 2015/412 of the European Parliament and of the Council of 11 March 2015 amending Directive 2001/18/EC as regards the Possibility for the Member States to Restrict or Prohibit the Cultivation of Genetically Modified Organisms (GMOs) in their Territory, *OJ* L68/1, 13.3.2015.

Directive (EU) 2015/1513
Directive (EU) 2015/1513 of the European Parliament and of the Council of 9 September 2015 amending Directive 98/70/EC relating to the Quality of Petrol and Diesel Fuels and amending Directive 2009/28/EC on the Promotion of the Use of Energy from Renewable Sources, *OJ* L239/1, 15.9.2015.

European Union Strategy for Sustainable Development COM(2001) 264 final
European Union Strategy for Sustainable Development European Commission, 'A Sustainable Europe for a Better World: a European Union Strategy for Sustainable Development' COM(2001) 264 final (Commission proposal to the Gothenburg European Council), Brussels, 15.5.2001.

Explanatory Notes (2015/C 076/01)
European Commission Explanatory Notes to the Combined Nomenclature of the European Union (2015/C 076/01) Publication made in accordance with Article 9(1) of Council Regulation (EEC) No 2658/87 of 23 July 1987 on the Tariff and Statistical Nomenclature and on the Common Customs Tariff, *OJ* C76/58, 04.03.2015.

General Guidance for Aircraft Operators MRR Guidance Document No. 2
European Commission Directorate-General Climate Action Directorate A - International and Climate Strategy CLIMA.A.3 - Monitoring, Reporting, Verification 1 Guidance Document The Monitoring and Reporting Regulation - General guidance for Aircraft Operators MRR Guidance document No. 2, Version of 16.07.2012.

Parliament Recommendation for Second Reading (26 February 2015)
Recommendation for Second Reading on the Council position at First Reading With A View to the Adoption of A Directive of the European Parliament and of the Council amending Directive 98/70/EC relating to the Quality of Petrol and Diesel Fuels and amending Directive 2009/28/EC on the promotion of the Use of Energy from Renewable Sources (10710/2/2014 - C8-0004/2015 - 2012/0288(COD)) Committee on the Environment, Public Health and Food Safety Rapporteur: Nils Torvalds A8-0025/2015, 26.2.2015.

Parliament Report A7-0279/2013
Report on the Proposal for a Directive of the European Parliament and of the Council amending Directive 98/70/EC relating to the Quality of Petrol and Diesel Fuels and Amending Directive 2009/28/EC on the Promotion of the Use of Energy from Renewable Sources (COM(2012)0595 - C7-0337/2012 - 2012/0288(COD)) Committee on the Environment, Public Health and Food Safety Rapporteur: Corinne Lepage A7-0279/2013, 26.7.2013.

Parliament Resolution on a 2030 Framework for Climate and Energy Policies
European Parliament Resolution of 5 February 2014 on a 2030 Framework for Climate and Energy Policies (2013/2135(INI)), 2014.

Presidency Conclusion (8-9 March 2007)
Presidency Conclusions of the Brussels European Council, 8-9 March 2007, 7224/1/07.

Regulation (EC) No 1782/2003
Council Regulation (EC) No 1782/2003 of 29 September 2003 establishing Common Rules for Direct Support Schemes under the Common Agricultural Policy and Establishing Certain Support Schemes for Farmers and amending Regulations (EEC) No 2019/93, *OJ* L270, 21.10.2003.

Regulation (EC) No 1829/2003
Regulation (EC) No 1829/2003 Of The European Parliament and of the Council of 22 September 2003 on Genetically Modified Food and Feed, *OJ* L268/1, 18.10.2003.

Regulation (EC) No 1830/2003
Regulation (EC) No 1830/2003 of the European Parliament and of the Council of 22 September 2003 concerning the Traceability and Labelling of Genetically Modified Organisms and the Traceability of Food and Feed Products Produced from Genetically Modified Organisms and amending Directive 2001/18/EC, *OJ* L268, 18.10.2003.

Regulation (EC) No 1946/2003
Regulation (EC) No 1946/2003 of the European Parliament and of the Council of 15 July 2003 on Transboundary Movements of Genetically Modified Organisms, *OJ* L287, 5.11.2003.

Regulation (EC) No 1013/2006
Regulation (EC) No 1013/2006 of the European Parliament and of the Council of 14 June 2006 on Shipments of Waste, *OJ* L190/1, 12.7.2006.

Regulation (EC) No 73/2009
Council Regulation (EC) No 73/2009 of 19 January 2009 Establishing Common Rules for Direct Support Schemes for Farmers under the Common Agricultural Policy and Establishing Certain Support Schemes For Farmers, amending Regulations (EC) No 1290/2005, *OJ* L30, 31.1.2009.

Regulation (EC) No 74/2009
Council Regulation (EC) No 74/2009 of 19 January 2009 Amending Regulation (EC) No 1698/2005 on Support for Rural Development by the European Agricultural Fund for Rural Development (EAFRD), *OJ* L30/100, 31.1.2009.

Regulation (EC) No 443/2009
Regulation (EC) No 443/2009 of the European Parliament and of the Council of 23 April 2009 setting Emission Performance Standards for New Passenger Cars as Part of the Community's integrated Approach to Reduce CO_2 Emissions from Light-duty Vehicles, *OJ* L140/1, 5.6.2009 (consolidated).

Regulation (EC) No 599/2009
Council Regulation (EC) No 599/2009 of 7 July 2009 Imposing a Definitive Anti-Dumping Duty and Collecting Definitively the Provisional Duty Imposed on Imports of Biodiesel Originating in the United States of America, *OJ* L179/26, 10.7.2009.

Regulation (EU) No 510/2011
Regulation (EU) No 510/2011 of the European Parliament and of the Council of 11 May 2011 setting emission performance standards for new light commercial vehicles as part of the Union's integrated approach to reduce CO2 emissions from light-duty vehicles, *OJ* L145/1, 31.5.2011.

Regulation (EU) No 525/2013
Regulation (EU) No 525/2013 of the European Parliament and of the Council of 21 May 2013 on a mechanism for monitoring and reporting greenhouse gas emissions and for reporting other information at national and Union level relevant to climate change and repealing Decision No 280/2004/EC, *OJ* L165/13, 18.6.2013.

Regulation (EU) No 1306/2013
Regulation (EU) No 1306/2013 of the European Parliament and of the Council of 17 December 2013 on the Financing, Management and Monitoring of the Common Agricultural Policy and repealing Council Regulations (EEC) No 352/78, *OJ* L347/549, 20.12.2013.

Regulation (EU) No 1307/2013
Regulation (EU) No 1307/2013 Of The European Parliament And Of The Council of 17 December 2013 establishing rules for direct payments to farmers under support schemes within the framework of the common agricultural policy and repealing Council Regulation (EC) No 637/2008 and Council Regulation (EC) No 73/2009, *OJ* L347/608, 20.12.2013.

Regulation (EU) No 1308/2013
Regulation (EU) No 1308/2013 Of the European Parliament and of the Council of 17 December 2013 establishing a Common Organisation of the Markets in Agricultural Products and Repealing Council Regulation (EEC) No 922/72, *OJ* L347/671, 20.12.2013.

Regulation (EU) No 253/2014
Regulation (EU) No 253/2014 of the European Parliament And Of The Council of 26 February 2014 amending Regulation (EU) No 510/2011 to define the modalities for reaching the 2020 target to reduce CO_2 emissions from new light commercial vehicles, *OJ* L84/38, 20.3.2014.

Regulation (EU) No 333/2014
Regulation (EU) No 333/2014 of the European Parliament and of the Council of 11 March 2014 amending Regulation (EC) No 443/2009 to define the Modalities for Reaching the 2020 Target to Reduce CO_2 Emissions from New Passenger Cars, *OJ* L103/15, 5.4.2014.

Regulation (EU) No 421/2014
Regulation (EU) No 421/2014 of the European Parliament and of the Council of 16 April 2014 amending Directive 2003/87/EC Establishing a Scheme for Greenhouse Gas Emission Allowance Trading Within the Community, in view of the Implementation By 2020 Of An International Agreement Applying A Single Global Market-Based Measure To International Aviation Emissions, *OJ* L129, 30.4.2014.

Regulations Council Regulation (EC) No 598/2009
Regulations Council Regulation (EC) No 598/2009 of 7 July 2009 Imposing a Definitive Countervailing Duty and Collecting Definitively the Provisional Duty Imposed on Imports Of Biodiesel Originating in The United States of America, *OJ* L179/1, 10.7.2009.

TEU
Consolidated version of the Treaty on European Union, *OJ* C115/13, 9. 5. 2008.

TFEU
Consolidated version of the Treaty on the Functioning of the European Union, *OJ* C326/47, 26.10.2012.

White Paper COM(2011) 144 final
White Paper, 'Roadmap to a Single European Transport Area – Towards a Competitive and Resource Efficient Transport System', COM(2011) 144 final, Brussels, 28.3.2011.

3. Relevant Documents in China

Administrative Measures on Grain and Oil Storage 2009
Administrative Measures on Grain and Oil Storage, promulgated by National Development and Reform Commission, 29 December 5 2009, 粮油仓储管理办法.

Administrative Regulation on Central Grain Reserves 2003
Administrative Regulation on the Central Grain Reserves, promulgated by State Council, 15 August 2003,中央储备粮管理条例.

Administrative Regulation on Grain Circulation 2004
Administrative Regulation on the Circulation of Grains, promulgated by State Council, promulgated 26 May 2004, revised 18 July 2013,粮食流通管理条例, 国务院令第407号.

Administrative Rule on Wetland Protection 2013
Administrative Rule on Wetlands Protection, promulgated by the State Forestry Administration, 28 March 2013, 《湿地保护管理规定》, 国家林业局令第32号.

Agriculture Law of China 1993
Agriculture Law of China, promulgated by Standing Committee of National People's Congress, 2 July 1993, revised 28 December 2002, 中华人民共和国农业法.

Air Pollution Prevention Action Plan 2013
Notice of Printing and Issuing Air Pollution Prevention Action Plan, promulgated by the State Council, 10 September 2013, 国务院关于印发大气污染防治行动计划的通知,国发[2013]37号.

Biodiesel Industry Development Policy 2014
Biodiesel Industry Development Policy, promulgated by the National Energy Administration, 28 November 2014, 生物柴油产业发展政策.

Biodiversity Conservation Strategy and Action Plan of China (2011-2030)
Biodiversity Conservation Strategy and Action Plan of China (2011-2030), promulgated by the Ministry of Environment, 17 September 2010, 关于印发《中国生物多样性保护战略与行动计划》（2011-2030年）的通知, 环发[2010]106号.

China's 21st Century Agenda 1994
The White Paper on China's Population, Environment and Development in the 21st Century, 中国21世纪议程：中国21世纪人口，环境与发展白皮书, 1994.

China's Fifth National Report to CBD 2014
China's Fifth National Report on the Implementation of the Convention on Biological Diversity, promulgated by Ministry of Environmental Protection, March 2014.

China's Peaceful Development 2011
China's Peaceful Development, issued by Information Office of the State Council, September 2011.

Constitution of China 1982
Constitution of China, promulgated by National People's Congress, 4 December 1982, revised 14 March 2004, 中华人民共和国宪法.

Environment Impact Assessment Law of China 2002
Environment Impact Assessment Law of China, promulgated by Standing Committee of the National People's Congress, 28 October 2002 Order of the President of the People's Republic of China (No. 77), 中华人民共和国环境影响评价法.

Environmental Protection Law of China 1989
Environmental Protection Law of China, promulgated by Standing Committee of the National People's Congress, 26 December 1989, revised 24 April 2014, 中华人民共和国环境保护法.

Forest Law of China 1998
Forest Law of China, promulgated by Standing Committee of the National People's Congress, 20 September 1984, revised 29 April 1998, 中华人民共和国森林法.

Guidelines for National GHG Emission Accounting and Reporting 2013
Notice on Printing and Issuing Guidelines (Trial) for Greenhouse Gases Emission Accounting and Reporting of Enterprises within the First 10 Sectors (Trial), promulgated by National Development and Reform Commission Office, 15 October 2013,
国家发展改革委办公厅关于印发首批10个行业企业温室气体排放核算方法与报告指南(试行)的通知, 发改办气候[2013]2526号.

Guidelines for Sustainable Cultivation of Energy Forestry 2011
Notice on Printing and Issuing Guidelines for Sustainable Cultivation of Energy Forestry and Jatropha Curcasi Forestry, Annex I Guidelines for Sustainable Cultivation of Energy Forestry, promulgated by State Forest Administration, 17 February
2011,国家林业局关于印发能源林及小桐子原料林可持续培育指南的通知能源林可持续培育指南, 林造发[2011]33号.

Guidelines on GHG Emission Accounting and Reporting of Shanghai (Trial) 2012
Greenhouse Gases Emission Accounting and Reporting Guidelines of Shanghai (Trial), promulgated by Shanghai Development and Reform Commission, 12 December 2012, SH/MRV-001-2012, 上海市温室气体排放核算与报告指南(试行).

Guidelines on Water Transport GHG Emission Accounting and Reporting of Shanghai (Trial) 2016
Guidelines on Water Transport GHG Emission Accounting and Reporting of Shanghai of Shanghai (Trial), promulgated by Shanghai Development and Reform Commission, 4 February 2016, SH/MRV-011-2016, 上海市水运行业温室气体排放核算与报告方法(试行).

Guiding Opinions on Promoting Forest Carbon Trading 2014
Guiding Opinions on Promoting Forest Carbon Trading, promulgated by State Forestry Administration, promulgated 29 April 2014, 国家林业局关于推进林业碳汇交易工作的指导意见, 林造发[2014]55号.

Hainan Biodiesel Promotion Work Plan 2009
Work Plan for Promoting Use of Biodiesel in the Market of Hainan Province, promulgated by the Hainan Provincial Government Office, 2 April 2009, 海南省生物柴油市场推广使用工作方案, 琼府办[2009]49号.

Implementing Regulation for Land Administration Law of China 1998
Implementing Regulation for Land Administration Law of China, promulgated by State Council, 27 December 1998, revised 8 January 2011, 中华人民共和国土地管理法实施条例, 国务院令第256号.

Interim Measures on Financial Support for Feedstock Bases 2007
Notice on Printing and Issuing Interim Measures on Financial Support for Bioenergy and Biochemistry Feedstock Bases, promulgated by Ministry of Finance, 20 September 2007, 财政部关于印发《生物能源和生物化工原料基地补助资金管理暂行办法》的通知, 财建[2007]435号.

Interim Measures on Financial Support for Energy Use of Crop Stalks 2008
Notice of Issuing and Printing Interim Measures on Administration of Financial Support for Energy Use of Crop Stalks, promulgated by Ministry of Finance, 30 October 2008, 关于印发《秸秆能源化利用补助资金管理暂行办法》的通知, 财建[2008]735号.

Interim Measures on Carbon Emission Trading Administration 2014
Interim Measures on Carbon Emission Trading Administration, promulgated by National Development and Reform Commission, 10 December 2014,碳排放权交易管理暂行办法, 国家发展和改革委员会令第17号.

Interim Regulation on Excise Duties 2008
Interim Regulation on Excise Duties of China, promulgated by the State Council, 10 November 2008, 中华人民共和国消费税暂行条例, 国务院令第539号.

Land Administration Law of China 1986
Land Administration Law of China, promulgated by Standing Committee of the National People's Congress, 25 June 1986, revised 28 August 2004, 中华人民共和国土地管理法.

Law on Grains (draft) 2012
Notice on Public Opinion Collection for 'Law on Grains (draft)', promulgated by the State Council Legal Office, 21 February 2012, 国务院法制办公室关于《粮食法（征求意见稿）》公开征求意见的通知.

Law on Legislation of China 2000
Law on Legislation of China, promulgated by the National People's Congress, 15 March 2000, revised 15 March 2015, 立法法.

Law on the Administration of Tax Collection of China 2001
Law of the People's Republic of China on the Administration of Tax Collection, promulgated by the Standing Committee of the National People's Congress, 28 April 2001, 中华人民共和国税收征收管理法.

Law on the Prevention and Control of Environmental Pollution by Solid Waste 2004
Law on the Prevention and Control of Environmental Pollution by Solid Waste, promulgated by Standing Committee of the National People's Congress, 29 December 2004, 中华人民共和国固体废物污染环境防治法.

Measures on Administration of Flexible Subsidy 2007
Notice on Printing and Issuing Measures on Administration of Flexible Subsidy for Bioethanol, promulgated by Ministry of Finance, 20 November 2007, 财政部关于印发《生物燃料乙醇弹性补贴财政财务管理办法》的通知, 财建[2007]724号.

Measures on Bioethanol Pilot Project 2002
Notice of Printing and Issuing 'Measures on Pilot Project of Use of Bioethanol for Automobiles' and 'Implementation Work Details on Pilot Project of Use of Bioethanol for Automobiles', promulgated by National Development and Reform Commission (ex. National Economic and Trade Commission), 22 March 2002, 关于印发《车用乙醇汽油使用试点方案》和《车用乙醇汽油使用试点工作实施细则》的通知, 国经贸技术[2002]174号.

Measures on Carbon Emissions Management of Shanghai 2013
Measures on Carbon Emissions Management of Shanghai (Trial), Shanghai Municipal Government, 6 November 2013, 上海市碳排放管理试行办法 (沪府令10号).

Measures on Expanding Bioethanol Pilot Project 2004
Notice of Printing and Issuing 'Measures on Expanding Pilot Projects on Use of Bioethanol for Automobiles' and 'Implementation Work Details on Expanding Pilot Projects on Use of Bioethanol for Automobiles', promulgated by National Development and Reform Commission, 10 February 2004, 关于印发《车用乙醇汽油扩大试点方案》和《车用乙醇汽油扩大试点工作实施细则》的通知, 发改工业[2004]230号.

Measures on Management of Waste Kitchen Oil Disposal of Shanghai 2012
Measures on Management of Waste Kitchen Oil Disposal of Shanghai, promulgated by Shanghai Municipal Government, 26 December 2012, 上海市餐厨废弃油脂处理管理办法, (沪府令97号).

Measures on Operation and Administration of CDM Projects 2011
Measures on Operation and Management of Clean Development Mechanism Projects, promulgated jointly by National Development and Reform Commission, Ministry of Foreign Affairs, Ministry of Science and Technology, and Ministry of Finance, 3 August 2011, 清洁发展机制项目运行管理办法.

National Plan on Climate Change of China (2014-2020)
Notice on Printing and Issuing the 'National Plan on Climate Change of China (2014-2020)', promulgated by the National Development and Reform Commission, 19 September 2014, 国家发展改革委关于印发国家应对气候变化规划 (2014-2020年) 的通知, 发改气候[2014]2347号.

Notice on Adjustment of Excise Duties 2014
Notice on Adjustment of Excise Duties, promulgated by Ministry of Finance and State Administration of Taxation, 25 November 2014, 国家税务总局关于调整消费税政策的通知, 财税[2014]93号.

Notice on Bioethanol Project Management 2006
Notice on Strengthening Bioethanol Project Management and Promotion of Industrial Healthy Development, promulgated by the National Development and Reform Commission and Ministry of Finance, 14 December 2006, 国家发展改革委、财政部关于加强生物燃料乙醇项目建设管理, 促进产业健康发展的通知, 发改工业[2006]2842号.

Notice on Carbon Emission Allowances Trading Pilot Schemes 2011
Notice on Carrying out Carbon Emission Allowances Trading Pilot Schemes, promulgated by National Development and Reform Commission Office, 29 September 2011, 国家发展改革委办公厅关于开展碳排放权交易试点工作的通知, 发改办气候[2011]2601号.

Notice on Excise Duty Policies 2008
Notice on Relevant Excise Duty Policies after Increasing Excise Duty Rates of Oil Products, promulgated by Ministry of Finance and State Administration of Taxation, 19 December 2008,关于提高成品油消费税税率后相关成品油消费税政策的通知, 财税[2008]168号.

Notice on Increasing Exerse Duty Rates of Oil Products 2008
Notice on Increasing Excise Duty Rates of Oil Products, promulgated by Ministry of Finance and State Administration of Taxation, 19 December 2008, 关于提高成品油消费税税率的通知, 财税[2008]167号.

Notice on Oil Products Taxation Reform 2008
Notice on Reform on Prices and Taxation of Oil Products, promulgated by the State Council, 18 December 2008, 国务院关于实施成品油价格和税费改革的通知, 国发[2008]37号.

Notice on Strengthening Synthesized Use of Agricultural Crop Stalks 2013
Notice on Strengthening Synthesized Use of Agricultural Crop Stalks and Prohibition on Burning, promulgated by National Development and Reform Commission, Ministry of Agriculture, and Ministry of Environmental Protection, 14 May 2013, 关于加强农作物秸秆综合利用和禁烧工作的通知, 发改环资[2013]930号.

Notice on the Excise Duty Exemption for Biodiesel (BD100) 2010
Notice on the Excise Duty Exemption for Biodiesel (BD100) Produced from Waste Animal Fat and Vegetable Oil, promulgated by Ministry of Finance and State Administration of Taxation, 17 December 2010, 关于对利用废弃的动植物油生产纯生物柴油免征消费税的通知,财税[2010]118号.

Notice on the First Group of Enterprises for Shanghai Pilot Scheme 2012
Notice on the List of Enterprises Participating in the Pilot Carbon Emission Trading Scheme (First Group), promulgated by Shanghai Development and Reform Commission, 29 November 2012, 上海市发展改革委关于公布本市碳排放交易试点企业名单(第一批)的通知, 沪发改环资(2012)172号.

Notice on the Scope of Excise Duty Exemption for Biodiesel (BD100) 2011
Notice on the Scope of Application of the Excise Duty Exemption for Biodiesel (BD100) Produced from Waste Animal Fat and Vegetable Oil, promulgated by Ministry of Finance and State Administration of Taxation, 15 June 2011, 关于明确废弃动植物油生产纯生物柴油免征消费税适用范围的通知, 财税[2011]46号.

Opinion on Accelerating Synthesized Utilisation of Agricultural Crop Stalks 2008
Opinion on Accelerating Promotion of Synthesized Utilisation of Agricultural Crop Stalks, promulgated by State Council Office, 27 July 2008, 国务院办公厅关于加快推进农作物秸秆综合利用的意见, 国办发[2008]105号.

Opinion on Fiscal Support Policy 2006
Opinion on Implementation of Fiscal Support Policy for Developing Biofuel and Biochemistry, promulgated by Ministry of Finance, National Development and Reform Commission, General Administration of Taxation, State Forestry Administration, 30 September 2006, 关于发展生物能源和生物化工财税扶持政策的实施意见, 财建[2006]702号.

Opinion on Implementation of Shanghai Carbon Emission Trading Pilot Scheme 2012
Opinion on Implementation of Shanghai Carbon Emission Trading Pilot Scheme, promulgated by Shanghai Municipal Government, 3 July 2012, 上海市人民政府关于本市开展碳排放交易试点工作的实施意见, 沪府发[2012]64号.

Opinion on Strengthening Crack-down on Gutter Oil 2010
Opinion on Strengthening Crack-down on Gutter Oil and Management of Kitchen Wastes, promulgated by State Council Office, 13 July 2010, 国务院办公厅关于加强地沟油整治和餐厨废弃物管理的意见, 国办发[2010]36号.

Programme of Investigating and Evaluating Marginal Land Resources 2007
Programme of Investigating and Evaluating Marginal Land Resources for Dedicated Energy Crops for Liquid Bio-fuel, promulgated by the Ministry of Agriculture, 20 April 2007, 关于开展对我国适宜种植能源作物边际土地资源进行调查评估的函, 农科教能函[2007]10 号.

Regulation on Basic Agricultural Land Protection 1998
Regulation on Basic Agricultural Land Protection, promulgated by State Council, 27 December 1998, 基本农田保护条例, 国务院令第257号.

Regulation on Implementation of Forest Law of China 2000
Regulation on Implementation of Forestry Law of China, promulgated by the State Council, 29 January 2000, revised 8 January 2011, 中华人民共和国森林法实施条例, 国务院令278号.

Regulation on Nature Reserves of China 1994
Regulation on Nature Reserves of China, promulgated by State Council, 9 October 1994, 中华人民共和国自然保护区条例.

Renewable Energy Law of China 2005
Renewable Energy Law of China, promulgated by Standing Committee of the National People's Congress, 28 February 2005, revised 26 December 2009, 可再生能源法.

Renewable Energy Medium- and Long-Term Plan 2007
Medium- and Long-Term Plan for China's Renewable Energy Development and Utilization , promulgated by National Development and Reform Commission, 31 August 2007, 可再生能源中长期发展规划, 发改能源[2007]2174号.

Report on National Soil Pollution 2014
Report on National Soil Pollution, promulgated by the Ministry of Environment Protection and Ministry of Land and Resources, 17 April 2014, 全国土壤污染状况调查公报.

The 11th Five Year Plan for Renewable Energy Development
Notice of Printing and Issuing the 11th Five Year Plan for Renewable Energy Development, promulgated by National Development and Reform Commission, 3 March 2008, 国家发展改革委关于印发可再生能源发展 "十一五" 规划的通知, 发改能源[2008]610号.

The 12th Five Year Plan for Energy Development
Notice of Printing and Issuing the 12th Five Year Guideline for Energy Development , promulgated by the State Council, 1 January 2013, 国务院关于印发能源发展"十二五"规划的通知, 国发 [2013] 2号.

The 12th Five Year Plan for National Economic and Social Development
The 12th Five-Year Guidelines for National Economic and Social Development, National People's Congress, 16 March 2011, 中华人民共和国国民经济和社会发展第十二个五年规划纲要.

The 12th Five Year Plan for Renewable Energy Development
Notice on Printing and Issuing the 12th Five Year Guideline for Renewable Energy Development, promulgated by the National Development and Reform Commission, 6 July 2012, 国家发展改革委关于印发可再生能源发展"十二五"规划的通知, 发改能源[2012]1207号.

The 12th Five Year Work Programme on Energy Saving and Emission Reduction 2011
Notice on Printing and Issuing Work Programme of Energy Saving and Emission Reduction during the 12th Five Year Period, promulgated by the State Council, 31 August 2011, 国务院关于印发 "十二五" 节能减排综合性工作方案的通知, 国发[2011]26号.

The 12th Five Year Work Programme on GHG Emission Control 2011
Notice on Printing and Issuing the 12th Five Year Work Programme for Greenhouse Gases Emissions Control, promulgated by State Council, 1 December 2011, 国务院关于印发"十二五"控制温室气体排放工作方案的通知, 国发[2011]41号.

The 12th Five Year Plan on Synthesized Utilisation of Agricultural Crop Stalks 2011
Notice on Printing and Issuing the 12th Five Year Plan on Synthesized Utilisation of Agricultural Crop Stalks, promulgated by the National Development and Reform Commission, Ministry of Agriculture, and Ministry of Finance, 29 November 2011, 关于印发"十二五"农作物秸秆综合利用实施方案的通知, 发改环资 [2011]2615号.

The 2013 White Paper on Climate Change Policy and Action in Forestry 2014
Notice on Printing and Issuing the 2013 White Paper on Climate Change Policy and Action in Forestry, promulgated by State Forestry Administration Office, 27 February 2014, 国家林业局办公室关于印发《2013年林业应对气候变化政策与行动白皮书》的通知, 办造字[2014]19号.

Urgent Notice on Corn Processing 2006
Urgent Notice on Strengthening Management of Corn Processing Projects' Construction, promulgated by National Development and Reform Commission, 8 December 2006, 关于加强玉米加工项目建设管理的紧急通知, 发改工业[2006]2781号.

Water and Soil Conservancy Law of China 1991
Water and Soil Conservancy Law of China, promulgated by Standing Committee of the National People's Congress, 29 June 1991, 水土保持法.

Water Pollution Prevention Law of China 1984
Water Pollution Prevention Law of China, promulgated by Standing Committee of the National People's Congress, 11 May 1984, revised 28 February 2008, 中华人民共和国水污染防治法.

Working Programme for Accelerating Oil Quality Upgrading 2015
Notice of Printing and Issuing Working Programme for Accelerating Oil Quality Upgrading, promulgated by the National Development and Reform Commission, 5 May 2015, 关于印发《加快成品油质量升级工作方案》的通知, 发改能源[2015]974号.

BIBLIOGRAPHY

Abbott 2012
Abbott, K.W., 'The Transnational Regime Complex for Climate Change', *Environment & Planning C: Government & Policy* 2012, Vol. 30(4), p. 571-590.

Abbott *et al.* **2012**
Abbott, K.W., Genschel, P., Zangl, B. & Snidal, D., 'International Organizations as Orchestrators', Paper presented at the APSA 2012 Annual Meeting Paper.

Abbott & Snidal 2010
Abbott, K.W. & Snidal, D., 'International Regulation Without International Government: Improving IO Performance Through Orchestration', *The Review of International Organizations* 2010, Vol. 5(3), p. 315-344.

Abrell & Weigt 2008
Abrell, J. & Weigt, H., 'The Interaction of Emissions Trading and Renewable Energy Promotion', Dresden University of Technology Working Paper, No. WP-EGW-05, 2008.

Ackrill & Kay 2011
Ackrill, R. & Kay, A., 'EU Biofuels Sustainability Standards and Certification Systems - How to Seek WTO Compatibility', *Journal of Agricultural Economics,* 2011, Vol. 62(3), p. 551-564.

Adam 2010
Adam, R., 'Missing the 2010 Biodiversity Target: A Wake-Up Call For The Convention On Biodiversity', *Colorado Journal of International Environmental Law and Policy,* 2010, Vol. 21, p. 123-166.

Afionis & Stringer 2012
Afionis, S. & Stringer, L.C., 'European Union Leadership in Biofuels Regulation: Europe as a Normative Power?', *Journal of Cleaner Production,* 2012, Vol. 32, p. 114-123.

Afionis & Stringer 2014
Afionis, S. & Stringer, L.C., 'The Environment as a Strategic Priority in the European Union–Brazil Partnership: Is the EU Behaving as a Normative Power or Soft Imperialist?', *International Environmental Agreements: Politics, Law and Economics,* 2014, Vol. 14(1), p. 47-64.

Ahamat & Rahman 2014
Ahamat, H. & Rahman, N., 'Restricting Biofuel Import in the Name of the Environment: How Does the Application of WTO Rules Affect Developing Countries?', *Journal of East Asia & International Law,* 2014, Vol. 7, p. 51-78.

Ajanovic 2011
Ajanovic, A., 'Biofuels Versus Food Production: Does Biofuels Production Increase Food Prices?', *Energy,* 2011, Vol. 36(4), p. 2070-2076.

Al-Riffai, Dimaranan & Laborde 2010
Al-Riffai, P., Dimaranan, B. & Laborde, D., 'Global Trade and Environmental Impact Study of the EU Biofuels Mandate', International Food Policy Institute (IFPRI), 2010.

Allwood *et al.* 2014
Allwood, J.M. *et al.*, 'Glossary', in Edenhofer, O. (ed.), *Climate Change 2014: Mitigation of Climate Change. Contribution of Working Group III to the Fifth Assessment Report of the Intergovernmental Panel on Climate Change,* Cambridge: Cambridge University Press, 2014, p. 1249-1279.

Altieri 2009
Altieri, M.A., 'The Ecological Impacts of Large-Scale Agrofuel Monoculture Production Systems in the Americas', *Bulletin of Science, Technology & Society,* 2009, Vol. 29(3), p. 236-244.

Amatayakul & Berndes 2012
Amatayakul, W. & Berndes, G., 'Determining Factor for the Development of CDM Biomass Power Projects', *Energy for Sustainable Development,* 2012, Vol. 16(2), p. 197-203.

Anderson-Sprecher & Jiang 2014
Anderson-Sprecher, A. & Jiang, J., 'Biofuels Annual. China's 2014 Fuel Ethanol Production is Forecast to Increase Six Percent', USDA Foreign Agricultural Service Global Agricultural Information Network (GAIN), 2014.

Andonova, Betsill & Bulkeley 2009
Andonova, L., Betsill, M. & Bulkeley, H., 'Transnational Climate Governance', *Global Environmental Politics,* 2009, Vol. 9(2), p. 52-73.

Anseeuw *et al.* 2012
Anseeuw, W. *et al.*, 'Transnational Land Deals for Agriculture in the Global South', Analytical Report Based on the Land Matrix Database, International Land Coalition, 2012.

Bäckstrand 2008
Bäckstrand, K., 'Accountability of Networked Climate Governance: The Rise of Transnational Climate Partnerships', *Global Environmental Politics,* 2008, Vol. 8(3), p. 74-102.

Baker 2006
Baker, S., *Sustainable Development,* Oxon: Routledge, 2006.

Bakker 2006
Bakker, S.J.K., 'CDM and biofuels: Can the CDM assist biofuel production and deployment?', Energy Research Centre of the Netherlands, ECN-E--06-033 2006, available at: <www.ecn.nl/docs/library/report/2006/e06033.pdf>.

Baldwin & Cave 1999
Baldwin, R. & Cave, M., *Understanding Regulation: Theory, Strategy and Practice,* Oxford: Oxford University Press, 1999.

Baldwin, Scott & Hood 1998
Baldwin, R, Scott, C. & Hood, C., *A Reader on Regulation,* Oxford: Oxford University Press, 1998.

Banuri, T. 2009
Banuri, T., 'Climate Change and Sustainable Development', Paper presented at the Natural Resources Forum, 2009.

Bao 2006
Bao, M., 'The Evolution of Environmental Policy and Its Impact in the People's Republic of China', *Conservation and Society,* 2006, Vol. 4(1), p. 36-54.

Barnard 2010
Barnard, C., *The Substantive Law of the EU: the Four Freedoms,* Oxford: Oxford University Press, 3rd edn., 2010.

Beck *et al.* 2014a
Beck, S. *et al.*, 'Towards a Reflexive Turn in the Governance of Global Environmental Expertise. The Cases of the IPCC and the IPBES', *GAIA-Ecological Perspectives for Science and Society,* 2014, Vol. 23(2), p. 80-87.

Beck *et al.* 2014b
Beck, S. *et al.*, 'Climate Change and the Assessment of Expert Knowledge: Does the IPCC Model Need Updating?', *Bridges,* 2014, Vol. 40, available at: <http://ostaustria.org/bridges-magazine/item/8244-climate-change-and-the-assessment-of-expert-knowledge-does-the-ipcc-model-need-updating> (last accessed 24 April 2016).

Beckman & Rasmussen 2010
Beckman, C. & Rasmussen, K., 'Peoples Republic of China: Readout from Sino-U.S.', Advanced Biofuels Forum Report Categories, Bio-Fuels GAIN Report Number: CH10035, USDA Foreign Agricultural Service Global Agricultural Information Network (GAIN), 2010.

Beg *et al.* 2002
Beg, N. *et al.*, 'Linkages between Climate Change and Sustainable Development', *Climate Policy,* 2002, Vol. 2(2-3), p. 129-144.

Benvenisti & Downs 2007
Benvenisti, E. & Downs, G.W., 'The Empire's New Clothes: Political Economy and the Fragmentation of International Law', *Stanford Law Review,* 2007, Vol. 60(2), p. 595-631.

Berndes & Hansson 2007
Berndes, G. & Hansson, J., 'Bioenergy Expansion in the EU: Cost-Effective Climate Change Mitigation, Employment Creation and Reduced Dependency on Imported Fuels', *Energy Policy,* 2007, Vol. 35(12), p. 5965-5979.

Bessou *et al.* 2011
Bessou, C. *et al.*, 'Biofuels, Greenhouse Gases and Climate Change. A Review', *Agronomy for Sustainable Development,* 2011, Vol. 31(1), p. 1-79.

Biermann & Bauer 2005
Biermann, F. & Bauer, S., *A World Environment Organization: Solution or Threat for Effective International Environmental Governance?,* Hants: Ashgate, 2005.

Biermann, Davies & Van der Grijp 2009
Biermann, F., Davies, O. & Van der Grijp, N., 'Environmental Policy Integration and the Architecture of Global Environmental Governance', *International Environmental Agreements: Politics, Law and Economics,* 2009, Vol. 9(4), p. 351-369.

Biermann, Pattberg & Van Asselt 2009
Biermann, F., Pattberg, P. & Van Asselt, H., 'The Fragmentation of Global Governance Architectures: A Framework for Analysis', *Global Environmental Politics,* 2009, Vol. 9(4), p. 14-40.

Black 2002
Black, J., 'Critical Reflections on Regulation', *Australian Journal of Legal Philosophy*, 2002, Vol. 27, p. 1-36.

Bodansky & Rajamani 2012
Bodansky, D. & Rajamani, L., 'The Evolution and Governance Architecture of the Climate Change Regime', in draft 28 October 2012, available at: <http://cprclimateworkshop.cprindia.org/uploads/2/3/7/5/23756750/bodansky_and_rajamani__evolution_and_governance_architecture_of_the_climate_change_regime.pdf> (last accessed 24 April 2016), (chapter in the book by Sprinz, D. & Luterbacher, U. (eds.), *International Law and Global Climate Change*, Cambridge, MA: MIT Press, 2nd revised edn., forthcoming 2016?)Sprinz, D. & Luterbacher, U. (eds.), *International Law and Global Climate Change*, Cambridge, MA: MIT Press, 2nd revised edn., forthcoming 2016.

Boele-Woelki 2008
Boele-Woelki, K., 'What Comparative Family Law Should Entail', *Utrecht Law Review*, 2008, Vol. 4(2), p. 1-24.

Borger 2012
Borger, G., 'All Things Not Being Equal: Aviation In The EU ETS', *Climate Law*, 2012, Vol. 3(3), p. 265-281.

Börjesson & Tufvesson 2011
Börjesson, P. & Tufvesson, L.M., 'Agricultural Crop-Based Biofuels - Resource Efficiency and Environmental Performance Including Direct Land Use Changes', *Journal of Cleaner Production*, 2011, Vol. 19(2), p. 108-120.

Bosselmann 2013
Bosselmann, K., *The Principle of Sustainability: Transforming Law and Governance*, Hampshire: Ashgate Publishing, 2013.

Bowman, Davies & Redgwell 2010
Bowman, M., Davies, P. & Redgwell, C., *Lyster's International Wildlife Law*, Cambridge, UK: Cambridge University Press, 2010.

Boyd *et al.* 2009
Boyd, E. *et al.* 'Reforming the CDM for Sustainable Development: Lessons Learned and Policy Futures', *Environmental Science & Policy*, 2009, Vol. 12(7), p. 820-831.

Boyle 2008
Boyle, A., 'Between Process and Substance; Sustainable Development in the Jurisprudence of International Courts and Tribunals', in Bugge, H.C. & Voigt, C. (eds.), *Sustainable Development in International and National Law What did the Brundtland Report do to Legal Thinking and Legal Development, and Where can we go From Here?*, Groningen: Europa Law Publishing, 2008, p. 203-219.

Boyle 2010
Boyle, A., 'Soft Law in International Law Making', in Evans, M.D. (ed.), *International Law*, New York: Oxford University Press, 2010, p. 122-140.

Boyle & Freestone 1999
Boyle, A.E. & Freestone, D., *International Law and Sustainable Development: Past Achievements and Future Challenges*, Oxford: University Press Oxford, 1999.

Bradford 2012
Bradford, A., 'The Brussels Effect', *Northwestern University Law Review*, 2012, Vol. 107(1), p. 1-68.

Bradley 2011
Bradley, K.S.C., 'Powers And Procedures in the EU Constitution: Legal Bases and the Court', in Craig, P. & De Búrca, G. (eds.), *The Evolution of EU Law*, Oxford: Oxford University Press, 2011, p. 85-110.

Brande 2009
Brande, E.V., 'Green Civilian Power Europe?', in Orbie, J. (ed.), *Europe's Global Role: External Policies of the European Union*, Farnham/Burlington: Ashgate, 2009, p. 157-180.

Broadhurst 2011
Broadhurst, T., 'Biofuels and Sustainability: A Case Study from Tanzania', PISCES Working brief, 2011.

Brunnée 2002
Brunnée, J., 'COPing with Consent: Law-Making under Multilateral Environmental Agreements', *Leiden Journal of International Law*, 2002, Vol. 15(1), p. 1-52.

Brunnée 2005
Brunnée, J., 'Reweaving the Fabric of International Law? Patterns of Consent in Environmental Framework Agreements', in Wolfrum, R. & Röben, V. (eds.), *Developments of International Law in Treaty Making*, Berlin: Springer, 2005, p. 101-126.

Bugge 2008
Bugge, H.C., '1987-2007: "Our Common Future" Revisited', in Bugge, H.C. & Voigt, C. (eds.), *Sustainable Development in Internatioanl and National Law What did the Brundtland Report do to Legal Thinking and Legal Development, and Where can We Go From Here?*, Groningen: Europa Law Publishing, 2008, p. 3-21.

Bulkeley *et al.* 2012
Bulkeley, H. *et al.*, 'Governing Climate Change Transnationally: Assessing the Evidence from a Database of Sixty Initiatives', *Environment and Planning C: Government and Policy*, 2012, Vol. 30(4), p. 591-612.

Cameron 2009
Cameron, E., 'Human Rights and Climate Change: Moving from an Intrinsic to an Instrumental Approach', *The Georgia Journal of International and Comparative Law,* 2009, Vol. 38, p. 673-716.

Carbon Market Watch 2014
'Role of International Offsets in the Fuel Quality Directive', *Carbon Market Watch,* 2014, available at: <http://carbonmarketwatch.org/report-the-role-of-international-offsets-in-the-fuel-quality-directive/>.

Carlarne 2008
Carlarne, C.P., 'Good Climate Governance: Only a Fragmented System of International Law Away?', *Law & Policy,* 2008, Vol. 30(4), p. 450-480.

Carlarne 2010
Carlarne, C.P., *Climate Change Law and Policy: EU and US Approaches,* Oxford: Oxford University Press, 2010.

Chang, Belis & Bruyninckx 2012
Chang, P.-F., Belis, D. & Bruyninckx, H., 'EU-China Climate Relations: The Clean Development Mechanism and Renewable Energy in China', in Wouters, J. *et al.* (eds.), *China, the European Union and Global Governance,* Cheltenham: Edward Elgar, 2012, p. 227-244.

Chang *et al.* 2012
Chang, S. *et al.*, 'Development of Biofuels in China: Technologies, Economics and Policies', The World Bank Development Research Group Environment and Energy Team, Policy Research Working Paper 2012.

Charles *et al.* 2013
Charles, C., 'Biofuels – At What Cost? A Review of Costs and Benefits of EU Biofuel Policies', International Institute for Sustainable Development (IISD), 2013.

Charnovitz 2002
Charnovitz, S., 'A World Environment Organization', *Columbia Journal of Environmental Law,* 2002, Vol. 27, p. 323-362.

Chen 2008
Chen, J., *Chinese Law: Context and Transformation,* Leiden: Martinus Nijhoff Publishers, 2008.

Cherubini *et al.* 2009
Cherubini, F. *et al.*, 'Energy- and Greenhouse Gas-Based LCA of Biofuel and Bioenergy Systems: Key issues, Ranges and Recommendations', *Resources, Conservation and Recycling,* 2009, Vol. 53(8), p. 434-447.

Chowdhury & Wessel 2012
Chowdhury, N. & Wessel, R.A., 'Conceptualising Multilevel Regulation in the EU: A Legal Translation of Multilevel Governance?', *European Law Journal*, 2012, Vol. 18(3), p. 335-357.

Chum *et al.* 2011
Chum, H. *et al.*, 'Bioenergy', in Edenhofer, O. *et al.* (eds.), *Renewable Energy Sources and Climate Change Mitigation: Special Report of the Intergovernmental Panel on Climate Change*, Cambridge/New York: Cambridge University Press, 2011, p. 209-332.

Churchill & Ulfstein 2000
Churchill, R.R. & Ulfstein, G., 'Autonomous Institutional Arrangements in Multilateral Environmental Agreements: A Little-Noticed Phenomenon in International Law', *American Journal of International Law*, 2000, Vol. 94(4), p. 623-659.

Cini, M. 2001
Cini, M., 'The Soft Law Approach: Commission Rule-Making in the EU's State Aid Regime', *Journal of European Public Policy*, 2001, Vol. 8(2), p. 192-207.

Cole 2011
Cole, D.H., 'From Global to Polycentric Climate Governance', *Climate Law*, 2011, Vol. 2(3), p. 395-413.

Condon 2009
Condon, B.J., 'Climate Change and Unresolved Issues in WTO Law', *Journal of International Economic Law*, 2009, Vol. 12(4), p. 895-926.

Connors & Schmidt 2010
Connors, J. & Schmidt, M., 'United Nations', in Moeckli, D., Shah, S. & Sivakumaran, S. (eds.), *International Human Rights Law*, New York: Oxford University Press, 2010, p. 391-432.

Corne 1995
Corne, P.H., 'Lateral Movements: Legal Flexibility and Foreign Investment Regulation in China', *Case Western Reserve Journal of International Law*, 1995, Vol. 27(2), p. 247-299.

Corne 2002
Corne, P.H., 'Creation and Application of Law in the PRC', *The American Journal of Comparative Law*, 2002, Vol. 50(2), p. 369-443.

Cottier & Matteotti-Berkutova 2009
Cottier, T. & Matteotti-Berkutova, S., 'International Environmental Law and the Evolving Concept of "Common Concern Of Mankind"', in Cottier, T., Nartova, O. & Bigdeli, S.Z. (eds.), *International Trade and the Mitigation of Climate Change*, Cambridge: Cambridge University Press, 2009, p. 21-47.

Cotula, Dyer & Vermeulen 2008
Cotula, L., Dyer, N. & Vermeulen, S., *Fuelling Exclusion?: The Biofuels Boom and Poor People's Access to Land*, London: International Institute for Environment and Development (IIED), 2008.

Cotula *et al.* 2009
Cotula, L. *et al.*, 'Land Grab or Development Opportunity?: Agricultural Investment and International Land Deals in Africa', FAO, IIED and IFAD, 2009.

Cramer Report 2007
'Testing Framework for Sustainable Biomass, Final Report from the Project Group "Sustainable production of biomass"', 2007.

Cremona 2011
Cremona, M., 'External Relations and External Competence of the European Union: The Emergence of an Integrated Policy', in Craig, P. & De Búrca, G. (eds.), *The Evolution of EU Law*, Oxford: Oxford University Press, 2011, p. 217-268.

Cremona 2012
Cremona, M., 'Coherence and EU External Environmental Policy', in Morgera, E. (ed.), *The External Environmental Policy of the European Union: EU and International Law Perspectives*, Cambridge: Cambridge University Press, 2012, p. 33-54.

Crowley & Howse 2014
Crowley, M. & Howse, R., 'Tuna-Dolphin II: A Legal and Economic Analysis of the Appellate Body Report', *World Trade Review*, 2014, Vol. 13(2), p. 321-355.

Curnow, Tait & Millar 2010
Curnow, P., Tait, L. & Millar, I., 'Financing Renewable Energy Projects in Asia: Barriers and Solutions', *Renewable Energy Law & Policy Review*, 2010, Vol. 1, p. 101-112.

Dai 2014
Dai, L., 'Tackling Diffuse Water Pollution from Agriculture in China: Drawing Inspiration from the European Union', *Utrecht Law Review*, 2014, Vol. 10(2), p. 136-154.

Dannemann 2006
Dannemann, G., 'Comparative Law: Study of Similarities or Differences?', in Reimann, M. & Zimmermann, R. (eds.), *The Oxford Handbook of Comparative Law*, Oxford: Oxford University Press, 2006, p. 383-419.

De Beer 2011
De Beer, J., 'Network Governance of Biofuels', Valgen Working Paper Series, 2011.

De Búrca 2000
De Búrca, G., 'Differentiation Within the Core: the Case of the Common Market', in De Búrca, G. & Scott, J. (eds.), *Constitutional Change in the EU: From Uniformity to Flexibility?*, Oxford: Hart Publishing, 2000, p. 133-172.

De Cendra de Larragán 2008
De Cendra de Larragán, J., 'Too Much Harmonization? An Analysis of the Commission's Proposal To Amend the EU ETS from the Perspective of Legal Principles', in Faure, M. & Peeters, M. (eds.), *Climate Change and European Emissions Trading. Lessons for Theory and Practice,* Cheltenham: Edward Elgar, 2008, p. 53-87.

De Cendra de Larragán 2012
De Cendra de Larragán, J., 'EU Climate and Energy Law: Challenges for Member States', in Peeters, M., Stallworthy, M. & De Cendra de Larragán, J. (eds.), *Climate Law in EU Member States: Towards National Legislation for Climate Protection,* Cheltenham: Edward Elgar, 2012, p. 39-66.

De Cruz 2010
De Cruz, P., *Family Law, Sex and Society: A Comparative Study of Family Law,* Oxon: Routledge, 2010.

De Gorter & Just 2009
De Gorter, H. & Just, D.R., 'The Economics of a Blend Mandate for Biofuels', *American Journal of Agricultural Economics,* 2009, Vol. 91(3), p. 738-750.

De Gorter & Tsur 2010
De Gorter, H. & Tsur, Y., 'Cost-Benefit Tests for GHG Emissions from Biofuel Production', *European Review of Agricultural Economics,* 2010, Vol. 37(2), p. 133-145.

De Man 2013
De Man, R., 'Private Sector Driven Sustainability Standards. How can They Promote Sustainability in Third States', in Douma, W.T. & Van der Velde, S. (eds.), *EU Environmental Norms and Third Countries: The EU as A Global Role Model?,* The Hague: Centre for the Law of EU External Relations, 2013, Vol. 5, p. 35-51.

De Sadeleer 2003
De Sadeleer, N., 'Procedures for Derogations from the Principle of Approximation of Laws under Article 95 EC', *Common Market Law Review,* 2003, Vol. 40(4) p. 889-915.

De Sadeleer 2014
De Sadeleer, N., *EU Environmental Law and the Internal Market,* Oxford: Oxford University Press, 2014.

De Schutter 2008
De Schutter, O., 'Background Note: Analysis of the World Food Crisis by the UN Special Rapporteur on the Right to Food', United Nations Office of the United Nations High Commissioner forHuman Rights, 2008.

De Schutter *et al.* 2012
De Schutter, O. *et al.*, 'Commentary to the Maastricht Principles on Extraterritorial Obligations of States in the Area of Economic, Social and Cultural Rights', *Human Rights Quarterly,* 2012, Vol. 34, p. 1084-1169.

De Vera 2007
De Vera, E.R., 'The WTO and Biofuels: The Possibility of Unilateral Sustainability Requirements', *Chicago Journal of International Law,* 2007, Vol. 8, p. 661-680.

De Witte 2012
De Witte, B., 'A Competence to Protect: The Pursuit of Non-Market Aims through Internal Market Legislation', in Syrpis, P. (ed.), *The Judiciary, the Legislature and the EU Internal Market,* Cambridge: Cambridge University Press, 2012, p. 25-46.

DeHaan *et al.* 2010
DeHaan, L.R., Weisberg, S., Tilman, D. & Fornara, D., 'Agricultural and Biofuel Implications of A Species Diversity Experiment with Native Perennial Grassland Plants', *Agriculture, Ecosystems & Environment,* 2010, Vol. 137(1), p. 33-38.

Dehue, Meyer & Hamelinck 2007
Dehue, B., Meyer, S. & Hamelinck, C., 'Towards A Harmonised Sustainable Biomass Certification Scheme', PBIONL072413, ECOFYS, 2007.

Dehue *et al.* 2011
Dehue, B. *et al.*, 'Inventory of Data Sources and Methodologies to Help Economic Operators Identify Land Status Relating to EU Sustainability Criteria for Biofuels and Bioliquids', PEGENL083540, ECOFYS, 2011,.

Deininger & Byerlee 2011
Deininger, K.W. & Byerlee, D., *Rising Global Interest in Farmland: Can It Yield Sustainable and Equitable Benefits?,* Washington DC: World Bank Publications, 2011.

Demirbas 2009
Demirbas, A., 'Political, Economic and Environmental Impacts of Biofuels: A Review', *Applied Energy,* 2009, Vol. 86, p. S108-S117.

Deshpande 2006
Deshpande, R., 'Biofuels and the WTO: An Emerging Context', *Asian Biotechnology and Development Review,* 2006, Vol. 8(2), p. 77-96.

Di Lucia, Ahlgren & Ericsoon 2012
Di Lucia, L., Ahlgren, S. & Ericsson, K., 'The Dilemma of Indirect Land-Use Changes in EU Biofuel Policy - An Empirical Study of Policy-Making in The Context of Scientific Uncertainty', *Environmental Science & Policy,* 2012, Vol. 16, p. 9-19.

Diaz-Chavez *et al.* 2013
Diaz-Chavez, R. *et al.*, 'Mandatory Requirements in Relation to Air, Soil, or Water Protection: Analysis of Need and Feasibility Final Report (Tasks 3 & 4)', ECOFYS, 2013.

Doherty, M. 2004
Doherty, M., 'Challenging A Commission Refusal to Allow the Introduction of More Stringent National Measures', *Environmental Law Review,* 2004, *Vol. 6,* p.120-126.

Dong, F. 2007
Dong, F., 'Food Security and Biofuels Development: The Case of China', Centre of Agricultural and Rural Development, Briefing Paper 07-BP 52, Iowa State University, 2007.

Douma 2010
Douma, W.T., 'Legal Aspects of the European Union's Biofuels Policy: Protection or Protectionism?', *German Yearbook of International Law,* 2010, Vol. 53, p. 371-420.

Douma 2014
Douma, W.T., 'The Sustainability of the EU's Biofuels Policy', in Squintani, L. *et al.*, *Sustainable Energy United in Diversity:Challenges and Approaches in Energy Transition in the EU,* Groningen: European Environmental Law Forum, Book Series, Vol. 1, p. 25-44.

Drenguis 2014
Drenguis, D.D., 'Reap what you sow: soil pollution remediation reform in China', *Pacific Rim Law & Policy Journal,* 2014, Vol. 23(1), p. 171-201.

Dresner 2008
Dresner, S., *The Principles of Sustainability,* London: Earthscan, 2008.

Drexhage & Murphy 2010
Drexhage, J. & Murphy, D., 'Sustainable Development: From Brundtland to Rio 2012', Background Paper Prepared for Consideration by the High Level Panel on Global Sustainability at Its First Meeting, 19 September 2010.

Driesen 2003
Driesen, D., 'Thirty Years of International Environmental Law: A Retrospective and Plea for Reinvigoration', *Syracuse Journal of International Law and Commerce,* 2003, Vol. 30, p. 353-368.

Drumbl 2010
Drumbl, M.A., 'Actors and Law-Making In International Environmental Law', in Fitzmaurice, M., Ong, D.M. & Merkouris, P. (eds.), *Research Handbook on International Environmental Law*, Cheltenham: Edward Elgar, 2010, p. 3-25.

Dufey 2006
Dufey, A., 'Biofuels Production, Trade and Sustainable Development: Emerging Issues', Sustainable Markets Discussion Paper Number 2, International Institute for Environment and Development (IIED), November 2006.

Duran & Morgera 2012
Duran, G.M., & Morgera, E., *Environmental Integration in the EU's External Relations: Beyond Multilateral Dimensions,* Oxford: Hart Publishing, 2012.

Echols 2009
Echols, M.A., 'Biofuels Certification and the Law of the World Trade Organization', Programme on Agricultural Trade and Sustainable Development, International Centre for Trade and Sustainable Development, Geneva, 2009.

Economy 2004
Economy, E.C., *The River Runs Black: The Environmental Challenge to China's Future,* Ithaca: Cornell University Press, 2004.

Edenhofer *et al.* 2014a
Edenhofer, O. *et al.*, 'IPCC, 2014: Summary for Policymakers', in Edenhofer, O. *et al.* (eds.), *Climate Change 2014: Mitigation of Climate Change. Contribution of Working Group III to the Fifth Assessment Report of the Intergovernmental Panel on Climate Change,* Cambridge/New York: Cambridge University Press, 2014, p. 4-30.

Edenhofer *et al.* 2014b
Edenhofer, O. *et al.*, 'Technical Summary', in Edenhofer, O. *et al.* (eds.), *Climate Change 2014: Mitigation of Climate Change. Contribution of Working Group III to the Fifth Assessment Report of the Intergovernmental Panel on Climate Change,* Cambridge/New York: Cambridge University Press, 2014, p. 33-107.

Edin 2003
Edin, M., 'State Capacity and Local Agent Control in China: CCP Cadre Management from A Township Perspective', *The China Quarterly,* 2003, Vol. 173, p. 35-52.

Eide 1987
Eide, A., 'The Right to Adequate Food as a Human Right', Report prepared by Asbjorn Eide, Final Report, E/CN.4/Sub.2/1987/23.

Eide 2008
Eide, A., *The Right to Food and the Impact of Liquid Biofuels (Agrofuels),* Rome: Food and Agriculture Organization of the United Nations, 2008.

Elbehri, Segerstedt & Liu 2013
Elbehri, A., Segerstedt, A. & Liu, P., *Biofuels and the Sustainability Challenge: A Global Assessment of Sustainability Issues, Trends and Policies for Biofuels and Related Feedstocks,* Rome: Food and Agriculture Organization of the United Nations, 2013.

Ellis & Kamel 2007
Ellis, J. & Kamel, S., 'Overcoming Barriers to Clean Development Mechanism Projects', *OECD Papers,* 2007, Vol. 7(1), p. 1-50.

Ellis & Moarif 2015
Ellis, J. & Moarif, S., 'Identifying and Addressing Gaps in the UNFCCC Reporting Framework' (OECD and IEA, COM/ENV/EPOC/IEA/SLT) 2015.

Endres 2010
Endres, J.M., 'Clearing the Air: the Meta-Standard Approach to Ensuring Biofuels Environmental and Social Sustainability', *Virginia Environmental Law Journal,* 2010, Vol. 28, p. 73-120.

Esty 2000
Esty, D., 'International Governance at the Global Level: The Value of Creating a Global Environmental Organization', *Environment Matters Annual Review of European, Comparative & International Environmental Law,* 1999-2000, p. 12-15.

European Renewable Energy Council 2013
European Renewable Energy Council, 'EU Tracking Roadmap 2013. Keeping Track of Renewable Energy Targets Towards 2020', 2013.

EEA 2013
European Environmental Agency (EEA), *EU Bioenergy Potential from A Resource Efficiency Perspective,* 2013.

EEA Scientific Committee 2011
European Environmental Agency Scientific Committee, *Opinion of the EEA Scientific Committee on Greenhouse Gas Accounting in Relation to Bioenergy,* 2011.

Falkner 2005
Falkner, G., *Complying with Europe: EU Harmonisation and Soft Law in the Member States,* Cambridge: Cambridge University Press, 2005.

Fargione, Plevin & Hill 2010
Fargione, J.E., Plevin, R.J. & Hill, J.D., 'The Ecological Impact of Biofuels', *Annual Review of Ecology, Evolution, and Systematics,* 2010, Vol. 41, p. 351-377.

Fargione *et al.* 2008
Fargione, J. *et al.,* 'Land Clearing and the Biofuel Carbon Debt', *Science,* 2008, Vol. 319(5867), p. 1235-1238.

Faria 2009
Faria, J.A.E., 'Future Directions of Legal Harmonisation and Law Reform: Stormy Seas or Prosperous Voyage', *Uniform Law Review,* 2009, Vol. 14(1-2), p. 5-34.

Faure 2000
Faure, M., 'The Harmonization, Codification and Integration of Environmental Law: A Search for Definitions', *European Environmental Law Review,* 2000, p. 174-182.

Faure & Peeters 2008
Faure, M.G. & Peeters, M., *Climate Change and European Emissions Trading: Lessons for Theory and Practice,* Cheltenham: Edward Elgar, 2008.

Field *et al.* 2014
Field, C.B. *et al.*, 'Technical Summary', in Field, C.B. *et al.* (eds.), *Climate Change 2014: Impacts, Adaptation, and Vulnerability. Part A: Global and Sectoral Aspects. Contribution of Working Group II to the Fifth Assessment Report of the Intergovernmental Panel on Climate Change,* Cambridge/New York: Cambridge University Press, 2014, p. 35-94.

Finlayson *et al.* 2011
Finlayson, C.M. *et al.*, 'The Ramsar Convention and Ecosystem-based Approaches to the Wise Use and Sustainable Development of Wetlands', *Journal of International Wildlife Law & Policy,* 2011, Vol. 14(3-4), p. 176-198.

Fletcher Jr *et al.* 2010
Fletcher Jr, R.J. *et al.*, 'Biodiversity Conservation in the Era of Biofuels: Risks and Opportunities', *Frontiers in Ecology and the Environment,* 2010, Vol. 9(3), p. 161-168.

Fonseca *et al.* 2010
Fonseca, M.B. *et al.*, 'Impacts of the EU Biofuel Target on Agricultural Markets and Land Use A Comparative Modelling Assessment', European Commission Joint Research Centre Institute for Prospective Technological Studies, EUR 24449 EN, 2010.

Fontaras *et al.* 2009
Fontaras, G. *et al.*, 'Effects of Biodiesel on Passenger Car Fuel Consumption, Regulated and Non-Regulated Pollutant Emissions over Legislated and Real-World Driving Cycles', *Fuel,* 2009, Vol. 88(9), p. 1608-1617.

Franconi 1996
Franconi, F., 'Extraterritorial Application of Environmental Law', in Meessen, K.M. (ed.), *Extraterritorial Jurisdiction In Theory And Practice,* London: Kluwer Law International, 1996, p. 122-124.

French 2010
French, D., 'Sustainable Development', in Fitzmaurice, M., Ong, D.M. & Merkouris, P. (eds.), *Research Handbook on International Environmental Law,* Cheltenham: Edward Elgar, 2010, p. 51-68.

Fritsche, Sims & Monti 2010
Fritsche, U.R., Sims, R.E. & Monti, A., 'Direct and Indirect Land-Use Competition Issues for Energy Crops and Their Sustainable Production - An Overview', *Biofuels, Bioproducts and Biorefining*, 2010, Vol. 4(6), p. 692-704.

Gaffney & Marley 2009
Gaffney, J.S. & Marley, N.A., 'The Impacts of Combustion Emissions on Air Quality and Climate - From Coal to Biofuels and Beyond', *Atmospheric Environment*, 2009, Vol. 43(1), p. 23-36.

Gaines 2014
Gaines, S.E., 'The Energy Revolution as Sustainable Development', in Squintani, L., Vedder, H.H.B., Reese, M. & Vanheusden, B. (eds.), *Sustainable Energy United in Diversity Challenges and Approaches in Energy Transition in the EU*, Groningen: European Environmental Law Forum Book Series, Vol. 1, p. 7-24.

Gan & Yu 2008
Gan, L. & Yu, J., 'Bioenergy Transition in Rural China: Policy Options and Co-benefits', *Energy Policy*, 2008, Vol. 36, p. 531-540.

Gardner 2003
Gardner, R.C., 'Perspectives On Wetlands And Biodiversity: International Law, Iraqi Marshlands, and Incentives For Restoration', *Colorado Journal of International Environmental Law & Policy*, 2003, Vol. 1, p. 1-117.

Gehring 2007
Gehring, T., 'Treaty-Making and Treaty Evolution', in Bodansky, D., Brunnée, J. & Hey, E. (eds.), *The Oxford Handbook of International Environmental Law*, Oxford: Oxford University Press, 2007, p. 467-497.

Gerbens-Leenes *et al.* 2012
Gerbens-Leenes, P.W. *et al.*, 'Biofuel Scenarios in A Water Perspective: The Global Blue and Green Water Footprint of Road Transport in 2030', *Global Environmental Change*, 2012, Vol. 22, p. 764-775.

German, Schoneveld & Pacheco 2011
German, L., Schoneveld, G.C. & Pacheco, P., 'Local Social and Environmental Impacts of Biofuels: Global Comparative Assessment and Implications for Governance', *Ecology and Society*, 2011, Vol. 16(4), p. 29-43.

Gevurtz 2014
Gevurtz, F.A., 'Determining Extraterritoriality', *William & Mary Law Review*, 2014, Vol. 56(2), p. 341-407.

Gillivray & Holder 2001
Gillivray, D.M., & Holder, J., 'Locating EC Environmental Law', *Yearbook of European Environmental Law*, 2001, Vol. 20(1), p. 139-171.

Gnansounou *et al.* 2009
Gnansounou, E. *et al.*, 'Life Cycle Assessment of Biofuels: Energy and Greenhouse Gas Balances', *Bioresource Technology*, 2009, Vol. 100(21), p. 4919-4930.

Goldmann 2010
Goldmann, M., 'Inside Relative Normativity: From Sources To Standard Instruments For The Exercise Of International Public Authority', *The Exercise of Public Authority by International Institutions*, 2010, Vol. 210, p. 661-711.

Gonzalez 2010
Gonzalez, C.G., 'The global food crisis: law, policy, and the elusive quest for justice', *Yale Human Rights and Development Law Journal*, 2010, Vol. 13(2), p. 462-479.

Gordon 1998
Gordon, J., 'The Concept of Human Rights: The History and Meaning of its Politicization', *Brooklyn Journal of International Law*, 1998, Vol. 23, p. 689-792.

Haberl 2012
Haberl, H., 'Correcting a Fundamental Error in Greenhouse Gas Accounting Related to Bioenergy', *Energy Policy*, 2012, Vol. 45, p. 18-23.

Hale & Roger 2014
Hale, T. & Roger, C., 'Orchestration and Transnational Climate Governance', *The Review of International Organizations*, 2014, Vol. 9(1), p. 59-82.

Hamelinck *et al.* 2011
Hamelinck, C. *et al.*, 'Biofuels Baseline 2008', ECOFYS, PECPNL084543, 2011.

Handl 2005
Handl, G., 'International Lawmaking by Conferences of Parties and Other Politically Mandated Bodies: Comment', in Wolfrum, R. & Röben, V. (eds.), *Developments of International Law in Treaty Making*, Berlin: Springer, 2005, p. 127-143.

Hansmann & Kraakman 2001
Hansmann, H. & Kraakman, R., 'The End of History for Corporate Law', *Georgetown Law Journal*, 2001, Vol. 89(2), p. 439-468.

Hao 2007
Hao, T. [郝铁川], 我国国民经济和社会发展规划具有法律约束力吗? [National Economic and Social Development Planning has Legal Force?], 2 *学习与探索* [Study & Exploration] 2007, p. 99-102 (in Chinese).

Harrison *et al.* 2010
Harrison, J.A. *et al.*, 'Mechanisms for Driving Sustainability of Biofuels in Developing Countries', *Renewable Energy Law & Policy Review*, 2010, Vol. 2, p. 197-211.

Hasenclever 1997
Hasenclever, A., *Theories of International Regimes,* Cambridge: Cambridge University Press, 1997.

Havlík *et al.* 2011
Havlík, P. *et al.*, 'Global Land-Use Implications of First and Second Generation Biofuel Targets', *Energy Policy,* 2011, Vol. 39(10), p. 5690-5702.

Heimer 2006
Heimer, M., 'The Cadre Responsibility System and the Changing Needs of the Party', in Brodsgaard, K.E. & Zheng, Y. (eds.), *The Chinese Communist Party in Reform,* Oxon: Routledge, 2006, p. 122-138.

Heinimö & Junginger 2009
Heinimö, J. & Junginger, M., 'Production and Trading of Biomass for Energy - An Overview of the Global Status', *Biomass and Bioenergy,* 2009, Vol. 33(9), p. 1310-1320.

Hellmann & Verburg 2010
Hellmann, F. & Verburg, P.H., 'Impact Assessment of the European Biofuel Directive on Land Use and Biodiversity', *Journal of Environmental Management,* 2010, Vol. 91(6), p. 1389-1396.

Hey 2009
Hey, E., 'Global Environmental Law: Common Interests and the (Re)Constitution of Public Space', *Iustum Aequum Salutare,* 2009, Vol. 1, p. 41-57.

Hill *et al.* 2009
Hill, J. *et al.*, 'Climate Change And Health Costs of Air Emissions From Biofuels and Gasoline', *Proceedings of the National Academy of Sciences,* 2009, Vol. 106(6), p. 2077-2082.

Hirschl *et al.* 2012
Hirschl, B. *et al.*, 'Science-Policy Interface and the Role of Impact Assessment in the Case of Biofuels', in Von Raggamby, A. & Rubik, F. (eds.), *Sustainable Development, Evaluation and Policy-Making Theory, Practise and Quality Assurance,* Cheltenham: Edward Elgar, 2012, p. 151-172.

Hoffmann 2014
Hoffmann, M.J., 'Global Climate Governance', in Payne, A. & Phillips, N. (eds.), *Handbook of the International Political Economy of Governance,* Cheltenham: Edward Elgar, 2014, p. 433-449.

Hooghe & Marks 2003
Hooghe, L. & Marks, G., 'Unraveling The Central State, But How? Types Of Multi-Level Governance', *American Political Science Review,* 2003, Vol. 97(2), p. 233-243.

Huang *et al.* 2009
Huang, J. *et al.*, 'Impacts of Bioethanol Development on China's Regional Agricultural Development', *Economics (Quarterly)*, 2009, Vol. 8(2), p. 727-742.

Huber 2013
Huber, S., 'The EU, International Aviation and Climate Change - A Case Study for the EU as A Global Role Model?', in Douma, W.T. & Van der Velde, S. (eds.), *EU Environmental Norms and Third Countries: The EU as A Global Role Model?*, The Hague: Centre for the Law of EU External Relations, 2013, Vol. 5, p. 83-93.

Humphreys 2010
Humphreys, S., 'Competing Claims: Human Rights and Climate Harms', in Humphreys, S. (ed.), *Human Rights and Climate Change*, Cambridge: Cambridge University Press, 2010, p. 37-68.

IEA 2011
International Energy Agency(IEA), *Technology Roadmap: Biofuels for Transport*, 2011.

IISD 2008
International Institute for Sustainble Development (IISD), 'Summary of the Second Meeting of the Ad Hoc Open-ended Working Group on Protected Areas and the Thirteenth Meeting of the Subsidiary Body on Scientific, Technical and Technological Advice of the Convention on Biological Diversity', *Earth Negotiations Bulletin*, 2008, Vol. 9, No. 427, p. 1-18.

ILA 2012
International Law Association (ILA), 'International Law on Sustainable Development', Sofia Conference Final Report, 2012.

ILC 2006
International Law Commission (ILC), 'Fragmentation of International Law: Difficulties Arising from the Diversification and Expansion of International Law', Fifty-eighth Session, Geneva, 2006.

Inkinen 2014
Inkinen, V., 'Application of Sustainability Criteria in the EU ETS', Master Thesis, Faculty of Law, University of Helsinki, 2014.

International Law Commission 2006
'Fragmentation Of International Law: Difficulties Arising from the Diversification And Expansion of International Law', Report of the Study Group of the International Law Commission, Finalized by Martti Koskenniemi, Fifty-Eighth Session, A/CN.4/L.682, Geneva, 2006.

IPCC 2007
Core Writing Team, 'Climate Change 2007: Synthesis Report', in Pachauri, R.K. & Reisinger, A. (eds.), *Contribution of Working Groups I, II and III to the Fourth Assessment Report of the Intergovernmental Panel on Climate Change*, Geneva: IPCC, 2007, p. 26-73.

Jacobsson 2004
Jacobsson, K., 'Soft Regulation and the Subtle Transformation of States: The Case of EU Employment Policy', *Journal of European Social Policy*, 2004, Vol. 14(4), p. 355-370.

Jans & Vedder 2012
Jans, J.H. & Vedder, H.H.B., *European Environmental Law After Lisbon*, Groningen: Europa Law Publishing, 2012.

Jansson & Kalimo 2014
Jansson, M.S. & Kalimo, H., 'On a Common Road towards Sustainable Biofuels? EU and U.S. Approaches to Regulating Biofuels', *Journal of Environmental and Public Health Law*, 2014, Vol. 8(2), p. 104-159.

Jansson *et al.* 2009
Jansson, C. *et al.*, 'Cassava, a Potential Biofuel Crop in (the) People's Republic of China', *Applied Energy*, 2009, Vol. 86 (Supplement 1), p. S95–S99.

Johnson 2009
Johnson, E., 'Goodbye to Carbon Neutral: Getting Biomass Footprints Right', *Environmental Impact Assessment Review*, 2009, Vol. 29(3), p. 165-168.

Johnson 2011
Johnson, F.X., 'Regional-Global Linkages in the Energy-Climate-Development Policy Nexus: The Case of Biofuels in the EU Renewable Energy Directive', *Renewable Energy Law and Policy Review*, 2011, Vol. 2, p. 91-106.

Joosten & Clarke 2002
Joosten, H. & Clarke, D., *Wise Use of Mires and Peatlands*, International Mire Conservation Group, International Peat Society, 2002.

Juhola & Westerhoff 2011
Juhola, S. & Westerhoff, L., 'Challenges of Adaptation to Climate Change across Multiple Scales: A Case Study of Network Governance in Two European Countries', *Environmental Science & Policy*, 2011, Vol. 14(3), p. 239-247.

Jull 2007
Jull, C., 'Recent trends in the Law and Policy of Bioenergy Production, Promotion and Use', Food and Agriculture Organisation 2007.

Junginger *et al.* 2011
Junginger, M. *et al.*, 'Opportunities and Barriers for International Bioenergy Trade', *Energy Policy,* 2011, Vol. 39(4), p. 2028-2042.

Kamdem 2008
Kamdem, I.F., 'Harmonisation, Unification et Uniformisation. Plaidoyer Pour un Discours Affiné sur les Moyens d'Intégration Juridique', *Uniform Law Review,* 2008, Vol. 13(3), p. 709-744.

Kane 1999
Kane, M., 'Sustainability Concepts: From Theory to Practice', in Köhn, J. (ed.), *Sustainability in Question: The Search for a Conceptual Framework,* Cheltenham: Edward Elgar, 1999, p. 15-32.

Kaphengst, Ma & Schlegel 2009
Kaphengst, T., Ma, M.S. & Schlegel, S., 'At a Tipping Point? How the Debate on Biofuel Standards Sparks Innovative Ideas for the General Future of Standardisation and Certification Schemes', *Journal of Cleaner Production,* 2009, Vol. 17, p. S99-S101.

Kaphengst, Wunder & Krista 2012
Kaphengst, T., Wunder, S. & Krista, T., *The Social Dimension of EU Biofuel Policy,* Ecologic Briefs on International Relations and Sustainable Development, Ecologic Briefs, 2012.

Kelemen 2010
Kelemen, R.D., 'Globalizing European Union Environmental Policy', *Journal of European Public Policy,* 2010, Vol. 17(3), p. 335-349.

Keller 1994
Keller, P., 'Sources of Order in Chinese Law', *The American Journal of Comparative Law,* 1994, Vol. 42(4), p. 711-759.

Kendall, Change & Sharpe 2009
Kendall, A, Chang, B. & Sharpe, B., 'Accounting for Time-Dependent Effects in Biofuel Life Cycle Greenhouse Gas Emissions Calculations', *Environmental Science & Technology,* 2009, Vol. 43(18), p. 7142-7147.

Kent 2013
Kent, A., 'WTO Law on Subsidies and Climate Change: Overcoming the Dissonance', *Trade, Law & Development,* 2013, Vol. 5(2), p. 344-382.

Keohane 1982
Keohane, R.O., 'The Demand for International Regimes', *International Organization,* 1982, Vol. 36(2), p. 325-355.

Keohane & Victor 2011
Keohane, R.O. & Victor, D.G., 'The Regime Complex for Climate Change', *Perspectives on Politics,* 2011, Vol. 9(1), p. 7-23.

Keohane, Revesz & Stavins 1998
Keohane,N.O., Revesz, R.L. & Stavins, R.N., ' The Choice of Regulatory Instruments in Environmental Policy', *Harvard Environmental Law Review,* 1998, Vol. 22(2), p. 313-367.

Kern & Bulkeley 2009
Kern, K. & Bulkeley, H., 'Cities, Europeanization and Multi-level Governance: Governing Climate Change through Transnational Municipal Networks', *Journal of Common Market Studies,* 2009, Vol. 47(2), p. 309-332.

Khanna *et al.* 2011
Khanna, M. *et al.*, 'Land Use and Greenhouse Gas Mitigation Effects of Biofuel Policies', *University of Illinois Law Review,* 2011, Vol. 2, p. 549-588.

Kingsbury *et al.* 2005
Kingsbury, B. *et al.*, 'The Emergence of Global Administrative Law', *Law and Contemporary Problems,* 2005, Vol. 68(3/4), p. 15-61.

Kiss 2005
Kiss, A., 'Public Lecture on International Environmental Law', in Häder, D.-P., Hemmersbach, R. & Lebert, M. (eds.), *The Law of Energy for Sustainable Development,* Cambridge: Cambridge University Press, 2005, p. 6-36.

Kiss & Shelton 2007
Kiss, A. & Shelton, D., *Guide to International Environmental Law,* Leiden: Martinus Nijhoff Publishers, 2007.

Klabbers 2009
Klabbers, J., *An Introduction to International Institutional Law,* New York: Cambridge University Press, 2009.

Knill 2005
Knill, C., 'Introduction: Cross-national Policy Convergence: Concepts, Approaches and Explanatory Factors', *Journal of European Public Policy,* 2005, Vol. 12(5), p. 764-774.

Knothe 2010
Knothe, G., 'Biodiesel and Renewable Diesel: A Comparison', *Progress in Energy and Combustion Science,* 2010, Vol. 36(3), p. 364-373.

Koh & Ghazoul 2008
Koh, L.P. & Ghazoul, J., 'Biofuels, Biodiversity, and People: Understanding the Conflicts and Finding Opportunities', *Biological Conservation,* 2008, Vol. 141(10), p. 2450-2460.

Kokkini-Iatridou 1986
Kokkini-Iatridou, D., 'Some Methodological Aspects of Comparative Law. The Third Part of a (Pre-) Paradigm', *Netherlands International Law Review,* 1986, Vol. 33(2), p. 143-194.

Konopacky 2012
Konopacky, J., 'Refueling Biofuel Legislation: Incorporating Social Sustainability Principles to Protect Land Rights', *Wisconsin International Law Journal,* 2012-2013, Vol. 30(2), p. 401-427.

Koskenniemi & Leino 2002
Koskenniemi, M. & Leino, P., 'Fragmentation of International Law? Postmodern Anxieties', *Leiden Journal of International Law,* 2002, Vol. 15(3), p. 553-579.

Kostka & Hobbs 2012
Kostka, G. & Hobbs, W., 'Local Energy Efficiency Policy Implementation In China: Bridging the Gap between National Priorities and Local Interests', *The China Quarterly,* 2012, Vol. 211, p. 765-785.

Krämer 2011
Krämer, L., *EU Environmental Law,* London: Sweet & Maxwell, 2011.

Krasner 1983
Krasner, S.D., 'Structural Causes and Regime Consequences: Regimes or Intervening Variables', in Krasner, S.D. (ed.), *International Regimes,* Ithaca: Cornell University Press, 1983, p. 1-22.

Kulovesi 2012
Kulovesi, K., 'Climate Change in the EU External Relations: Please Follow My Example (or I Might Force You to)', in Morgera, E. (ed.), *The External Environmental Policy of the European Union: EU and International Law Perspectives,* New York: Cambridge University Press, 2012, p. 115-148.

Kulovesi, Morgera & Muñoz 2011
Kulovesi, K., Morgera, E. & Muñoz, M., 'Environmental Integration and Multifaceted International Dimensions of EU Law: Unpacking the EU's 2009 Climate and Energy Package', *Common Market Law Review,* 2011, Vol. 48(3), p. 829-891.

Laborde 2011
Laborde, D., 'Assessing the Land Use Change Consequences of European Biofuel Policies', International Food Policy Institute (IFPRI), 2011.

Laliberte & Lanteigne 2008
Laliberte, A. & Lanteigne, M., 'The Issue of Challenges to the Legitimacy of CCP Rule', in Laliberte, A. & Lanteigne, M., *The Chinese Party-State in the 21st Century: Adaptation and the Reinvention of Legitimacy*, Oxon: Routledge, 2008, p. 1-21.

Lamers *et al.* 2011
Lamers, P. *et al.*, 'International Bioenergy Trade — A Review Of Past Developments In The Liquid Biofuel Market', *Renewable and Sustainable Energy Reviews,* 2011, Vol. 15(6), p. 2655-2676.

Lan *et al.* 2006
Lan, X. *et al.*, 'Environmental Governance in China', Report of the Task Force on Environmental Governance to the China Council for International Cooperation on Environment and Development (CCICED), 2006.

Lan 2011
Lan, H., 'Comments on EU Aviation ETS Directive and EU - China Aviation Emission Dispute', *Revue Juridique Themis,* 2011, Vol. 45(3), p. 589-608.

Langlais 2012
Langlais, A., 'Biomass or the Story of an Unfinished Coming Together of Energy and Waste: Observations on the EU and French Legal Approach', in Peeters, M., Stallworthy, M. & De Cendra de Larragán, J. (eds.), *Climate Law in EU Member States Towards National Legislation for Climate Protection,* Cheltenham: Edward Elgar, 2012, p. 277-293.

Larson 2006
Larson, E.D., 'A Review of Life-Cycle Analysis Studies on Liquid Biofuel Systems for the Transport Sector', *Energy for Sustainable Development,* 2006, Vol. 10(2), p. 109-126.

Latner, O'Kray & Jiang 2006
Latner, K., O'Kray, C. & Jiang, J., 'Biofuels: An Alternative Future for Agriculture', GAIN Report CH6049, USDA-FAS, 2006.

Lavranos 2002
Lavranos, N., 'Multilateral Environmental Agreements: Who Makes the Binding Decisions?', *European Energy and Environmental Law Review,* 2002, Vol. 11(2), p. 44-50.

Lee 2011
Lee, S., 'In the Prison of the Mind: Punishment, Social Order, and Self-Regulation', in Sarat, A., Douglas, L. & Umphrey, M.M. (eds.), *Law as Punishment/Law as Regulation*, Stanford: Stanford University Press, 2011, p. 124-154.

Lee *et al.* 2011
Lee, J.S.H. *et al.*, 'No Farmer Left Behind in Sustainable Biofuel Production', *Biological Conservation,* 2011, Vol. 144(10), p. 2512-2516.

Lendle & Schaus 2010
Lendle, A. & Schaus, M., 'Sustainability Criteria in the EU Renewable Energy Directive: Consistent with WTO Rules?', International Centre for Trade and Sustainable Development, 2010.

Leng 2008
Leng, R., 'Life Cycle Inventory and Energy Analysis of Cassava-based Fuel Ethanol In China', *Journal of Cleaner Production,* 2008, Vol. 16(3), p. 374-384.

Levidow & Paul 2010
Levidow, L. & Paul, H., 'Global Agrofuel Crops as Contested Sustainability, Part I: Sustaining What Development?', *Capitalism Nature Socialism,* 2010, Vol. 21(2), p. 64-86.

Levin *et al.* 2009
Levin, K., Cashore, B. & Koppell, J., 'Can Non-State Certification Systems Bolster State-Centered Efforts to Promote Sustainable Development through the Clean Development Mechanism', *Wake Forest Law Review,* 2009, Vol. 44, p. 777-798.

Lewandowski & Faaij 2006
Lewandowski, I. & Faaij, A.P., 'Steps towards the Development of a Certification System for Sustainable Bio-Energy Trade', *Biomass and Bioenergy,* 2006, Vol. 30(2), p. 83-104.

Lewis 2004
Lewis, B.C., 'It's a Small World After All: Making the Case for the Extraterritorial Application of the National Environmental Policy Act', *Cardozo Law Review,* 2004, Vol. 25, p. 2143-2188.

Lewis 2010
Lewis, J.I., 'The Evolving Role of Carbon Finance in Promoting Renewable Energy Development in China', *Energy Policy,* 2010, Vol. 38(6), p. 2875-2886.

Li & Chan-Halbrendt 2009
Li, S.Z. & Chan-Halbrendt, C., 'Ethanol Production in China: Potential and Technologies', *Applied Energy,* 2009, Vol. 86 (Supplement 1), p. S162-S169.

Liang *et al.* 2013
Liang, S. *et al.*, 'Waste Oil Derived Biofuels in China Bring Brightness for Global GHG Mitigation', *Bioresource Technology,* 2013, Vol. 131, p. 139-145.

Lieberthal 1995
Lieberthal, K., *Governing China: From Revolution through Reform,* London: W.W. Norton, 1995.

Lieberthal 1997
Lieberthal, K., 'China's Governing System and Its Impact on Environmental Policy', *China Environment Series*, 1997, Vol. 1, p. 3-8.

Lima 2009
Lima, M.G.B., 'Biofuel Governance and International Legal Principles: Is It Equitable and Sustainable?', *Melbourn Journal of International Law*, 2009, Vol. 10, p. 470-492.

Lin 1997
Lin, F., 'From "Harmonious Development" to "Sustainable Development" – A Strategic Move in China's Environmental Protection?', *Summer Asia Pacific Law Review*, 1997, Vol. 5(2), p. 25-43.

Lin 2011a
Lin, J., 'Environmental Regulation of Biofuels: Limits of the Meta-Standard Approach', *Carbon & Climate Law Review*, 2011, Vol. 5(1), p. 34-43.

Lin 2011b
Lin, J., 'Environmental Law and Policy in China: Responding to Climate Change', in Yu, G. (ed.), *The Development of the Chinese Legal System: Change and Challenges*, Oxon: Routledge, 2011, p. 295-310.

Lin 2012a
Lin, J., 'Governing Biofuels: A Principal-Agent Analysis of the European Union Biofuels Certification Regime and the Clean Development Mechanism', *Journal of Environmental Law*, 2012, Vol. 24(1), p. 43-73.

Lin 2012b
Lin, J., 'Climate Governance in China: Using the "Iron Hand"', in Richardson, B.J. (ed.), *Local Climate Change Law: Environmental Regulation in Cities and Other Localities*, Cheltenham: Edward Elgar, 2012, p. 300-323.

Liu 2012
Liu, X. [刘小敏], 国家"十二五"规划中能源强度与碳排放强度约束性指标的比较研究. [Comparative Study on Energy Intensity Mandatory Target and Carbon Intensity Mandatory Target in the '12th Five Year' Guideline], 5 *金融评论* [Financial Review] 2012, p. 58-69 (in Chinese).

Lo & Wang 2015
Lo, K. & Wang, M.Y., 'Low Carbon Policies and Programs in China', in Han, S.S., Green, R. & Wang, M.Y. (eds.), *Towards Low Carbon Cities in China: Urban Form and Greenhouse Gas Emissions*, Oxon: Routledge, 2015, p. 21-44.

Lu *et al.* 2014
Lu, L. *et al.*, 'Evaluating Energy Benefit of Pistacia Chinensis Based Biodiesel in China', *Renewable and Sustainable Energy Reviews*, 2014, Vol. 35, p. 258-264.

Lubman 2000
Lubman, S., 'Bird in a Cage: Chinese Law Reform after Twenty Years', *Northwestern Journal of International Law & Business,* 2000, Vol. 20, p. 383-423.

Lydgate 2012a
Lydgate, E.B., 'Sustainable Development in the WTO: From Mutual Supportiveness to Balancing', *World Trade Review,* 2012, Vol. 11(4), p. 621-639.

Lydgate 2012b
Lydgate, E.B., 'Biofuels, Sustainability, and Trade-Related Regulatory Chill', *Journal of International Economic Law,* 2012, Vol. 15(1), p. 157-180.

Lydgate 2013
Lydgate, E.B., 'The EU, the WTO and Indirect Land-Use Change', *Journal of World Trade,* 2013, Vol. 47(1), p. 159-186.

Machado-Filho 2008
Machado-Filho, H., 'Climate Change and the International Trade of Biofuels', *Carbon & Climate Law Review,* 2008, Vol. 1, p. 67-77.

Magraw & Hawke 2007
Magraw, D.B. & Hawke, L.D., 'Sustainable Development', in Bodansky, D., Brunnee, J. & Hey, E. (eds.), *The Oxford Handbook of International Environmental Law,* Oxford: Oxford University Press, 2007, p. 613-638.

Maletic 2013
Maletic, I., *The Law And Policy of Harmonization in Europe's Internal Market,* Cheltenham: Edward Elgar, 2013.

Markesinis 2003
Markesinis, B.S., *Comparative Law in the Courtroom and Classroom: The Story of the Last Thirty-Five Years,* Oregon: Hart Publishing, 2003.

Mathews 2007
Mathews, J.A., 'Biofuels: What a Biopact between North and South could Achieve', *Energy Policy,* 2007, Vol. 35(7), p. 3550-3570.

Mbengue 2011
Mbengue, M.M., 'The Intergovernmental Panel on Climate Change (IPCC): A Singular Model of Expertise at the International Level', in Kerbrat, Y. & Maljean-Dubois, S. (eds.), *The Transformation of International Environmental Law,* Oxford/Paris: Editions A. Pedone and Hart Publishing, 2011, p. 97-116.

McElwee 2011
McElwee, C., *Environmental Law in China: Mitigating Risk and Ensuring Compliance,* Oxford: Oxford University Press, 2011.

McGraw 2002
McGraw, D.M., 'The CBD – Key Characteristics And Implications for Implementation', *Review of European Community & International Environmental Law*, 2002, Vol. 11(1), p. 17-28.

McMichael 2010
McMichael, P., 'Agrofuels in the Food Regime', *The Journal of Peasant Studies*, 2010, Vol. 37(4), p. 609-629.

Messerlin 2012
Messerlin, P.A., 'Climate and Trade Policies: From Mutual Destruction to Mutual Support', *World Trade Review*, 2012, Vol. 11(1), p. 53-80.

Michaels 2006
Michaels, R., 'The Functional Method of Comparative Law', in Reimann, M. & Zimmermann, R. (eds.), *The Oxford Handbook of Comparative Law*, Oxford: Oxford University Press, 2006, p. 339-382.

Michaels 2012
Michaels, R., 'Comparative Law', in Basedow, J.K.J. *et al.* (eds.), *Max Planck Encyclopedia of European Private Law*, Oxford: Oxford University Press, 2012, p. 298-302.

Mitchell & Tran 2010
Mitchell, A.D. & Tran, C., 'The Consistency of the EU Renewable Energy Directive with the WTO agreements: the Case of Biofuels', *Renewable Energy Law & Policy Review*, 2010, Vol. 1, p. 33-44.

Mol 2010
Mol, A.P., 'Environmental Authorities and Biofuel Controversies', *Environmental Politics*, 2010, Vol. 19(1), p. 61-79.

Molony & Smith 2010
Molony, T. & Smith, J., 'Biofuels, Food Security, and Africa', *African Affairs*, 2010, Vol. 109(436), p. 489-498.

Momirov & Fourie 2009
Momirov, A. & Fourie, A.N., 'Vertical Comparative Law Methods: Tools for Conceptualising the International Rule of Law', *Erasmus Law Review*, 2009, Vol. 2, p. 291-309.

Monateri 2012
Monateri, P.G., *Methods of Comparative Law*, Cheltenham: Edward Elgar Publishing, 2012.

Montini 2008
Montini, M., 'Sustainable Development within the Climate Change Regime', in Bugge, H.C. & Voigt, C. (eds.), *Sustainable Development in International and National Law,* Groningen: Europa Law Publishing, 2008, p. 523-546.

Morgera 2011
Morgera, E., 'Far Away, So Close: A Legal Analysis of the Increasing Interactions Between the Convention on Biological Diversity and Climate Change Law', *Climate Law,* 2011, Vol. 2(1), p. 85-115.

Morgera 2013
Morgera, E., 'No Need to Reinvent the Wheel for A Human Rights-Based Approach to Tackling Climate Change: The Contribution of International Biodiversity Law', in Hollo, E., Kulovesi, K. & Mehling, M. (eds.), *Climate Change and the Law: A Global Perspective,* Dordrecht: Springer, 2013), p. 359-390.

Morgera, Kulovesi & Gobena 2009
Morgera, E., Kulovesi, K. & Gobena, A., 'Case Studies on Bioenergy Policy and Law: Options for Sustainability', 102 FAO Legislative Study, 2009.

Motaal 2008
Motaal, D.A., 'The Biofuels Landscape: Is There A Role for the WTO?', *Journal of World Trade,* 2008, Vol. 42(1), p. 61-86.

Msangi & Ewing 2008
Msangi, S. & Ewing, M., 'Food, Feed, or Fuel: Examining Linkages between Biofuels and Agricultural Market Economies', *Georgetown Journal of International Affairs,* 2008, Vol. 9(1), p. 17-24.

Mueller, Anderson & Wallington 2011
Mueller, S.A., Anderson, J.E. & Wallington, T.J., 'Impact of Biofuel Production and Other Supply and Demand Factors on Food Price Increases in 2008', *Biomass and Bioenergy,* 2011, Vol. 35(5), p. 1623-1632.

Murphy, De Vit & Nolet 2009
Murphy, D., De Vit, C. & Nolet, J., 'Climate Change Mitigation through Land Use Measures in the Agriculture and Forestry Sectors', IISD, 2009.

Murphy *et al.* 2011
Murphy, R. *et al.*, 'Global Developments in the Competition for Land from Biofuels', *Food Policy,* 2011, Vol. 36, p. S52-S61.

Nagle 1996
Nagle, J.C., 'The Missing Chinese Environmental Law Statutory Interpretation Cases', *New York University Environmental Law Journal,* 1996, Vol. 5, p. 517-555.

Narula 2006
Narula, S., 'The Right to Food: Holding Global Actors Accountable under International Law', *Columbia Journal of Transnational Law,* 2006, Vol. 44, p. 691-800.

Naylor *et al.* 2007
Naylor, R.L. *et al.*, 'The Ripple Effect: Biofuels, Food Security, and the Environment', *Environment: Science and Policy for Sustainable Development,* 2007, Vol. 49(9), p. 30-43.

O'Sullivan & Sheffrin 2003
O'Sullivan, A. & Sheffrin, S.M., *Economics: Principles in Action,* Pearson: Prentice Hall, 2003.

Oberthür 2009
Oberthür, S., 'Interplay Management: Enhancing Environmental Policy Integration among International Institutions', *International Environmental Agreements: Politics, Law and Economics* 2009, Vol. 9(4), p. 371-391.

Oliver *et al.* 2009
Oliver, H.H. *et al.*, 'China's Fuel Economy Standards for Passenger Vehicles: Rationale, Policy Process, and Impacts', *Energy Policy,* 2009, Vol. 37(11), p. 4720-4729.

Olsen 2007
Olsen, K.H., 'The Clean Development Mechanism's Contribution to Sustainable Development: A Review of the Literature', *Climatic Change,* 2007, Vol. 84(1), p. 59-73.

Olsen & Fenhann 2008
Olsen, K.H. & Fenhann, J., 'Sustainable Development Benefits of Clean Development Mechanism Projects: A New Methodology for Sustainability Assessment Based on Text Analysis of The Project Design Documents Submitted For Validation', *Energy Policy,* 2008, Vol. 36(8), p. 2819-2830.

Ostrom 2014
Ostrom, E., 'A Polycentric Approach for Coping with Climate Change', *Annals of Economics And Finance,* 2014, Vol. 15, p. 97-134.

Ottinger 2009
Ottinger, R.L., 'Biofuels – Potential, Problems & Solutions', *Fordham Environmental Law Review,* 2009, Vol. 19, p. 253-263.

Ottinger & Miller 2010
Ottinger, R.L. & Miller, S.E., 'Bioenergy in Developing Countries: Potential and Risks', *Renewable Energy Law and Policy Review,* 2010, Vol. 1, p. 23-32.

Overmars *et al.* 2011
Overmars, K.P., Stehfest, E., Ros, J.P. & Prins, A.G., 'Indirect Land Use Change Emissions Related to EU Biofuel Consumption: An Analysis Based on Historical Data', *Environmental Science & Policy,* 2011, Vol. 14(3), p. 248-257.

Pallemaerts 2006
Pallemaerts, M., 'The EU and Sustainable Development: An Ambiguous Relationship', in allemaerts, M. & Azmanov, A. (eds.), *The European Union and Sustainable Development: Internal and External Dimensions,* Brussels: Brussels University Press, 2006, p. 19-52.

Parrish 2009
Parrish, A.L., 'Reclaiming International Law from Extraterritoriality', *Minnesota Law Review,* 2009, Vol. 93, p. 815-874.

Parry *et al.* 2011
Parry, A.S. *et al.*, 'Automobile Fuel Economy Standards: Impacts, Efficiency, and Alternatives', *Review of Environmental Economics and Policy,* 2011, Vol. 5, p. 89-108.

Pattberg & Stripple 2008
Pattberg, P. & Stripple, J., 'Beyond the Public and Private Divide: Remapping Transnational Climate Governance in the 21st Century', *International Environmental Agreements: Politics, Law and Economics,* 2008, Vol. 8(4), p. 367-388.

Pauwelyn 2003
Pauwelyn, J., 'Bridging Fragmentation and Unity: International Law as a Universe of Inter-connected Islands', *Michigan Journal of International Law,* 2003, Vol. 25, p. 903-916.

Payne 2008
Payne, C., 'Local Meets Global: The Low Carbon Fuel Standard And The WTO', *North Carolina Journal of International Law and Commercial Regulation,* 2008, Vol. 34, p. 891-918.

Pedersen 2011
Pedersen, O.W., 'The Janus-Head of Human Rights and Climate Change: Adaptation and Mitigation', *Nordic Journal of International Law,* 2011, Vol. 80(4), p. 403-423.

Peeters 2011
Peeters, M., 'The EU ETS and the Role of the Courts: Emerging Contours in the Case of Arcelor', *Climate Law,* 2011, Vol. 2(1), p. 19-36.

Peeters 2014
Peeters, M., 'Instrument Mix or Instrument Mess? The Administrative Complexity of the EU Legislative Package for Climate Change', in Peeters, M. & Uylenburg, R. (eds.), *EU Environmental Legislation: Legal Perspectives on Regulatory Strategies*, Cheltenham: Edward Elgar, 2014, p. 173-192.

Peeters & Nóbrega 2014
Peeters, M. & Nóbrega, S., 'Climate Change-related Aarhus Conflicts: How Successful are Procedural Rights in EU Climate Law?', *Review of European, Comparative & International Environmental Law*, 2014, Vol. 23(3), p. 354-366.

Peeters & Schomerus 2014
Peeters, M. & Schomerus, T., 'Regional Renewable Energy: A String of Legal And Financial Challenges', in Peeters, M. & Schomerus, T. (eds.), *Renewable Energy Law in the EU: Legal Perspectives on Bottom-up Approaches*, Cheltenham: Edward Elgar, 2014, p. 281-316.

Peeters & Weishaar 2009
Peeters, M. & Weishaar, S., 'Exploring Uncertainties in the EU ETS: Learning by Doing Continues beyond 2012', *Carbon & Climate Law Review*, 2009, Vol. 1, p. 88-101.

Pelsy 2008
Pelsy, F., 'The European Commission 2008 Directive Proposal on Biofuels: A Critique', *Law, Environment and Development Journal*, 2008, Vol. 4(2) , p. 119-134.

Peters *et al.* 2012
Peters, D. *et al.*, 'Analysis of Member State RED implementation Final Report' (Task 2), BIONL11469, ECOFYS, 2012.

Pittock 2010
Pittock, J., 'A Pale Reflection of Political Reality: Integration of Global Climate, Wetland, And Biodiversity Agreements', *Climate Law* 2010, Vol. 1(3), p. 343-373.

Prabhakar & Elder 2009
Prabhakar, S. & Elder, M., 'Biofuels and Resource Use Efficiency in Developing Asia: Back to Basics', *Applied Energy*, 2009, Vol. 86, p. S30-S36.

Prost & Clark 2006
Prost, M. & Clark, P.K., 'Unity, Diversity and the Fragmentation of International Law: How Much Does the Multiplication of International Organizations Really Matter?', *Chinese Journal of International Law*, 2006, Vol. 5(2), p. 341-370.

Puddle 2007
Puddle, K., 'Unilateral Trade Measures to Combat Climate Change: A Biofuels Case Study', *New Zealand Journal of Environmental Law*, 2007, Vol. 11, p. 99-140.

Qi & Cheng 2015
Qi, S. & Cheng, S., Comparative Study on Pilot Emission Trading Schemes of China [中国碳排放权试点比较研究], 清华大学-中国低碳报告编写组[Low Carbon Developmemt Research Group, Tsinghua University] 2015 (in Chinese).

Qi *et al.* 2008
Qi, Y. *et al.*, 'Translating a Global Issue into Local Priority China's Local Government Response to Climate Change', *The Journal of Environment & Development,* 2008, Vol. 17(4), p. 379-400.

Qiu *et al.* 2010
Qiu, H. *et al.*, 'Bioethanol Development in China and the Potential Impacts on Its Agricultural Economy', *Applied Energy,* 2010, Vol. 87, p. 76-83.

Qiu *et al.* 2012
Qiu, H. *et al.*, 'Liquid Biofuels in China: Current Status, Government Policies, and Future Opportunities and Challenges', *Renewable and Sustainable Energy Reviews,* 2012, Vol. 16, p. 3095-3104.

Rached 2014
Rached, D.H., 'The Intergovernmental Panel on Climate Change: Holding Science and Policy-Making to Account', *Yearbook of International Environmental Law,* 2014, Vol. 24(1), p. 3-36.

Rajagopal & Zilberman 2007
Rajagopal, D. & Zilberman, D., *Review of Environmental, Economic and Policy Aspects of Biofuels,* Washington DC: World Bank Publications, Vol. 4341, 2007.

Ramsar Convention Secretariat 2010
Ramsar Convention Secretariat, 'Wise Use of Wetlands: Concepts and Approaches for the Wise Use Of Wetlands', in Pritchard, D. (ed.), *Ramsar Handbooks For The Wise Use Of Wetlands,* Gland: Ramsar Convention Secretariat, 2010, p. 8-56.

Ramsar Convention Secretariat 2013
Ramsar Convention Secretariat, *The Ramsar Convention Manual: A Guide to the Convention on Wetlands* (Ramsar, Iran, 1971), Gland: Ramsar Convention Secretariat, 2013.

Rancourt 2009
Rancourt, M.-E., 'Promoting Sustainable Biofuels under the WTO Legal Regime', *McGill International Journal of Sustainable Development Law and Policy,* 2009, Vol. 5, p. 73-114.

Rao 2004
Rao, P.S., 'Multiple International Judicial Forums: A Reflection of the Growing Strength Of International Law or its Fragmentation', *Michigan Journal of International Law,* 2004, Vol. 25, p. 929-962.

Rathmann *et al.* 2010
Rathmann, R. *et al.*, 'Land Use Competition for Production of Food and Liquid Biofuels: An Analysis of the Arguments in the Current Debate', *Renewable Energy,* 2010, Vol. 35(1), p. 14-22.

Regan 2009
Regan, D.H., 'How to Think about PPMs (and Climate Change)', in Cottier, T., Nartova, O. & Bigdeli, S.Z. (eds.), *International Trade and the Mitigation of Climate Change,* Cambridge: Cambridge University Press, 2009, p. 97-123.

Report on the Policies for Biofuel Industry 2014
Report on the Policies for Biofuel Industry in the Major Countries [世界主要国家生物液体燃料产业政策研究报告], 中国农业大学国家能源非粮生物质原料研发中心 [National Energy R&D Centre for Non-Food Biomass, China Agricultural University] 2014 (in Chinese).

Resnik 2006
Resnik, J., 'Law's Migration: American Exceptionalism, Silent Dialogues, and Federalism's Multiple Ports of Entry', *The Yale Law Journal,* 2006, Vol. 115, p. 1564-1670.

Ribeiro 2013
Ribeiro, B.E., 'Beyond Commonplace Biofuels: Social Aspects of Ethanol', *Energy Policy,* 2013, Vol. 57, p. 355-362.

Richardson 2014
Richardson, B., 'The Governance of Primary Commodities: Biofuel Certification in the European Union', in Payne, A. & Phillips, N. (eds.), *Handbook of the International Political Economy of Governance,* Cheltenham: Edward Elgar, 2014, p. 201-219.

Röben 2013
Röben, V., Conference (Meeting) of States Parties, Max Planck Institute for Comparative Public Law and International Law, 2013.

Robertson *et al.* 2008
Robertson, G.P. *et al.*, 'Sustainable Biofuels Redux', *Science,* 2008, Vol. 322(5898), p. 49-50.

Roht-Arriaza 2010a
Roht-Arriaza, N., 'Human Rights in the Climate Change Regime', *Journal of Human Rights and the Environment,* 2010, Vol. 1(2), p. 211-235.

Roht-Arriaza 2010b
Roht-Arriaza, N., '"First, Do No Harm": Human Rights and Efforts to Combat Climate Change', *Georgia Journal of International & Comparative Law,* 2010, Vol. 38(3), p. 593-612.

Romppanen 2012a
Romppanen, S., 'Regulating Better Biofuels for the European Union', *European Energy and Environmental Law Review,* 2012, Vol. 21(3), p. 123-141.

Romppanen 2012b
Romppanen, S., 'The EU's Biofuels: Certified as Sustainable?', *Renewable Energy Law & Policy Review,* 2012, Vol. 3, p. 173-186.

Romppanen 2013
Romppanen, S., 'The Role and Relevance of Private Actors in EU Biofuel Governance', *Review of European, Comparative & International Environmental Law,* 2013, Vol. 22(3), p. 340-353.

Romppanen 2014
Romppanen, S., 'Legitimacy and EU Biofuel Governance: In Search of Greater Coherence', *Climate Law,* 2014, Vol. 4(3-4), p. 239-266.

Rousseaux 2005
Rousseaux, S., 'Carbon Sinks in the Kyoto Protocol's Clean Development Mechanism: An Obstacle to the Implementation of the Convention on Biological Diversity?', *Environmental Law Review,* 2005, Vol. 7(1), p. 1-11.

Runge & Senauer 2007
Runge, C.F. & Senauer, B., 'How Biofuels Could Starve the Poor', *Foreign Affairs,* 2007, Vol. 86(3), p. 41-53.

Ruthner *et al.* 2011
Ruthner, L. *et al.*, 'Study on the Integrity of the Clean Development Mechanism (CDM)', Final Report CLIMA.B.3/ETU/2010/0020r, AEA Technology plc, 2011.

Sagemuller 2006
Sagemuller, I., 'Forest Sinks under the United Nations Framework Convention on Climate Change and the Kyoto Protocol: Opportunity or Risk for Biodiversity', *Columbia Journal of Environmental Law,* 2006, Vol. 31, p. 189-242.

Sampson 2005
Sampson, G.P., *The WTO and Sustainable Development,* Hong Kong: United Nations University Press, 2005.

Sanchez *et al.* 2012
Sanchez, S.T. *et al.*, 'Accounting for Indirect Land-Use Change in the Life Cycle Assessment of Biofuel Supply Chains', *Journal of the Royal Society Interface,* 2012, Vol. 9, p. 1105-1119.

Sand 2011
Sand, P.H., 'The Role of Environmental Agreements' Conferences of the Parties', in Kerbrat, Y. & Maljean-Dubois, S. (eds.), *The Transformation of International Environmental Law*, Oxford/Paris: Editions A. Pedone and Hart Publishing, 2011, p. 89-96.

Sands & Peel 2012
Sands, P. & Peel, J., *Principles of International Environmental Law*, New York: Cambridge University Press, 2012.

Savaresi 2011
Savaresi, A., 'Reducing Emissions from Deforestation in Developing Countries under the UNFCCC: Caveats and Opportunities for Biodiversity', *Yearbook of International Environmental Law*, 2011, p. 1-33.

Scarlat & Dallemand 2011
Scarlat, N. & Dallemand, J.-F., 'Recent Developments of Biofuels/Bioenergy Sustainability Certification: A Global Overview', *Energy Policy*, 2011, Vol. 39(3), p. 1630-1646.

Schermers & Blokker 2011
Schermers, H.G. & Blokker, N.M., *International Institutional Law: Unity within Diversity*, Leiden: Martinus Nijhoff Publishers, 2011.

Schleifer 2013
Schleifer, P., 'Orchestrating Sustainability: The Case of European Union Biofuel Governance', *Regulation & Governance*, 2013, Vol. 7(4), p. 533-546.

Schneider 2009
Schneider, L., 'Assessing the Additionality of CDM Projects: Practical Experiences and Lessons Learned', *Climate Policy*, 2009, Vol. 9(3), p. 242-254.

Schrijver 2008
Schrijver, N., 'Development – The Neglected Dimension in the Post-Rio International Law of Sustainable Development', in Bugge, H.C. & Voigt, C. (eds.), *Sustainable Development in International and National Law*, Groningen: Europa Law Publishing, 2008, p. 223-248.

Schwaiger *et al.* 2012
Schwaiger, H. *et al.*, 'The Future European Emission Trading Scheme and Its Impact on Biomass Use', *Biomass and Bioenergy*, 2012, Vol. 38, p. 102-108.

Scott 2009
Scott, J., 'From Brussels with Love: The Transatlantic Travels of European Law and the Chemistry of Regulatory Attraction', *The American Journal of Comparative Law*, 2009, Vol. 57(4), p. 897-942.

Scott 2011a
Scott, J., 'The Multi-Level Governance of Climate Change', in Craig, P. & De Burca, G. (eds.), *The Evolution of EU Law,* Oxford: Oxford University Press, 2011, p. 805-835.

Scott 2011b
Scott, J., 'The Multi-Level Governance of Climate Change', *Carbon & Climate Law Review,* 2011, Vol. 5(1), p. 25-33.

Scott 2014
Scott, J., 'Extraterritoriality and Territorial Extension in EU Law', *American Journal of Comparative Law,* 2014, Vol. 62(1), p. 87-125.

Scott & Jiang 2013
Scott, R. & Jiang, J., 'China People's Republic Biofuels Annual', USDA Foreign Agriculture Service Global Agriculture Information Network (GAIN), 2013.

Searchinger *et al.* 2008
Searchinger, T. *et al.*, 'Use of US Croplands for Biofuels Increases Greenhouse Gases through Emissions from Land-Use Change', *Science,* 2008, Vol. 319(5867), p. 1238-1240.

Searchinger *et al.* 2009
Searchinger T. *et al.*, 'Fixing a Critical Climate Accounting Error', *Science,* 2009, Vol. 326(5952), p. 527-528.

Segger 2008
Segger, M.-C.C., 'Sustainable Development in International Law', in Bugge, H.C. & Voigt, C. (eds.), *Sustainable Development in International and National Law: What did the Brundtland Report do to Legal Thinking and Legal Development, and Where Can We Go from Here?*, Groningen: Europa Law Publishing, 2008, p. 87-199.

Senden 2004
Senden, L., *Soft Law in European Community Law,* Oxford: Hart Publishing, 2004.

Sevenster 2000
Sevenster, H., 'The Environmental Guarantee after Amsterdam: Does the Emperor Have New Clothes', in Somsen, H. (ed.), *Yearbook of European Environmental Law,* Oxford: Oxford University Press, 2000, p. 291-310.

Sexton 2015
Sexton, R., *Exploring Economics,* Boston: Cengage Learning, 7th edn., 2015.

Sexton *et al.* 2008
Sexton, S. *et al.*, 'Food Versus Fuel: How Biofuels Make Food More Costly and Gasoline Cheaper', *Agricultural and Resource Economics Update,* 2008, Vol. 12(1), p. 1-6.

Shaffer & Bodansky 2012
Shaffer, G. & Bodansky, D., 'Transnationalism, Unilateralism and International Law', *Transnational Environmental Law*, 2012, Vol. 1(1), p. 31-41.

Shelton 2010
Shelton, D., 'International Law and "Relative Normativity"', in Evans, M.D. (ed.), *International Law*, New York: Oxford University Press, p. 141-171.

Siang 2007
Siang, C.C., 'China's Medium to Long-term Renewable Energy Development Plan: Promotion of Bio-ethanol Introduction and Future Strategic Developments', New Energy and Coal Group, Strategy and Industry Research Unit, The Institute of Energy Economics, 2007.

Sims *et al.* 2014
Sims, R. *et al.*, 'Transport', in Edenhofer O. *et al.* (eds.), *Climate Change 2014: Mitigation of Climate Change. Contribution of Working Group III to the Fifth Assessment Report of the Intergovernmental Panel on Climate Change*, Cambridge: Cambridge University Press, 2014, p. 599-670.

Smith *et al.* 2014
Smith, P. *et al.*, 'Agriculture, Forestry and other Land Use (AFOLU)', in Edenhofer, O. *et al.* (eds.), *Climate Change 2014: Mitigation of Climate Change. Contribution of Working Group III to the Fifth Assessment Report of the Intergovernmental Panel on Climate Change*, Cambridge: Cambridge University Press, 2014, p. 811-922.

Snyder 1994
Snyder, F., *Soft Law and Institutional Practice in the European Community*, Dordrecht: Springer, 1994.

Soimakallio & Koponen 2011
Soimakallio, S. & Koponen, K., 'How to Ensure Greenhouse Gas Emission Reductions by Increasing the Use of Biofuels? – Suitability of the European Union Sustainability Criteria', *Biomass and Bioenergy*, 2011, Vol. 35(8), p. 3504-3513.

Song & Lei 2014
Song, R. & Lei, H., 'Emissions Trading in China: First Reports from the Field', World Resources Institute, 2014.

Sorda, Banse & Kemfert 2010
Sorda, G., Banse, M. & Kemfert, C., 'An Overview of Biofuel Policies across the World', *Energy Policy*, 2010, Vol. 38(11), p. 6977-6988.

Sorrell & Sijm 2003
Sorrell, S. & Sijm, J., 'Carbon Trading in the Policy Mix', *Oxford Review of Economic Policy*, 2003, Vol. 19(3), p. 420-437.

Staffas, Gustavsson & McCormick 2013
Staffas, L., Gustavsson, M. & McCormick, K., 'Strategies and Policies for the Bioeconomy and Bio-Based Economy: An Analysis of Official National Approaches', *Sustainability,* 2013, Vol. 5, p. 2751-2769.

Steiner, Alston & Goodman 2008
Steiner, H.J., Alston, P. & Goodman, R., *International Human Rights In Context: Law, Politics, Morals: Text And Materials,* Oxford: Oxford University Press, 2008.

Sutter & Parreño 2007
Sutter, C. & Parreño, J.C., 'Does the Current Clean Development Mechanism (CDM) Deliver its Sustainable Development Claim? An Analysis of Officially Registered CDM Projects', *Climatic Change,* 2007, Vol. 84(1), p. 75-90.

Swart *et al.* 2003
Swart, R., Robinson, J. & Cohen, S., 'Climate Change and Sustainable Development: Expanding the Options', *Climate Policy,* 2003, Vol. 3 (Supplement 1), p. S19-S40.

Swinbank 2009
Swinbank, A., 'EU Policies on Bioenergy and their Potential Clash with the WTO', *Journal of Agricultural Economics,* 2009, Vol. 60(3), p. 485-503.

Swinbank & Daugbjerg 2013
Swinbank, A. & Daugbjerg, C., 'Improving EU Biofuels Policy? Greenhouse Gas Emissions, Policy Efficiency, and WTO Compatibility', *Journal of World Trade,* 2013, Vol. 47(4), p. 813-834.

Switzer 2007
Switzer, S., 'International Trade Law and the Environment: Designing a Legal Framework to Curtail the Import of Unsustainably Produced Biofuel', *Journal of International Trade Law and Policy,* 2007, Vol. 6(1), p. 30-44.

Switzer & McMahon 2011
Switzer, S. & McMahon, J.A., 'EU Biofuels Policy - Raising the Question of WTO Compatibility', *International and Comparative Law Quarterly,* 2011, Vol. 60(3), p. 713-736.

Tarasofsky & Hoare 2004
Tarasofsky, R.G. & Hoare, A.L., *Implications of A UNEO for the Global Architecture of the International Environmental Governance System,* Paris: Institute for Sustainable Development and International Relations, 2004.

Taylor 2013
Taylor, M., 'Ethanol at the Crossroads: Fuelling the Potential of International Trade in Ethanol', *Renewable Energy Law & Policy Review,* 2013, Vol. 3, p. 161-174.

Teng & Gu 2007
Teng, F. & Gu, A., 'Climate Change: National and Local Policy Opportunities in China', Fondazione Eni Enrico Mattei Working Papers 2007.

Teubner 1996
Teubner, G., 'Global Bukowina: Legal Pluralism in the World-Society', in Teubner, G. (ed.), *Global Law without A State*, Dartsmouth: Ashgate Publishing, 1996, p. 3-28.

Teubner & Fischer-Lescano 2004
Teubner, G. & Fischer-Lescano, A., 'Regime-Collisions: The Vain Search For Legal Unity in the Fragmentation of Global Law', *Michigan Journal of International Law*, 2004, Vol. 25(4), p. 999-1046.

Tilman *et al.* 2009
Tilman, D. *et al.*, 'Beneficial Biofuels - The Food, Energy, and Environment Trilemma', *Science*, 2009, Vol. 325(5938), p. 270-271.

Tladi 2012
Tladi, D., 'Principles of Sustainable Development in the Case Concerning Pulp Mills on the River Uruguay', International Development Law Organization Legal Working Paper, 2012.

Trouwborst 2009
Trouwborst, A., 'International Nature Conservation Law and the Adaptation of Biodiversity to Climate Change: A Mismatch?', *Journal of Environmental Law*, 2009, Vol. 21(3), p. 419-442.

Trubek & Trubek 2005
Trubek, D.M. & Trubek, L.G., 'Hard and Soft Law in the Construction of Social Europe: the Role of the Open Method of Coordination', *European Law Journal*, 2005, Vol. 11(3), p. 343-364.

Tuerk, Cowie & Leopold 2011
Tuerk, A., Cowie, A. & Leopold, A., 'The Influence of Emissions Trading Schemes on Bioenergy Use', Paper presented at the IEA Bioenergy Task 38, 2011.

Tung 2009
Tung, C., 'Carbon Law and Practice in China', in Freestone, D. & Streck, C. (eds.), *Legal Aspects of Carbon Trading Kyoto, Copenhagen and Beyond*, Oxford: Oxford University Press, 2009, p. 488-516.

Tung 2013
Tung, C., 'Sustainable Development and Climate Policy and Law in China', in Hollo, E.J., Kulovesi, K. & Mehling, M. (eds.), *Climate Change and the Law*, Dordrecht: Springer, 2013, p. 597-604.

Ulfstein 2007
Ulfstein, G., 'International Environmental law: Treaty Bodies', in Bodansky, D., Brunnée, J. & Hey, E. (eds.), *The Oxford Handbook of International Environmental Law*, Oxford: Oxford University Press, 2007, p. 876-889.

Ulfstein 2010
Ulfstein, G., 'International Framework for Environmental Decision-making', in Fitzmaurice, M., Ong, D.M. & Merkouris, P. (eds.), *Research Handbook on International Environmental Law*, Cheltenham: Edward Elgar, 2010, p. 26-47.

UNFCCC Secretary 2014
UNFCCC Secretary, 'Handbook on Measurement, Reporting and Verification for Developing Country Parties', Bonn: UNFCCC Secretary, 2014.

Van Asselt 2011a
Van Asselt, H., 'Managing the Fragmentation of International Environmental Law: Forests at the Intersection of the Climate and Biodiversity Regimes', *New York University International Law & Policy*, 2011, Vol. 44, p. 1205-1278.

Van Asselt 2011b
Van Asselt, H., 'Integrating Biodiversity in the Climate Regime's Forest Rules: Options and Tradeoffs in Greening REDD Design', *Review of European Community & International Environmental Law*, 2011, Vol. 20(2), p. 139-149.

Van Asselt & Zelli 2014
Van Asselt, H. & Zelli, F., 'Connect the Dots: Managing the Fragmentation of Global Climate Governance', *Environmental Economics and Policy Studies*, 2014, Vol. 16(2), p. 137-155.

Van Asselt, Sindico & Mehling 2008
Van Asselt, H., Sindico, F. & Mehling, M.A., 'Global Climate Change and the Fragmentation of International Law', *Law & Policy*, 2008, Vol. 30(4), p. 423-449.

Van Dam, Junginger & Faaij 2010
Van Dam, J., Junginger, M. & Faaij, A.P., 'From the Global Efforts on Certification Of Bioenergy towards An Integrated Approach Based on Sustainable Land Use Planning', *Renewable and Sustainable Energy Reviews*, 2010, Vol. 14(9), p. 2445-2472.

Van de Staaij *et al.* 2012
Van de Staaij, J. *et al.*, 'Analysis of the Operation of the Mass Balance System and Alternatives', ECOFYS Final Report (Task 1), BIONL11469, 30 November 2012.

Van den Bossche, Schrijver & Faber 2007
Van den Bossche, P., Schrijver, N. & Faber, G., *Unilateral Measures Addressing Non-Trade Concerns: A Study on WTO Consistency, Relevance of Other International Agreements, Economic Effectiveness and Impact on Developing Countries of Measures Concerning Non-Product-Related Processes and Production Methods,* The Hague: Ministry of Foreign Affairs of The Netherlands, 2007.

Van Holten & Van Rijswick 2014
Van Holten, S. & Van Rijswick, M., 'The Governance Approach in European Union Environmental Directives and its Consequences for Flexibility, Effectiveness and Legitimacy', in Peeters, M. & Uylenburg, R. (eds.), *EU Environmental Legislation: Legal Perspectives on Regulatory Strategies New Horizons in Environmental and Energy Law series,* Cheltenham: Edward Elgar, 2014, p. 13-47.

Van Rooij 2006
Van Rooij, B., 'Implementation of Chinese Environmental Law: Regular Enforcement and Political Campaigns', *Development and Change,* 2006, Vol. 37(1), p. 57-74.

Vedder 2010
Vedder, H.H.B., 'The Treaty of Lisbon and European Environmental Policy', *Journal of Environmental Law,* 2010, Vol. 22(2), p. 285-299.

Verbruggen, Moomaw & Nyboer 2011
Verbruggen, A., Moomaw, W. & Nyboer, J., 'Annex I: Glossary, Acronyms, Chemical Symbols and Prefixes', in Edenhofer O. *et al.* (eds.), *IPCC Special Report on Renewable Energy Sources and Climate Change Mitigation,* Cambridge: Cambridge University Press, 2011, p. 954-972.

Vogel 1997
Vogel, D., *Trading Up: Consumer and Environmental Regulation in A Global Economy,* Cambridge: Harvard University Press, 1997.

Vogel 2012
Vogel, D., *The Politics of Precaution: Regulating Health, Safety, and Environmental Risks in Europe and the United States,* Oxfordshire: Princeton University Press, 2012.

Vogler 2014
Vogler, J., 'The International Politics of Sustainable Development', in Atkinson, G. *et al.* (eds.), *Handbook of Sustainable Development,* Cheltenham: Edward Elgar, 2014, p. 430-446.

Voigt 2009
Voigt, C., *Sustainable Development as A Principle of International Law: Resolving Conflicts between Climate Measures and WTO Law,* Leiden/Boston: Martinus Nijhoff Publishers, 2009.

Von Blottnitz & Curran 2007
Von Blottnitz, H. & Curran, M.A., 'A Review of Assessments Conducted on Bio-ethanol as A Transportation Fuel from a Net Energy, Greenhouse Gas, and Environmental Life Cycle Perspective', *Journal of Cleaner Production,* 2007, Vol. 15(7), p. 607-619.

Von Doudda, Corkery & Chartres 2007
Von Doudda, J., Corkery, A. & Chartres, R., 'Human Rights and Climate Change', *Australian International Law Journal,* 2007, Vol. 14, p. 161-183.

Von Glahn & Taulbee 2007
Von Glahn, G. & Taulbee, J.L., *Law Among Nations: An Introduction to Public International Law,* Boston: Pearson, 2007.

Vos 2001
Vos, E., 'Differentiation, Harmonisation and Governance', in De Witte, B., Hanf, D. & Vos, E. (eds.), *The Many Faces of Differentiation in EU Law,* Antwerp/Oxford/New York: Intersentia, 2001, p. 145-180.

Vos & Weimer 2016 (forthcoming)
Vos, E. & Weimer, M., 'Differentiated Integration or Uniform Regime? Searching for the Soul of the Internal Market with A View to National Derogations', in Ott, A., Vos, E. & De Witte, B. (eds.), *Between Flexibility and Disintegration: The State of EU Law Today,* Cheltenham: Edward Elgar, 2016, forthcoming.

Vuichard, Ciais & Wolf 2009
Vuichard, N., Ciais, P. & Wolf, A., 'Soil Carbon Sequestration or Biofuel Production: New Land-Use Opportunities for Mitigating Climate over Abandoned Soviet Farmlands', *Environmental Science & Technology,* 2009, Vol. 43(22), p. 8678-8683.

Wagner 2010
Wagner, M., *Non-State Actors,* New York: Max Planck Encyclopedia of Public International Law, 2010.

Wagner, An & Wang 2009
Wagner, D.V., An, F. & Wang, C., 'Structure and Impacts of Fuel Economy Standards for Passenger Cars in China', *Energy Policy,* 2009, Vol. 37(10), p. 3803-3811.

Wang 2011
Wang, H., 'Building a Regulatory Framework for Biofuels Governance in China: Legislation as the Starting Point', *Natural Resources Forum,* 2011, Vol. 35(3), p. 201-212.

Wang & Chang 2014
Wang, N. & Chang, Y.-C., 'The Development of Policy Instruments in Supporting Low-carbon Governance in China', *Renewable and Sustainable Energy Reviews,* 2014, Vol. 35, p. 126-135.

Wang & Tian 2011
Wang, Q. & Tian, Z., 'Biofuels and the Policy Implications for China', *Asian-Pacific Economic Literature,* 2011, Vol. 25(1), p. 161-168.

Wang, Xiong & Liu 2009
Wang, F., Xiong, X.-R. & Liu, C.-Z., 'Biofuels in China: Opportunities and Challenges', *Vitro Cellular & Developmental Biology-Plant,* 2009, Vol. 45(3), p. 342-349.

Wara 2008
Wara, M.W., 'Measuring the Clean Development Mechanism's Performance and Potential', *UCLA Law Review,* 2008, Vol. 55(6), p. 1759-1804.

Watson 1993
Watson, A., *Legal Transplants: An Approach to Comparative Law,* Athens, Georgia: University of Georgia Press, 1993.

Weatherill 2006
Weatherill, S., 'Supply of and Demand for Internal Market Regulation: Strategies, Preferences and Interpretation', in Shuibhne, N.N. (ed.), *Regulating the Internal Market,* Cheltenham: Edward Elgar, 2006, p. 29-60.

Weigel 2008
Weigel, W., *Economics of the Law: A Primer,* London/New York: Routledge, 2008.

Weil 1983
Weil, P., 'Towards Relative Normativity in International Law', *American Journal of International Law,* 1983, Vol. 77(3), p. 413-442.

Wessel 2011
Wessel, R.A., 'Informal International Law-making as a New Form of World Legislation', *International Organisations Law Review,* 2011, Vol. 8, p. 253-265.

White, & Dasgupta 2010
White, B. & Dasgupta, A., 'Agrofuels Capitalism: A View from Political Economy', *The Journal of Peasant Studies,* 2010, Vol. 37(4), p. 593-607.

Whiting 2004
Whiting, S.H., 'The Cadre Evaluation System at the Grass Roots: The Paradox of Party Rule', in Naughton, B.J. & Yang, D.L. (eds.), *Holding China Together,* Cambridge: Cambridge University Press, 2004, p. 101-119.

Whitman 2007
Whitman, J.Q., 'Consumerism Versus Producerism: A Study in Comparative Law', *The Yale Law Journal*, 2007, Vol. 117, p. 340-406.

Wickeri & Kalhan 2010
Wickeri, E. & Kalhan, A., 'Land Rights Issues in International Human Rights Law', *Malaysian Journal on Human Rights*, 2010, Vol. 4(10), p. 16-25.

Wiener 2001
Wiener, J.B., 'Something Borrowed for Something Blue: Legal Transplants and the Evolution of Global Environmental Law', *Ecology Law Quarterly*, 2001, Vol. 27 p. 1295-1372.

Wiener 2004
Wiener, J.B., 'Convergence, Divergence, and Complexity in US and European Risk Regulation', in Vig, N.J. & Faure, M.G. (eds.), *Green Giants? Environmental Policies of the United States and the European Union*, Cambridge: MIT Press, 2004, p. 73-110.

Wiener 2011
Wiener, J.B., 'The Real Pattern of Precaution', in Wiener, J.B. *et al.* (eds.), *The Reality of Precaution: Comparing Risk Regulation in the United States and Europe*, Washington/London: Earthscan, 2011, p. 519-565.

Wiens, Fargione & Hill 2011
Wiens, J., Fargione, J. & Hill, J., 'Biofuels and Biodiversity', *Ecological Applications*, 2011, Vol. 21(4), p. 1085-1095.

Wiersema 2009
Wiersema, A., 'The New International Law-Makers? Conferences of the Parties to Multilateral Environmental Agreements', *Michigan Journal of International Law*, 2009, Vol. 31, p. 231-287.

Wiesenthal *et al.* 2009
Wiesenthal, T. *et al.*, 'Biofuel Support Policies in Europe: Lessons Learnt for the Long Way Ahead', *Renewable and Sustainable Energy Reviews*, 2009, Vol. 13(4), p. 789-800.

Winter 2008
Winter, G., 'A Fundament and Two Pillars: The Concept of Sustainable Development 20 Years after the Brundtland Report', in Bugge, H.C. & Voigt, C. (eds.), *Sustainable Development in International and National Law What did the Brundtland Report Do to Legal Thinking and Legal Development, and Where can we go From Here?*, Groningen: Europa Law Publishing, 2008, p. 25-45.

Witt 2010
Witt, A.B., 'Biofuels and Invasive Species from an African Perspective – A Review', *Global Change Biology Bioenergy*, 2010, Vol. 2(6), p. 321-329.

Wouters *et al.* 2012
Wouters, J. *et al.* (eds.), *China, the European Union, and Global Governance,* Cheltenham: Edward Elgar, 2012.

Wu & Huo 2014
Wu, L. & Huo, H., 'Energy Efficiency Achievements in China's Industrial and Transport Sectors: How Do They Rate?', *Energy Policy,* 2014, Vol. 73, p. 38-46.

Wyatt 2008
Wyatt, A.M., 'The Dirt on International Environmental Law Regarding Soils: Is the Existing Regime Adequate?', *Duke Environmental Law & Policy Forum,* 2008, Vol. 19, p. 165-208.

Xie 2012
Xie, G. [谢光辉], 非粮生物质原料体系研发进展及方向. [Progress and direction of non-food biomass feedstock supply research and development in China], 17(6) *中国农业大学学报* [Journal of China Agriculture University] 2012, p. 1-19 (in Chinese).

Xu 2012
Xu, M. [徐孟洲], 论经济社会发展规划与规划法制建设 [Study on Economic and Social Development Planning and Planning Law Making], 1(2) *法学家 [The Jurist]* 2012, p. 43-55 (in Chinese).

Xu 2011
Xu, Y., 'Addressing the Hidden Costs of Automobile Use in China: The Potential Role of Tax', *Australian Tax Forum,* 2011, Vol. 26, p. 693-718.

Yang 2008
Yang, T., 'Implementation Challenge of Mitigating China's Greenhouse Gas Emissions', *The Georgetown International Environmental Law Review,* 2008, Vol. 20, p. 681-701.

Yang & Percival 2009
Yang, T. & Percival, R.V., 'The Emergence of Global Environmental Law', *Ecology Law Quarterly,* 2009, Vol. 36, p. 615-664.

Yang *et al.* 2009
Yang, H., Zhou, Y. & Liu, J., 'Land and Water Requirements of Biofuel and Implications for Food Supply and the Environment in China', *Energy Policy,* 2009, Vol. 37, p. 1876-1885.

Yeh 2007
Yeh, S.Y., 'Is China's Development Path Sustainable – An Overview on the Legal Policy Framework in Relation to Environmental Protection in China', *Asian Journal of WTO & International Health Law and Policy,* 2007, Vol. 2(2), p. 399-456.

Young 2011
Young, M.A., 'Climate Change Law and Regime Interaction', *Carbon and Climate Law Review,* 2011, Vol. 2, p. 147-157.

Zahar 2010a
Zahar, A., 'Verifying Greenhouse Gas Emissions of Annex I Parties: Methods We Have and Methods We Want', *Climate Law,* 2010, Vol. 1(3), p. 409-427.

Zahar 2010b
Zahar, A., 'Does Self-Interest Skew State Reporting of Greenhouse Gas Emissions? A Preliminary Analysis Based on the First Verified Emissions Estimates under the Kyoto Protocol', *Climate Law,* 2010, Vol. 1(2), p. 313-324.

Zang 2009a
Zang, D., 'Green from Above: Climate Change, New Developmental Strategy, and Regulatory Choice in China', *Texas International Law Journal,* 2009, Vol. 45, p. 201-232.

Zang 2009b
Zang, D., 'From Environment to Energy: China's Reconceptualization of Climate Change', *Wisconsin International Law Journal,* 2009, Vol. 27,p. 543-574.

Zerk 2010
Zerk, J. A., 'Extraterritorial Jurisdiction: Lessons for the Business and Human Rights Sphere from Six Regulatory Areas', Report for the Harvard Corporate Social Responsibility Initiative to Help Inform the Mandate of the UNSG's Special Representative on Business and Human Rights, Working Paper No. 59, 2010.

Zhang 2010
Zhang, Z., 'China in the Transition to A Low-carbon Economy', *Energy Policy,* 2010, Vol. 38(11), p. 6638-6653.

Zhang 2011
Zhang, H., 'China's Low Carbon Strategy: The Role of Renewable Energy Law in Advancing Renewable Energy', *Renewable Energy Law & Policy Review,* 2011, Vol. 2(2), p. 133-146.

Zhang & Crooks 2012
Zhang, Q. & Crooks, R., 'Toward an Environmentally Sustainable Future: Country Environmental Analysis of the People's Republic of China', Asian Development Bank, 2012.

Zhang *et al.* 2003
Zhang, C. *et al.*, 'Life Cycle Economic Analysis of Fuel Ethanol Derived from Cassava in Southwest China', *Renewable and Sustainable Energy Reviews,* 2003, Vol. 7(4), p. 353–366.

Zhang *et al.* 2009
Zhang, P. *et al.*, 'Bioenergy Industries Development in China: Dilemma and Solution', *Renewable and Sustainable Energy Reviews,* 2009, Vol. 13(9), p. 2571-2579.

Zhu 2005
Zhu, J., 'Agricultural Renewable Energy and its Management in China', in Bradbrook, A.J. *et al.* (eds.), *The Law of Energy for Sustainable Development,* Cambridge: Cambridge University Press, 2005, p. 156-180.

Zilberman *et al.* 2012
Zilberman, D. *et al.*, 'The Impact of Biofuels on Commodity Food Prices: Assessment of Findings', *American Journal of Agricultural Economics,* 2012, p. 1-7.

Zleptnig 2010
Zleptnig, S., *Non-economic Objectives in WTO Law: Justification Provisions of GATT, GATS, SPS, and TBT Agreements,* Leiden: Martinus Nijhoff Publishers, 2010.

Zweigert & Kötz 1998
Zweigert, K. & Kötz, H., *An Introduction to Comparative Law,* New York: Oxford University Press, 1998.

CURRICULUM VITAE

Taotao Yue (born in Wuhan, China in 1985) studied at the Zhongnan University of Economics and Law where she obtained her Bachelor of Law and Master of International Law. She participated in an international exchange programme during 2009-2010 and studied at the University of Cergy-Pontoise in Paris. From September 2011, she joined the Maastricht Institute for Transnational Legel Research (METRO) at Maastricht University to conduct her doctoral research.

IUS COMMUNE EUROPAEUM

A peer-reviewed book series in which the common foundations of the legal systems of the Member States of the European Union are the central focus.

The *Ius Commune Europaeum* series includes horizontal comparative legal studies as well as studies on the effect of treaties within the national legal systems. All the classic fields of law are covered. The books are published in various European languages under the auspices of METRO, the Institute for Transnational Legal Research at the Maastricht University.

Recently published:

Volume 138: *The Nature of Mutual Recognition in European Law. Re-examining the Notion from an Individual Rights Perspective with a View to Its Further Development in the Criminal Justice Area,* W. VAN BALLEGOOIJ
Volume 139: *Evidence in Contemporary Civil Procedure. Fundamental Issues in a Comparative Perspective,* C.H. VAN RHEE and A. UZELAC (eds.)
Volume 140: *The Effect of D&O Insurance on Managerial Risk Taking,* M. GABER
Volume 141: *Information and Notification Duties,* I. SAMOY and M.B. LOOS (eds.)
Volume 142: *The Application of the Theory of Efficient Breach in Contract Law. A Comparative Law and Economics Perspective,* W. LIAO
Volume 143: *A European Legal Framework for Nuclear Liability. Rethinking Current Approaches,* T. HELDT
Volume 144: *The Development of European Private Law in a Multilevel Legal Order,* E.A.G. VAN SCHAGEN
Volume 145: *Property Law Perspectives IV,* D. GRUYAERT, E. RAMAEKERS and L. ROSTILL (eds.)
Volume 146: *The Citizen in European Private Law: Norm-setting, Enforcement and Choice,* C. CAUFFMAN and J. SMITS (eds.)
Volume 147: *Legal Certainty in Real Estate Transactions: A Comparison of England and France,* B. DU MARAIS and D. MARRANI (eds.)
Volume 148: *Scarcity and the State I. The Allocation of Limited Rights by the Administration,* P. ADRIAANSE, F. VAN OMMEREN, W. DEN OUDEN and J. WOLSWINKEL (eds.)
Volume 149: *Optional Instruments of the European Union. A Definitional, Normative and Explanatory Study,* W.A. BULL
Volume 150: *The European Union and National Civil Procedure. A Smooth Process or a Rocky Road for Member States?,* A. NYLUND and B.H. KRANS (eds.)

www.ingramcontent.com/pod-product-compliance
Ingram Content Group UK Ltd.
Pitfield, Milton Keynes, MK11 3LW, UK
UKHW051925100926
8556IPUK00022B/1374

9 781780 684130